Cyclodextrins for Chemosensing

Sivakumar Krishnamoorthy
Editor

Cyclodextrins for Chemosensing

Springer

Editor
Sivakumar Krishnamoorthy
Department of Chemistry
Faculty of Science
Sri Chandrasekharendra Saraswathi Viswa
Mahavidyalaya (SCSVMV) Deemed to be
University
Kanchipuram, Tamil Nadu, India

ISBN 978-3-031-89298-1 ISBN 978-3-031-89299-8 (eBook)
https://doi.org/10.1007/978-3-031-89299-8

This Springer imprint is published by the registered company Springer Nature Switzerland AG
The registered company address is: Gewerbestrasse 11, 6330 Cham, Switzerland

To my beloved wife, Bhuvaneswari,
and to my daughter, Indhumathi,
for the unwavering support and endless patience.

Preface

The field of chemosensing, the science of detecting chemical substances through molecular recognition, has experienced significant advancements over the past few decades. Among the many tools available for chemosensing, cyclodextrins (CDs) have emerged as powerful and versatile molecules due to their unique structural properties and ability to form stable inclusion complexes with a wide variety of guest molecules. This book, *Cyclodextrins for Chemosensing*, explores the multi-faceted roles of cyclodextrins in the development of chemosensors, focusing on their application in environmental monitoring, healthcare diagnostics, and industrial processes.

Cyclodextrins are cyclic oligosaccharides, traditionally derived from starch, with hydrophobic cavities that can encapsulate a variety of guest molecules, altering their chemical and physical properties. This ability has positioned CDs as an essential component in the development of sensors for detecting metal ions, small organic molecules, and environmental pollutants. Over the years, CDs have been modified and functionalized to enhance their performance, leading to a wide range of applications in fluorescence-based, colorimetric, electrochemical, and potentiometric sensing techniques.

The versatility of cyclodextrins, coupled with the rapid advances in nanotechnology, has revolutionized the field, enabling the creation of more sensitive, selective, and cost-effective chemosensors. This book brings together cutting-edge research from leading scientists in the field, showcasing state-of-the-art technologies and future directions in cyclodextrin-based chemosensing. From traditional designs to innovative nanomaterials, the chapters presented here provide an in-depth exploration of the principles, mechanisms, and applications of cyclodextrins in chemosensing.

It is my hope that this compilation of knowledge will serve as a valuable resource for researchers, practitioners, and students interested in the evolving field of chemosensors. As the world continues to face environmental and health challenges, the role of cyclodextrins in enhancing the sensitivity, selectivity, and affordability of sensors will play a pivotal part in tackling these issues in the coming years.

Prof. Dr. Sivakumar Krishnamoorthy
Director—Research and Development Cell
Sri Chandrasekharendra Saraswathi
Viswa Mahavidyalaya (SCSVMV)
Deemed to be University
Kanchipuram, Tamil Nadu, India

Acknowledgements I extend my heartfelt gratitude to the entire staff at Springer for their invaluable support throughout this project. In particular, I wish to thank Ms. Charlotte Hollingworth, Editorial Director, for her meticulous attention to refining the book's content, and Ms. Sabrina Reba, Associate Editor, for her consistent guidance and unwavering support.

Contents

About the Editor

Prof. Dr. Sivakumar Krishnamoorthy serves as Professor in the Department of Chemistry, Faculty of Science, and as the Director of the Research and Development Cell at Sri Chandrasekharendra Saraswathi Viswa Maha Vidyalaya (SCSVMV), a Deemed-to-be-University in Kanchipuram, Tamil Nadu, India. Inspired by the versatility of cyclodextrins, his research focuses on developing innovative application materials through cyclodextrin inclusion complexes. He holds one granted Indian patent on cyclodextrin application and has filed five patent applications. He is a recipient of a research fellowship from the prestigious Indian Academy of Sciences and has authored over 115 research papers. He has secured funding from various Government of India agencies, including DST, CSIR, UGC, DAE, the Indian Academy of Sciences, and the Tamil Nadu State Council for Science and Technology, for his research projects and programs.

Chapter 1
Cyclodextrins for Chemosensing: *Principles and Advances*

Sivakumar Krishnamoorthy

Abstract Chemosensors play a vital role in detecting and quantifying various chemical species with high sensitivity and selectivity. This chapter, titled *Cyclodextrins for Chemosensing: Principles and Advances,* provides a comprehensive exploration of chemosensing, its underlying processes, mechanisms, and the molecular forces that govern sensor performance. Beginning with an overview of chemosensors and the chemosensing process, the chapter elucidates the interplay between sensing and signaling, supported by advanced analytical techniques. A focal point of the chapter is the integration of cyclodextrins into chemosensing systems. Cyclodextrins, with their unique molecular architecture and ability to form inclusion complexes, have emerged as indispensable tools in amplifying sensor performance. The chapter discusses the key contributions of cyclodextrins in chemosensing, comparing cyclodextrin-based chemosensors and cyclodextrin-modified systems. It highlights their versatile roles in enhancing selectivity, sensitivity, and signal amplification, addressing the optimization of chemosensing systems. Additionally, the chapter delves into the detection of specific chemical species, ranging from ions and small molecules to complex biomolecules, and their applications across diverse fields such as environmental monitoring, biomedical diagnostics, and industrial processes. A critical analysis of literature data underscores the advancements in cyclodextrin-based chemosensors, providing insights into current trends and challenges. Concluding with a summary and an overview of the book's content, this chapter serves as a valuable resource for researchers and practitioners, emphasizing the pivotal role of cyclodextrins in advancing the field of chemosensing.

Keywords Chemosensors · Cyclodextrins · Chemosensing mechanism · Cyclodextrins in chemosensing · Molecular forces

S. Krishnamoorthy (✉)
Department of Chemistry, Faculty of Science, Sri Chandrasekharendra Saraswathi Viswa Mahavidyalaya (Deemed to be University) [SCSVMV], Enathur, Kanchipuram, Tamilnadu, India
e-mail: sivakumar.k@kanchiuniv.ac.in; chemshiva@gmail.com

S. Krishnamoorthy (ed.), *Cyclodextrins for Chemosensing*,
https://doi.org/10.1007/978-3-031-89299-8_1

1.1 Introduction: Chemosensors and Chemosensing

A ***chemosensor*** is a molecule with the ability to detect and respond to specific analytes, which may be molecules or ions. Upon interacting with the analyte, the chemosensor produces a measurable signal—such as fluorescence, color change, or an electrochemical response—that signifies the analyte's presence. This entire process of interaction and signal generation is known as **chemosensing**.

Chemosensing Process: *A Gateway to Molecular Detection*

Chemosensing is the process of detecting and responding to chemical stimuli through specific molecular interactions. This highly specialized mechanism involves the use of chemosensors—molecules or molecular assemblies that undergo a measurable change, such as a shift in fluorescence, color, or electrochemical properties, upon interaction with a target analyte.

Chemosensing Mechanism: *Signal Transduction from Molecular Recognition* The sensing process begins with the recognition of the analyte through weak interactions like hydrogen bonding, π-π stacking, van der Waals forces, or hydrophobic effects. This recognition event triggers a transduction mechanism, converting the molecular interaction into a detectable signal [Fig. 1.1].

Sensing and Signaling: *The Dual Core of Chemosensory Molecules*

A chemosensory molecule operates through two fundamental components: the sensing unit and the signaling unit [1].

Sensing Unit: This part of the molecule is responsible for recognizing the target analyte. It employs specific interactions such as hydrogen bonding, electrostatic forces, van der Waals forces, or coordination bonds to ensure selective binding. The design of the sensing unit is critical, as it determines the chemosensor's specificity and affinity toward the analyte.

Signaling Unit: Once the sensing unit interacts with the analyte, the signaling unit translates this interaction into a detectable output. Common outputs include

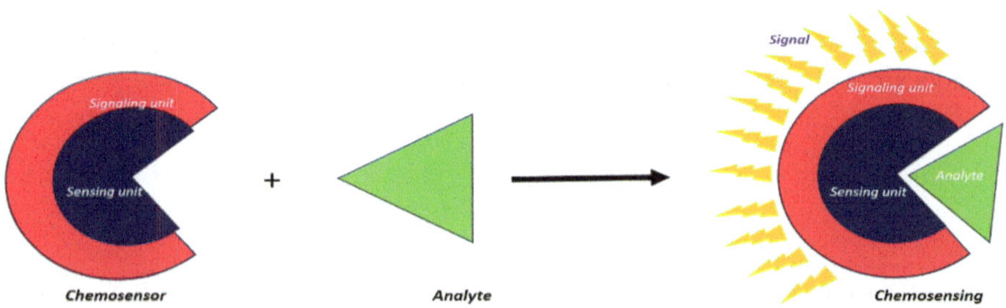

Fig. 1.1 Schematic representation of the interaction of a chemosensor with analyte, demonstrating the *chemosensing* process

changes in fluorescence, color, or electrochemical properties. This unit serves as the communication bridge, making the molecular recognition event observable and quantifiable.

The harmonious interplay of these two units enables chemosensors to act as precise molecular detectors, driving advancements in diagnostics, environmental monitoring, and chemical analysis.

1.2 Analytical Techniques for Chemosensing *Tools for Decoding Molecular Interactions*

In general, optical and electrochemical methods are widely used in chemosensing. However, various analytical methods widely reported for chemosensing in the literature [1–6] are listed below.

Optical Methods: Colorimetric detection, UV–Visible Spectroscopy, Fluorescence Spectroscopy, Raman Spectroscopy and Surface-Enhanced Raman Scattering (SERS), Infrared (IR) Spectroscopy.

Electrochemical Methods: Potentiometry, Cyclic Voltammetry (CV), Differential Pulse Voltammetry (DPV), Electrochemical Impedance Spectroscopy (EIS).

Chromatographic Techniques: Gas Chromatography (GC), High-Performance Liquid Chromatography (HPLC), Thin-Layer Chromatography (TLC).

Mass Spectrometry (MS): MS is often coupled with other methods such as gas chromatography (GC–MS) or liquid chromatography (LC–MS) to enhance the separation and detection of analytes.

Microscopy-Based Techniques: Fluorescence Microscopy, Scanning Electron Microscopy (SEM), Atomic Force Microscopy (AFM).

Thermal Analysis: Differential Scanning Calorimetry (DSC), Thermogravimetric Analysis (TGA).

Photoelectrochemical Methods: Photoelectrochemical (PEC) methods are an advanced sensing strategy that combines light-induced chemical reactions with electrochemical detection using photoactive materials.

Capillary Electrophoresis (CE): Capillary Electrophoresis (CE) is a powerful analytical technique used in chemosensing for the separation and detection of analytes based on their size-to-charge ratio under the influence of an electric field.

Nanotechnology-Based Techniques: Nanoparticle-Based Sensors, Quantum Dot Sensors, Surface-Enhanced Raman Scattering (SERS).

Biosensors: Biosensors are analytical devices that integrate a biological recognition element with a transducer to detect chemical or biological analytes.

1.3 Molecular Forces in Chemosensing: *The Power Behind Detection*

Chemosensing relies on a delicate dance of molecular forces that enable sensors to "feel" and selectively interact with their targets [1–6]. Here are the key players:

Van der Waals Forces: The subtle pull between molecules, helping sensors detect the faintest interactions.

Hydrogen Bonding: The "friendly handshake" that ensures strong and specific binding, especially in biological sensing.

π-π Interactions: Aromatic rings teaming up for highly selective binding, essential for organic compound detection.

Dipole–Dipole Interactions: The alignment of polar molecules, boosting accuracy in detecting charged or polar analytes.

Covalent Bonds: The strong, lasting connections, locking sensors to their targets for irreversible detection.

Coordination Bonds: Metal ions forming a secure bond with ligands, crucial in detecting metals and complex molecules.

Hydrophobic Interactions: The "water-avoiding" attraction that's vital for detecting non-polar, organic compounds.

These forces, working together, enable chemosensors to identify and measure substances with precision, power, and specificity.

1.4 Species Detected in Chemosensing: *Target Molecules and Analytes of Interest*

Chemosensors detect a wide variety of species, including ions, small molecules, and biological markers. Here are examples of species commonly detected:

Ions: Cations and Anions. Cations: Na^+, K^+, Ca^{2+}, Mg^{2+}, Hg^{2+}, Cd^{2+}, Pb^{2+}, Cu^{2+}. *Anions*: Cl^-, Br^-, I^-, CN^-, NO^{3-}, $SO4^{2-}$, $PO4^{3-}$

Small Organic Molecules: Volatile Organic Compounds (VOCs): acetone, ethanol, toluene, benzene, Biologically Relevant Molecules: glucose, urea, lactate, dopamine, serotonin.

Gases: CO_2, O_2, NO, SO_2, H_2, CH_2, NH_2.

Biomolecules: Nucleotides: ATP, GTP, Proteins, and Peptides: insulin, antibodies, lipids, and metabolites.

Pollutants and Environmental Toxins: Heavy metals: As, Hg, Cd, Pesticides: organophosphates, carbamates, Industrial Pollutants: dioxins, polycyclic aromatic hydrocarbons (PAHs).

Pathogens: Detection of bacterial or viral components, like lipopolysaccharides (LPS) or viral RNA, in biosensing applications.

Drugs and Pharmaceuticals: Detection of therapeutic drugs, antibiotics, or recreational drugs.

1.5 Applications of Chemosensing: *From Molecules to Solutions*

Chemosensing is a crucial mechanism in analytical chemistry pivotal in diverse fields such as environmental monitoring, biomedical diagnostics, and industrial quality control. Innovations in designing chemosensors with enhanced sensitivity, selectivity, and rapid response are driving advancements in precise chemical detection.

1.6 Cyclodextrins in Chemosensing: *Masters of Molecular Recognition*

Cyclodextrins (CDs) are cyclic oligosaccharides with a hydrophobic inner cavity and a hydrophilic outer surface (Fig. 1.2a). The inner cavity is formed by 1, 4 linked α-D-glucopyranose units (Fig. 1.2b and Fig. 1.2c) that give a truncated cone shape to the CDs. This unique molecular structure enables them to act as versatile hosts for a wide range of guest molecules, making them highly valuable in chemosensing applications [1, 2].

Cyclodextrins (CDs) are categorized based on the number of glucose units they contain, which influences their size and properties. The main types of cyclodextrins are the following:

- *Alpha-Cyclodextrin (α-CD)*: Composed of six glucose units, α-CD has a relatively small cavity size (about 5.0 Å). It is primarily used for encapsulating smaller molecules, such as some organic compounds and small guest molecules.
- *Beta-Cyclodextrin (β-CD)*: Composed of seven glucose units, β-CD has a larger cavity size (about 6.0 Å), making it more versatile for accommodating a wide range of guest molecules, including drugs, fragrances, and environmental pollutants. It is one of the most widely used cyclodextrins in various applications. Hence, the shape and specifications of the *β-CD* molecule are demonstrated in Fig. 1.2d and Fig. 1.2e.
- *Gamma-Cyclodextrin (γ-CD)*: Composed of eight glucose units, γ-CD has an even larger cavity (about 7.5 Å) and can accommodate larger molecules or complex

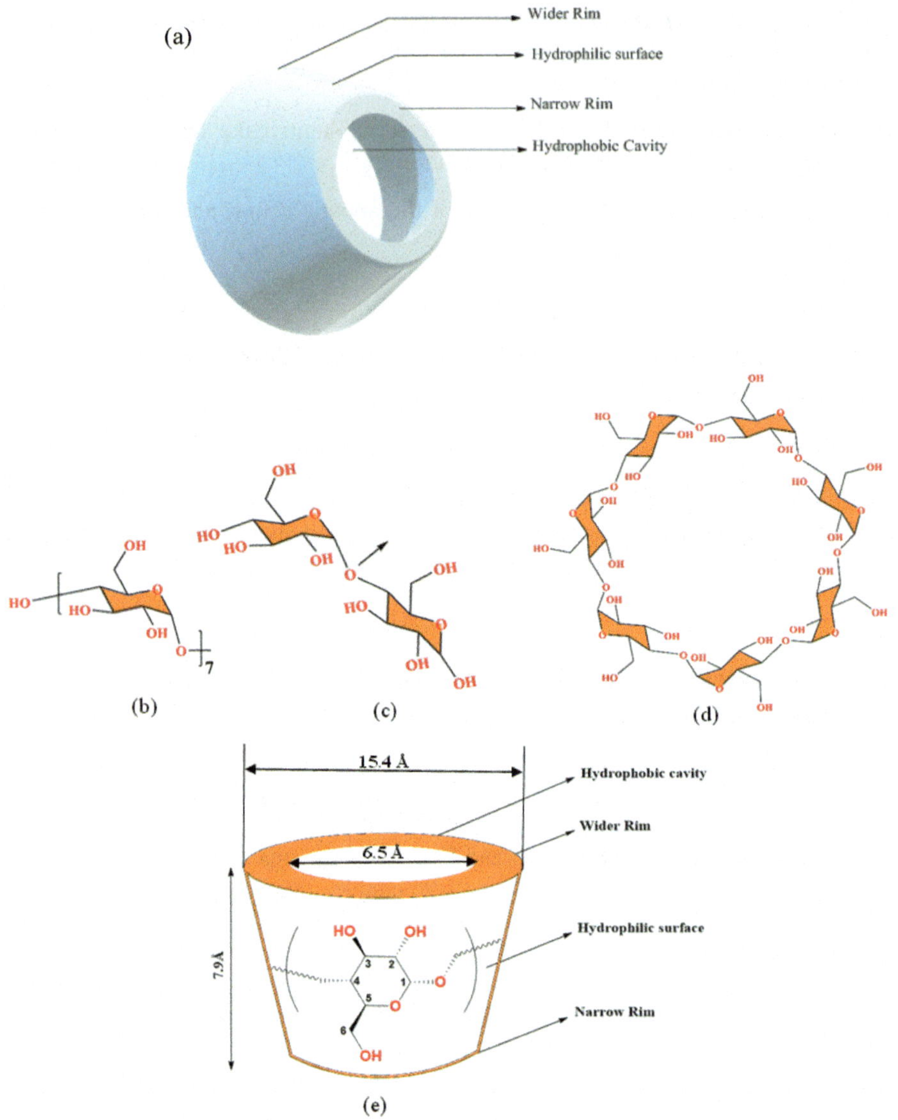

Fig. 1.2 **a** Conical shape of cyclodextrins **b** Monosaccharide unit **c** α-1,4-linked dextrose units **d** Structure of β-Cyclodextrin **e** Truncated conical shape of β-Cyclodextrin. [Reproduced with permission from Sivakumar et al. [1]. Copyright, Springer Nature]

guests. It is useful in encapsulating bulkier molecules such as certain lipophilic substances.

While the primary natural types of cyclodextrins (α-CD, β-CD, and γ-CD) are the most well-known, there are indeed over 100 different types of cyclodextrins available commercially today. These types include not only naturally occurring cyclodextrins but also a wide range of **derivatives** and **modified forms** that offer enhanced properties for specific applications. Some examples of modifications and specialized types include:

- *Hydroxypropyl-β-cyclodextrin (HPβCD)*: Offers improved water solubility, commonly used in pharmaceuticals and food products.
- *Methyl-β-cyclodextrin (MβCD)*: Enhances solubility for hydrophobic compounds and is useful in disrupting lipid membranes.
- *Sulfo-β-cyclodextrin (SβCD)*: Sulfonated derivatives that offer higher solubility in water and can be used in a variety of sensing and drug delivery applications.
- *Carboxymethyl-β-cyclodextrin (CMβCD)*: A carboxylated form used for improving the solubility and stability of guest molecules in water.
- *Dodecyl-β-cyclodextrin*: Modified with hydrophobic groups to improve its ability to encapsulate lipophilic molecules.

These modifications allow cyclodextrins to be tailored for specific needs in various industries, including pharmaceuticals, food, agriculture, environmental monitoring, and chemical sensing. Companies such as **CycloLab** [www.cyclolab.hu] provide a comprehensive list of over 100 types of cyclodextrins, demonstrating the broad range of possibilities these compounds offer in terms of functionality and application. This diversity enhances the versatility of cyclodextrins, making them crucial components in a wide variety of technologies and industries.

1.7 Key Contributions of Cyclodextrins for Chemosensing

Rigid Fit to the analytes: Cyclodextrins enhance chemosensing by providing a **rigid fit** to the guest compound upon encapsulation within their hydrophobic cavity. This structural stabilization improves both **selectivity** and **sensitivity**, enabling the cyclodextrin-chemosensor complex to interact more effectively with the target analyte, thereby enhancing detection accuracy.

Molecular Recognition: Cyclodextrins (CDs) form inclusion complexes with analytes, improving selectivity and sensitivity. Host–guest interactions are driven by hydrophobic forces, hydrogen bonding, and van der Waals interactions.

Signal Amplification: CDs enhance the detection signals in optical and electrochemical-based sensors by stabilizing or concentrating the analyte near the sensor's active site.

Improved Solubility: CDs increase the solubility of hydrophobic analytes in aqueous systems, enabling their detection in water-based environments.

Functionalization: CDs can be chemically modified with functional groups (e.g., thiols, amines, or fluorescent tags) to improve their binding affinity or incorporate additional sensing capabilities.

Selective Binding: CDs show size and shape selectivity for guest molecules, making them ideal for detecting specific analytes such as toxins, pollutants, or biomolecules.

1.8 The Versatile Role of Cyclodextrins in Amplifying the Sensor's Performance: *Selectivity, Sensitivity, Stability, and Reversibility*

Cyclodextrins (CDs) significantly enhance the ***selectivity, sensitivity, stability, and reversibility of sensing systems*** due to their unique molecular structure and host–guest chemistry. Their hydrophobic cavities enable selective recognition of specific analytes, improving selectivity. The ability to form stable inclusion complexes boosts sensitivity by amplifying detection signals. CDs also stabilize sensing materials by protecting sensitive components from environmental degradation, such as moisture or light. Additionally, their reversible binding nature allows for repeated use in dynamic sensing applications. The tunability of CDs through chemical modifications further enhances these properties, making them versatile tools in various chemosensing applications. This synergistic role optimizes sensor performance across diverse analyte detection scenarios. Figure 1.3 demonstrates the enhancement in the signal due to the impact of *β-cyclodextrin based chemosensory* systems in detecting different heavy metal ions using spectrophotometric and spectrofluorometric techniques [1, 2].

Optimizing Chemosensing: *The Role of Matrix and Medium in Cyclodextrin-Based Detection Systems*

The matrix and medium used in sensing with cyclodextrins (CDs) play a crucial role in enhancing the performance of chemosensors [1, 2].

Matrix: The matrix refers to the material that supports or incorporates the cyclodextrin-based sensing system. It can be organic (e.g., polymers, hydrogels) or inorganic (e.g., nanoparticles, metal oxides) and serves as a solid support to embed the cyclodextrins. The matrix influences the stability, mechanical properties, and ease of handling of the sensor. It can also be engineered to ensure that the cyclodextrin interacts effectively with the target analytes while maintaining sensor durability.

Medium: The medium is the environment in which the sensing process occurs, typically the solution or gas phase containing the target analyte. Cyclodextrins can be utilized in aqueous or non-aqueous solutions, depending on the application. The

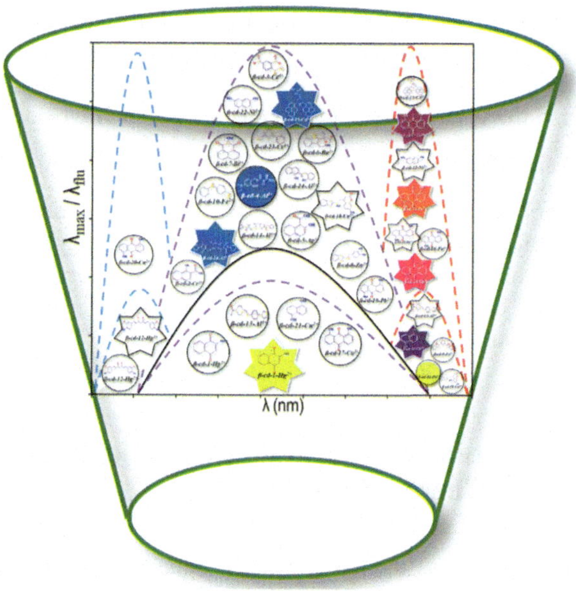

Fig. 1.3 β-cyclodextrin inclusion complex of heavy metal sensors reported in the literature. Star and circle shapes represent absorbance shift and fluorescence shifts of the sensors respectively. Sensors which exhibit color are given their respective color. [Reproduced with permission from Sivakumar et al. [1]. Copyright, Springer Nature]

solvent or medium affects the binding efficiency of cyclodextrins with the analyte and the overall performance of the sensor. In some cases, the medium is modified to optimize the host–guest interaction or to improve the selectivity and sensitivity of the sensor.

Together, the matrix and medium determine the effectiveness of cyclodextrin-based sensing systems, impacting factors such as response time, stability, and reversibility of the sensing process.

1.9 Cyclodextrin Chemosensors Versus Cyclodextrin-Based Chemosensors

Cyclodextrin Chemosensors: These sensors rely on cyclodextrins (CDs) as the primary sensing elements, where cyclodextrin directly interacts with the target analyte to trigger a measurable response. The sensor's performance depends on the ability of CDs to form host–guest complexes, affecting optical, electrochemical, or mechanical properties. Typically, natural CDs (α-, β-, γ-CD) or their simple derivatives are used as recognition units. *Example:* A sensor where β-cyclodextrin forms a complex with an organic molecule, causing a fluorescence or absorbance change.

Cyclodextrin-Based Chemosensors: These sensors incorporate cyclodextrins into a larger system, often embedded in a matrix (polymers, nanoparticles) or modified chemically to enhance their sensing capabilities. Cyclodextrins act as recognition elements, while the overall sensor's performance is improved by the integration of additional materials or technologies, enhancing selectivity, sensitivity, and stability. Example: A sensor where modified cyclodextrins are embedded in a polymer matrix, and the response is detected electrochemically or optically.

Key Differences

Cyclodextrin Chemosensors focus on cyclodextrins alone as sensing agents. Cyclodextrin-Based Chemosensors integrate cyclodextrins with other materials to optimize sensor performance. While both leverage cyclodextrin's host–guest chemistry, cyclodextrin-based sensors are more complex and tailored for enhanced real-world applications.

1.10 Literature Data on Cyclodextrins-Based Chemosensing

In order to assess the literature data on "*Cyclodextrins based Chemosensing*" a comprehensive set of keyword combinations were derived and searched in AI-based literature search engine https://app.dimensions.ai/. The set of keyword combinations used for assessing the literature on "*Cyclodextrins for Chemosensing*" is listed Table 1.1.

Along with the set of keyword combinations, Table 1.1 details the number of published papers and patents hits obtained in the AI-based literature search engine https://app.dimensions.ai/.

Literature Data Analysis (on "*Cyclodextrin for Chemosensing*")

In order to assess the literature data on "*Cyclodextrins for Chemosensing*" the keyword combination; ***Cyclodextrin AND Chemosensor*** for the *Title and abstract* search in the https://app.dimensions.ai/ yielded 133 hits (Table 1.1) of published papers.

The first scientific paper with these keywords has appeared in the year 1995. Since then, from 1995 to 2024, the https://app.dimensions.ai/ yields 133 (Fig. 1.4a) papers in 10 different subject categories (Fig. 1.4b).

Table 1.1 Literature data on *Cyclodextrins for Chemosensing*

Keyword combinations	No. of hits	
	Papers	Patents
Cyclodextrin AND Chemosensor	133	12
Cyclodextrin AND Colorimetric Sensor	73	15
Cyclodextrins AND Fluorescence Sensor/Sensing	348	24
Cyclodextrin AND Electrochemical Sensor	676	83

(a)

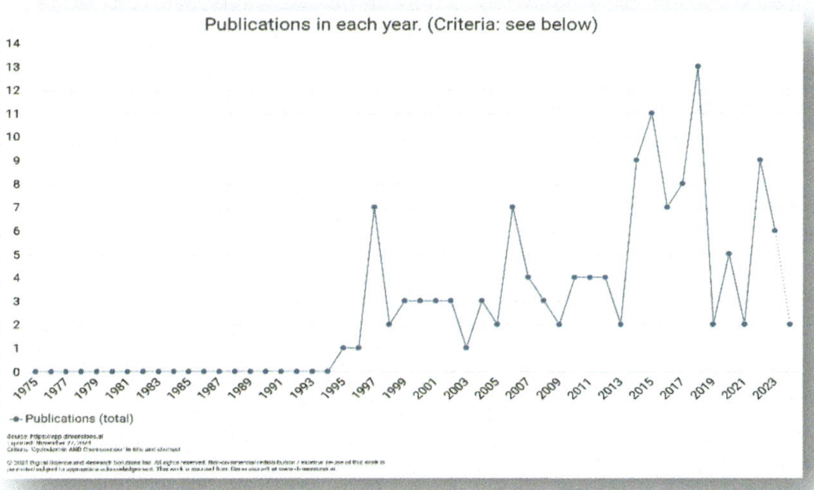

(b)

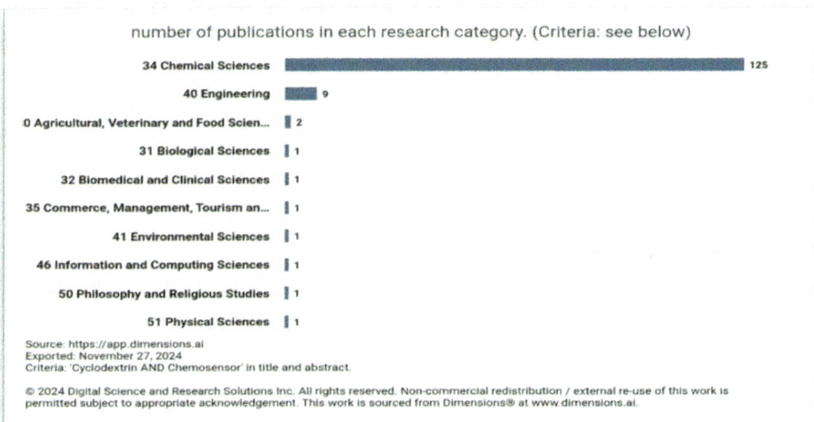

Fig. 1.4 Articles on "*Cyclodextrins for Chemosensing*" from 1975 to 2024; **a** year-wise publication of articles and **b** subject categories of publications.; www.dimensions.ai

The limited number of publications, with fewer than 1000 hits for a *title and abstract* search on 'Cyclodextrins for Chemosensing' (as depicted in Table 1.1 and Fig. 1.4) underscores the niche and emerging nature of this field. This scarcity of comprehensive literature not only reflects the untapped potential of cyclodextrins in chemosensing but also reinforces the need for this focused monograph, aiming to consolidate existing knowledge and inspire further research.

1.11 An Overview of the Content of the Book

Chapter 1 Cyclodextrins for Chemosensing: *Principles and Advances*—This chapter provides a comprehensive overview of the fundamental principles underlying the use of cyclodextrins in chemosensing, along with an analysis of the recent advancements and existing literature.

Chapter 2 Cyclodextrins as Molecular Hosts in Chemosensing: A Comprehensive Overview of Mechanisms—This chapter reviews mechanisms of CDs-guest interactions, highlighting advancements in CDs-based sensor design.

Chapter 3 Role of Weak Interactions and Steric Effect in Chemosensing Using Cyclodextrins—Cyclodextrins (CDs), with hydrophobic cavities, utilize weak non-covalent interactions like hydrogen bonding and hydrophobic forces to selectively bind diverse guest molecules. This chapter also reviews the steric effects, influenced by the fit between host and guest, which play a critical role in optimizing sensitivity and specificity in chemosensing applications.

Chapter 4 The Versatile Roles of Cyclodextrins in Colorimetric and Spectrophotometric Detection of Metal Ions—Cyclodextrins (CDs) enhance metal ion detection through chelation complexes, inclusion complexes, or nanoparticle functionalization, leveraging their unique binding properties. These interactions induce measurable spectral or colorimetric changes, enabling selective sensing.

Chapter 5 The Chromophore Appended Cyclodextrin Inclusion Complexes as Chemical Sensor for Organic Molecules—CDs impact on chromophoric groups, such as naphthalene and pyrene, and useful in identifying steroids, pesticides, and nitro aromatics, offering valuable insights for modern researchers.

Chapter 6 Intriguing Role of Cyclodextrin in Fluorescence-based Metal Ion Sensors—CDs enhance selectivity and anti-interference properties in fluorescence-based detection systems. Advanced designs, including CD dimers, polymers, and nanomicelle-incorporated polymer dots, leverage mechanisms like PET, FRET, and AIE for ‘turn-on’ and ‘turn-off’ sensing. These innovations highlight the potential of CDs in advancing metal ion detection technologies.

Chapter 7 Cyclodextrins as Sensors in Metals Determination: An Exploration Through Fluorometry—This chapter gives the reader an insight into the use of CDs as chemosensors for detecting metals through fluorometry. CDs are versatile supramolecular agents that form inclusion complexes with metal ions, enabling selective and sensitive detection. Their use in fluorescent probes allows accurate quantification of metal ion concentrations via changes in fluorescence intensity. This method is highly adaptable, offering significant advantages such as sensitivity, selectivity, and reliability in complex matrices. Applications span environmental monitoring, industrial processes, pharmaceutical analysis, and biological research, highlighting CDs pivotal role in advancing analytical chemistry.

Chapter 8 Fluorescence Sensing of Metal ions using Cyclodextrins—This chapter explores the principles and innovations in using modified CDs for fluorescence-based metal sensing. CDs enhance sensor performance by forming host–guest complexes and nano assemblies, improving solubility, selectivity, and tunability. These structures, such as polymer dots and micelles, exhibit sensitivity to environmental factors like pH and display unique properties when interacting with metal ions. The integration of CDs with quantum dots further supports efficient metal ion adsorption and fluorescence detection. The chapter provides a detailed overview of CDs transformative role in environmental, analytical, and biomedical applications.

Chapter 9 Electrochemical Detection of Metals Using Cyclodextrins—This chapter highlights the advancements in CDs-based electrochemical sensing platforms for detecting heavy metal ions. It addresses the urgent need for cost-effective, sensitive, and rapid detection methods to tackle pollution from non-biodegradable metals like lead, cadmium, and mercury. CDs enhance sensor performance through their ability to form inclusion complexes, often combined with nanostructured materials like carbon nanotubes, graphene, and conducting polymers. Applications in biosensors, chiral recognition, and pollutant detection are explored alongside recent trends in trace metal speciation and electroanalytical techniques. The chapter also evaluates these methods in the context of environmental quality standards and water safety directives.

Chapter 10 Electrochemical Detection of Pollutants and Biomolecules Using Cyclodextrins—This chapter reviews the potential of CDs-based functional materials for electrochemical sensing applications. CDs form host–guest inclusion complexes due to their hydrophobic cavities and hydrophilic surfaces, making them ideal for pollutant and biomolecule detection. Their integration with nanomaterials like graphene, metal nanoparticles, and conducting polymers enhances electrode performance with high conductivity and surface area. Key electrochemical techniques, including cyclic voltammetry and amperometry, are highlighted for their precision in detecting organic and inorganic pollutants. These advancements pave the way for real-time environmental monitoring, wearable diagnostics, and multi-analyte sensing technologies.

Chapter 11 Chemosensing of Small Molecules using Cyclodextrins—This chapter reviews the potential of CDs-based functional material. This chapter highlights the versatile role of CDs in developing sensors for detecting biologically significant small molecules. CDs, known for their biocompatibility, water solubility, and functionalization ease, form effective supramolecular architectures for molecular recognition-based sensing. These sensors have transformative applications in healthcare, environmental monitoring, and other fields. The chapter concludes with insights into future directions, showcasing the evolving potential of CD-based supramolecular sensors.

Chapter 12 Cyclodextrins: Chemical Modifications and Applications in Sensing—This chapter examines CDs as versatile chemical sensors for ions and small molecular species through molecular recognition. CDs are valued for their flexibility, hydrophilicity, and ability to form host–guest complexes. Though CDs lack intrinsic UV–Vis absorption, structural modifications enhance their spectroscopic sensitivity and adaptability for diverse applications. The chapter highlights synthetic strategies to optimize CDs' steric and chemical properties for sensing. Recent advancements in CD-based sensors are reviewed, showcasing their potential in synthetic organic and material chemistry.

Chapter 13 Cyclodextrin Nanoparticles: A New Frontier in Molecular Detection—This chapter explores the advancements in CDs-based nanoparticles (CD-NPs) for sensing applications, with a focus on small molecule detection. CDs, enhancing selectivity and sensitivity in electrochemical, potentiometric, and fluorescent sensing. These nanoparticles are pivotal for detecting biomolecules, environmental pollutants, and physiologically significant substances through host–guest interactions. Their applications span drug delivery, biosensing, disease detection, and therapeutic monitoring. The chapter highlights CD-NPs as versatile platforms for innovative detection systems in fields like pharmaceutical analysis, environmental monitoring, and chiral sensing.

Chapter 14 Applications of β-Cyclodextrin-linked Carbon Nanodots and Future Perspectives—This chapter highlights the importance of macrocyclic molecules like cyclodextrins and their role in supramolecular chemistry and functional materials. CDs, especially β-cyclodextrin-based carbon nanodots (CNDs), are valued for their biocompatibility, non-toxicity, and synthesis simplicity. These CNDs find applications in biomedical diagnostics, drug delivery, environmental monitoring, and food safety. Their low toxicity and brightness make them ideal for sensing, catalysis, and optoelectronics, with potential in medical imaging and intelligent material development. Future advancements in β-cyclodextrins and CNDs are expected to drive innovation in drug formulation, nanomedicine, and responsive materials. Continued nanotechnology research promises novel applications across medicine, environmental science, and other industries.

Chapter 15 Cyclodextrin-Based Sensory Molecules for Diverse Sensing Platforms—This chapter explores the use of CDs in developing chemical sensors using diverse platforms. The chapter highlights recent advancements in using CDs for colorimetric, fluorescence, electrochemical, and potentiometric sensing. It also discusses the unique optical features of CDs in monomeric, dimeric, clustered, and nano-assembled forms. The potential of CD-based sensors in detecting a wide range of compounds is thoroughly examined.

Chapter 16 Cyclodextrins as Fluorescent Chemosensors for the heavy metal ions and Volatile Organic Carbons—This chapter discusses the role of CDs in detecting volatile organic compounds (VOCs) and toxic heavy metal ions using their host–guest chemistry. VOCs and metal ions interact with CD, enabling selective and sensitive detection through fluorescence. This area of research is gaining significant attention for environmental and health monitoring.

Chapter 17 Advances in cyclodextrin-based sensors for recognition of small molecules—This chapter highlights the growing role of CDs in chemical sensors for recognizing small biologically relevant molecules. CDs are utilized in molecular recognition techniques such as electroanalysis, spectroscopy, and clinical diagnostics. The chapter focuses on advancements in fluorescence, electrochemical, and potentiometric response methods for sensing. It emphasizes chromophore-linked CDs, metallocyclodextrins, and polymer-based CDs, as well as their integration into nanomaterial-based sensor systems. The chapter also discusses the advantages and limitations of CD-functionalized materials in electrochemical sensing. Recent developments in CD-based electrochemical sensors are also explored.

Chapter 18 Sensing reactive oxygen and nitrogen species with cyclodextrins and their nanostructured systems—This chapter discusses the critical balance of reactive oxygen species (ROS) and nitrogen species (RNS) in living organisms, which act as messengers but can be toxic at excessive levels, leading to diseases like neurological and cardiovascular disorders. It highlights the importance of developing detection systems for ROS and RNS. CDs, with their ability to encapsulate molecules and enhance their properties, are ideal for designing such sensing systems. The chapter explores various CDs-based systems, from simple derivatives to complex structures incorporating polymers and nanoparticles, for monitoring oxidative species. These systems offer promising approaches for the precise detection and regulation of ROS and RNS. Representative examples of these innovative sensing technologies are also discussed.

1.12 Summary

Cyclodextrins (CDs) are a class of versatile molecules widely used in chemosensing due to their unique ability to form host–guest complexes. They are employed in a range of sensing applications, including detecting small biologically relevant molecules, metal ions, and environmental pollutants through various techniques such as fluorescence, electrochemistry, and potentiometry. CDs enhance sensor selectivity and sensitivity by forming inclusion complexes with target molecules, which can be monitored by changes in optical or electrochemical properties. Additionally, CDs' ability to encapsulate reactive oxygen and nitrogen species (ROS and RNS) is explored for detecting oxidative stress, a crucial factor in numerous diseases. The chapter further discusses the development of CD-based nanoparticles, which offer enhanced sensitivity and versatility in biosensing and environmental monitoring.

Advances in modifying CDs, such as integrating them with nanomaterials and functionalizing them with chromophores or metal ions, are also highlighted. These modifications improve sensor performance, enabling more accurate and specific detection in complex environments. CDs' biocompatibility and ease of functionalization make them ideal candidates for applications in pharmaceuticals, environmental monitoring, and diagnostics. The chapters also emphasize the future potential of CDs in nanomedicine, drug delivery, and the development of responsive materials, paving the way for innovative solutions to pressing global challenges.

References

1. Sivakumar K, Parinamachivayam G (2022) Review on β-cyclodextrin inclusion complex based chemosensors for heavy metals. J Incl Phenom Macrocycl Chem 102:603–618. https://doi.org/10.1007/s10847-022-01153-2
2. Sivakumar K, Krishna Chaitanya G (2023) α-Cyclodextrin based chemosensors: a Review. Comments Inorg Chem 43:177–201. https://doi.org/10.1080/02603594.2022.2121277
3. Sivakumar K, Nichodemus M, Sankaran KR (2014) Steric effect is a decisive factor on the orientation preference of guest into host: evidence from 5-indanol/ β-CD model. Mol Phys 112(14):1879–1891. https://doi.org/10.1080/00268976.2013.868057
4. Sivakumar K, Parinamachivayam G, MuraliKrishnan M, Ragavendran V, Stalin T, Premkumar K, SujayChakravarty, Bharathi A (2021) Reinforcement of 'imine-hydroxyl chelation pocket' by encapsulating into the β-CD cavity for the sterically protective detection of Al^{3+}. J Mol Liquid 323.https://doi.org/10.1016/j.molliq.2020.114949
5. Sivakumar K, Parinamachivayam G, Murali Krishnan M, Chakravarty S, Bharathi A (2018) Preparation, characterization and molecular modeling studies of the beta-cyclodextrin inclusion complex with benzoguanamine and its analytical application as chemosensor for the selective sensing of Ce4+. Spectrochim Acta Part A Mol Biomol Spectrosc 200:212–225. https://doi.org/10.1016/j.saa.2018.04.034
6. Sivakumar K, Ganesan V, Munisamy M, Shanmugasundaram E, Senthilvelan S, Paramasivaganesh K, Stalin T (2021) Electrochemical sensing of N-phenyl-1-naphthylamine using the MWCNT/β-CD through 'host scavenger–guest pollutant' mechanism. Chem Pap 75:1421–1430. https://doi.org/10.1007/s11696-020-01394-w

Prof. Dr. Sivakumar Krishnamoorthy serves as Professor in the Department of Chemistry, Faculty of Science, and as the Director of the Research and Development Cell at Sri Chandrasekharendra Saraswathi Viswa Mahavidyalaya (SCSVMV), a Deemed-to-be University in Kanchipuram, Tamil Nadu, India. Inspired by the versatility of cyclodextrins, his research focuses on developing innovative application materials through cyclodextrin inclusion complexes. He holds one granted Indian patent on cyclodextrin application and has filed five patent applications. He is a recipient of a research fellowship from the prestigious Indian Academy of Sciences and has authored over 115 research papers. He has secured funding from various Government of India agencies, including DST, CSIR, UGC, DAE, the Indian Academy of Sciences, and the Tamil Nadu State Council for Science and Technology, for his research projects and programs.

Chapter 2
Cyclodextrins as Molecular Hosts in Chemosensing: A Comprehensive Overview *of Mechanisms*

Priyanka Chetia, Runa Begum, and Pranjal Saikia

Abstract Cyclodextrins (CD), cyclic oligosaccharides composed of glucopyranose units, have garnered considerable attention as molecular hosts in the realm of chemosensing due to their unique structural features and molecular recognition capabilities. This review focuses on exploring the fundamental principles and recent advances of CDs in the field by examining the mechanisms governing CD-guest interactions and the design considerations for CD-based sensor platforms. Through a critical analysis of current challenges and emerging trends, it outlines the roadmap for future advancements in CD-mediated chemosensing, paving the way for enhanced detection capabilities and broader applications.

Keywords Cyclodextrins · Host–guest Interactions · Sensing · Molecules · Ions

2.1 Introduction

In various fields of chemistry and materials science, there has been a recent surge of interest in the self-assembly of molecular systems into meticulously ordered nanostructured functional assemblies, spurred by supramolecular-driven interactions [1, 2]. Macrocyclic hosts, such as crown ethers, cucurbiturils, cryptands, and calixarenes, have emerged as key players in this endeavor, attracting considerable attention. These hosts have been extensively utilized as fundamental building blocks for self-assembled structures, often serving as highly efficient host molecules in host–guest chemistry [3–6]. Among these hosts, cyclodextrins (CDs) stand out as particularly vital and promising due to their inherent characteristics: affordability, water solubility, non-toxicity, commercial availability, and ease of functionalization [7, 8]. Composed of α-D-glucopyranose units connected by 1,4-type glycosidic

P. Chetia · R. Begum · P. Saikia (✉)
Department of Applied Sciences (Chemical Science Division), Gauhati University, Guwahati, Assam, India
e-mail: psjorhat@gmail.com

S. Krishnamoorthy (ed.), *Cyclodextrins for Chemosensing*,
https://doi.org/10.1007/978-3-031-89299-8_2

linkages, cyclodextrins are cyclic oligosaccharides known for their ability to encapsulate hydrophobic molecules, a property extensively utilized in the pharmaceutical industry for solubilizing and delivering drugs [9–11].

In the annals of scientific exploration, cyclodextrins (CDs) emerged as a revelation in 1891, first discerned by Villiers amidst the labyrinthine pathways of starch [8, 12]. Subsequently, Schardinger bestowed upon them the titles of α-, β-, and γ-CDs, unveiling their regal stature [13, 14]. Pringsheim's pioneering work illuminated their extraordinary capacity to form steadfast alliances with pharmaceutical compounds, while Higuchi and Connors crafted an elegant symphony of mathematical precision, elucidating the ballet of inclusion complexation. The grand debut of CDs in therapeutics arrived with Prostaglandin E2/βCD, gracing the world stage as ethereal sublingual tablets in 1976 [8, 15]. Since then, CDs have adorned the realms of gastronomy and aesthetics, embodying the epitome of scientific elegance.

Cyclodextrins, cyclic oligosaccharides composed of six or more D-(+)-glucopyranose units linked by α-1,4-glycosidic bonds, manifest in an array of forms dictated by their glucopyranose count [16, 17]. While α-, β-, and γ-CDs house six, seven, and eight units respectively, larger variants like ζ-CD, η-CD, and θ-CD boast even greater numbers [18]. Despite their grandeur, larger cyclodextrins demonstrate lower yields and diminished complexing properties compared to their more diminutive counterparts, being less significant pharmaceutically. Furthermore, their collapsed structure yields a smaller cavity than γ-CD, hindering their utility [19]. Produced from renewable sources, α-, β-, and γ-CDs exhibit a cone-like shape, with increasing glucopyranose units correlating to a larger cavity diameter. This hydrophobic interior and hydrophilic exterior enable CDs to form molecular inclusion complexes, epitomizing their versatile and enduring utility in solution and solid states [20].

CDs, with their characteristic truncated cone structure, harbor a hydrophobic cavity due to the presence of C-3 and C-5 hydrogens alongside a glycosidic oxygen atom, while the exterior presents a hydrophilic façade [8, 21]. Along the broader secondary rim, hydroxyl groups grace C-2 and C-3, while the narrower primary rim boasts a lone C-6 hydroxyl group (Fig. 2.1). These structural nuances endow CDs with versatility across myriad chemical and analytical applications, from sample preparation to spectroscopic analysis. Notably, CDs exhibit a unique prowess in selectively forming noncovalent host–guest complexes, guided by molecular and chiral recognition processes [22]. Within these interactions, hydrogen bonding, van der Waals interactions, electrostatic forces, and hydrophobic interactions intertwine, driven by the pursuit of conformational harmony and energy relief [14, 23]. However, the comparison of noncovalent interactions between gas and condensed phases warrants caution, as polar interactions dominate in the former, while hydrophobic interactions weaken due to the absence of solvent shielding in the latter. With their distinctive structure—a hollow truncated cone featuring a hydrophilic outer surface and a lipophilic inner cavity—cyclodextrins interact effectively with hydrophobic molecules, enhancing their solubility in water [24]. This characteristic makes cyclodextrins invaluable in various applications across biomedical, environmental, biosensing, and food science fields [25, 26]. Furthermore, CDs exhibit optical transparency in the ultraviolet and

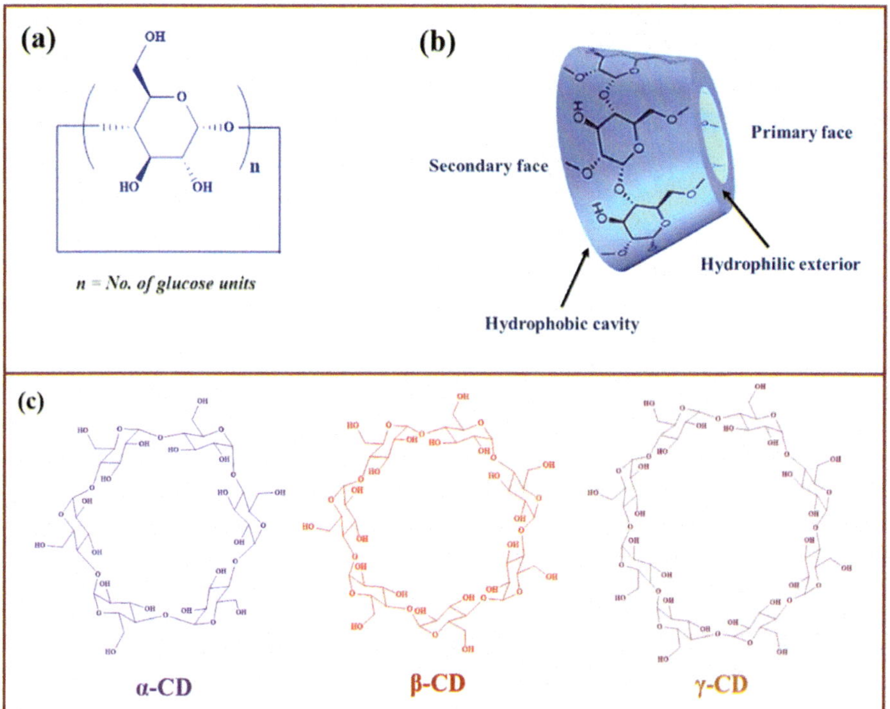

Cyclodextrin	α	β	γ
Glucopyranose units	6	7	8
Formulae	$C_{36}H_{60}O_{30}$	$C_{42}H_{70}O_{35}$	$C_{48}H_{80}O_{40}$
Molecular weight (g/mol)	972.9	1135.0	1297.1
Central cavity diameter: external/internal (Å)	5.3/4.7	6.5/6.0	8.3/7.5

Fig. 2.1 Structures of cyclodextrins (CDs): **a** the monomer unit; **b** the overall form; **c** the α-, β-, as well as γ-CD structures (n = 6, 7, and 8, respectively)

visible regions of the spectrum, thereby circumventing any potential interference with light signals during imaging procedures. Additionally, unlike many other macrocyclic compounds that require intricate multistep synthesis, CDs are readily available commercially, serving as convenient "ready-made" molecular entities [27, 28]. Moreover, CDs are inherently safe and biocompatible in their native forms, with a well-established track record of directly enhancing imaging signal intensity. These favorable attributes of CDs have spurred significant interest in CD-based "host–guest chemistry" within sensing research [29–31]. Their ability to act as host molecules for a wide range of substances, facilitating stabilization and solubilization, positions cyclodextrins as attractive candidates for the development of functional materials.

Table 2.1 Dimensions of cyclodextrin molecules

CDs	Height of torus (Å)	Cavity volume ($Å^3$)	Inner diameter (Å)	Outer diameter (Å)	Melting point (°C)	Specific rotation
α-CD	7.9	174	5.7	13.7	278	+146° to + 153°
β-CD	7.9	262	7.8	15.3	290–300	+160° to + 164°
γ-CD	7.9	427	9.5	16.9	267	+174° to + 180°

The nuances among cyclodextrin types hinge upon their size, notably reflected in the inner diameter of their cavities, spanning from 4.7 to 8.3 Å. As glucose units increase, so does the cavity size (Table 2.1). This dimensionality dictates the encapsulation potential for diverse entities such as drug molecules, proteins, oligonucleotides, or ions, with binding affinity primarily influenced by the cavity's depth [19, 24]. Furthermore, the solubility order of native cyclodextrins (CDs) at 25 °C follows $\beta < \alpha < \gamma$, with β-CD exhibiting significantly lower solubility than α-CD and γ-CD. This disparity arises due to the "flip-flop type intramolecular H-bonding" in β-CD, which creates a rigid structure by forming intramolecular H-bonds between the -OH groups of the 2° C-2 and C-3 carbons [10, 14]. This rigidity impedes β-CD's interaction with water molecules, resulting in its diminished solubility compared to α-CD and γ-CD. While all three types of CDs possess intramolecular H-bonding, the permanent rigid intramolecular H-bonding loop formed in β-CD due to the "flip-flop" pattern contributes to its reduced solubility. Researchers often resort to chemical modification approaches to enhance the solubility of native CDs, thereby augmenting their selectivity and sensing capabilities [8, 32]. Consequently, the greater solubility of α-CD over β-CD constitutes an advantageous feature, facilitating broader utility in sensing and molecular recognition applications.

2.2 Mechanisms in Cyclodextrins-Based Chemosensing

2.2.1 Mechanism in CD-Based Metal Complexes

In the realm of materials science, noble metal nanomaterials like gold (AuNMs), silver (AgNMs), and palladium (PdNMs) boast biocompatibility, stability, and modifiability [33, 34]. Their surfaces host unique properties such as catalytic prowess, optical allure, and electrochemical finesse. Combined with diverse molecular functionalities, they yield novel composites with fluorescence, phosphorescence, and surface-enhanced Raman scattering (SERS). Noble nanomaterials, when adorned with β-CDs, find utility in environmental analyses, including electrochemical, colorimetric, fluorescence, and SERS assays, heralding a new frontier in scientific exploration. Chen et al. delved into a pioneering electrochemical sensor employing AgNMs adorned with β-CD for nitroaromatic compound detection. This innovative sensor proved effective in detecting nitroaromatic compounds in natural water samples, boasting a remarkable detection limit of 50 nM [35].

The utilization of CD-based metal complex systems extends across chiral selection and chemical sensing domains. For instance, the dinuclear $Cu(II)_2$-β-CD complex emerges as a potent chiral selector, enhancing the enantio-separation of aromatic α-hydroxycarboxylic acids. This heightened enantio-separation capability is attributed to ion-pairing interactions between the deprotonated carboxyl group of the α-hydroxycarboxylic acid and the Cu(II) ion within $Cu(II)_2$-β-CD. Meanwhile, the $[Pb_{14}(\beta\text{-}CD)_2]0.35H_2O$ complex serves as a fluorescent probe, enabling selective detection of H_2S within living cells [36]. This sensing mechanism relies on the coordination of H_2S with Pb(II) ions, thereby significantly amplifying fluorescence emission. Moreover, the $[Pb_{14}(\beta\text{-}CD)_2]0.35H_2O$ complex facilitates the creation of visual H_2S test papers, boasting excellent sensitivity and selectivity to H_2S [37]. Furthermore, this complex demonstrates reducing capabilities towards $AuCl_4^-$, resulting in the formation of Au@Pb-β-CD NPs that exhibit exceptional electrochemiluminescent behavior. This property is exploited in the development of an immunosensor for insulin detection, where energy transfer from Au@Pb-β-CD NPs to CRuSi NPs [38]. The fabricated immunosensor, coated with insulin antigen solution and CRuSi NPs secondary antibody (Ab_2-CRuSi NPs), provides a remarkable electrochemiluminescent response, facilitating accurate insulin quantification.

2.2.2 Mechanism in CD-Based MOF Hybrids

A myriad of sensors have emerged for detecting various species, leveraging the fluorescence properties of certain CD-MOFs. Their operational mechanisms, albeit less intricate, mirror those of the aforementioned hybrid materials. One such sensor utilizes a γ-CD-MOF infused with tetraphenylethylene (TPE) molecules within its framework. Remarkably, TPE@K-γ-CD-MOF displayed heightened fluorescence compared to TPE alone. Upon exposure to nitro-aromatic compounds, TPE@K-γ-CD-MOF exhibited fluorescence quenching, endowing it with sensing capabilities [39]. Another fluorescence-centric sensor was devised by Xin et al. for H_2S detection [36]. Employing a CD-based metal–organic nanotube harboring a metallamacrocycle, the sensor demonstrated increased fluorescence intensity upon H_2S coordination to central Pb(II) ions as the auxochromic group. Farha's research team uncovered that CD-functionalized MOFs (CDs/MOFs) contain numerous free hydroxyl groups capable of reacting with carbon dioxide, rapidly forming alkyl carbonates within the nanomaterials. This property renders them ideal for detecting carbon dioxide [40]. Consequently, they introduced a sensing approach for CO_2 detection in CO_2/N_2 and CO_2/air mixtures, leveraging these nanomaterials.

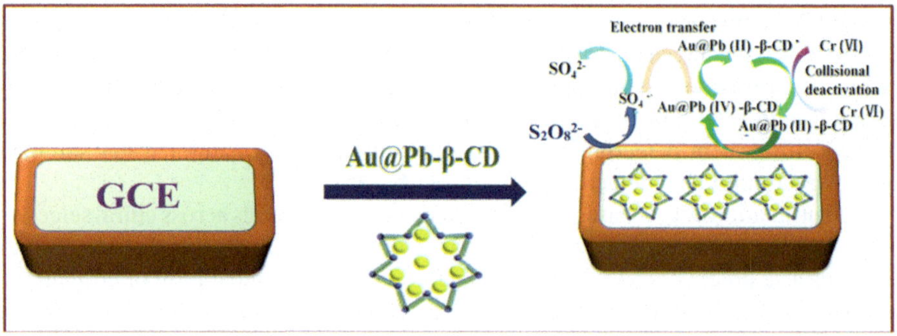

Fig. 2.2 Detection of toxic Cr(VI) ions via Au@Pb-β-CD hybrid material [41]

Ma et al. developed a biosensor for detecting toxic Cr(VI) ions using a novel Au@Pb-β-CD hybrid material with enhanced electrochemiluminescence (ECL) properties [41]. The sensor's high sensitivity relied on the ECL response of Pb-β-CD-MOF, which increases upon immobilization of Ag or Au in the framework. This phenomenon facilitated effective Cr(VI) sensing (Fig. 2.2). The proposed mechanism suggests that during the oxidation–reduction reaction between Pb-β-CD-MOF and $S_2O_8^{2-}$, the former transitions into an excited state, deactivated upon collision with Cr(VI) ions. This process diminishes ECL intensity, enabling Cr(VI) ion detection. The sensor's selectivity was assessed in the presence and absence of ethylenediamine tetraacetic acid (EDTA), which selectively masked cations other than Cr(VI), enhancing detection sensitivity.

2.2.3 Mechanism in CD-Based Quantum Dot Hybrids Top of Form

Quantum dots (QDs) represent a class of materials with remarkable properties, including tunable bandgap and vibrant color emission [42]. To broaden their utility, CD molecules have been introduced to QD surfaces, forming host–guest complexes for sensing applications. For instance, the integration of carbon quantum dots within a net-polymer matrix comprising α-cyclodextrin and citric acid facilitates the extraction and identification of toluene and xylene from water-based solutions, opening avenues for the advancement of novel optical sensing tools [43]. Similarly, α–CD-functionalized Ag nanoparticles facilitated the detection of o–phenylenediamine via surface-enhanced Raman spectroscopy (SERS) with optimized parameters [44].

In a recent study by Dong and co-workers, modified CDs were employed in conjunction with graphene quantum dots to successfully separate tyrosine enantiomers [45]. Initially, β-CD quantum dot derivatives were synthesized for integration into an electrode setup, facilitating the separation process. Additionally, an electrochemical sensor utilizing magnetic graphene oxide/β–CD/Au nanoparticles displayed high sensitivity in detecting chrysoidine in water [46]. Despite initial constraints, the synergistic combination of CDs and QDs, exemplified by QD@CD inclusion compounds, holds significant promise for future applications. R. Sangubotla and J. Kim fabricated blue-fluorescent carbon nanoparticles (CDs) that were synthesized via the combination of melamine and 3-carboxyphenylboronic acid (3-CPBA), tailored with β-cyclodextrin to produce β-MPBCDs, employed for the specific identification of dopamine [47]. Following the addition of PYL, the fluorescence emission of β-MPBCDs experienced augmentation, leading to the development of β-MPBCDs-PYL probe, showcasing elevated fluorescence sensitivity towards dopamine at varying concentrations.

2.2.4 Mechanism in CD-Based Carbon Nanotubes

Carbon nanotubes (CNTs) offer excellent potential as versatile platforms for sensor development due to their facile chemical modification, robust mechanical strength, and high electrical conductivity [48]. β-CD-modified CNTs have emerged as particularly promising candidates for detecting pesticides, pharmaceuticals, and other organic pollutants in the environment [49, 50]. For instance, Liu and co-workers successfully immobilized β-CD/multiwalled CNTs on carbon paper to create an efficient electrochemical sensor for detecting methyl parathion, a pesticide [51]. Additionally, Kong and the research team employed *p*-phenylenediamine as a linker to tether β-CD onto single-walled CNTs (SWCNTs), enabling the sensing of persistent organic pollutants through host–guest interactions [52]. Howlader's group explored various methods to modify multiwalled CNTs (MWCNTs) with β-CD, ultimately finding that Steglich esterification was the preferred approach [53]. This method yielded MWCNT-β-CD composites with significant β-CD content, offering large effective surface areas and efficient electron transfer capabilities. Such composite materials served as the basis for novel electrochemical sensors capable of accurately detecting pharmaceuticals in real water and urine samples.

Li and collaborators devised a method to create β-CD covalently functionalized SWCNTs nanohybrids through an amidation reaction involving mono-6-deoxy-6-hexanediamino-β-CD (MH-β-CD) and SWCNTs-COOH [54]. Employing these MH-β-CD/SWCNTs, they developed an electrochemical sensor for bisphenol A. Results demonstrated that the MH-β-CD/SWCNTs modified GCE exhibited robust catalytic activity in oxidizing bisphenol A. Zhao and Li's group pioneered the development of disulfide-bridged β-CD dimer-functionalized MWCNTs (DBβ-CD/MWCNTs) nanohybrids for the simultaneous detection of three phenols (4-chlorophenol, 4-aminophenol, 4-nitrophenol) [55]. Their modified electrode, incorporating DBβ-CD/MWCNTs, exhibited significantly enhanced electrochemical signals compared to electrodes with mono-β-CD/MWCNTs or MWCNTs alone, demonstrating superior performance in detecting all three target phenols.

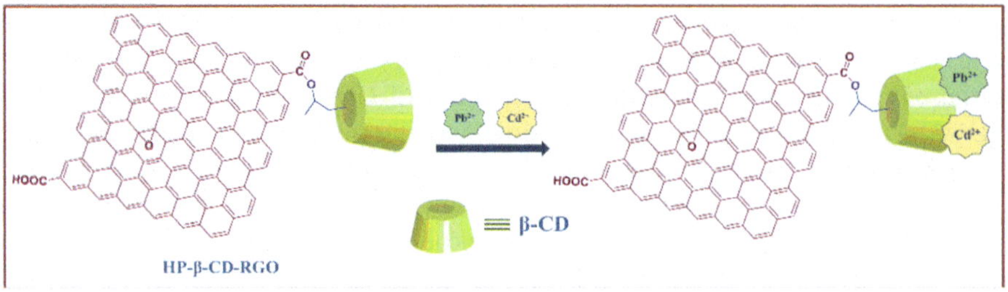

Fig. 2.3 Schematic representation of HP-β-CD-RGO hybrid nanomaterial and their interaction with Pb^{2+} and Cd^{2+}ions [59]

2.2.5 Mechanism in CD-Based Graphene Materials

Graphene and graphene oxide (GO) stand out among carbon nanomaterials for their superior conductivity, electrocatalytic activity, and large surface area [56]. They serve as excellent electrode materials for constructing electrochemical sensors or biosensors. Their versatility allows for integration with various functional materials, enabling promising applications across nanoelectronics, nanocomposites, supercapacitors, and sensor technologies.

Guo et al. introduced a sensitive sensing platform for detecting carbendazim using CDs-graphene-modified glassy carbon electrode (CDs-graphene/GCE). Their study showed enhanced electrochemical performance compared to graphene/GCE and bare GCE, indicating superior molecular recognition properties of CDs and excellent electrochemical properties of graphene [57]. Meanwhile, Wang and colleagues utilized microwave-assisted decoration of reduced GO with hydroxypropyl β-CD, exhibiting robust electrochemical response to phenolic organic pollutants [58]. Additionally, when modified onto GCE with Nafion, the sensor displayed improved sensitivity, stability, and reproducibility for $Pb^{2+}$and Cd^{2+} detection (Fig. 2.3) [59]. In a recent study, Guo's team developed novel composite materials by combining copper nanospheres (CuNSs) with β-CD-SH-immobilized reduced graphene oxide (SH-β-CD-RGO/CuNSs). Utilizing a hydrothermal method, they synthesized the SH-β-CD-RGO [60]. Functionalization of CuNSs significantly enhanced the reduction of 4-nitrophenol (4-NP). Their electrochemical sensor for 4-NP demonstrated excellent sensitivity within the range of 0.05–25 μM and 25–100 μM, boasting a low detection limit of 20 nM. This sensor proved effective in detecting 4-NP in environmental water samples.

2.2.6 Mechanism in CD-Based Conducting Polymer Hybrids

The combination of conducting polymers (CPs) and cyclodextrins (CDs) to create CP-CD hybrids presents a promising strategy for developing novel materials that capitalize on the strengths of both components, offering versatile practical applications [61, 62]. CPs exhibit inherent advantages, particularly in electrical signal transport, making them highly attractive for sensor applications [63]. Given the affinity between CPs and CD derivatives, their synergy has been actively explored across various fields. Among the widely used CPs like polypyrrole (PPy) and polyaniline (PANI), hybrid formations with CDs have been extensively investigated, showcasing their potential for innovative material design and application development [64–66].

In pioneering work by Breslin's research group, a novel hybrid material was synthesized, comprising a polypyrrole (PPy) film doped with sulfonated β-CD, demonstrated through cyclic voltammetry scans [67]. This innovative PPy-β-CD hybrid exhibited remarkable selectivity in detecting dopamine even in the presence of competitive compounds like ascorbic acid [68]. Moreover, employing a similar methodology, they developed a urea sensor with an extraordinary sensitivity of 0.1 nM. These advancements underscore the potential of hybrid materials for high-performance sensing applications. Luong's team devised a boron-doped diamond electrode coated with an electropolymerized composite film of sulfobutylether-β-CD-doped poly(N-acetyltyramine) and polypyrrole, achieving selective nanomolar detection of dopamine [69]. Incorporating N-acetyltyramine aimed to enhance surface hydrophobicity, reducing surface fouling. The inclusion of negatively charged sulfobutylether-β-CD selectively absorbed dopamine while repelling interferents like uric acid and ascorbic acid. Another study by Chen et al. developed β-CD/poly(N-acetylaniline)/CNTs composite-modified carbon paste electrodes, enabling a dual-signaling electrochemical sensing strategy for 1-aminopyrene (1-AP) based on competitive host–guest interactions [70]. Yashima et al. synthesized helical polyacetylene-carrying CD moieties, termed Polym-2 [71, 72]. Polym-2 exhibited a visible color alteration due to dynamic changes in the conjugated backbone caused by the inclusion of a complex into the CD cavity. Furthermore, the helicity of Polym-2 could be manipulated by complexing with chiral guests. Interaction with (S)-1-phenylethylamine induced a color shift from yellow to red, whereas (R)-1-phenylethylamine addition did not result in any observable color change [73].

2.2.7 Mechanism in Aryl-Substituted CDs

The endeavor to develop aryl-substituted cyclodextrins (CDs) seeks to create photochemical molecular devices for detecting spectroscopically inactive molecules through fluorescence. Substituted CDs undergo altered spectroscopic properties upon forming host–guest complexes, leading to changes in their fluorescent spectrum. This, coupled with CDs' capability to discern guest molecule characteristics, renders them versatile molecular sensors [74, 75].

Ueno et al. pioneered the synthesis of aryl-substituted CDs, demonstrating significant fluorescence changes in pyrene-modified γ-CDs with suitable analytes [76, 77]. This foundational work laid the groundwork for CD-based chemosensors, with subsequent advancements including Aoyagi's improved 'turn-off' chemosensors, enhancing sensing robustness through intramolecular excimer formation [78]. However, challenges persist in synthesizing regioisomerically pure disubstituted CDs. Additionally, the linker flexibility between the CD core and fluorescent substituent influences intramolecular complexation. Mallard et al. illustrated how linker flexibility affects the inclusion ability of guest molecules [79]. Beyond chemosensing, anthrylamine-substituted CDs hold promise for DNA-targeted drug delivery, leveraging their intercalating properties and CD cavity binding ability. Recent developments have yielded sensors capable of distinguishing adenosine phosphates based on the self-inclusion phenomenon of anthrylamine-substituted CDs [80].

2.2.8 Mechanism in Dansyl-Substituted CDs

CDs modified with a dansyl (DN) moiety constitute a fascinating class of fluorescent chemosensors, renowned for their efficacy in detecting molecules in aqueous environments [81]. Their operational principle hinges upon the remarkable fluorescence changes exhibited by the DN moiety in response to the polarity of the surrounding medium. Notably, the DN emission experiences a significant enhancement within hydrophobic microenvironments, including enzyme binding pockets or even the confined spaces within CD molecules themselves. It has been elucidated that covalently linked DN-β-CD conjugates effectively encapsulate the DN moiety within the CD cavity, thereby substantially augmenting the fluorophore's quantum yield [82]. Upon the inclusion of guest molecules, a notable decrease in fluorescence intensity is observed, thereby classifying DN-based sensors as 'turn-off' chemosensors. These foundational insights into DN-CD chemosensing were first reported by Ueno and team, who ingeniously linked 6-monoamino-β-CD to DN-glycine, thereby enhancing the performance of Pyr-γ-CD chemosensors [83]. Unlike their Pyr-γ-CD counterparts, DN-*β-CD* operates in a monomolecular fashion, exhibiting exceptional sensitivity in detecting steroidal compounds. Further advancements have been made through the preparation of regioisomers of monosubstituted DN-γ-CD [84] and DN-*β*-CD, demonstrating distinct sensitivities towards guest molecules [85]. This underscores the critical importance of achieving regioisomeric purity, not only for molecular recognition but also for applications relying on the fluorescence emission of fluorophore-tagged CDs.

The influence of Cu(II) ions on altering DN moiety fluorescence has been meticulously studied. Interestingly, while Cu(II) ions showed no quenching effect on ethylenediamine-bridged DN-*β-CD* in its self-included state, quenching occurred upon expulsion of DN by competitive guest molecules. This phenomenon is attributed to the conformational constraints of the conjugate, which impede Cu(II) ion complexation in the self-included species [82]. However, with a longer and more flexible ethylenetriamine linker, the self-included species allowed metal binding, with notable selectivity for Cu(II) ions. Furthermore, incorporating a chiral center into the linker via an amino acid motive has facilitated the creation of powerful fluorescent enantioselective sensors. These sensors leverage the formation of ternary diastereomeric complexes between the DN-AA-*β-CD* conjugate, Cu(II) ions, and the analyte [86]. As these complexes exhibit varying stability, alterations in fluorescence intensity or Cu(II) quenching effects differ, rendering the enantioselective process detectable via fluorescence. The extensive research conducted on DN-CD-based chemosensors, particularly their sensing ability towards steroidal compounds, has propelled these systems towards practical applications [87]. This is driven by the pressing need for fast and reliable cholesterol sensing in medicine and daily life. Recent advancements include the construction of sensory devices based on DN-glycine-modified *β-CD* immobilized on cellulose membranes, along with the development of molecularly imprinted chemosensors using DN-modified *β-CD* as a template molecule [88]. These findings, coupled with diverse fabrication techniques, open up exciting possibilities for disposable sensor chips or papers, ensuring rapid and precise detection of biologically essential molecules.

2.2.9 Mechanism in Xanthene-Dye-Tagged CDs

The initial exploration of xanthene dye modification on CDs aimed to develop chemosensors with additional capabilities compared to aryl-CDs. Fluorescein (Flu), a xanthene dye derivative, exhibits variable ionization states (cationic, anionic, and neutral) depending on pH [89, 90]. Wang et al. synthesized mono-6-fluoresceinyl-*β-CD* (Flu-*β-CD*) and highlighted the significant influence of Flu moiety ionization on guest molecule inclusion [91]. This charge-changeable receptor effectively discriminated guests based on electrostatic interactions, with the cationic form showing superior binding to adamantane carboxylic acid compared to the anionic form. Intriguingly, Flu-*β-CD* could also differentiate uncharged guests, suggesting the role of charge effects in binding such molecules. Molecular orbital calculations supported these findings, indicating electrostatic potential gradients within the CD cavity upon guest complexation.

Subsequent efforts by Hamada et al. led to the synthesis of mono-6-fluorescein-*γ*-CD (Flu-*γ*-CD), demonstrating pH-tunable complexation properties distinct from Flu-*β*-CD [92]. Despite potential sensing abilities towards aliphatic compounds, the Flu-*γ-CD* analogue remains unexplored. Xanthene dyes like Flu and rhodamine B (RhB) are extensively used in biological imaging, necessitating stable conjugation strategies with CDs to explore CD-based drug delivery in biological settings [93]. However, previous Flu-CD chemosensors utilized ester bonds, limiting biomedical applications due to enzymatic degradation. To address this, Harada's group developed stable RhB-α-CD conjugates through amide formation, enabling visualization of CD-based rotaxanes on glass surfaces [94]. Similarly, Hasegawa et al. synthesized RhB-linked *β-CD* derivatives using amide bonds, leveraging pH-sensitive probes to visualize acidic cellular compartments [95]. The pH sensitivity of RhB-linked CDs, combined with aqueous solubility and low cytotoxicity, facilitated their application in biological fields, marking a significant advancement in Rh*B*-*CD* research within biological matrices.

In a pioneering endeavor, Mourtzis et al. [96] introduced a novel synthetic approach involving the attachment of fluorescein isothiocyanate (FITC) to per-6-guanidinoalkylamino- and per-6-aminoalkylamino-*β*- and *γ*-*CD*s [96]. This method yielded derivatives featuring a robust thioureido bond between the probe and the CD core, ensuring enzymatic stability. While the primary objective was biological visualization, particularly examining the DNA transfection process on HeLa cells using fluorescent microscopy, the detailed synthesis offered a promising avenue for future investigations [96]. This seminal work marked the inception of in-vitro exploration into fluorophore-modified CDs, showcasing their potential for diverse applications. Following this effort, Wenz et al. adopted a similar strategy to prepare FITC-tagged mono-6-amino-*β*-*CD*. Incorporating this conjugate into a polymeric backbone alongside unmodified α-CD resulted in a polyrotaxene, enabling visualization through fluorescence microscopy and atomic force microscopy [97]. Significantly, both research groups, led by Harada and Wenz, renowned for their contributions to rotaxane molecules, utilized xanthene-dye-appended CDs to gain insights into molecular structures [94, 97]. This indirect contribution highlights the versatility of fluorophore-tagged cyclodextrins, expanding their relevance beyond traditional applications. The fusion of synthetic chemistry and biological visualization not only advanced the understanding of molecular structures but also paved the way for innovative applications.

2.2.10 Mechanism in Porphyrinoid- and Phthalocyanoid-Tagged CDs

Porphyrins (Por) and phthalocyanines (Pc) have garnered significant attention as fluorescent molecules in conjunction with cyclodextrins (CDs) [98–100]. The concept of linking these macrocycles with CDs dates back to 1984 when Gonzalez et al. pioneered the preparation of porphyrinoid-CDs (Por-CDs) to simulate light-induced electron transfer in photosynthesis [101]. Subsequently, numerous research groups explored various Por-CD systems due to their structural resemblance to enzymes, with the Por moiety mirroring the active site and the CD moiety serving as the binding site [100]. Various Por-CD architectures have been constructed, including Por-capped α- and *β*-*CD*s, Por-linked *β*-*CD* dimers, and Por-linked *γ*-*CD* tetramers, mimicking P450 enzyme models and facilitating photoinduced electron transfer. These constructs exhibited remarkable regio- and stereoselectivity, making them valuable for catalytic applications, such as epoxidation of alkenes and selective cleavage of double bonds in carotenoids.

A novel porphyrin assembly on β-Cyclodextrin supramolecular sensitizer targeting zinc ions had been conceptualized by Yang and co-workers, leveraging the ratio of dual fluorescence emissions from the porphyrin [102]. This inclusion complex holds promising potential for biomedical applications. Fan *et al.* devised a novel biomimetic configuration by coupling per-O-methylated cyclodextrin dimer with cobalt proto-porphyrin (Co^{III}-PPIX@Py2CD), demonstrating an expansive linear detection range for hydrogen peroxide, underscoring its economical prowess as a catalyst for (bio)analytical applications [103]. Zhang's team synthesized a novel porphyrin compound, α, β, γ, δ-tetrakis(3,5-dibromo-2-hydroxylphenyl)porphyrin (TDBHPP), investigating its interaction with β-CD and exploring the fluorescence sensing capability of the resultant inclusion complex towards Pb^{2+} ions in aqueous solutions [104]. The heightened fluorescence exhibited by the complex undergoes rapid and complete quenching upon exposure to lead ions at room temperature. In the pursuit of constructing a pesticide sensor aimed at detecting boscalid within solutions, the adaptation of a perfluorinated copper phthalocyanine ($F_{16}CuPc$) organic field-effect transistor (OFET) with β-cyclodextrin (β-CD), as devised by Liu et al., facilitated the enhancement of sensing capabilities by virtue of pioneering molecular interactions [105]. The construction of the target drug delivery system ZnPc-$(PEG)_5$:Ac-CD:DOX involved the self-assembly of acetalated β-CD and DOX, along with the coating of amphiphilic phthalocyanine ZnPc-$(PEG)_5$, showcasing remarkable tumor-targeting capabilities in animal studies [106]. This strategy formulated by Zheng and coworkers employed for developing drug delivery systems, incorporating pH-responsive release and combination therapy utilizing CDs, offers valuable insights into the realm of CD chemistry and holds promise for application to a wide array of antitumor drugs.

A detailed comparison of various modified cyclodextrins (CDs), highlighting their unique structural adaptations and versatile applications, is elegantly presented in Table 2.2.

2.2.11 Mechanism in CD-Based Other Supramolecules

Khan et al. demonstrated a novel inclusion complex of a coumarin derivative within β-CD for sensing Cu^{2+}. Molecular modeling revealed that the benzothiazole group of the complex is encapsulated inside the β-CD cavity, while the coumarin component protrudes outside. This complex selectively detects Cu^{2+} ions over other cations and exhibits quenching of the 465 nm emission band upon Cu^{2+} addition. Computational studies suggested that Cu^{2+} coordinates with specific functional groups of β-CD and the coumarin derivative, forming a stable 1:1 complex. Confocal microscopy confirmed the efficacy of this complex as an 'on–off' sensor for intracellular Cu^{2+} in doped HeLa cells [110]. Additionally, Liu et al. developed a water-soluble β-CD for fluorescence sensing of Zn^{2+} in biological systems. The probe selectively binds Zn^{2+}ions, resulting in a significant fluorescence enhancement and blue shift in emission wavelength. This sensor effectively penetrated onion epidermal cells' membranes in fluorescence microscopy studies, exhibiting noticeable changes in emission profile upon Zn^{2+} treatment [111].

Table 2.2 Comparison of different types of modified cyclodextrins (CDs)

Modified CDs	Key Characteristics	Notable applications	Refs
Aryl-Substituted CDs	Alteration of spectroscopic properties upon complex formation; versatile molecular sensors for detecting spectroscopically inactive molecules	Detection of inert molecules through fluorescence; versatile sensing	[82, 107]
Dansyl-Substituted CDs	Effective in aqueous environments; fluorescence changes based on polarity; 'turn-off' chemosensing mechanism; enhanced sensitivity to steroidal compounds	Detection in hydrophobic microenvironments; DNA-targeted drug delivery	[86, 108]
Xanthene-Dye-Tagged CDs	Variable ionization states dependent on pH; discrimination of guests based on electrostatic interactions; potential for biomedical imaging and gene therapy	Visualizing CD-based structures on surfaces; biomedical imaging; gene therapy	[93]
Porphyrinoid and Phthalocyanoid-Tagged CDs	Simulating light-induced electron transfer; catalytic applications with remarkable selectivity; enhanced water solubility and aggregation behavior	Catalysis in epoxidation and selective cleavage; biodistribution studies; chiral sensing; tumor-targeted delivery; cancer therapy	[100, 109]

Similarly, a triazole-tetraphenylethylene (TPE) substituted β-CD was reported as a highly sensitive sensor for Cd^{2+}, with selectivity in water/DMSO solution [112]. The sensing mechanism relies on aggregation-induced emission (AIE), where Cd^{2+} binding disrupts the aggregation state of the ligand, leading to fluorescence enhancement. The 2:1 complex formation between Cd^{2+} and the ligand was confirmed by Job's plot analysis, involving coordination to the triazole N_2 nitrogen and a primary hydroxyl group of the β-CD from two ligand molecules (Fig. 2.4). These studies highlight the versatility of β-CD-based complexes as fluorescence sensors for metal ions, offering selective detection and potential applications in biological imaging and environmental monitoring. The combination of molecular modeling, computational simulations, and experimental validation enhances our understanding of the sensing mechanisms and paves the way for the development of more advanced sensing platforms.

The adamantyl group's high affinity for β-CD has been utilized in sensing $H_2PO_4^-$ ions by incorporating a sensor with β-CD [113]. The synthesized probe exhibited strong fluorescence in response to $H_2PO_4^-$, showing selectivity over other anions. The formation of a complex via anthracene fluorophore stacking between these

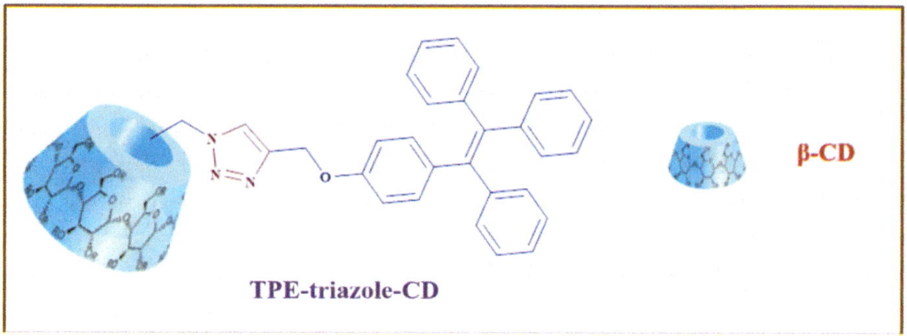

Fig. 2.4 Structure of triazole-tetraphenylethylene (TPE) substituted β-CD [112]

units was suggested by fluorescence and ^{1}H-NMR spectroscopic data. The addition of β-CD to the complex disrupted anthracene unit stacking, leading to exciplex band quenching and emission wavelength blue shift. Anthracene-substituted β-CD sensor was reported as a Pb^{2+} sensor by Antony and team members [114]. Fluorescence enhancement of the 537 nm emission band was observed only after Pb^{2+} addition, likely due to charge transfer from nitrogen atoms to the anthracene fluorophore. Competitive binding experiments revealed strong binding of Pb^{2+} to the sensor, forming a 1:1 complex confirmed by Job's plot analysis. Kanagaraj et al. reported Hg^{2+} sensing using naphthamide β-CD inclusion complex via colorimetric detection and fluorescence [115]. The designed sensor, comprising a 3-hydroxy-N-phenyl-2-naphthamide within per-6-amino-6-β-CD, showed high selectivity for Hg^{2+} over other cations. The binding of Hg^{2+} resulted in absorption profile changes, fluorescence emission blue shift, and intensity quenching. The fluorescence quenching effect was attributed to disrupted excited-state intramolecular proton transfer (ESIPT) as Hg^{2+} coordinates with the sensor, disrupting ESIPT and causing fluorescence quenching. Environmental applications of the sensor were demonstrated with doped Hg^{2+} in real-world water samples. These studies illustrate the diverse applications of β-CD-based sensors in detecting metal ions with high selectivity and sensitivity, providing insights into their mechanisms and potential for environmental monitoring and biological sensing.

The diverse applications of complexes functionalized with cyclodextrins (CDs) in ion/molecule detection are showcased in Table 2.3, emphasizing synthetic approaches along with their sensitivity and selectivity.

2.3 Conclusion and Perspectives

Over the past three decades, the domain of chemosensors has burgeoned into a vibrant research arena, witnessing the publication of numerous books and reviews. A chemosensor encompasses any organic or inorganic molecule that undergoes changes

Table 2.3 Applications of some complexes decorated with CDs in ion/molecule detection

Complex of CDs	Synthetic approaches	Sensing platform/ method	Sensing target	Linear range	Refs
β-CD-RGO/CuNSs	Hydrothermal method	GCE/Electrocatalytic	4-nitrophenol	0.05–100 μM	[60]
HP-β-CDs/RGO	Microwave assist	GCE/Electrochemical	Pd^{2+} Cd^{2+}	0.1–9 nM 0.5–9 nM	[59]
β-CDs/GO	Mechanical mixing	SPCEs/Electrochemical	Nitrobenzene	0.5–1000 μM	[116]
β-CDs polymer/rGO	Hydrothermal method	CPE/Electrochemical	Dopamine	2.2–80 mM	[117]
β-CDs/ MWCNTs/CP	Drop-coating method	CP/Electrochemical	Methyl parathion	0.01–13 μg/mL	[51]
Graphene-CDs-cMWCNT	Hydrothermal method	GCE/Electrocatalytic	4-aminophenol 4-chlorophenol 4-nitropheno	0.2–125 μM 0.025–350 μM 0.3–50 μM	[118]
CDs/carbon dots	Hydrothermal method	-/Fluorescence	Cr^{2+}	–	[119]
β-CDs/AuNMs	Hydrothermal method	Gold microelectrode/ Electrical	PCBs	–	[120]
DMAB- α-CD	–	-/Enhanced emission	n-pentanol n-hexanol	–	[121]
NC0α-CD	–	-/Fluorescence	CBr_4 CCl_4	–	[122]
Fe_3O_4@SiO_2-azobenzene-rh - α-CD	Hydrothermal method	-/Enhanced emission	Cu^{2+}	–	[123]
PAN-co-AA/α-CD	–	-/Enhanced emission	F^-	–	[124]
PdNPs-β-CDs/rGO	Ultrasonic assist	GCE/Electrocatalytic	Hydrazine	0.05–1600 μM	[125]

in its native optical properties upon interacting with an analyte, such as a metal, within a given matrix. These alterations are attributed to the presence of optically active constituents within the sensor molecules, responsible for both baseline optical signals and variations induced by analyte interactions. In optical sensing techniques relying on UV–Visible or fluorescence methods, the detection process hinges on the involvement of chromophores, fluorophores, and auxochromes, which facilitate analyte recognition and signal modulation. This review undertakes a comprehensive exploration of the multifaceted realm of cyclodextrin-based chemosensing, encompassing a wide array of interconnected topics. It meticulously delves into the intricate structural attributes of cyclodextrins that endow them with unparalleled efficacy as molecular hosts, intricately analyzing their variegated cavity sizes, geometric configurations, and the nuanced dynamics of host–guest interactions they engender. Moreover, the review investigates an extensive spectrum of stratagems devised for the functionalization of cyclodextrins, each meticulously tailored to augment their sensing prowess. This encompasses a thorough examination of the incorporation of bespoke functional groups with the covalent and non-covalent modifications to impart a heightened discriminative acumen to these molecular sentinels. Additionally, the review ventures into the nascent frontier of integrating cyclodextrins into the vanguard of nanomaterials, a pioneering endeavor that holds immense promise in ushering forth a new epoch of advanced sensing modalities. Furthermore, it cogently expounds upon the attendant challenges and conjectures on the future trajectory of the field, thereby furnishing invaluable insights destined to illuminate the path forward for discerning researchers and practitioners actively engaged in the vanguard of molecular sensing technologies.

Despite their promising applications, cyclodextrins face notable challenges in chemosensing that hinder their practical implementation. One significant issue is their susceptibility to environmental variations, which can compromise sensor functionality and reliability. Furthermore, the inherent sluggishness of host–guest interactions often results in delayed response times, limiting their effectiveness in dynamic sensing scenarios. Achieving consistent and reliable functionalization of cyclodextrins also remains a hurdle, complicating their deployment across various applications. Additionally, the risk of cyclodextrins leaching from sensor systems raises concerns about the reproducibility of results, potentially affecting the accuracy of measurements.

Looking to the future, the potential for cyclodextrins in chemosensing is expansive. Research efforts could concentrate on developing innovative formulations and composite materials to amplify their stability and enhance response times significantly. Investigating novel functionalization techniques is crucial for widening their application scope to detect a broader array of specific analytes effectively. Moreover, the integration of cyclodextrins with state-of-the-art technologies, such as nanomaterials and microfluidics, holds the promise of creating highly sensitive and selective sensors. This integration could revolutionize their application in critical areas such as healthcare diagnostics, environmental monitoring, and food safety, unlocking new possibilities for advanced sensing solutions.

References

1. Busseron E, Ruff Y, Moulin E, Giuseppone N (2013) Supramolecular self-assemblies as functional nanomaterials. Nanoscale 5:7098–7140
2. Li Y, Liu T, Liu H, Tian MZ, Li Y (2014) Self-assembly of intramolecular charge-transfer compounds into functional molecular systems. Acc Chem Res 47:1186–1198
3. Martí-Centelles V, Pandey MD, Burguete MI, Luis SV (2015) Macrocyclization reactions: the importance of conformational, configurational, and template-induced preorganization. Chem Rev 115:8736–8834
4. Barrow SJ, Kasera S, Rowland MJ, del Barrio J, Scherman OA (2015) Cucurbituril-based molecular recognition. Chem Rev 115:12320–12406
5. Krakowiak KE, Bradshaw JS, Zamecka-Krakowiak DJ (1989) Synthesis of Aza-crown Ethers. Chem Rev 89:929–972
6. Nimse SB, Kim T (2013) Biological applications of functionalized calixarenes. Chem Soc Rev 42:366–386
7. Szejtli J (1998) Introduction and general overview of cyclodextrin chemistry. Chem Rev 98:1743–1754
8. Crini G (2014) Review: a history of cyclodextrins. Chem Rev 114:10940–10975
9. Schöffer JDN, Matte CR, Charqueiro DS, de Menezes EW, Costa TMH, Benvenutti EV, Rodrigues RC, Hertz PF (2017) Effects of immobilization, pH and reaction time in the modulation of α-, β- or γ-cyclodextrins production by cyclodextrin glycosyltransferase: batch and continuous process. CarbohydrPolym 169:41–49
10. Li S, Purdy WC (1992) Cyclodextrins and their applications in analytical chemistry. Chem Rev 92:1457–1470
11. Costa MAS, Anconi CPA, Santos HFD, Almeida WBD, Nascimento CS (2015) Inclusion process of tetracycline in β and γ-cyclodextrins: a theoretical investigation. Chem Phys Lett 626:80–84
12. Szente L, Szemán J, Sohajda T (2016) Analytical characterization of cyclodextrins: history, official methods and recommended new techniques. J Pharm Biomed Anal 130:347–365
13. Eastburn SD, Tao BY (1994) Applications of modified cyclodextrins. Biotechnol Adv 12:325–339
14. Loftsson T, Duchene D (2007) Cyclodextrins and their pharmaceutical applications. Int J Pharm 329:1–11
15. Peng L, Liu S, Feng A, Yuan J (2017) Polymeric nanocarriers based on cyclodextrins for drug delivery: host–guest interaction as stimuli responsive linker. Mol Pharmaceutics 14:2475–2486
16. Zhang J, Ma PX (2010) Host–guest interactions mediated nano-assemblies using cyclodextrin-containing hydrophilic polymers and their biomedical applications. Nano Today 5:337–350
17. Dinar K, Sahra K, Seridi A, Kadri M (2014) Inclusion complexes of N-sulfamoyloxazolidinones with β-cyclodextrin: a molecular modelling approach. Chem Phys Lett 595–596:113–120
18. Roy A, Saha S, Roy MN (2016) Study to explore host-guest inclusion complexes of cyclodextrins with biologically active molecules in aqueous environment. Fluid Phase Equilib 425:252–258
19. Jansook P, Ogawa N, Loftsson T (2018) Cyclodextrins: structure, physicochemical properties and pharmaceutical applications. Int J Pharm 535:272–284
20. Petitjean M, García-Zubiri IX, Isasi JR (2021) History of cyclodextrin-based polymers in food and pharmacy: a review. Environ Chem Lett 19:3465–3476
21. Crumling MA, King KA, Duncan RK (2017) Cyclodextrins and iatrogenic hearing loss: new drugs with significant risk. Front Cell Neurosci 11:355
22. Przybyla MA, Yilmaz G, Becer CR (2020) Natural cyclodextrins and their derivatives for polymer synthesis. Poly Chem 11:7582–7602
23. Singh M, Sharma R, Banerjee UC (2002) Biotechnological applications of cyclodextrins. Biotechnol Adv 20:341–359

24. Das S, Nath S, Singh TS, Chattopadhyay N (2020) Cavity size dependent stoichiometry of probe–cyclodextrin complexation: experimental and molecular docking demonstration. J Photochem Photobiol A Chem 388:11215
25. Matencio A, Navarro-Orcajada S, García-Carmona F, López-Nicolás JM (2020) Applications of cyclodextrins in food science. A review. Trends Food Sci Technol 104:132–143
26. Crini G, Fourmentin S, Fenyvesi É, Torri G, Fourmentin M, Morin-Crini N (2018) Cyclodextrins, from molecules to applications. Environ Chem Lett 16:1361–1375
27. Connors KA (1997) The stability of cyclodextrin complexes in solution. J Am Chem Soc 97:1325–1358
28. Chattopadhyay N, Chakraborty T, Nag A, Chowdhury M (1990) Effect of inclusion of cyclodextrin on excited state proton transfer: carbazole−γ-cyclodextrin. J Photochem Photobiol A Chem 52:199–204
29. Xu M, Wu S, Zeng F, Yu C (2010) Cyclodextrin supramolecular complex as a water-soluble ratiometric sensor for ferric ion sensing. Langmuir 26:4529–4534
30. Alam AU, Deen MJ (2020) Bisphenol A electrochemical sensor using graphene oxide and β-cyclodextrin-functionalized multi-walled carbon nanotubes. Anal Chem 92:5532–5539
31. Li C, Wu Z, Yang H, Deng L, Chen X (2017) Reduced graphene oxide-cyclodextrin-chitosan electrochemical sensor: effective and simultaneous determination of o-and p-nitrophenols. Sens Actuators B Chem 251:446–454
32. Braga SS, Lysenko K, El-Saleh F, PazF AA (2021) Cyclodextrin-Efavirenz complexes investigated by solid state and solubility studies. Proceedings 78:15
33. El-Sayed MA (2001) Some interesting properties of metals confined in time and nanometer space of different shapes. Acc Chem Res 34:257–264
34. Wang J, Yang L, Boriskina S, Yan B, Reinhard BM (2011) Spectroscopic ultra-trace detection of nitroaromatic gas vapor on rationally designed two-dimensional nanoparticle cluster arrays. Anal Chem 83:2243–2249
35. Chen X, Cheng X, Gooding JJ (2012) Detection of trace nitroaromatic isomers using indium tin oxide electrodes modified using betacyclodextrin and silver nanoparticles. Anal Chem 84:8557–8563
36. Xin X, Wang J, Gong C, Xu H, Wang R, Ji S, Dong H, Meng Q, Zhang L, Dai F, Sun D (2016) Cyclodextrin-based metal-organic nanotube as fluorescent probe for selective turn-on detection of hydrogen sulfide in living cells based on H_2S-involved coordination mechanism. Sci Rep 6:21951
37. Xin X, Dai F, Li F, Jin X, Wang R, Sun D (2017) A visual test paper based on Pb(II) metal−organic nanotubes utilized as a H_2S sensor with high selectivity and sensitivity. Anal Methods 9:3094–3098
38. Ma H, Li X, Yan T, Li Y, Liu H, Zhang Y, Wu D, Du B, Wei Q (2016) Sensitive insulin detection based on electrogenerated chemiluminescence resonance energy transfer between $Ru(bpy)_3{}^{2+}$ and Au nanoparticle-Doped β-Cyclodextrin-Pb (II) metal−organic framework. ACS Appl Mater Interfaces 8:10121–10127
39. Qiu Z-J, Fan S-T, Xing C-Y, Song M-M, Nie Z-J, Xu L, Zhang S-X, Wang L, Zhang S, Li B-J (2020) Facile fabrication of an AIE-active metal–organic framework for sensitive detection of explosives in liquid and solid phases. ACS Appl Mater Interfaces 12:55299–55307
40. Farha OK, Özgür Yazaydın A, Eryazici I, Malliakas CD, Hauser BG, Kanatzidis MG, Nguyen ST, Snurr RQ, Hupp JT (2010) De novo synthesis of a metal–organic framework material featuring ultrahigh surface area and gas storage capacities. Nat Chem 2:944–948
41. Ma H, Li X, Yan T, Li Y, Liu H, Zhang Y, Wu D, Du B, Wei Q (2016) Electrogenerated Chemiluminescence Behavior of Au nanoparticles-hybridized Pb (II) metal-organic framework and its application in selective sensing hexavalent chromium. Sci Rep 6:22059
42. Chen F, Pei H, Jia Q, Guo W, Zhang X, Guo R, Liu N, Mo Z (2021) Construction of cyclodextrin functionalized nitrogen-doped graphene quantum dot electrochemical sensing interface for recognition of tryptophan isomers. Mater Chem Phy 273:125086
43. Silva EC, Pina J, PereiraRFP MurtinhoD, Valente AJM, Fajardo AR (2024) Synthesis and characterization of fluorescent poly(α-cyclodextrin)/carbon quantum dots composite for efficient removal and detection of toluene and xylene from aqueous media. Environ Pollut 347:123778

44. Ma P, Liang F, Wang D, Yang Q, Cao B, Song D, Gao D, Wang X (2015) Selective determination of o-phenylenediamine by surface-enhanced Raman spectroscopy using silver nanoparticles decorated with α-cyclodextrin. Microchim Acta 182:167–174
45. Dong S, Bi Q, Qiao C, Sun Y, Zhang X, Lu X, Zhao L (2017) Electrochemical sensor for discrimination tyrosine enantiomers using graphene quantum dots and β-cyclodextrins composites. Talanta 173:94–100
46. Wang X, Li X, Luo C, Sun M, Li L, Duan H (2014) Ultrasensitive molecularly imprinted electrochemical sensor based on magnetism graphene oxide/β-cyclodextrin/Au nanoparticles composites for chrysoidine analysis. Electrochim Acta 130:519–525
47. Sangubotla R, Kim J (2023) Ultra-sensitive turn-on fluorescent sensor based on β-cyclodextrin-stabilized melamine-derived carbon dots for the highly selective dual-detection of dopamine. Ceram Int 49:16272–16282
48. Qin Y, Kwon H-J, Subrahmanyam A, Howlader MMR, Selvaganapathy PR, Adronov A, Deen MJ (2016) Inkjet-printed bifunctional carbon nanotubes for pH sensing. Mater Lett 176:68–70
49. Ran X, Yang L, Zhang J, Deng G, Li Y, Xie X, Zhao H, Li C-P (2015) Highly sensitive electrochemical sensor based on β-cyclodextrin– gold@3, 4, 9, 10-perylene tetracarboxylic acid functionalized single-walled carbon nanohorns for simultaneous determination of myricetin and rutin. Anal Chim Acta 892:85–94
50. Yu X, Chen Y, Chang L, Zhou L, Tang F, Wu X (2013) β-cyclodextrin non-covalently modified ionic liquid-based carbon paste electrode as a novel voltammetric sensor for specific detection of bisphenol A. Sens Actuators B 186:648–656
51. Liu R, Wang Y, Li D, Dong L, Li B, Liu B, Ma H, Li F, Yin X, Chen X (2019) A simple, low-cost and efficient ß-CD/MWCNTs/CP-based electrochemical sensor for the rapid and sensitive detection of methyl parathion. Int J Electrochem Sci 14:9785–9795
52. Kong L, Wang J, Meng F, Chen X, Jin Z, Li M, Liu J, Huang X-J (2011) Novel hybridized SWCNT–PCD: synthesis and host–guest inclusion for electrical sensing recognition of persistent organic pollutants. J Mater Chem 21:11109–11115
53. Alam AU, Qin Y, Catalano M, Wang L, Kim MJ, Howlader MMR, Hu NX, Deen MJ (2018) Tailoring MWCNTs and β-cyclodextrin for sensitive detection of acetaminophen and estrogen. ACS Appl Mater Interfaces 10:21411–21427
54. Gao Y, Cao Y, Yang D, Luo X, Tang Y, Li H (2012) Sensitivity and selectivity determination of bisphenol A using SWCNT–CD conjugate modified glassy carbon electrode. J Hazard Mater 199:111–118
55. Li X, Liu D, Wang Z (2011) Highly selective recognition of naphthol isomers based on the fluorescence dye-incorporated SH-β-cyclodextrin functionalized gold nanoparticles. Biosens Bioelectron 26:2329–2333
56. Bai H, Li C, Wang X, Shi G (2010) A pH-sensitive graphene oxide composite hydrogel. Chem Commun 46:2376–2378
57. Guo Y, Guo S, Li J, Wang E, Dong S (2011) Cyclodextrin–graphene hybrid nanosheets as enhanced sensing platform for ultrasensitive determination of carbendazim. Talanta 84:60–64
58. Xu C, Wang J, Wan L, Lin J, Wang X (2011) Microwave-assisted covalent modification of graphene nanosheets with hydroxypropyl-βcyclodextrin and its electrochemical detection of phenolic organic pollutants. J Mater Chem 21:10463–10471
59. Lv M, Wang X, Li J, Yang X, Zhang C, Yang J, Hu H (2013) Cyclodextrinreduced graphene oxide hybrid nanosheets for the simultaneous determination of lead(II) and cadmium(II) using square wave anodic stripping voltammetry. Electrochim Acta 108:412–420
60. Liu Z, Guo Y (2018) Sensitive determination of trace 4-nitrophenol in water based on thio-β-cyclodextrin functionalized graphene/copper nanospheres. J Electroanal Chem 825:57–64
61. Fan X, Geng J, Wang Y, Gu H (2022) PVA/gelatin/β-CD-based rapid self-healing supramolecular dual-network conductive hydrogel as bidirectional strain sensor. Polymer 246:124769
62. Ke L, Li Z, Fan X, Loh XJ, Cheng H, Wu Y, Li Z (2021) Cyclodextrin-based hybrid polymeric complex to overcome dual drug resistance mechanisms for cancer therapy. Polymers 13:1254
63. Niu X, Mo Z, Yang X, Sun M, Zhao P, Li Z, Ouyang M, Liu Z, Gao H, Guo R, Liu N (2018) Advances in the use of functional composites of β-cyclodextrin in electrochemical sensors. Microchim Acta 185:328

64. Zia J, Riyazuddin M, Jackson S, Taylor DK, Riaz U (2023) In-vitro release studies of ciprofloxacin drug using polyaniline/polypyrrole-encapsulated β-cyclodextrin nanocomposites. Polym Bull 80:12557–12576
65. Wang B, He J, Sun D, Zhang R, Han B, Huang Y, Yang G (2005) Preparation of β-cyclodextrin–polyaniline complex in supercritical CO_2. Eur Polym J 41:2483–2487
66. Gu ZJ, Wang JT, Li LL, Chen LF, Shen Q (2014) Formation of polyaniline nanotubes with different pore shapes using α-, β-and γ-cyclodextrins as templates. Mater Lett 117:66–68
67. Hamilton A, Breslin CB (2014) The development of a highly sensitive urea sensor due to the formation of an inclusion complex between urea and sulfonated-β-cyclodextrin. Electrochim Acta 125:250–257
68. Harley CC, Rooney AD, Breslin CB (2010) The selective detection of dopamine at a polypyrrole film doped with sulfonated β-cyclodextrins. Sens Actuators B Chem 150:498–504
69. Shang F, Zhou L, Mahmoud KA, Hrapovic S, Liu Y, Moynihan HA, Glennon JD, Luong JHT (2009) Selective nanomolar detection of dopamine using a boron-doped diamond electrode modified with an electropolymerized sulfobutylether-β-cyclodextrin-doped poly (N-acetyltyramine) and polypyrrole composite film. Anal Chem 81:4089–4098
70. Zhu G, Wu L, Zhang X, Liu W, Zhang X, Chen J (2013) A new dual-signalling electrochemical sensing strategy based on competitive host–guest interaction of a β-cyclodextrin/poly (N-acetylaniline)/graphene-modified electrode: sensitive electrochemical determination of organic pollutants. Chem Eur J 19:6368–6373
71. Onouchi H, Hasegawa T, Kashiwagi D, Ishiguro H, Maeda K, Yashima E (2006) Chirality sensing of various biomolecules with helical poly (phenylacetylene) s bearing acidic functional groups in water. J Polym Sci Part A Polym Chem 44:5039–5048
72. Yashima E, Maeda K, Sato O (2001) Switching of a macromolecular helicity for visual distinction of molecular recognition events. J Am Chem Soc 123:8159–8160
73. Maeda K, Mochizuki H, Watanabe M, Yashima E (2006) Switching of macromolecular helicity of optically active poly (phenylacetylene) s bearing cyclodextrin pendants induced by various external stimuli. J Am Chem Soc 128:7639–3650
74. Adam W, Lukacs Z, Viebach K, Humpf HU, Saha-Möller CR, Schreier P (2000) Microscale determination of the absolute configuration of α-aryl-substituted alcohols by the CD exciton chirality method. J Org Chem 65:186–190
75. Liu Y, You C-C, Zhang H-Y, Zhao Y-L (2003) Enantioselective recognition of aliphatic amino acids by β-cyclodextrin derivatives bearing aromatic organoselenium moieties on the primary or secondary side. Eur J Org Chem 2003:1415–1422
76. Ueno A, Suzuki I, Osa T (1988) Formation of pyrene excimer in an association dimer of pyrene-appended γ-cyclodextrin. J Chem Soc Chem Commun 20:1373–1374
77. Ueno A, Suzuki I, Osa T (1990) Host-guest sensory systems for detecting organic compounds by pyrene excimer fluorescence. Anal Chem 62:2461–2466
78. Aoyagi T, Ikeda H, Ueno A (2001) Fluorescence properties, induced-fit guest binding and molecular recognition abilities of modified γ-cyclodextrins bearing two pyrene moieties. Bull Chem Soc Jpn 74:157–164
79. Mallard I, Landy D, Bouchemal N, Fourmentin S (2011) Synthesis and inclusion ability of anthracene appended β-cyclodextrins: unexpected effect of triazole linker. Carbohydr Res 346:35–42
80. Chen HY, Zhao M, Li Y, Liu GF, Ji LN, Mao ZW (2014) An anthracene-modified β-cyclodextrin that distinguishes adenosine phosphates fluorescently. Tetrahedron Lett 55:1802–1805
81. Cardona CM, Alvarez J, Kaifer AE, McCarley TD, Pandey S, Baker GA, Bonzagni NJ, Bright FV (2000) Dendrimers functionalized with a single fluorescent dansyl group attached off center: synthesis and photophysical studies. J Am Chem Soc 122:6139–6144
82. Corradini R, Dossena A, Marchelli R, Panagia A, Sartor G, Saviano M, Lombardi A, Pavone V (1996) A modified cyclodextrin with a fully encapsulated dansyl group: self-inclusion in the solid state and in solution. Chem Eur J 2:373–381

83. Ueno A, Minato S, Suzuki I, Fukushima M, Ohkubo M, Osa T, Hamada F, Murai K (1990) Host-guest sensory system of dansyl-modified BETA-cyclodextrin for detecting steroidal compounds by dansyl fluorescence. Chem Lett 4:605–608
84. Wang Y, Ikeda T, Ueno A, Toda F (1992) Syntheses and molecular recognition abilities of 6-O-, 2-O-, and 3-O-dansyl-γ-cyclodextrins. Chem Lett 21:863–866
85. Wang Y, Ikeda T, Ikeda H, Ueno A, Toda F (1994) Dansyl-β-cyclodextrins as fluorescent sensors responsive to organic compounds. Bull Chem Soc Jpn 67:1598–1607
86. Ikeda H, Li Q, Ueno A (2006) Chiral recognition by fluorescent chemosensors based on N-dansyl-amino acid-modified cyclodextrins. Bioorg Med Chem Lett 16:5420–5423
87. Tanabe T, Touma K, Hamasaki K, Ueno A (2001) Immobilized fluorescent cyclodextrin on a cellulose membrane as a chemosensor for molecule detection. Anal Chem 73:3126–3130
88. Cheng Y, Jiang P, Lin S, Li Y, Dong X (2014) An imprinted fluorescent chemosensor prepared using dansyl-modified β-cyclodextrin as the functional monomer for sensing of cholesterol with tailor-made selectivity. Sens Actuator B Chem 193:838–843
89. Lucas-Abellán C, Mercader-Ros MT, Zafrilla MP, Fortea MI, Gabaldón JA, Núñez-Delicado E (2008) ORAC-fluorescein assay to determine the oxygen radical absorbance capacity of resveratrol complexed in cyclodextrins. J Agric Food Chem 56:2254–2259
90. Zhu X, Hu Y, Gong A (2007) Investigation on the interaction of tetrachloride fluorescein–bovine serum albumin-β-cyclodextrin and the determination of protein by flow injection analysis. Anal Chim Acta 592:24–29
91. Wang Y, Ikeda T, Ueno A, Toda F (1993) Fluorescein modified β-cyclodextrin as a charge-changeable receptor. Tetrahedron Lett 34:4971–4974
92. Hamada F, Ishikawa K, Higuchi Y, Akagami Y, Ueno A (1996) Strong binding between acidic guests and fluorescein modified-γ-Cyclodextrin via hydrogen bonding. J Incl Phenom Mol Recognit Chem 25:283–294
93. Peng M, Kaczmarek AM, Hecke KV (2022) Ratiometric thermometers based on rhodamine B and fluorescein dye-incorporated (nano) cyclodextrin metal–organic frameworks. ACS Appl Mater Interfaces 14:14367–14379
94. Nishimura D, Takashima Y, Aoki H, Takahashi T, Yamaguchi H, Ito S, Harada A (2008) Single-molecule imaging of rotaxanes immobilized on glass substrates: observation of rotary movement. Angew Chem Int Ed 47:6077–6079
95. Hasegawa T, Kondo Y, Koizumi Y, Sugiyama T, Takeda A, Ito S, Hamada F(2009) A highly sensitive probe detecting low pH area of HeLa cells based on rhodamine B modified β-cyclodextrins. Bioorg Med Chem 17:6015–6019
96. Mourtzis N, Paravatou M, Mavridis IM, Roberts ML, Yannakopoulou K (2008) Synthesis, characterization, and remarkable biological properties of cyclodextrins bearing guanidinoalkylamino and aminoalkylamino groups on their primary side. Chem Eur J 14:4188–4200
97. Wenz G, Kräuter I, Sackmann E (2009) Assembly of a fluorescent cyclodextrin polyrotaxane and its detection by fluorescence microscopy. J Polym Sci A Polym Chem 47:6223–6230
98. Ermilov EA, Menting R, Lau JTF, Leng X, Röder B, Ng DKP (2011) Switching the photoinduced processes in host–guest complexes of β-cyclodextrin-substituted silicon (iv) phthalocyanines and a tetrasulfonated porphyrin. Phys Chem Chem Phys 13:17633–17641
99. Kryjewski M, Goslinski T, Mielcarek J (2015) Functionality stored in the structures of cyclodextrin–porphyrinoid systems. Coord Chem Rev 300:101–120
100. Mavridis IM, Yannakopoulou K (2020) Porphyrinoid-Cyclodextrin assemblies in biomedical research: an update. J Med Chem 63:3391–3424
101. Gonzalez MC, McIntosh AR, Bolton JR, Weedon AC (1984) Intramolecular photochemical electron transfer to acceptors in a β-cyclodextrin linked to a porphyrin. J Chem Soc Chem Commun 1138–1140
102. Yang R, Li K, Wang K, Zhao F, Li N, Liu F (2003) Porphyrin assembly on β-cyclodextrin for selective sensing and detection of a zinc ion based on the dual emission fluorescence ratio. Anal Chem 75:612–621
103. Fan MF, Wang HM, Nan LJ, Wang AJ, Luo X, Yuan P-X, Feng J-J (2020) The mimetic assembly of cobalt prot-porphyrin with cyclodextrin dimer and its application for H_2O_2 detection. Anal Chim Acta 1097:78–84

104. Zhang Y, Xiang WC, Yang RH, Liu F, Li KA (2005) Highly selective sensing of lead ion based on α-, β-, γ-, and δ-tetrakis (3, 5-dibromo-2-hydroxylphenyl) porphyrin/β-CD inclusion complex. J Photochem Photobiol A Chem 173:264–270
105. Zhang Y, Kuang J, Dong J, Shi L, Li Q, Zhang B, Shi W, Huang X, Zhu Z, Ma Y, Guo Y, Liu Y (2021) Ultra-sensitive boscalid sensors based on a β-cyclodextrin modified perfluorinated copper phthalocyanine field-effect transistor. J Mater Chem C 9:12877–12883
106. Zheng K, Liu X, Liu H, Dong D, Li L, Jiang L, Huang M, Ding C (2021) Novel pH-triggered doxorubicin-releasing nanoparticles self-assembled by functionalized β-cyclodextrin and amphiphilic phthalocyanine for anticancer therapy. ACS Appl Mater Interfaces 13:10674–10688
107. Suresh P, Azath IA, Pitchumani K (2010) Naked-eye detection of Fe^{3+} and Ru^{3+} in water: Colorimetric and ratiometric sensor based on per-6-amino-β-cyclodextrin/p-nitrophenol. Sens Actuators B Chem 146:273–277
108. Jiang B, Liu Y, Zhao L, Zhao L, Wang C, Liu C, Xu B (2021) Construction of a pH-sensitive self-assembly in aqueous solutions based on a dansyl-modified β-cyclodextrin. Soft Matter 17:7516–7523
109. Galstyan A, Kauscher U, Block D, Ravoo BJ, Strassert CA (2016) Silicon (IV) phthalocyanine-decorated cyclodextrin vesicles as a self-assembled phototherapeutic agent against MRSA. ACS Appl Mater Interfaces 8:12631–12637
110. Khan RI, Pitchumani K (2016) β-Cyclodextrin included coumarin derivatives as selective fluorescent sensors for Cu^{2+} ions in HeLa cells. R Soc Chem Adv 6:20269–20275
111. Liu Z, Yang W, Li Y, Tian F, Zhu W (2015) A facile synthesis of a highly water-soluble and selective fluorescent sensor towards zinc ions derived from β-cyclodextrin based on an unexpected sensing process. R Soc Chem Adv 5:100482–100487
112. Zhang L, Hu W, Yu L, Wang Y (2015) Click synthesis of a novel triazole bridged AIE active cyclodextrin probe for specific detection of Cd^{2+}. Chem Commun 51:4298–4301
113. Zhang L, He Y, Zhang N, Liu D, Han J, Gong W (2016) Construction of a novel INHIBIT logic gate through a fine-tuned assembly of anthryl fluorophores via selective anion recognition and host-guest interactions. R Soc Chem Adv 6:805–809
114. Antony EJ, Raj M, Paulpandi RQ, Paulraj MS, Enoch IVMV (2015) A highly selective fluorescent sensor for Pb^{2+} based on a modified β-cyclodextrin. J Fluoresc 25:1031–1036
115. Kanagaraj K, Bavanidevi K, Chow TJ, Pitchumani K (2014) Selective turn-off fluorescent sensing of mercury ions using aminocyclodextrin:3-hydroxy-N-phenyl-2-naphthamide complex in aqueous solution. R Soc Chem Adv 4:11714–11722
116. Velmurugan M, Karikalan N, Chen S-M, Dai Z-C (2017) Studies on the influence of β-cyclodextrin on graphene oxide and its synergistic activity to the electrochemical detection of nitrobenzene. J Colloid Interface Sci 490:365–371
117. Heydari A, Sheibani H (2016) Facile polymerization of β-cyclodextrin functionalized graphene or graphene oxide nanosheets using citric acid crosslinker by in situ melt polycondensation for enhanced electrochemical performance. RSC Adv 6:9760–9771
118. Gao J, Liu M, Song H, Zhang S, Qian Y, Li A (2016) Highly-sensitive electrocatalytic determination for toxic phenols based on coupled cMWCNT/cyclodextrin edge-functionalized graphene composite. J Hazard Mater 318:99–108
119. Wang J, Qiu F, Wu H, Li X, Zhang T, Niu X, Yang D, Pan J, Xu J (2017) Fabrication of fluorescent carbon dots-linked isophorone diisocyanate and β-cyclodextrin for detection of chromium ions. Spectrochim Acta Part A 179:163–170
120. Yu Y, Chen X, Wei Y, Liu JH, Huang XJ (2012) Strategy for polychlorinated biphenyl detection based on specific inhibition of charge transport using a nanogapped gold particle film. Anal Chem 84:9818–9824
121. Hamasaki K, Ueno A, Toda F (1993) A fluorescent α-cyclodextrin as a sensor for detecting aliphatic alcohols by dual fluorescence arising from normal planar and twisted intramolecular charge transfer excited states. J Chem Soc Chem Commun 331–333
122. Ikeda H, Ueno A (2009) Fluorescent α-cyclodextrin as a chemosensor for halomethanes. Chem Commun 28:4281–4283

123. Wang W, Wong NK, Sun M, Yan C, Ma S, Yang Q, Li Y (2015) Regenerable fluorescent nanosensors for monitoring and recovering metal ions based on photoactivatable monolayer self-assembly and host–guest interactions. ACS Appl Mater Interfaces 7:8868–8875
124. Zhang H, Zhou C, Sun J, Li X, Tang J (2021) A nanofibrous membrane fluorescent sensor for fluoride ions prepared by electrospinning and host-guest interaction. Matéria (Rio J) 26
125. Sakthinathan S, Kubendhiran S, Chen SM, Sireesha P, Karuppiah C, Su C (2017) Functionalization of reduced graphene oxide with β-cyclodextrin modified palladium nanoparticles for the detection of hydrazine in environmental water samples. Electroanal 29:587–594

Priyanka Chetia is a Ph.D Research Scholar in the Department of Applied Sciences (Chemical Science Division) at Gauhati Uniersity, Guwahati, Assam. She joined in 2023 under the supervision of Prof. Pranjal Saikia. Her research interest is in Nanomaterials synthesis, Supercapacitors and Environmental research.

Runa Begum is a Ph.D Research Scholar in Chemistry under the supervision of Prof. Pranjal Saikia in the Department of Applied Sciences (Chemical Science division) at Gauhati University, Guwahati, Assam. She has completed B.Sc. with major in Chemistry and M.Sc. with specialization in inorganic chemistry from Cotton University, Guwahati. Her research interest is Synthesis of nanomaterials and their Applications (Energy and Environment).

Pranjal Saikia is a professor in the Department of Applied Sciences at Gauhati University, Guwahati, Assam. His areas of expertise are nanoscience, energy, and environmental research. He earned his PhD in Chemistry at the Indian Institute of Chemical Technology (IICT), Hyderabad. He has conducted postdoctoral research at the University of Cincinnati, USA. He has received many accolades, such as the Hindustan Platinum Award from the Catalysis Society of India and prestigious fellowships like the DST-DAAD Exchange Fellowship. He has published a good number of research articles, book chapter and a edited book with a h-index of 23. He is a Fellow of the Royal Society of Chemistry (FRSC). Six students have earned PhD under his guidance and eight are pursuing.

Chapter 3
Role of Weak Interactions and Steric Effect in Chemosensing Using Cyclodextrins

Ayfer Koyuncu and Fatih Ciftci

Abstract Cyclodextrins (CDs), a family of cyclic oligosaccharides, have garnered significant attention in chemosensing applications due to their unique ability to form inclusion complexes with various guest molecules. The hydrophobic inner cavity and hydrophilic outer surface of CDs enable selective encapsulation of various guest molecules, including small organic compounds, ions, and biological entities. This selective binding is primarily driven by weak non-covalent interactions, such as hydrogen bonding, van der Waals forces, hydrophobic interactions, and occasionally π-π stacking. These interactions make CDs an excellent molecular scaffold for designing chemosensors with enhanced sensitivity and specificity. Chemosensing applications utilizing CDs exploit their ability to act as molecular hosts, which modulate optical or electrochemical signals upon guest binding. For example, CDs are often incorporated into fluorescence-based sensors, where the inclusion of a target molecule within the CD cavity can lead to changes in fluorescence intensity or wavelength. Similarly, CDs are used in electrochemical sensors to detect redox-active species by facilitating the selective binding of analytes in complex sample matrices. CDs also enhance the detection of pollutants, drugs, or biomarkers in environmental, clinical, and industrial settings through such interactions. A critical factor influencing the efficiency of these chemosensing systems is the steric effect, which refers to the spatial arrangement and size of guest molecules relative to the CD cavity. The three main types of cyclodextrins such as α-CD, β-CD, and γ-CD differ in cavity size (with 6, 7, and 8 glucopyranose units, respectively), allowing for selective encapsulation based on the size and shape of the guest. Steric hindrance can either enhance or hinder the inclusion process depending on the degree of fit between the host and guest. Larger or irregularly shaped molecules may experience steric clashes with the CD cavity walls, reducing binding affinity or preventing complexation altogether. On the

A. Koyuncu
Graduate School of Science and Engineering, Bioengineering Department, Hacettepe University, Beytepe, Ankara, Türkiye

F. Ciftci (✉)
Department of Biomedical Engineering, Fatih Sultan Mehmet Vakif University, Istanbul, Türkiye
e-mail: faciftcii@gmail.com; fciftci@fsm.edu.tr

Technology Transfer Office, Fatih Sultan Mehmet Vakif University, Istanbul, Türkiye

S. Krishnamoorthy (ed.), *Cyclodextrins for Chemosensing*,
https://doi.org/10.1007/978-3-031-89299-8_3

contrary, a well-fitting guest molecule can lead to strong binding and high selectivity. Moreover, the functionalization of cyclodextrins with various chemical groups can modulate their steric and electronic properties, enabling fine-tuning of guest selectivity and sensor performance. For instance, the introduction of bulky substituents or charged groups can create additional steric barriers or enhance electrostatic interactions, respectively, thus influencing the strength and specificity of binding. These steric considerations are crucial in designing CDs for chemosensing applications, as they govern both the recognition and transduction mechanisms underlying sensor functionality. In conclusion, cyclodextrin-based chemosensors are versatile platforms that rely on weak non-covalent interactions and steric effects to achieve high selectivity and sensitivity in the detection of a wide range of analytes. By optimizing the size and functionalization of CDs, researchers can develop advanced chemosensors with tailored properties for applications in environmental monitoring, biomedical diagnostics, and industrial processes. This chapter explains the above-mentioned in a comprehensive manner.

Graphical Abstract

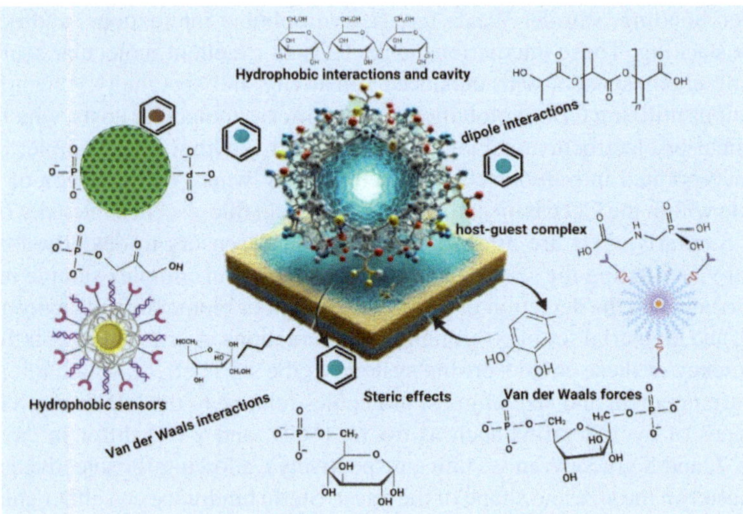

The graphical abstract illustrating the role of weak interactions and steric effects in chemosensing using cyclodextrins.

Keywords Cyclodextrin · Analytical chemistry · Interactions · Selective binding · Chemosensor

3.1 Introduction

Chemosensors are analytical devices designed to detect and respond to specific chemical substances by generating a measurable signal, often in the form of optical or electrochemical changes. These sensors have found widespread applications in fields ranging from environmental monitoring to biomedical diagnostics. A key challenge in chemosensor design is achieving high selectivity and sensitivity, particularly when detecting trace amounts of analytes in complex matrices. To address this challenge, host–guest chemistry has emerged as a powerful strategy, wherein a host molecule selectively binds a target analyte (guest) through non-covalent interactions. Among the various molecular hosts, cyclodextrins (CDs) have gained significant attention due to their ability to form stable inclusion complexes with a wide range of molecules [1–3].

Cyclodextrins are cyclic oligosaccharides composed of six to eight glucose units (α-CD, β-CD, γ-CD) that form a truncated cone-like structure with a hydrophobic inner cavity and a hydrophilic exterior. This unique architecture allows cyclodextrins to encapsulate guest molecules within their cavities, primarily driven by weak interactions such as hydrogen bonding, van der Waals forces, and hydrophobic interactions [5, 6, 8]. The encapsulation process not only facilitates the selective recognition of analytes but also induces significant changes in the physicochemical properties of both the cyclodextrin and the guest, making CDs ideal candidates for chemosensing applications [3, 4, 7].

In addition to weak interactions, the steric effect plays a pivotal role in the host–guest complexation process. The size and shape of both the cyclodextrin cavity and the guest molecule dictate the efficiency of binding and sensor response. For example, the three major cyclodextrins (α-CD, β-CD, γ-CD) offer varying cavity sizes, allowing for size-selective recognition of different analytes. Steric hindrance can either enhance the specificity of the sensor by excluding non-target molecules or limit the efficiency of guest binding due to spatial constraints. Furthermore, the functionalization of cyclodextrins with different chemical groups can introduce additional steric and electronic effects, further modulating the selectivity and sensitivity of the sensor [1, 3].

This paper explores the dual roles of weak interactions and steric effects in cyclodextrin-based chemosensing systems. It discusses how these factors influence the formation of host–guest complexes, their impact on sensor performance, and the strategies employed to optimize cyclodextrins for specific sensing applications. By understanding the interaction between molecular recognition and steric constraints, it is anticipated that designing more efficient and selective chemosensors with potential applications in environmental, industrial, and medical fields will become easier with developing technology in the coming years.

3.2 Structure and Unique Properties of Cyclodextrins

Cyclodextrins are a family of cyclic oligosaccharides composed of α-D-glucose units linked by α-1,4-glycosidic bonds and are produced from starch via enzymatic degradation, typically involving the enzyme cyclodextrin glycosyltransferase (CGTase). CDs are characterized by their toroidal (doughnut-like) shape, with a hydrophobic internal cavity and a hydrophilic external surface, giving them unique molecular recognition capabilities [2, 8, 9]. So, the unique chemical properties of cyclodextrins allow them to selectively encapsulate molecules with distinct molecular structures that form a cone or doughnut-like shape with a hydrophobic interior and a hydrophilic exterior, making them extremely versatile in host–guest chemistry [5, 8, 10, 11].

3.3 Host–Guest Chemistry: The Mechanism Behind Cyclodextrin Functionality

Host–guest chemistry is a fundamental concept supporting the functionality of CDs. This concept involves the formation of non-covalent inclusion complexes in which cyclodextrins act as "hosts" and other molecules (guests) are encapsulated within their hydrophobic cavities. The mechanism of host–guest interactions in cyclodextrins is given in Table 3.1 [3, 10–12].

3.4 The Role of Weak Interactions: How Cyclodextrins Achieve Selective Recognition

Cyclodextrins are widely used in selective recognition due to their ability to form host–guest inclusion complexes with a variety of molecules. This selective recognition is largely driven by weak non-covalent interactions, including hydrophobic interactions, hydrogen bonding, van der Waals forces, and, in some cases, electrostatic interactions. These weak forces collectively allow cyclodextrins to selectively bind specific guest molecules, which is critical in applications such as chemical sensing, drug delivery, and separation technologies [10, 13]. There are various non-covalent interactions between CDs and guest molecules that play an important role in the binding and recognition process. These weak interactions essential in Cyclodextrin recognition are given in Table 3.2 [3, 10–12].

Selective recognition mechanisms are used to selectively bind or recognize specific guest molecules from a mixture of similar compounds. Selective recognition mechanisms are fundamental in chemistry and biology, particularly in areas where specificity is critical, such as drug delivery, sensors, and catalysis.

Table 3.1 The mechanism of host–guest interactions in cyclodextrins

Structural basis of inclusion	• CDs are shaped like truncated cones with a hydrophobic inner cavity and a hydrophilic outer surface composed of hydroxyl groups. This dual nature allows CDs to interact differently with various molecules: hydrophobic interactions dominate within the cavity, while hydrogen bonding and van der Waals forces can occur at the cavity entrance or on the exterior • The size of the cavity varies with the type of CD: α-cyclodextrin (smallest), β-cyclodextrin (medium), and γ-cyclodextrin (largest), which dictates the size range of guest molecules that can be encapsulated
Driving forces for complexation	• Hydrophobic Interactions: The primary driving force for the inclusion of guest molecules is the hydrophobic effect. When a hydrophobic guest enters the hydrophobic cavity of CD, it displaces water molecules that were previously occupying the cavity, resulting in an energetically favorable state • Van der Waals Forces: These weak, non-covalent forces contribute to the stabilization of the host–guest complex. The fit between the CD cavity and the guest molecule allows for optimal van der Waals contacts, enhancing binding affinity • Hydrogen Bonding: Guest molecules with functional groups capable of hydrogen bonding (like hydroxyl, amine) can form hydrogen bonds with the hydroxyl groups on the CD rim, further stabilizing the complex • Electrostatic Interactions: If the guest molecule carries a charge, electrostatic interactions between the charged guest and the polar CD rim can also play a role, especially in chemically modified CDs
Selectivity and specificity	• The specificity of CDs for certain guest molecules is determined by the size and shape of the CD cavity relative to the guest. This size compatibility ensures selective encapsulation, allowing CDs to act as highly specific molecular receptors • Modifications to the CD structure (like as methylation, acetylation) can fine-tune the host–guest interactions, enhancing selectivity and binding strength for particular guests
Dynamic nature of inclusion complexes	• Host–guest complexes with CDs are typically dynamic and reversible. The guest can enter and exit the CD cavity depending on environmental conditions (like pH, temperature, solvent composition), which can be exploited in applications such as controlled release systems or responsive sensors • The reversibility of these interactions is crucial in applications like drug delivery, where the release of the active compound from the CD complex needs to occur under physiological conditions
Impact of solvent and environment	• The surrounding environment significantly affects the host–guest interaction. In aqueous solutions, the hydrophobic effect is particularly strong, driving guest inclusion into the CD cavity. Solvent polarity, temperature, and pH can all influence the strength and stability of the inclusion complex

Table 3.2 Weak interactions essential in cyclodextrin recognition

Hydrophobic interactions	• The primary driving force for guest inclusion in CDs is the hydrophobic effect. CDs have a hydrophobic internal cavity that can encapsulate non-polar guest molecules, while the hydrophilic exterior interacts favorably with the aqueous environment • The encapsulation process is entropically favorable as it leads to the release of structured water molecules from the hydrophobic cavity, increasing the overall disorder in the system. This interaction is particularly strong in aqueous environments where the tendency to minimize hydrophobic surface exposure drives the guest molecule into the CD cavity • Hydrophobic interactions are crucial for recognizing and binding non-polar and slightly polar guests, such as aromatic compounds, steroids, and certain drugs
Hydrogen bonding	• CDs possess multiple hydroxyl groups on the rim of their structure, which can engage in hydrogen bonding with guest molecules. The hydrogen bonding capability depends on the functional groups present on the guest and their orientation relative to the CD • These interactions contribute significantly to the binding specificity and stability of the host–guest complex. For instance, guest molecules with hydroxyl, carbonyl, or amine groups can form hydrogen bonds with the CD, enhancing the strength of the interaction • Hydrogen bonding also helps in differentiating guests with similar hydrophobicity but different polar functional groups, thereby enhancing the selectivity of CDs
Van der waals forces	• Van der Waals forces, including dipole–dipole interactions and London dispersion forces, are weak but play a significant role in stabilizing the inclusion complex. These forces arise from transient dipoles that occur when molecules are in close proximity • Although individually weak, Van der Waals forces can cumulatively contribute to the overall stability of the host–guest complex. The fit between the guest molecule and the CD cavity affects the extent of these interactions, with better fits leading to stronger overall binding • This aspect is crucial for recognizing guests that do not strongly interact via hydrogen bonding but can still fit snugly within the CD cavity
Steric efects and size selectivity	• The size and shape of the CD cavity impose steric constraints that influence guest inclusion. The three common cyclodextrins-α, β, and γ-have cavities of different sizes, allowing for size-selective recognition of guests • Steric effects play a critical role in excluding molecules that are too large or mismatched in shape, thus providing a form of size and shape selectivity. For example, β-cyclodextrin, with an intermediate cavity size, is ideal for encapsulating medium-sized aromatic compounds but may exclude larger molecules like steroids, which can fit better into γ-cyclodextrin • Modifications to the CD structure, such as adding bulky substituents, can further refine the selectivity by introducing additional steric hindrance
Electrostatic interactions	• While CDs themselves are neutral, their modified forms can include charged groups that enable electrostatic interactions with charged guest molecules. This modification allows for the selective binding of guests with complementary charges • Electrostatic interactions enhance the selectivity and binding affinity for charged guests, such as ionic drugs or dyes, particularly in applications like controlled release or targeted delivery

The primary mechanisms that contribute to selective recognition are as follows [3, 11, 14, 15]:

(i) **Size and Fit Matching**:

- The first level of selectivity is achieved through size matching between the cyclodextrin cavity and the guest molecule. The correct size and shape of the guest allow it to fit snugly into the cyclodextrin, maximizing van der Waals interactions and minimizing steric clashes.

(ii) **Complementary Interactions**:

- Beyond size and shape, cyclodextrins achieve selectivity through complementary interactions between the host and guest. The combination of hydrophobic effects, hydrogen bonding, and Van der Waals forces allows cyclodextrins to distinguish between closely related molecules.
- For example, subtle differences in the guest's functional groups can alter hydrogen bonding patterns, leading to the selective binding of one molecule over another.

(iii) **Dynamic Binding and Reversibility**:

- Cyclodextrin inclusion complexes are typically dynamic and reversible. This allows for a continuous exchange of guests based on the equilibrium between the host, guest, and the surrounding environment.
- The reversibility of the binding allows cyclodextrins to act as reusable sensing elements or carriers that can release the guest under specific conditions, such as changes in pH, temperature, or competitive binding with other molecules.

3.5 Steric Effects and Selectivity: Targeted Sensing with Cyclodextrins

As seen in Fig. 3.1, steric effects are crucial in the selective recognition and binding of guest molecules by cyclodextrins, playing a pivotal role in targeted sensing applications. Cyclodextrins, with their well-defined cyclic structure and hydrophobic cavity, rely on steric compatibility to selectively host–guest molecules. This steric complementarity ensures that only guests of the appropriate size, shape, and orientation are included within the cyclodextrin cavity, providing a mechanism for precise molecular recognition [16].

Steric effects arise from the physical size and spatial arrangement of atoms within molecules, influencing how these molecules interact with each other. In the context of cyclodextrin inclusion complexes, steric effects determined are given in Table 3.3 [17, 18].

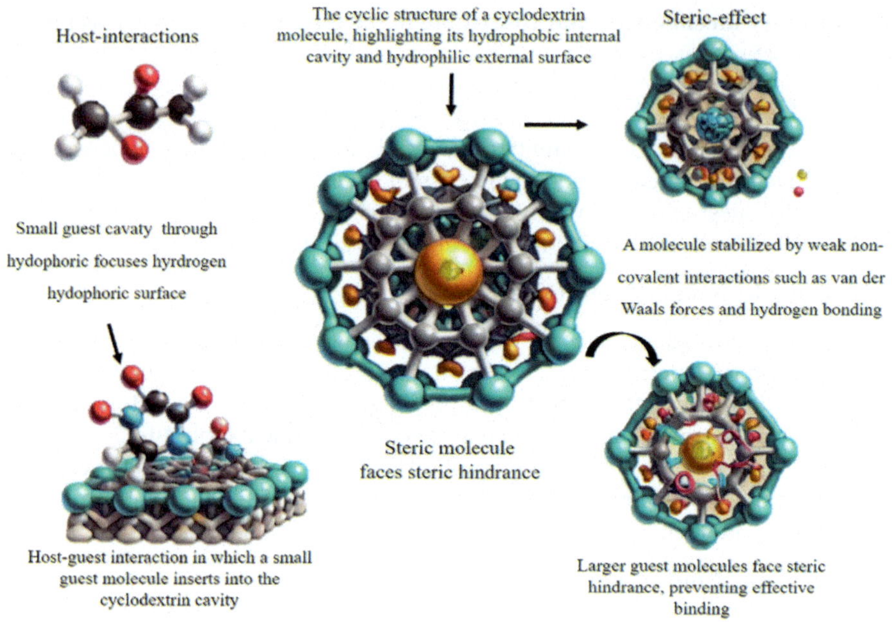

Fig. 3.1 The role of weak interactions and steric effects in chemosensing using cyclodextrins

Table 3.3 Steric effects interactions

Fit and exclusion	• The size of the CD cavity dictates which molecules can physically fit inside. CDs come in different sizes-α-cyclodextrin (smallest cavity), β-cyclodextrin (medium cavity), and γ-cyclodextrin (largest cavity)-each suited to host guests of corresponding sizes • If a guest molecule is too large or has bulky substituents, it will be sterically hindered from entering the cavity. Conversely, if the guest is too small, it may not interact sufficiently with the cavity walls, leading to weak binding
Orientation and conformation	• The three-dimensional shape of both the CD and the guest molecule affects how well the guest can align within the cavity. Steric hindrance can prevent the optimal orientation of a guest molecule, reducing binding affinity or selectivity • The conformation of the guest can also play a role; flexible molecules might adapt to fit, whereas rigid molecules must match the CD cavity precisely
Selective bnding through steric complementarity	• CDs achieve selectivity not just by the size of their cavities but also through steric complementarity. Molecules with shapes that match the internal contour of the CD cavity are more likely to form stable inclusion complexes • This steric matching allows CDs to differentiate between structurally similar molecules, providing a high degree of specificity

In cyclodextrins (CDs), steric effects play a significant role in determining how well a guest molecule fits into the cyclodextrin cavity, how stable the inclusion complex is, and how cyclodextrins function in various applications [1, 10, 17, 19]**. For instance**;

In Chemical Sensing

- In chemical sensing, steric effects are exploited to detect specific target molecules among a mixture of similar compounds. The selectivity afforded by steric compatibility ensures that only the target analytes, which match the cyclodextrin's cavity size and shape, are included and detected.
- For instance, cyclodextrins can be used in sensors for detecting specific pollutants, drugs, or biological markers by selectively binding only those molecules that fit their cavities without steric hindrance.

In Chiral Recognition and Enantiomeric Separation

- Cyclodextrins are particularly valuable in chiral recognition, where they can distinguish between enantiomers of a compound. Enantiomers have identical chemical structures but differ in spatial arrangement, leading to distinct steric interactions with the cyclodextrin.
- The ability of cyclodextrins to differentiate between enantiomers is crucial in pharmaceuticals, where the biological activity of each enantiomer can vary significantly. Steric effects ensure that only the desired enantiomer is selectively bound and separated.

In Drug Delivery

- In drug delivery, cyclodextrins are used to encapsulate and stabilize drug molecules, enhancing their solubility and bioavailability. Steric effects play a role in determining which drugs can be effectively encapsulated.
- Cyclodextrins can be designed or modified to preferentially bind specific drug molecules based on their size and shape, thereby enhancing targeted delivery and reducing off-target effects.

In Environmental Remediation

- Cyclodextrins can be used to selectively capture and remove pollutants from the environment. Steric effects ensure that only pollutants of a certain size and shape are encapsulated, preventing the binding of non-target molecules.
- This selectivity is advantageous in complex environmental matrices, where numerous substances may be present, allowing for the targeted removal of specific contaminants.

In Food and Cosmetic Industries

i. In food and cosmetics, cyclodextrins are employed to stabilize flavors, fragrances, and active ingredients. Steric effects contribute to the selective encapsulation of desired molecules, enhancing product stability and performance.
ii. For example, cyclodextrins can selectively encapsulate volatile compounds, protecting them from degradation and controlling their release in response to specific triggers, such as temperature changes.

Some adaptations can be made to increase selectivity with steric modifications [10, 12, 17]. For example;

(i) **Chemical Modification of Cyclodextrins**

- The selectivity of cyclodextrins can be enhanced by chemical modifications that alter the size and shape of the cavity or add sterically bulky groups to the rims of cyclodextrin.
- These modifications can introduce additional steric hindrance or facilitate a better fit for specific guest molecules, further refining the selectivity and specificity of the cyclodextrin.

(ii) **Tailoring Cyclodextrins for Specific Applications**

- By choosing the appropriate cyclodextrin (α, β, or γ) or chemically modifying it, the steric properties of the cyclodextrin can be tailored to target specific molecules.
- This tailoring allows for the design of cyclodextrins that are highly selective for particular applications, such as detecting trace amounts of a pollutant in a complex mixture or delivering a drug to a specific site in the body.

(iii) **Optimization of Inclusion Complex Formation**

- The formation of inclusion complexes can be optimized by adjusting conditions such as pH, temperature, and solvent, which can affect the steric fit and binding affinity of the guest molecules.
- Understanding and controlling these conditions allows for more precise tuning of cyclodextrin complexes to achieve the desired selectivity and stability.

Steric effects are fundamental to the selective binding capabilities of cyclodextrins, enhancing their utility in targeted sensing and recognition applications. By leveraging the size, shape, and conformational properties of both cyclodextrins and their guest molecules, these interactions allow for precise molecular recognition, enabling the development of highly selective sensors, delivery systems, and separation technologies. The ability to exploit and manipulate steric effects underscores the versatility and effectiveness of cyclodextrins as molecular hosts in a wide array of fields. However, there are some difficulties and situations that need to be taken into consideration. For example, (i) **Balancing Steric Hindrance and Binding Affinity**, and *(ii)* **Dynamic Nature of Steric Interactions**.

In Balancing Steric Hindrance and Binding Affinity; While steric effects contribute to selectivity, excessive steric hindrance can prevent binding altogether. Therefore, a careful balance must be achieved to optimize both selectivity and binding affinity.

In Dynamic Nature of Steric Interactions; The dynamic and reversible nature of steric interactions can be both an advantage and a limitation. While it allows for flexible binding and release, it can also lead to competition between similar guests, necessitating further refinements for ultra-selective applications [17, 20].

3.6 Size and Shape Compatibility: Determinants of Binding Efficiency

Steric effects, driven by size and shape compatibility, are crucial determinants of binding efficiency in host–guest interactions and molecular recognition processes. The fit between the host and guest determines how well they interact, with steric hindrance limiting access or preventing binding, and a good steric fit maximizing contact and stabilizing the complex. These principles are widely applied in fields like drug delivery, catalysis, and molecular recognition [19].

For efficient binding, the size of the guest molecule must closely match the dimensions of the host's binding cavity. If the guest is too small, it may not form enough stabilizing interactions (like van der Waals forces, hydrogen bonding) with the host, leading to a weak complex. Conversely, if the guest is too large, steric hindrance may prevent it from fitting into the cavity, reducing or eliminating binding. For instance, Cyclodextrins come in different sizes (α-, β-, γ-cyclodextrin), with each type having a cavity of specific dimensions. The guest molecule needs to have a size that matches the cavity for efficient binding. For example, **β-cyclodextrin** (with a moderate-sized cavity) binds well with molecules like aromatic compounds or steroids but may not accommodate larger molecules that would better fit in a **γ-cyclodextrin** [1, 3].

In addition to size, the shape of the guest molecule must complement the host's binding site. Binding efficiency is maximized when the guest's shape allows it to closely interact with the inner walls of the cavity. For example, a linear molecule may fit poorly in a spherical cavity, while a more compact, spherical molecule may have a higher binding affinity due to better shape compatibility. For instance, Enzyme active sites are highly selective for substrates that have the right shape to fit into the active site. The "lock-and-key" model illustrates this principle, where only substrates with a complementary shape to the enzyme's active site can bind efficiently and undergo a chemical reaction [17, 21].

The ability of cyclodextrins to form stable complexes with guest molecules depends significantly on how well the size and shape of the guest match the cavity dimensions listed. These properties are given in Table 3.4 [17, 19].

Size and shape compatibility are fundamental determinants of binding efficiency in cyclodextrin inclusion complexes, such as (***i***). By ensuring a good fit between the

Table 3.4 The ability of cyclodextrins to form stable complexes with guest molecules

Size matching	• For optimal binding, the guest molecule must fit snugly within the CD cavity. If the guest is too large, steric hindrance prevents inclusion, while if it is too small, insufficient contact with the cavity walls results in weak binding due to reduced van der Waals and hydrophobic interactions • Size compatibility is a primary determinant of binding efficiency, as it ensures maximal interaction between the CD and the guest, enhancing the overall stability of the complex
Shape matching	• In addition to size, the shape of the guest molecule plays a crucial role. Linear or slightly curved molecules may fit well into α-cyclodextrin, whereas more complex, planar, or spherical guests might align better with β- or γ-cyclodextrins • Shape compatibility ensures that the guest molecule is properly oriented within the cavity, allowing optimal interaction with the cyclodextrin's hydrophobic interior and the hydroxyl groups at the cavity rims
Conformational fexibility	• Guests that are flexible in conformation may adapt their shape to better fit the CD cavity, improving binding efficiency. However, highly rigid or bulky guests that cannot conform to the cavity shape may experience reduced binding • CDs themselves can exhibit some flexibility, allowing slight adjustments in cavity size and shape to better accommodate the guest, although this flexibility is limited compared to the adaptability of more flexible molecules
Influence of substituents	• The presence of substituents on the guest molecule, such as bulky side chains or functional groups, can enhance or hinder binding depending on how they affect the overall fit within the CD cavity • CDs can also be chemically modified with substituents that alter cavity dimensions or introduce specific interactions, thereby tuning the size and shape compatibility for particular guests

cyclodextrin cavity and the guest molecule, these parameters govern the strength, stability, and specificity of the host–guest interactions, such as (***ii***). This understanding allows for the strategic design and application of cyclodextrins in diverse fields, where precise molecular recognition and targeted sensing are essential, such as (***iii***) [1, 3].

(i) **High Binding Efficiency**:

- When the size and shape of the guest closely match the cyclodextrin cavity, the binding efficiency is maximized due to optimal van der Waals forces, hydrophobic interactions, and potential hydrogen bonding.
- This high efficiency is critical in applications like chemical sensing, where strong and selective binding of the target analyte is necessary for accurate detection.

(ii) **Enhanced Selectivity**:

- Cyclodextrins can differentiate between molecules of similar chemistry but different sizes or shapes, allowing selective binding of target molecules. This selectivity is particularly useful in enantiomeric separation, drug formulation, and environmental sensing, where precise molecular recognition is required.

- Size and shape selectivity can also minimize interference from non-target molecules, enhancing the specificity of cyclodextrin-based sensors and separation processes.

(iii) **Fine-Tuning Through Cyclodextrin Modification**:

- Cyclodextrins can be modified to adjust cavity dimensions or introduce functional groups that favor the binding of specific guest shapes or sizes. For example, methylation or other substitutions can expand the cavity slightly, accommodating larger guests or enhancing the fit for certain shapes.
- These modifications enable the design of cyclodextrins that are tailored for specific applications, providing a high degree of control over binding efficiency and selectivity.

Besides these, some challenges and optimization strategies such as Balancing Fit and Flexibility, Designing for Specificity, and Adjusting Environmental Conditions should be taken. For instance [8, 17];

(i) **In Balancing Fit and Flexibility**:

- Striking the right balance between snug fit and conformational flexibility is key to optimizing binding efficiency. Overly rigid systems may exclude slightly mismatched guests, while overly flexible systems may lack selectivity.

(ii) **In Designing for Specificity**:

- Cyclodextrins can be chemically modified to create cavities that are precisely tuned for the size and shape of specific target molecules. This approach is valuable in designing highly selective sensors, delivery systems, and separation techniques.

(iii) **In Adjusting Environmental Conditions**:

- Environmental factors such as pH, temperature, and solvent can influence the size and shape compatibility between cyclodextrins and their guests. Optimizing these conditions can further enhance binding efficiency and selectivity.

3.7 Electrostatic Interactions: Tuning Selectivity for Enhanced Performance

Electrostatic interactions, such as (*i*) ionic bonds, (*ii*) dipole–dipole interactions, and (*iii*) hydrogen bonding, play a significant role in the binding and selectivity of cyclodextrin inclusion complexes. These interactions can be harnessed to fine-tune the selectivity and performance of cyclodextrin-based systems in various applications, such as chemical sensing, drug delivery, and molecular recognition [10, 13, 17].

(i) **Ionic Interactions**:

- **Ion–Dipole Interactions**: They occur between charged guest molecules (ions) and the polar hydroxyl groups on the cyclodextrin's outer surface. These interactions are significant when the guest has a charged or polar functional group.
- **Salt Bridges**: They are formed between oppositely charged groups on the guest and on the cyclodextrin. This can occur if the cyclodextrin is modified with charged groups (like sulfonate or quaternary ammonium groups).

(ii) **Dipole–Dipole Interactions**:

- **Permanent Dipoles**: These interactions occur between guest molecules with permanent dipoles and the dipoles of the cyclodextrin's hydroxyl groups. These interactions can enhance the specificity of binding.
- **Induced Dipoles**: They occur when the cyclodextrin's polar groups induce a dipole in a non-polar guest molecule, facilitating weak but significant interactions.

(iii) **Hydrogen Bonding**:

- Although not purely electrostatic, hydrogen bonding often overlaps with electrostatic interactions. It involves the interaction between hydrogen atoms bound to electronegative atoms (like oxygen) and lone pairs on other electronegative atoms.

Enhancing selectivity through electrostatic interactions refers to improving the selective binding properties of cyclodextrins (a family of cyclic oligosaccharides) by leveraging electrostatic interactions. This requires (*i*) **Designing for Specific Binding,** (*ii*) **Optimizing Binding Affinity, and** (*iii*) **Improving Complex Stability** to be done [1, 13].

(i) **Designing for Specific Binding**:

- **Charge Complementarity**: Cyclodextrins can be modified to include charged groups on their rims or within their structure. These modifications enable them to selectively bind guest molecules with complementary charges, enhancing the specificity of the inclusion complex.
- **Functionalization**: Cyclodextrins can be functionalized with ionic or polar groups (like sulfonic acid, amino groups) to interact with specific ionic or polar guest molecules. This functionalization improves the ability of cyclodextrins to selectively capture target analytes in a mixture.

(ii) **Optimizing Binding Affinity**:

- **Electrostatic Tuning**: The strength of electrostatic interactions can be adjusted by altering the charge density on the cyclodextrin or guest molecule. For example, increasing the number of charged groups on cyclodextrin can enhance its interaction with oppositely charged guest molecules.
- **pH Dependence**: The ionization state of charged groups can be influenced by pH. By adjusting the pH of the solution, the electrostatic interactions can be fine-tuned to enhance or reduce binding affinity depending on the target application.

(iii) **Improving Complex Stability**:

- **Stabilization Through Ionic Bonds**: Strong ionic interactions can stabilize the inclusion complex, making it less prone to dissociation. This stabilization is particularly important in applications requiring long-term stability, such as drug delivery systems.
- **Buffer Systems**: In chemical sensing and environmental applications, the use of buffer systems can help maintain optimal pH and ionic strength, ensuring that electrostatic interactions remain effective.

Electrostatic Interactions are exploited in various scientific, industrial, and technological fields. This improves efficiency, specificity, and control in different applications. These areas are as follows [10, 17]:

In Chemical Sensing:

- **Selective Detection**: Cyclodextrins with charged or polar functional groups can selectively detect specific ionic or polar analytes. For instance, a cyclodextrin functionalized with carboxyl groups might selectively bind to positively charged metal ions.
- **Enhanced Sensitivity**: Electrostatic interactions can improve the sensitivity of sensors by increasing the binding affinity of target molecules, leading to more pronounced changes in sensor signals.

In Drug Delivery:

- **Targeted Delivery**: Cyclodextrins can be engineered to have specific electrostatic interactions with drugs or biological targets. For example, drugs with charged functional groups can be targeted more effectively using cyclodextrins with complementary charges.
- **Controlled Release**: Electrostatic interactions can also be used to control the release rate of drugs from cyclodextrin complexes, by modulating the strength of binding through changes in ionic strength or pH.

In Chiral Separation:

- **Enantiomeric Differentiation**: Electrostatic interactions can enhance the separation of chiral compounds. Cyclodextrins modified with charged groups can selectively interact with one enantiomer over the other, improving the resolution of chiral separation techniques.

In Environmental Remediation:

- **Selective Capture**: Cyclodextrins modified with charged groups can be used to selectively capture and remove specific pollutants, such as heavy metals or ionic contaminants, from environmental samples.
- **Enhanced Removal Efficiency**: The increased binding affinity resulting from electrostatic interactions can improve the efficiency of pollutant removal processes.

In Food and Cosmetic Industries:

- **Stabilization of Active Ingredients**: Electrostatic interactions can be utilized to stabilize and control the release of active ingredients in food and cosmetic products. This ensures that the ingredients remain effective and are released in a controlled manner.
- **Selective Encapsulation**: Cyclodextrins with electrostatic interactions can selectively encapsulate charged or polar ingredients, improving the formulation of products with specific performance characteristics.

Electrostatic interactions are a powerful tool for tuning the selectivity and performance of cyclodextrin inclusion complexes. By strategically modifying cyclodextrins and optimizing environmental conditions, electrostatic interactions can be harnessed to enhance binding efficiency, improve stability, and achieve high levels of specificity in various applications. This capability underscores the versatility of cyclodextrins as molecular hosts and their potential in advancing technologies in sensing, drug delivery, and environmental management. In addition to all these positive aspects, there are also challenges and optimization strategies and necessary precautions should be taken for this [13, 22]. For instance;

(1) **Balancing Electrostatic and Hydrophobic Interactions**:

   - While electrostatic interactions are important, they must be balanced with hydrophobic interactions to achieve optimal binding. Overemphasis on electrostatic interactions may reduce the role of hydrophobic effects, which are also critical for cyclodextrin binding.

(2) **Controlling Environmental Conditions**:

   - Environmental factors such as ionic strength, pH, and temperature can influence electrostatic interactions. Careful control of these conditions is necessary to maintain effective binding and selectivity.

(3) **Designing for Specific Applications**:

   - Tailoring cyclodextrins with specific electrostatic properties for particular applications requires a deep understanding of the target molecules and the desired interaction. Customizing the electrostatic environment of cyclodextrins can enhance their performance in diverse applications.

3.8 The Future of Chemical Sensing: Innovations and Applications with Cyclodextrins

Cyclodextrins have long been recognized for their unique ability to form inclusion complexes with a wide range of guest molecules, making them valuable tools in chemical sensing. As research and technology advance, cyclodextrins continue to evolve, offering new possibilities for innovative sensing applications. The future of chemical sensing with cyclodextrins promises enhanced sensitivity, selectivity, and versatility, driven by ongoing innovations and emerging trends. **There are some emerging innovations in Cyclodextrin-Based chemical sensing** [23, 24]. **For example**;

Smart Sensors and Responsive Systems:

- **Stimuli-Responsive Cyclodextrins**: Cyclodextrins are being integrated into smart sensors that respond to specific stimuli, such as pH changes, temperature variations, or light exposure. These responsive systems can provide real-time monitoring and dynamic adjustments in sensing applications.
- **Mechanically Flexible Sensors**: Incorporating cyclodextrins into flexible and stretchable substrates enables the development of sensors that can conform to various surfaces, including wearable technology and dynamic environments.

Nanotechnology and Miniaturization:

- **Nanocyclodextrins**: Advances in nanotechnology have led to the creation of nanometer-sized cyclodextrin particles or coatings. These nanoscale materials enhance the surface area and binding capacity of cyclodextrins, leading to improved sensitivity and faster response times.
- **Hybrid Nanomaterials**: Combining cyclodextrins with nanoparticles or nanomaterials can create hybrid sensing platforms with enhanced optical, electronic, or catalytic properties, offering new opportunities for detecting low concentrations of analytes.

Advanced Functionalization and Modification:

- **Targeted Functional Groups**: The development of cyclodextrins with tailored functional groups allows for precise interaction with specific analytes, including the ability to bind and sense complex molecules or mixtures.
- **Multifunctional Cyclodextrins**: Cyclodextrins can be engineered to perform multiple functions simultaneously, such as sensing and catalysis, enabling more versatile and integrated sensing platforms.

Integration with Digital Technologies:

- **Internet of Things (IoT) Integration**: Cyclodextrin-based sensors are being integrated with IoT platforms for remote monitoring and data collection. This integration allows for real-time data analysis and communication, enhancing the usability and effectiveness of sensing systems.

- **Smartphone and Portable Devices**: Advances in smartphone technology enable the development of portable cyclodextrin-based sensors that can be used for on-the-go chemical analysis and environmental monitoring.

Bio-Compatible and Green Chemistry Approaches:

- **Biocompatible Sensors**: Research is focusing on developing cyclodextrin-based sensors that are safe for biological applications, such as in vivo sensing and medical diagnostics.
- **Sustainable Materials**: Efforts are being made to use eco-friendly and renewable resources in the production of cyclodextrins and their derivatives, promoting greener and more sustainable sensing technologies.

The future of chemical sensing with cyclodextrins is bright, with ongoing innovations and advancements poised to enhance the capabilities and applications of these versatile molecules [1, 17]. By leveraging new technologies, functionalizations, and integration with digital systems, cyclodextrins will continue to play a pivotal role in advancing sensing technologies across a wide range of fields. However, there are some challenges that need to be taken into consideration [1, 17, 24–26, 28–30]. These are;

(i) **Scalability and Cost**:

- Developing cost-effective and scalable manufacturing processes for advanced cyclodextrin-based sensors is crucial for widespread adoption and commercialization.

(ii) **Long-Term Stability and Reusability**:

- Ensuring the long-term stability and reusability of cyclodextrin-based sensors is essential for practical applications. Research will focus on improving the durability and performance of these sensors over time.

(iii) **Complexity of Analyte Mixtures**:

- Addressing challenges related to sensing complex mixtures and achieving high selectivity in the presence of interfering substances will be an ongoing area of research.

(iv) **Regulatory and Safety Considerations**:

- Ensuring that cyclodextrin-based sensors meet regulatory standards and safety requirements for medical, environmental, and industrial applications is essential for successful implementation.

3.9 Conclusion

Cyclodextrins (CDs) have emerged as powerful and versatile molecules in the field of chemical sensing, offering significant advantages due to their unique ability to form inclusion complexes with a broad range of guest molecules. Through the careful exploitation of weak interactions—such as hydrophobic interactions, hydrogen bonding, Van der Waals forces, and electrostatic effects—cyclodextrins provide high selectivity, sensitivity, and stability in sensing systems. Furthermore, their tunable steric properties allow for enhanced specificity in detecting target analytes, making them ideal candidates for applications ranging from environmental monitoring to medical diagnostics, food safety, and industrial process control.

As research in this field progresses, innovations in the functionalization of cyclodextrins, nanotechnology integration, and the development of responsive and smart sensing systems are unlocking new possibilities. Cyclodextrins are not only improving existing sensing technologies but are also paving the way for novel applications in advanced drug delivery, personalized medicine, wearable and IoT-enabled sensors, and sustainable environmental solutions.

Looking forward, cyclodextrin-based chemical sensing is set to play a crucial role in addressing future challenges in public health, environmental protection, and industrial efficiency. Continued interdisciplinary collaboration and commercialization efforts will further expand their potential, positioning cyclodextrins as key components in the next generation of chemical sensing technologies. With their wide range of applications and capacity for innovation, cyclodextrins will continue to be at the forefront of advancements in molecular recognition and targeted detection, driving progress across multiple scientific and industrial domains.

References

1. Del Valle EM (2004) Cyclodextrins and their uses: a review. Process Biochem 39(9):1033–1046. https://doi.org/10.1016/S0032-9592(03)00258-9
2. Benaissa A, Bouhadiba A, Naili N, Chekkal F, Khelfaoui M, Bouras I, Madi F (2023) Computational investigation of dimethoate and β-cyclodextrin inclusion complex: molecular structures, intermolecular interactions, and electronic analysis. Struct Chem 34(3):1189–1204. https://doi.org/10.1007/s11224-023-02162-8
3. Esteso MA, Romero CM (2024) Cyclodextrins: properties and applications. Int J Mol Sci 25(8):4547. https://doi.org/10.3390/ijms25084547
4. Morabito K, Shapley NC, Steeley KG, Tripathi A (2011) Review of sunscreen and the emergence of non-conventional absorbers and their applications in ultraviolet protection. Int J Cosmet Sci 33(5):385–390. https://doi.org/10.1111/j.1468-2494.2011.00654.x
5. Celebioglu A, Uyar T (2012) Electrospinning of nanofibers from non-polymeric systems: polymer-free nanofibers from cyclodextrin derivatives. Nanoscale 4(2):621–631. https://doi.org/10.1039/C1NR11364J
6. Cova TF, Murtinho D, Pais AA, Valente AJ (2018) Combining cellulose and cyclodextrins: fascinating designs for materials and pharmaceutics. Front Chem 6:271. https://doi.org/10.3389/fchem.2018.00271

7. Hassabo AG, Zayed M, Bakr M, Othman H (2022) An overview of carpet manufacture: design, dyeing, printing and finishing. J Text Color Polym Sci 19(2):269–290. https://doi.org/10.21608/JTCPS.2022.150669.1129
8. Kali G, Haddadzadegan S, Bernkop-Schnürch A (2024) Cyclodextrins and derivatives in drug delivery: new developments, relevant clinical trials, and advanced products. Carbohyd Polym 324:121500. https://doi.org/10.1016/j.carbpol.2023.121500
9. Loftsson T, Másson M, Brewster ME (2004) Self-association of cyclodextrins and cyclodextrin complexes. J Pharm Sci 93(5):1091–1099. https://doi.org/10.1002/jps.20047
10. Bezerra FM, Lis MJ, Firmino HB, Dias da Silva JG, Curto Valle RDCS, Borges Valle JA, Tessaro AL (2020) The role of β-cyclodextrin in the textile industry. Molecules 25(16):3624. https://doi.org/10.3390/molecules25163624
11. Alrbaihat MR (2023) Mechanochemical synthesis of host-guest ınclusion complexes of cyclodextrin: a review. Biomed Chem Sci 2(2):102–108. https://doi.org/10.48112/bcs.v2i2.287
12. Kou X, Zhang Y, Su D, Wang H, Huang X, Niu Y, Meng Q (2023) Study on host-guest interaction of aroma compounds/γ-cyclodextrin inclusion complexes. LWT 178:114589. https://doi.org/10.1016/j.lwt.2023.114589
13. Ross PD, Rekharsky MV (1996) Thermodynamics of hydrogen bond and hydrophobic interactions in cyclodextrin complexes. Biophys J 71(4):2144–2154. https://doi.org/10.1016/S0006-3495(96)79415-8
14. Tian B, Xiao D, Hei T, Ping R, Hua S, Liu J (2020) The application and prospects of cyclodextrin inclusion complexes and polymers in the food industry: a review. Polym Int 69(7):597–603. https://doi.org/10.1002/pi.5992
15. Morin-Crini N, Fourmentin S, Fenyvesi É, Lichtfouse E, Torri G, Fourmentin M, Crini G (2021) 130 years of cyclodextrin discovery for health, food, agriculture, and the industry: a review. Environ Chem Lett 19(3):2581–2617. https://doi.org/10.1007/s10311-020-01156-w
16. Li X, Chen S, Hu X, Zi F (2024) Efficient and selective recovery of gold from thiosulfate leaching solution using functionalized β-cyclodextrins synthesized by a steric hindrance strategy. Adv Func Mater 34(1):2307472. https://doi.org/10.1002/adfm.202307472
17. Lichtfouse E, Schwarzbauer J, Robert D (eds) (2011) Environmental chemistry for a sustainable world: volume 2: remediation of air and water pollution, vol 2, Springer Science & Business Media
18. Peluso P, Landy D, Nakhle L, Dallocchio R, Dessì A, Krait S, Scriba GK (2023) Isothermal titration calorimetry and molecular modeling study of the complex formation of daclatasvir by γ-cyclodextrin and trimethyl-β-cyclodextrin. Carbohyd Polym 313:120870. https://doi.org/10.1016/j.carbpol.2023.120870
19. Harada A, Takashima Y, Nakahata M (2014) Supramolecular polymeric materials via cyclodextrin–guest interactions. Acc Chem Res 47(7):2128–2140. https://doi.org/10.1021/ar500109h
20. Guo X, Wang J, Li L, Pham DT, Clements P, Lincoln SF, Prud'Homme RK (2010) Steric effects and competitive intra-and intermolecular host-guest complexation between beta-cyclodextrin and adamantyl substituted poly (acrylate) s in water: a 1H NMR, rheological and preparative study. J Polym Sci Part B Polym Phys 48(16):1818–1825. https://doi.org/10.1002/polb.22052
21. Wenz G, Strassnig C, Thiele C, Engelke A, Morgenstern B, Hegetschweiler K (2008) Recognition of ıonic guests by ıonic β-cyclodextrin derivatives. Chem A Eur J 14(24):7202–7211. https://doi.org/10.1002/chem.200800295
22. Geue N, Alcázar JJ, Campodónico PR (2023) Influence of β-cyclodextrin methylation on host-guest complex stability: a theoretical study of intra-and intermolecular interactions as well as host dimer formation. Molecules 28(6):2625. https://doi.org/10.3390/molecules28062625
23. Velázquez-Contreras F, Zamora-Ledezma C, López-González I, Meseguer-Olmo L, Núñez-Delicado E, Gabaldón JA (2021) Cyclodextrins in polymer-based active food packaging: a fresh look at nontoxic, biodegradable, and sustainable technology trends. Polymers 14(1):104. https://doi.org/10.3390/polym14010104
24. Visakh PM (2022) Nanomaterials and nanotechnology in medicine: medical applications: challenges and opportunities. Nanomater Nanotechnol Med 1–25. https://doi.org/10.1002/9781119558026.ch1

25. Cabrales L, Abidi N, Hammond A, Hamood A (2012) Cotton fabric functionalization with cyclodextrins. Surfaces 6(8). ISSN: 2028–2508 CODEN: JMESCN
26. Saffarionpour S (2019) Nanoencapsulation of hydrophobic food flavor ingredients and their cyclodextrin inclusion complexes. Food Bioprocess Technol 12(7):1157–1173. https://doi.org/10.1007/s11947-019-02285-z
27. Nagy ZM, Molnár M, Fekete-Kertész I, Molnár-Perl I, Fenyvesi É, Gruiz K (2014) Removal of emerging micropollutants from water using cyclodextrin. Sci Total Environ 485:711–719. https://doi.org/10.1016/j.scitotenv.2014.04.003
28. Liu Q, Zhou Y, Lu J, Zhou Y (2020) Novel cyclodextrin-based adsorbents for removing pollutants from wastewater: a critical review. Chemosphere 241:125043. https://doi.org/10.1016/j.chemosphere.2019.125043
29. Aiassa V, Garnero C, Zoppi A, Longhi MR (2023) Cyclodextrins and their derivatives as drug stability modifiers. Pharmaceuticals 16(8):1074. https://doi.org/10.3390/ph16081074
30. Desoky MM, Caldera F, Brunella V, Ferrero R, Hoti G, Trotta F (2023) Cyclodextrins for lithium batteries applications. Materials 16(16):5540. https://doi.org/10.3390/ma16165540

Ayfer Koyuncu received her Bachelor of Science (BS) and Master of Science (MS) degrees from Kırıkkale University, Kırıkkale, Türkiye, Department of Biology. She completed her Doctor of Philosophy (Ph.D.) in 2021 at Hacettepe University, Ankara, Turkey, Department of Bioengineering. Her research interests are focused on the synthesis of nanomaterials, exploring their biological effects, and their potential applications in biomedicine.

Dr. Fatih Ciftci is a Assistant Professor at Fatih Sultan Mehmet Vakıf University, Department of Biomedical Engineering (ISTANBUL, TURKEY). He obtained his Ph.D. in the field of Bioengineering in 2021. He has published 41 research papers and 5 Turkey patent grant. His research interests include biomaterials, biopolymers, bioceramics, hard-soft tissue engineering applications, 3D transdermal tissue scaffold, MXene-MOF-Graphene nanomaterials, wearable technologies, biosensors and rare diseases. Dr. Ciftci actively participates in numerous national and international collaborative projects and serves as a reviewer for various scientific journals.

Chapter 4
The Versatile Roles of Cyclodextrins in Colorimetric and Spectrophotometric Detection of Metal Ions

Anupama Vijayan, C. K. Najiya, and John Prakash

Abstract Cyclodextrins are multi-faceted molecules capable of assuming versatile roles in the detection of metal ions through electronic absorption spectroscopy and colorimetric methods. In spectrophotometry-based metal sensing, cyclodextrins have been employed via three main approaches: (1) The formation of chelation complexes between metal ions and structurally modified cyclodextrins, where the modification enhances the binding affinity for specific ions. (2) The creation of inclusion complexes, where cyclodextrins encapsulate guest molecules that serve as signaling groups for metal ions. (3) Functionalisation of nanoparticles with cyclodextrins. In all three approaches, the interaction between metal ions and cyclodextrins—whether through direct chelation, inclusion complex formation, or nanoparticle functionalisation—induces observable changes in the absorption spectrum of the sensor. This chapter examines the role of cyclodextrins in each of these three approaches, with a focus on how specific chemical modifications are introduced to enable the selective detection of metal ions. The chapter also correlates the resulting spectral changes and colorimetric responses, to the underlying mechanistic processes that occur during metal ion interaction, providing a comprehensive understanding of how these modifications influence the detection performance of cyclodextrin-based sensors.

Keywords Metal sensing · Colorimetric detection · Modified cyclodextrin · Nanoparticles · Inclusion complex · Absorption spectroscopy

---

A. Vijayan · C. K. Najiya · J. Prakash (✉)
Department of Chemistry, School of Basic and Applied Sciences,
Central University of Tamil Nadu, Thiruvarur, India
e-mail: johnprakash@cutn.ac.in

S. Krishnamoorthy (ed.), *Cyclodextrins for Chemosensing*,
https://doi.org/10.1007/978-3-031-89299-8_4

4.1 Introduction: Metal Ion Sensing

Metal ions play fundamentally important roles in biological processes and industrial applications. Metals such as chromium (Cr), copper (Cu), iron (Fe), calcium (Ca), magnesium (Mg), zinc (Zn) and manganese (Mn) are part of the human diet and physiological functioning. However, excessive amounts of these metals in the same or different oxidation states (as in the case of chromium) can be toxic to the human body [1]. The metals such as aluminium (Al), arsenic (As), cadmium (Cd), gold (Au), lead (Pb), lithium (Li), mercury (Hg), nickel (Ni), platinum (Pt), silver (Ag), tin (Sn), uranium (U) and several others do not play recognised roles in physiological processes. They are typically considered harmful contaminants rather than essential components for life [2]. While these metals naturally occur in the Earth's crust, the delicate balance of their distribution can be disrupted by anthropogenic activities as well as certain natural processes. Human-intervention such as mining, smelting operations, industrial production, agricultural practices and natural processes such as weathering and volcanic eruptions are crucial pathways that mobilise and introduce metal contamination into air, water, and soil [2]. According to a UN report, at least 80% of wastewater is released into the ecosystem without pre-treatment or with minimal pre-treatment [3]. Bioaccumulation of metal ions in the food chain takes place through processes such as plant uptake from soil and fish uptake from water [4]. Once they enter the body of mammals such as human beings, they are largely incapable of metabolising the majority of these metals. The need for efficient and precise metal ion sensing arises from the potential threats related to their accumulation, which can lead to environmental pollution, compromised water quality, and detrimental health effects [5]. Table 4.1 provides a list of chemicals, their possible sources of exposure and related health issues.

4.2 Supramolecular-Spectrophotometric Approach for Metal Ion Sensing

In recent years, there has been a significant increase in the studies related to detecting and tracking metal ions. This heightened interest has led to extensive research focused on the selective and sensitive detection of specific ions. Various testing methods, including electrochemistry, atomic absorption spectrometry, atomic emission spectrometry, inductively coupled plasma mass spectroscopy, X-ray fluorescence spectrometry, flame atomic absorption spectrometry and X-ray photoelectron spectroscopy have been widely employed for their detection. However, these techniques necessitate sophisticated equipment, limiting their applicability in resource-poor settings [19]. In contrast to other sensing methods, the utilisation of colorimetric chemosensing techniques grounded in supramolecular chemistry has emerged as a pivotal research tool for detecting metal ions. Supramolecular host–guest chemistry, integral to this technique, imparts the much-needed specificity required for accurate

Table 4.1 Some toxic common metals, their biological relevance, sources and related health issues in human beings

Metal	Biological relevance	Sources of exposure	Health issues
Mg	Nerve function, muscle contractions, bone formation, blood pressure regulation	Excessive dietary magnesium	Cramping, nausea, depression, and low blood pressure [6]
Al	Non-essential	Drinking water, cooking utensils, vaccines, antacids, personal care products	Alzheimer's disease, neurotoxicity, osteomalacia [7]
Ca	Critical for bone health, nerve function, muscle contractions	Excessive dietary intake	Hypercalcemia, kidney stones with excessive intake [8]
Fe	Oxygen transport, enzyme cofactor, red blood cell formation, enzyme catalysis	Excess dietary intake, contaminated drinking water	Oxidative stress and fibrosis of various organs [9]
Co	Component of red blood cells and vitamin B12	Contaminated soil and water	Respiratory, cardiovascular issues, potential carcinogenicity [10]
Ni	Enzyme cofactor	Inhalation, ingestion of contaminated water/food	Respiratory problems, skin irritation, gastrointestinal issues, and potential carcinogenicity [11]
Cu	Iron transport, enzyme cofactor	Diet, plumbing materials, supplements	Wilson disease, anaemia, cardiovascular issues [12]
Zn	Enzyme cofactor, neurotransmission, signal transduction, and gene expression	Contaminated water, industrial emissions, and certain consumer products	Abdominal pain, intense vomiting and liver damage [13]
As	Non-essential	Groundwater, soil contamination, airborne emissions	Skin lesions, cancers cardiovascular diseases, and neurological effects [14]
Ag	Non-essential	Industrial discharge, medical equipment, jewellery	Argyria (skin discolouration) [15]
Cd	Non-essential	Air, water, food contamination, tobacco smoke	Itai-itai disease, hepatic damage, respiratory problems, carcinogenicity, testicular damage and osteomalacia [16]
Hg	Non-essential	Fish consumption, dental amalgam, industrial discharges	Neurological, cardiovascular, developmental, kidney, respiratory, and skin issues [17]
Pb	Non-essential	Lead-based paints, contaminated soil, dust, lead-acid batteries	Mental retardation, haematological, gastrointestinal, cardiovascular issues, renal disorders, muscle paralysis [18]

metal ion identification. Colorimetric detection can enable naked-eye detection, and the use of absorption spectroscopy can provide a sensitive and cost-effective tool for the quantitative determination of metal ions. The marriage of supramolecular chemistry and absorption spectroscopy stands out as a promising avenue due to its inherent versatility, user-friendly application, rapid response, and suitability for on-site detection without necessitating sophisticated equipment or extensive sample preparation. Cyclodextrins (CD) represent a class of important macromolecules used in analytical chemistry for colorimetric detection of metal ions. CDs, being cyclic oligosaccharides with well-defined structures, facilitate selective interactions with metal ions, enhancing the precision of detection methods. This chapter will explore the use of CDs in versatile roles for the detection of cationic species (metal ions) and the associated mechanisms that allow quantification via Ultra Violet-Visible (UV–Visible) absorption spectroscopy.

4.3 CD-Based Metal Ion Sensing

Native CDs represent a class of unmodified forms of CDs, mostly alpha (α), beta (β) and gamma (γ) CDs, that consist of six, seven and eight α-D-glucopyranose units, respectively. Native CDs show metal ion complexation in an alkaline medium on deprotonation of alcohol functionality [20]. The lack of selectivity and inability to be observed using UV–Visible absorption spectroscopy renders this approach inapt for this discussion. However, CDs' versatility has driven their integration into advanced colourimetric metal sensors, thanks to their appealing properties for metal sensing applications.

- *Water Solubility*: The water solubility of CDs is a crucial attribute for metal sensing applications, especially in aqueous environments commonly encountered in biological systems and environmental monitoring. However, it is to be kept in mind that 14 g of α-CD or 23 g of γ-CD are soluble in 100 mL of water at 25 °C. In contrast, only 1.8 g of β-CD can be dissolved under the same conditions [21].
- *Low Toxicity*: CDs have low cellular toxicity, which makes them suitable for designing sensors for use in living organisms or biomedical samples. In fact, the cell permeability of a probe is greatly benefitted through the interaction of CD with the phospholipids and cholesterol on the cell membrane [22].
- *Encapsulation Capacity*: The hydrophobic cavities of CDs allow them to encapsulate guest molecules.
- *Chemical Modification*: CDs can undergo functionalisation and structural modifications tailored to the needs of sensing specific ions.

Mainly CDs are employed in the following ways in metal ion sensing applications:

1. Derivatised forms of CDs
2. Formation of inclusion complexes with probe molecules
3. In combination with nanomaterials.

These approaches collectively address the limitations associated with native CDs, offering promising avenues for the development of highly selective and sensitive metal ion sensing via spectrophotometric techniques.

4.4 Structurally Modified CDs in Metal Sensing

4.4.1 Modification of CDs

Native CDs inherently lack chromogenic groups, and hence electronic absorption in the UV–Visible region ranges from 200–800 nm [25]. It becomes imperative to introduce strategic structural modifications to harness the capability of CDs as spectrophotometric sensors. The functionalisation of CDs can impart absorption–emission features, enabling their investigation through spectrophotometric techniques like absorption and emission spectroscopy. Moreover, the judicious introduction of modifications to the native CD structure provides a valuable tool for tuning the selectivity of these molecules towards specific metal ions. Thus, by carefully modifying the structure of CDs, we may fully realise the spectrophotometric potential CDs offer as versatile and selective sensors for metal ions. The coordination of metal with a modified CD system can be investigated using spectrophotometric titration. The C2, C3, and C6 hydroxyl groups are potential structural alteration sites (Fig. 4.1A). While the hydroxyl groups in C_3 are much less reactive than those in C_2, the hydroxyl groups in C_6 are the most reactive [26].

4.4.1.1 Sensing Mechanism and Spectral Changes

In numerous instances, the complexation of a metal ion with a modified CD has been observed to induce a hypsochromic shift. This spectral shift is ascribed to the lowering of electron density on the modified CD due to the presence of electron-deficient cationic metal species. This shift is accompanied by the emergence of a distinct peak, indicative of the formation of the metal-modified CD complex.

For transition metal complexes, metal has d-orbitals, and the peak may correspond to d-d transitions. It involves the promotion of electrons from lower energy d-orbitals to the higher energy ones formed by the ligand field splitting of metal d-orbitals. Alternatively, new peaks can also arise due to charge transfer transitions involving the movement of electrons from one part of a molecule to another. There are two main types of charge transfer: Metal-to-Ligand Charge Transfer (MLCT) and Ligand-to-Metal Charge Transfer (LMCT). LMCT involves the transfer of electronic charge from a ligand to a metal ion within a coordination complex whereas MLCT refers to the transfer of electronic charge from a metal ion to a ligand [27].

The intensity of this new peak exhibits a progressive augmentation with the gradual addition of the metal ion until reaching a saturation point, signifying the

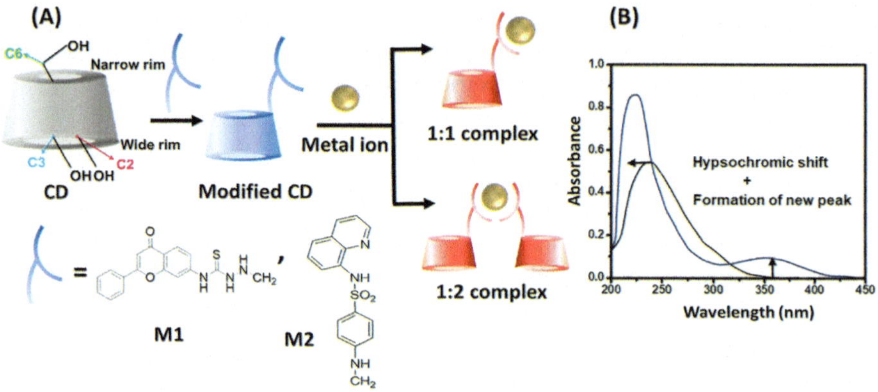

Fig. 4.1 **A** Modification of cyclodextrin (CD) is introduced through reactive sites at C2, C3 and C6 positions (as shown in β-CD). The modified CD interacts with metal ions, forming 1:1 or 1:2 metal-CD complexes. M1 and M2 represent chromophore appendages introduced during the modification of CD, where M1 is 7-aminoflavone [23] and M2 is N-(8-quinolyl)-p-aminobenzene-sulfonamide [24]. **B** Spectral changes during the complexation of modified cyclodextrin (CD) with metal ions. The plot illustrates a notable blue shift in the spectral band associated with the modified CD, alongside the emergence of a new peak resulting from the interaction with the metal ion

coordination of the maximum metal ions with the CD. Beyond this point, there is no further increase in absorbance, suggesting the completion of the coordination process. The determination of the coordination stoichiometry is accomplished through the molar ratio method [24].

In 7-aminoflavone-modified β-CD complex (refer M1 in Fig. 4.1A), the addition of Cu (II) prompted a blue shift in the absorption band (266 nm to 253 nm) and a decrease in absorbance was reported. Simultaneously, a novel band surfaced, corresponding to the d-d transition of the Cu (II)-Flav-β-CD complex. The presence of a singular isosbestic point in the absorbance titration affirmed a straightforward, one-step transformation during the copper complexation process [23]. A similar mechanism was observed for the N-(8-quinolyl)-p-aminobenzene-sulfonamide modified β-CD for sensing of Zn (II) [24] (refer M2 in Fig. 4.1A). Figure 4.1(B) shows the absorption spectrum of a modified CD before (black) and after metal ion addition (blue), highlighting a blue shift and the formation of a new spectral band.

4.5 CD Inclusion Complexes in Metal Ion Sensing

CDs possess an amphiphilic nature due to their hydrophobic inner cavity and hydrophilic outer surface, allowing them to form inclusion complexes with guest molecules. This complex formation is selective and depends on factors such as the size and geometrical fit of the guest molecules to the CD cavity [21]. The size of the cavity for α-, β-, and γ-CDs increases with increasing number of glucopyranose repeating units, and the cavity diameter is about 0.49, 0.62, and 0.80 nm, respectively [28]. The driving force involved in the complex formation can be attributed to reasons such as the replacement of water molecules by guest molecules from the CD cavity, release of conformational strain, and charge transfer interactions, out of which the replacement of the high enthalpy water molecules by a suitable "guest" molecule plays an important role. In an aqueous environment, the non-polar CD cavity houses polar water molecules, which is an energetically unfavoured interaction. As a result, these molecules of water can be spontaneously substituted by appropriate guest molecules having a more non-polar nature than water [26].

Various methods, such as co-precipitation, kneading, spray-drying, and freeze-drying, are used to prepare CD-guest complexes [29]. The inclusion complex formation significantly alters the physicochemical properties of guest molecules, enhancing their aqueous solubility, modifying their spectral features (e.g., increased fluorescence intensity), and reducing their diffusion and volatility. CDs can also act as synthetic enzymes, accelerating reactions and altering pathways [26].

The formation of inclusion complexes can be confirmed using UV–Visible spectroscopy, which often shows a blue shift in spectral bands due to changes in the guest's microenvironment [30]. CDs typically form a 1:1 stoichiometric complex with guest molecules, though other ratios like 1:2 or 2:1 are possible. Figure 4.2 shows the formation of inclusion complexes and the structures of guest molecules reported in the literature.

The stoichiometry of the complex can be determined using Job's method (continuous variation method), and the association constant is calculated using the Benesi-Hildebrand method based on spectrophotometric data [31].

4.5.1 Spectral Changes During Metal Ion-Inclusion Complex Interaction

In this section, we shall see the spectral changes caused by the interaction of metal ions with the inclusion complex. Although CD imparts changes in the chemical properties of the inclusion complex, the notable changes in the UV–Visible absorption spectrum are observed primarily due to the interaction between the guest molecule and metal ion. Spectral changes such as hyperchromic, and hypochromic shifts along with combinations of hypsochromic and bathochromic shifts are possible depending on the characteristics of the guest molecule. A general idea of what may influence these spectral changes is necessary to better interpret the UV–Visible absorption spectra. A red or bathochromic shift to the inclusion complex peak can be the result of increased conjugation/electron density on the sensor molecule brought about by the metal ion. A blue/hypsochromic shift is brought about by decreased conjugation/electron density [27]. A few specific cases will be discussed for better understanding.

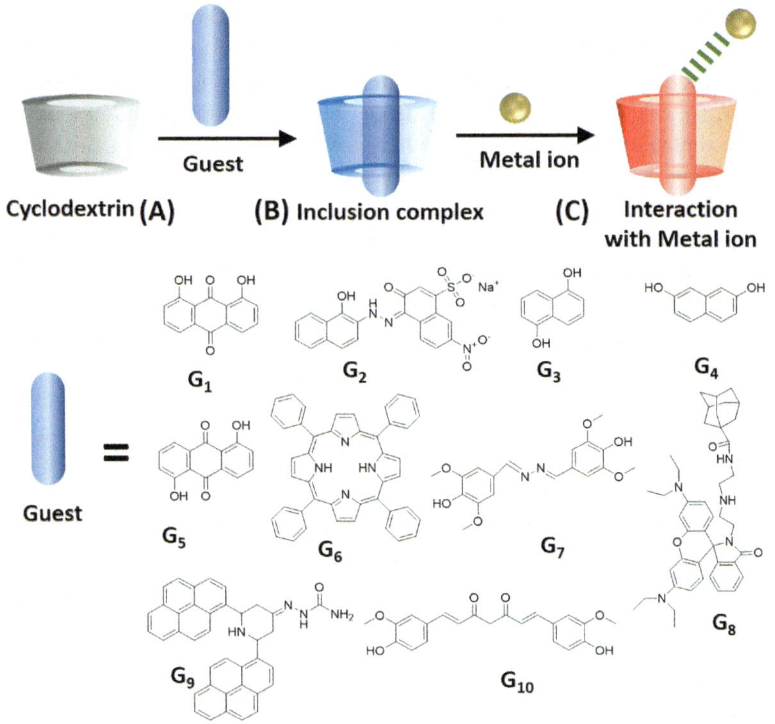

Fig. 4.2 The scheme demonstrates the formation of inclusion complex: The first step involves the complexation of CD with guest(G) and the subsequent step illustrates the interaction of its interaction with metal ions. The dashed line in the third step represents the interaction between the inclusion complex and metal ion. The interaction can lead to complexation or charge transfer. Certain reported guests are given, labelled G1-G10 (G1: 1,8-DHAQ, G2: EBT, G3: 1,5-ND, G4-2,7-ND, G5:1,5-DHAQ, G6:H2ttP, G7:SyA, G8:spirolactum rhodamine moiety, G9:Pyrene appended piperidone derivative, G10:curcumin)

4.5.1.1 Hyperchromic and Hypochromic Shifts

In this case, the λ_{max} of the inclusion complex (sensor) undergoes a hypochromic shift as it is used up due to the formation of a complex with the metal ion. Simultaneously, the peak corresponding to the formation of the metal-sensor complex increases in intensity with a rise in the concentration of the metal ion (Fig. 4.3B). When the 1,8-dihydroxyanthraquinone-β-CD (1,8 DHAQ- β-CD) inclusion complex interacted with Cu (II), the absorption peak corresponding to the inclusion complex at 500 nm decreased gradually and a new peak corresponding to metal complexation emerged at 546 nm [32]. This facilitated colorimetric sensing as a colour change occurred from citrus red to purple-red. An analogous scenario was recorded in the complexation reaction of the Eriochrome Black T(EBT-β-CD) system for the sensing of Ca (II) and Mg (II) at varying pH levels [33]. Figure 4.3A illustrates the gradual decrease

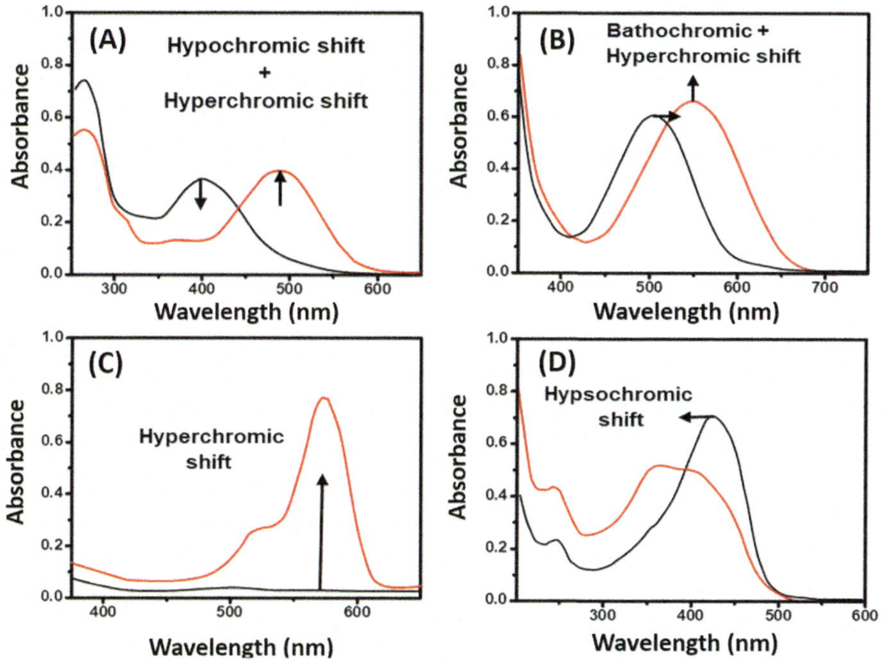

Fig. 4.3 **A** Addition of metal ions induces a hypochromic shift in the inclusion complex spectral band, accompanied by the formation of a new peak (hyperchromic shift). **B** The inclusion complex spectral band exhibits a bathochromic shift and increased absorbance following metal ion addition. **C** A hyperchromic shift is observed due to structural changes resulting from metal ion binding. **D** The spectral band of the inclusion complex demonstrates a hypochromic shift upon the addition of metal ions. [In the absorption spectra, the black line represents CD-NPs alone, while the red line represents CD-NPs after interaction with metal ions.]

in absorbance of the spectral band associated with the inclusion complex and the emergence of a new peak corresponding to complexation.

4.5.1.2 Hyperchromic and Bathochromic Shift

The 1,5-naphthalene diol (1,5-ND) incorporated with β-CD exhibits a characteristic bathochromic shift with enhancement in absorbance (hyperchromic shift) with the addition of Cu (II) ions in an aqueous solution (colour changes from pink to purple). 2,7-ND caused similar changes in the absorbance spectrum selectively in the presence of Ni (II) ions [34]. The absorption spectra of β-CD/1,5-DHAQ: Fe (III) exhibited a red shift in λ_{max} along with a hyperchromic shift in accordance with an increase in the concentration of Fe (III) [35]. CD/porphyrin (H_2ttp) supramolecular sensitiser exhibited a bathochromic shift of the *Soret* band of porphyrin on the addition of Zn (II) ions along with the appearance of a new band in the place of prominent *Q*-band of the porphyrin. The new band was attributed to the formation of the zinc/porphyrin complex [36].

The inclusion complex of syringaldehyde (SyA) di-Schiff azine with β-CD exhibited dual sensing for Cu (II) and Pb (II). In the presence of Cu (II), the colour of the complex changes from colourless to deep red. This is due to the formation of two peaks: a strong 526 nm band corresponding to the charge transfer from non-bonding electrons of nitrogen to the metal and a shoulder at 489 nm signifies metal-induced charge transfer. The Pb (II) ions induced a red shift in the SyA band, corresponding to π-π^* transition, also associated with charge transfer [37]. The expected spectral change in such examples has been depicted in Fig. 4.3B.

4.5.1.3 Hyperchromic Shift Due to Specific Structural Changes Induced by the Metal Ion

The absorption spectrum of an inclusion complex formed by a combination of the dansyl-linked β-CD and adamantyl-linked spirolactum rhodamine, upon addition of Fe (III) exhibits a hyperchromic shift. This is due to the formation of the pink-coloured open-ring state of the spirolactum rhodamine moieties on the addition of Fe (III) ions, as the rhodamine absorbs in the open-ring state [38]. Figure 4.3C shows an illustrative absorption spectrum depicting the phenomena. Similarly, a pyrene-appended piperidone derivative β-CD complex for Cd(II) sensing exhibits a hyperchromic shift of its prominent absorption band selectively for Cd (II) ion, while a hypochromic shift for other interfering ions. This is due to the structural modification of the inclusion complex caused by the binding with Cd (II) [39].

4.5.1.4 Hypsochromic Shift

Chelation of metal ions with the inclusion complex can also induce a blue shift in certain cases. Figure 4.3D shows an illustrative spectrum of this particular case. It has been seen that curcumin-β-CD complex on chelation with Hg (II) through the O^- atom of curcumin and OH group of β-CD causes a major blue shift of absorbance maxima from 433 to 379 nm [40]. In this case, a colorimetric response from yellow to colourless was also observed.

4.6 CD-Based Nanoparticles (NPs) for Metal Ion Sensing

NPs are a class of optical probes that showcase various appealing features such as bright coloured, high sensitivity, photostability, and biocompatibility. Nanomaterials are defined by dimensions typically ranging from 1 to 100 nm and include several categories including NPs, nanotubes, nanorods, nanocomposites, nanofibers, quantum dots, graphene, fullerenes, nanocapsules and so on. Quantum confinement, a phenomenon prevalent at the nanoscale, bestows nanomaterials with size-dependent electronic and optical characteristics [41]. One of the advantages of employing nanomaterials for metal sensing is that they provide abundant active sites for interactions with metal ions due to their high surface area. Nanomaterials also have a key role in addressing challenges related to selectivity in metal sensing.

4.6.1 Use of CD as a Stabilising Agent

Stabilisers such as polymers, proteins, surfactants and macromolecules are integral and indispensable components of nanoparticles. Komiyana reported the first instance of metal NPs stabilised by native CD in [42]. Chemical reduction, laser-induced ablation, microwave irradiation and ball milling techniques have been utilised for the synthesis of CD-based NPs [28]. Stabilisers including cyclodextrins are utilised for important reasons as follows:

- *Prevention of Agglomeration or Aggregation*: NPs have high surface energy, which makes them prone to agglomeration or aggregation. Stabilisers form a protective layer around the NPs, preventing them from sticking together.
- *Enhanced Stability and Shelf Life*: Stabilisers act as a protective barrier, preventing direct exposure of NPs to external factors such as oxygen or moisture, thus preserving their integrity and functionality.
- *Improved Solubility*: Many NPs are hydrophobic or poorly soluble in certain solvents, and the right choice of stabilisers can improve the solubility of NPs.
- *Controlled Synthesis and Size Distribution*: Stabilisers can play a crucial role in controlling the size and morphology of NPs during their synthesis. By influencing the nucleation and growth processes, stabilisers help achieve a more controlled and narrow size distribution of NPs.
- *Facilitation of Functionalisation*: Stabilisers can provide functional groups or reactive sites on the surface of the NPs, facilitating further modification or functionalisation.

What makes CDs stand out is their unique cavity-like geometry that complements their ability to improve solubility in aqueous medium and biocompatibility. Furthermore, the capacity to incorporate functional groups by alteration of the CD can enhance the selectivity of NPs towards metal ions [43]. Research suggests that the particle size of NPs depends on the concentration and form of CD used, and CD has the capability to influence the morphology of NPs. Additionally, there is evidence suggesting that CD molecules are multifunctional, capable of serving as both stabilising and reducing agents in an alkaline medium [28]. This versatility in enhancing key characteristics makes CD-functionalised NPs an excellent choice in selective metal ion sensing applications [44].

4.6.2 Surface Plasmon Resonance (SPR) and Spectral Features of NPs

The absorption spectra of NPs are influenced by a phenomenon known as SPR. When NPs, typically made of noble metals like Au or Ag, are exposed to light, the conduction electrons at the surface of the nanoparticle collectively oscillate in response to the incident electromagnetic field. Resonance takes place when the frequency of the incident light coincides with the natural frequency of the surface plasmons. The absorption spectra of the nanoparticle exhibit a distinct absorption band due to the absorption of light attributed to the resonance of surface plasmons. This band is known as the SPR band, and its position is highly sensitive to the size, shape, and surrounding environment of the NPs [45]. The position of the SPR band is closely associated with the size of the NPs. The size of the NPs influences the resonance frequency of the surface plasmons, affecting the position of the SPR band. Smaller NPs exhibit SPR bands at shorter wavelengths (blue region), while larger NPs have SPR bands at longer wavelengths (red region). Note that while SPR involves the collective oscillations of electrons on a macroscopic metal film and is widely used in biosensing,

Localised Surface Plasmon Resonance (LSPR) pertains to nanoscale metallic NPs, with applications in nanotechnology, sensing, and imaging [46]. The nature of the SPR band indicates the morphological features of NPs. This can be explained in accordance with Mie's theory i.e., for spherical NPs, only a single SPR band can be observed whereas for anisotropic particles, two or more SPR bands may be seen, depending on their shape [47]. Plasmonic metal NPs, particularly Ag and Au NPs, are good choices for use in optical sensors [48]. Colorimetric sensors detect changes in colour resulting from alterations to the UV–Visible spectroscopy employed for investigating the SPR bands of NPs. It is also utilised to analyse spectral changes resulting from CD functionalisation and, moreover, it may be utilised to sense metal ions using CD NPs.

In the following sections, we will explore: (1) spectral changes during CD functionalisation and (2) detection of metal ions using CD NPs in the context of UV–Visible absorption spectroscopy.

4.6.3 Spectral Changes During CD Functionalisation

The stabilisation of NPs by CD can be confirmed by the changes in the absorption spectrum. The stabilisation is attributed to the strong interaction between NPs and CD due to the reductant nature of CD i.e., the primary hydroxyl groups present at the narrow rim of CD get oxidised to carboxylic acids in the presence of metal precursor. These carboxylic acid groups are involved in the stabilisation of metal NPs [49].

In general, functionalisation by the introduction of CD prevents particle growth or leads to reduced particle size due to the effective capping of CD. In UV–Visible spectra, as the concentration of CD increases, a hyperchromic shift occurs till it reaches the optimum concentration of CD, after which a hypochromic shift is observed. When CD induces a reduction in particle size, a blue shift in the SPR band is expected [44]. Despite the reduction in NP size induced by CDs, as confirmed by Scanning Electron Microscopy and Transmission Electron Microscopy, there are indications in reports that introducing CD causes a red shift [47]. It is suggested that the observed red shift may result from alterations in the dielectric nature of the NPs' surface induced by CD; however, conclusive evidence is lacking [50]. This intriguing phenomenon warrants further exploration in nanomaterial research.

4.6.4 Detection of Metal Ions Using CD-Functionalised NPs

When a chemical species attaches to the surface of NPs, it induces a discernible change in their SPR, facilitating the determination of specific substances [46]. Various ways of sensing can be proposed based on the type of NP composite and the specific metal ion they interact with.

4.6.4.1 Aggregation-Based Sensing

When metal ions interact with the colloidal system of NPs, the inter-particle distance might decrease, resulting in aggregation. This results in a noticeable change in the SPR band of the NPs, for instance, a red shift in the SPR band signifies an increase in the size of NPs and a hypochromic shift of the SPR band is indicative of aggregation. Through proper calibration, the concentration of metallic ions in the analyte can be quantified using UV–Visible spectroscopy [46]. β-CD functionalised AuNPs responded to the addition of Cu (II) and underwent a gradual decrease of the SPR band at 546 nm. Simultaneously, a novel band emerged at 781 nm, indicative of aggregation of CD-NPs caused by the coordination of Cu (II) ions [50]. Hydroxypropyl-β-CD functionalised AuNPs showed aggregation in the presence of Pb (II) ions which was evaluated by monitoring the ratio of the absorption band at 650 nm corresponding to aggregated AuNPs to the absorbance of plasmon band at 525 nm, which represents the free AuNPs [51]. A comparable outcome was seen in an alternative study, in which Pb (II) ions were added to the AuNP solution, causing a progressive rise in intensity and a red shift from 528 to 550 nm. The chelation of Pb (II) was facilitated by the deprotonated -OH of β-CD [44]. Refer Fig. 4.4B for the schematic representation and illustration of spectral changes seen in aggregation-based sensing of metal ions by CD-NPs.

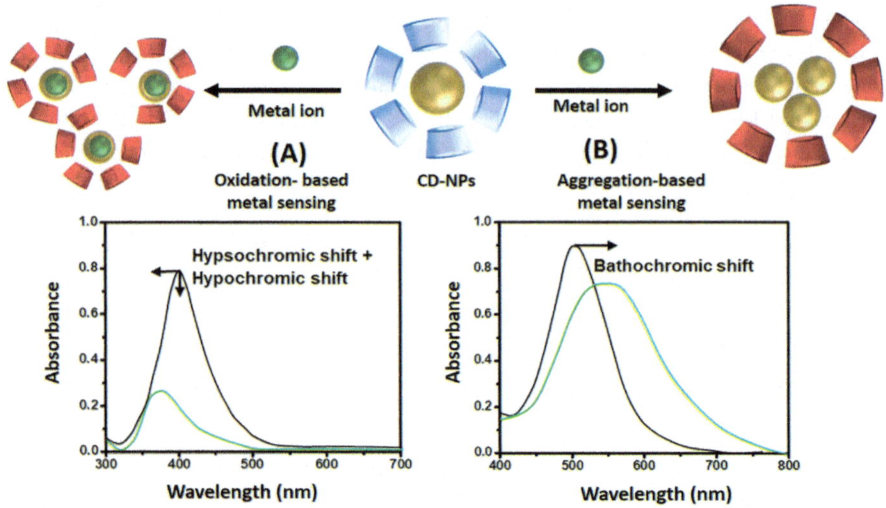

Fig. 4.4 **A** Oxidation-based sensing: The schematic illustrates the interaction of metal ions with CD-NPs resulting in oxidation of NPs and subsequent amalgamation of the metal NPs with interacting metal ions. The absorption spectrum shows a hypsochromic shift accompanied by a decrease in absorbance as oxidation progresses. **B** Aggregation-based sensing: The schematic represents the interaction of metal ions with CD-NPs, which induces aggregation of the nanoparticles. The absorption spectrum shows a bathochromic shift upon interaction with metal ions, indicating nanoparticle aggregation. [In the absorption spectra, the black line represents the CD-NPs alone, while the green line represents the CD-NPs after interaction with metal ions.]

4.6.4.2 Oxidation-Based Sensing

The introduction of specific metal ions into an NP solution induces the oxidation of NPs, particularly AgNPs, leading to a decrease in the number of metal NPs. Consequently, observable alterations in both colour and absorbance of SPR bands occur. This mechanistic pathway is particularly popular in sensing of Hg (II) using AgNPs. The reduction of Hg (II) to Hg (0) at the cost of Ag (0) oxidation to Ag (I) results in the formation of Ag-Hg amalgams on the surface of AgNPs, which dramatically alters the SPR signals and may induce a visual colour transformation as well [46]. Incremental addition of Hg (II) to AgNPs shows a hypsochromic shift and a decrease in absorbance (hypochromic shift) of the SPR band has been reported. This phenomenon is attributed to the creation of Ag-Hg amalgam when Hg (II) is adsorbed onto the surface of β-CD AgNPs, followed by the subsequent oxidation of Ag (0) to Ag (I) [52, 53]. Figure 4.4A shows a schematic representation and illustration of spectral changes associated with oxidation-based sensing of metal ions.

4.6.4.3 Sensing-Based on Morphological Changes

Metal ions can oftentimes induce a morphological change in nanostructures such as the conversion of one-dimensional nanorods and nanowires into NPs. In such cases, the disappearance of existing SPR bands and the appearance of new ones is expected [46].

4.7 Summary

In conclusion, this chapter has explored the role of CDs in the realm of metal ion sensing. CDs demonstrate remarkable versatility in metal ion detection through UV–Visible absorption spectroscopy. As low-toxicity macrocycles, CDs can encapsulate guest molecules and nanoparticles or be functionalized to selectively encapsulate metal ions. Their ability to enhance the aqueous solubility of sensing systems leads to more efficient metal ion detection in aqueous environments.

The different approaches and the multiple uses of CDs in spectrophotometric determination of metal ions are discussed by categorising them into three separate sections that explore the very distinct roles of CDs in the sensing systems. Native CDs lack chromogenic groups, necessitating structural modifications to enhance their capabilities as spectrophotometric sensors. By functionalising CDs, they can gain absorption features, improving their selectivity towards specific metal ions and enabling investigation through spectrophotometric techniques. The coordination of metal ions with modified CDs often induces a spectral shift, and emergence of new peaks due to electronic transitions and underlying mechanisms such as d-d transitions or charge transfer processes. CDs also have the ability to form inclusion complexes with specific guest molecules. These CD-guest systems interact with metal ions inducing spectral changes based on the different binding dynamics. Finally, CD-based NPs are another category of CD-based sensing systems that showcase the ability to quantitatively determine metal concentration based on the interaction of CD-NPs with the metal ion and subsequent alteration of the SPR bands.

Looking ahead, the strength of CD-based spectrophotometric sensors lies in their ability to enable *on- site*, 'naked-eye' detection of metal ions through observable color changes (qualitative analysis). Additionally, the potential for low-cost miniaturising absorption spectrometers enables portable, quantitative analysis. Combining the strengths of CDs with absorption spectroscopy, makes CD-based sensing systems ideal for on-site monitoring of metal ions in water bodies and water purification systems.

Acknowledgements Anupama Vijayan acknowledges DST-INSPIRE for providing research fellowship. John Prakash thanks DST-FIST (SR/FST/CS-I/2021/215) for the financial assistance.

References

1. Ahmad W, Alharthy RD, Zubair M et al (2021) Toxic and heavy metals contamination assessment in soil and water to evaluate human health risk. Sci Rep 11:17006. https://doi.org/10.1038/s41598-021-94616-4
2. Tchounwou PB, Yedjou CG, Patlolla AK, Sutton DJ (2012) Heavy metal toxicity and the environment, pp 133–164
3. Sheth Y, Dharaskar S, Chaudhary V et al (2022) Prospects of titanium carbide-based MXene in heavy metal ion and radionuclide adsorption for wastewater remediation: a review. Chemosphere 293:133563. https://doi.org/10.1016/j.chemosphere.2022.133563
4. Moiseenko TI, Gashkina NA (2020) Distribution and bioaccumulation of heavy metals (Hg, Cd and Pb) in fish: influence of the aquatic environment and climate. Environ Res Lett 15:115013. https://doi.org/10.1088/1748-9326/abbf7c
5. Cui W, Ren Z, Song Y, Ren CL (2022) Development and potential for point-of-care heavy metal sensing using microfluidic systems: a brief review. Sens Actuators A Phys 344:113733. https://doi.org/10.1016/j.sna.2022.113733
6. Pickering G, Mazur A, Trousselard M et al (2020) Magnesium status and stress: the vicious circle concept revisited. Nutrients 12:3672. https://doi.org/10.3390/nu12123672
7. Krupińska I (2020) Aluminium drinking water treatment residuals and their toxic impact on human health. Molecules 25:641. https://doi.org/10.3390/molecules25030641
8. Witkowska D, Słowik J, Chilicka K (2021) Heavy metals and human health: possible exposure pathways and the competition for protein binding sites. Molecules 26:6060. https://doi.org/10.3390/molecules26196060
9. Gattermann N, Muckenthaler MU, Kulozik AE et al (2021) The evaluation of iron deficiency and iron overload. Dtsch Arztebl Int. https://doi.org/10.3238/arztebl.m2021.0290
10. Banerjee P, Bhattacharya P (2021) Investigating cobalt in soil-plant-animal-human system: dynamics, Impact and management. J Soil Sci Plant Nutr 21:2339–2354. https://doi.org/10.1007/s42729-021-00525-w
11. Begum W, Rai S, Banerjee S et al (2022) A comprehensive review on the sources, essentiality and toxicological profile of nickel. RSC Adv 12:9139–9153. https://doi.org/10.1039/D2RA00378C
12. Chen J, Jiang Y, Shi H et al (2020) The molecular mechanisms of copper metabolism and its roles in human diseases. Pflugers Arch 472:1415–1429. https://doi.org/10.1007/s00424-020-02412-2
13. Okereafor U, Makhatha M, Mekuto L et al (2020) Toxic metal implications on agricultural soils, plants, animals, aquatic life and human health. Int J Environ Res Public Health 17:2204. https://doi.org/10.3390/ijerph17072204
14. Rahaman MdS, Rahman MdM, Mise N et al (2021) Environmental arsenic exposure and its contribution to human diseases, toxicity mechanism and management. Environ Pollut 289:117940. https://doi.org/10.1016/j.envpol.2021.117940
15. Tortella GR, Rubilar O, Durán N et al (2020) Silver nanoparticles: toxicity in model organisms as an overview of its hazard for human health and the environment. J Hazard Mater 390:121974. https://doi.org/10.1016/j.jhazmat.2019.121974
16. Genchi G, Sinicropi MS, Lauria G et al (2020) The effects of cadmium toxicity. Int J Environ Res Public Health 17:3782. https://doi.org/10.3390/ijerph17113782
17. Yang L, Zhang Y, Wang F et al (2020) Toxicity of mercury: molecular evidence. Chemosphere 245:125586. https://doi.org/10.1016/j.chemosphere.2019.125586
18. Kumar A, Kumar A, MMS C-P, et al (2020) Lead toxicity: health hazards, influence on food chain, and sustainable remediation approaches. Int J Environ Res Public Health 17:2179. https://doi.org/10.3390/ijerph17072179
19. Batool M, Junaid HM, Tabassum S et al (2022) Metal ion detection by carbon dots—a review. Crit Rev Anal Chem 52:756–767. https://doi.org/10.1080/10408347.2020.1824117
20. Norkus E (2009) Metal ion complexes with native cyclodextrins. an overview. J Incl Phenom Macrocycl Chem 65:237–248. https://doi.org/10.1007/s10847-009-9586-x

21. Sivakumar K, Chaitanya GK (2023) α- cyclodextrin based chemosensors: a review. Comments Inorg Chem 43:177–201. https://doi.org/10.1080/02603594.2022.2121277
22. Zhang Y-H, Zhang Y-M, Chen Y et al (2014) Phenanthroline bridged bis(β-cyclodextrin)s/adamantane-carboxylic acid supramolecular complex as an efficient fluorescence sensor to Zn2+. Org Chem Front 1:355. https://doi.org/10.1039/c3qo00054k
23. Azath IA, Pitchumani K (2013) Flavone modified-β-cyclodextrin as a highly selective and efficient fluorescent chemosensor for Cu2+ ions and l-histidine. Sens Actuators B Chem 188:59–64. https://doi.org/10.1016/j.snb.2013.07.009
24. Liu Y, Zhang N, Chen Y, Wang L-H (2007) Fluorescence sensing and binding behavior of aminobenzenesulfonamidoquinolino- β -cyclodextrin to Zn 2+. Org Lett 9:315–318. https://doi.org/10.1021/ol062816w
25. Jiang H, Xu Y, Na L et al (2008) UV-vis spectral analysis of inclusion complexes between -cyclodextrin and aromatic/aliphatic guest molecules. Curr Drug Discov Technol 5:173–176. https://doi.org/10.2174/157016308784746283
26. Szejtli J (1998) Introduction and general overview of cyclodextrin chemistry. Chem Rev 98:1743–1754. https://doi.org/10.1021/cr970022c
27. Kaur B, Kaur N, Kumar S (2018) Colorimetric metal ion sensors–a comprehensive review of the years 2011–2016. Coord Chem Rev 358:13–69. https://doi.org/10.1016/j.ccr.2017.12.002
28. Prochowicz D, Kornowicz A, Lewiński J (2017) Interactions of native cyclodextrins with metal ions and inorganic nanoparticles: fertile landscape for chemistry and materials science. Chem Rev 117:13461–13501. https://doi.org/10.1021/acs.chemrev.7b00231
29. Cid-Samamed A, Rakmai J, Mejuto JC et al (2022) Cyclodextrins inclusion complex: preparation methods, analytical techniques and food industry applications. Food Chem 384:132467. https://doi.org/10.1016/j.foodchem.2022.132467
30. Periasamy R, Nayaki SK, Sivakumar K, Ramasamy G (2020) Synthesis and characterization of host-guest inclusion complex of β-cyclodextrin with 4,4′-methylenedianiline by diverse methodologies. J Mol Liq 316:113843. https://doi.org/10.1016/j.molliq.2020.113843
31. Paul J (1928) Formation and stability of inorganic complexes in solution. Ann Chim (Paris) 9:113–203
32. Wang Y, Wang L, Shi LL et al (2012) Colorimetric and fluorescence sensing of Cu2+ in water using 1,8-dihydroxyanthraquinone-β-cyclodextrin complex with the assistance of ammonia. Talanta 94:172–177. https://doi.org/10.1016/j.talanta.2012.03.013
33. Afkhami A, Khalafi L (2008) Investigation of the effect of inclusion of eriochrome black T with β-cyclodextrin on its complexation reaction with Ca 2+ and Mg 2+ using rank annihilation factor analysis. Supramol Chem 20:579–586. https://doi.org/10.1080/10610270701522906
34. Kavitha R, Stalin T (2015) Naphthalenediols: a new class of novel fluorescent chemosensors for selective sensing of Cu2+ and Ni2+ in aqueous solution. J Lumin 158:313–321. https://doi.org/10.1016/j.jlumin.2014.10.029
35. Mohandoss S, Maniyazagan M, Stalin T (2015) A highly selective dual mode detection of Fe3+ ion sensing based on 1,5-dihydroxyanthraquinone in the presence of β-cyclodextrin. Mater Sci Eng C 48:94–102. https://doi.org/10.1016/j.msec.2014.11.060
36. Yang R, Li K, Wang K et al (2003) Porphyrin assembly on β-cyclodextrin for selective sensing and detection of a zinc ion based on the dual emission fluorescence ratio. Anal Chem 75:612–621. https://doi.org/10.1021/ac020467n
37. Narayanan V, Ganesan V, Shanmugasundram E et al (2023) Spectral studies of synthetic symmetric syringaldehyde di-Schiff azine with β-cyclodextrin inclusion complexes as a fluorescent probe for the detection of metal ions. J Photochem Photobiol A Chem 445:115069. https://doi.org/10.1016/j.jphotochem.2023.115069
38. Xu M, Wu S, Zeng F, Yu C (2010) Cyclodextrin supramolecular complex as a water-soluble ratiometric sensor for ferric ion sensing. Langmuir 26:4529–4534. https://doi.org/10.1021/la9033244
39. Poomalai S, Govindaraj TS, Soundrapandian S et al (2018) A new fluorescent chemosensor for cadmium(II) based on a pyrene-appended piperidone derivative and its β-cyclodextrin complex. Luminescence 33:538–544. https://doi.org/10.1002/bio.3443

40. Prabu S, Mohamad S (2020) Curcumin/beta-cyclodextrin inclusion complex as a new "turn-off" fluorescent sensor system for sensitive recognition of mercury ion. J Mol Struct 1204:127528. https://doi.org/10.1016/j.molstruc.2019.127528
41. Asha AB, Narain R (2020) Nanomaterials properties. In: Polymer science and nanotechnology. Elsevier, pp 343–359
42. Komiyama M, Hirai H (1983) Colloidal rhodium dispersions protected by cyclodextrins. Bull Chem Soc Jpn 56:2833–2834. https://doi.org/10.1246/bcsj.56.2833
43. Denicourt-Nowicki A, Roucoux A (2010) Noble metal nanoparticles stabilized by cyclodextrins: a pertinent partnership for catalytic applications. Curr Org Chem 14:1266–1283. https://doi.org/10.2174/138527210791616911
44. Aswathy B, Avadhani GS, Suji S, Sony G (2012) Synthesis of β-cyclodextrin functionalized gold nanoparticles for the selective detection of Pb2+ ions from aqueous solution. Front Mater Sci 6:168–175. https://doi.org/10.1007/s11706-012-0165-5
45. Hyder A, Buledi JA, Nawaz M et al (2022) Identification of heavy metal ions from aqueous environment through gold, silver and copper nanoparticles: an excellent colorimetric approach. Environ Res 205:112475. https://doi.org/10.1016/j.envres.2021.112475
46. Amirjani A, Haghshenas DF (2018) Ag nanostructures as the surface plasmon resonance (SPR)-based sensors: a mechanistic study with an emphasis on heavy metallic ions detection. Sens Actuators B Chem 273:1768–1779. https://doi.org/10.1016/j.snb.2018.07.089
47. Akhondi M, Jafari AH, Jamalizadeh E (2020) Selective colorimetric detection of HgII using silver nanoparticles modified with apple and nigella sativa seed extracts and β-Cyclodextrin. J Environ Chem Eng 8:103566. https://doi.org/10.1016/j.jece.2019.103566
48. Wang X, Wang Y, Ying Y (2021) Recent advances in sensing applications of metal nanoparticle/metal–organic framework composites. TrAC Trends Anal Chem 143:116395. https://doi.org/10.1016/j.trac.2021.116395
49. Amanulla B, Subbu HKR, Ramaraj SK (2018) A sonochemical synthesis of cyclodextrin functionalized Au-FeNPs for colorimetric detection of Cr6+ in different industrial waste water. Ultrason Sonochem 42:747–753. https://doi.org/10.1016/j.ultsonch.2017.12.041
50. Bindhu MR, Saranya P, Sheeba M et al (2021) Functionalization of gold nanoparticles by β-cyclodextrin as a probe for the detection of heavy metals in water and photocatalytic degradation of textile dye. Environ Res 201:111628. https://doi.org/10.1016/j.envres.2021.111628
51. Kou Y, Zhao Q, Wang X, Liu Y (2020) Rapid colorimetric detection of Pb2+ with β-cyclodextrin-functionalized nanocrystalline gold. Mater Chem Phys 243:122168. https://doi.org/10.1016/j.matchemphys.2019.122168
52. Manivannan S, Ramaraj R (2013) Silver nanoparticles embedded in cyclodextrin–silicate composite and their applications in Hg(ii) ion and nitrobenzene sensing. Analyst 138:1733. https://doi.org/10.1039/c3an36488g
53. Rajamanikandan R, Ilanchelian M (2018) β-cyclodextrin functionalised silver nanoparticles as a duel colorimetric probe for ultrasensitive detection of Hg2+ and S2− ions in environmental water samples. Mater Today Commun 15:61–69. https://doi.org/10.1016/j.mtcomm.2018.02.024

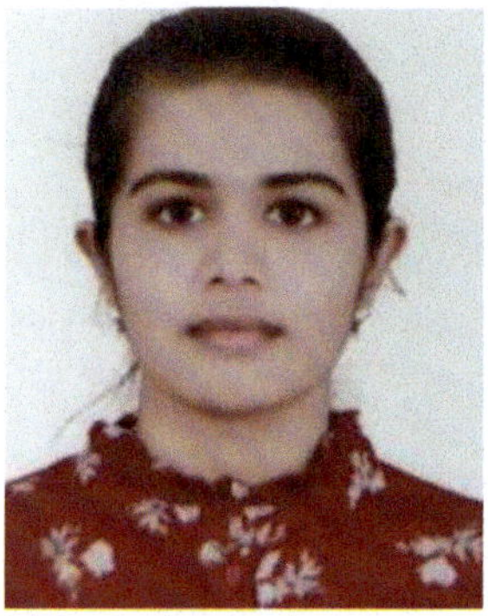

Anupama Vijayan is a DST-INSPIRE Fellow at the Central University of Tamil Nadu, Thiruvarur. She completed her 5-year integrated M.Sc. degree from the same institution. Her research primarily focuses on unconventional steady-state fluorescence spectroscopic techniques and has introduced a new method called Angular Sweep Total Fluorescence Spectroscopy.

Najiya C. K. completed his Integrated Master's in Chemistry at the Central University of Tamil Nadu, Thiruvarur. She is an DST-INSPIRE scholarship awardee. Najiya's current research involves microwave-assisted synthesis of fluorescent sensors.

Dr. John Prakash is an Assistant Professor at the Central University of Tamil Nadu. He earned his Ph.D. from the Indian Institute of Technology, Madras, and was awarded the prestigious Eli Lilly Outstanding Thesis Award in 2013. His research interests encompass analytical fluorimetry, the development of modular spectroscopic instrumentation for environmental and biologically relevant systems, and the application of chemometric tools to spectroscopic data.

Chapter 5
Chromophore-Appended Cyclodextrin Inclusion Complexes as Chemical Sensor for Organic Molecules

Rima Das

Abstract Chemical sensors have garnered significant interest for their role in detecting and overseeing potential environmental threats. Chromophore moieties are the groups that absorb certain wavelengths of visible light and reflect or transmit others, resulting in the perceived colour. Chromophores are often found in organic compounds, such as pigments and dyes, and they contain a conjugated system of alternating single and multiple bonds, which allow the absorption of light. Host guest complexation with chromophore moieties is an important principle for developing chemical sensors. Due to the versatile characteristics of native cyclodextrins (CDs), they can effectively sense various organic molecules by forming inclusion complexes. The CD host molecule detects small organic guest molecules directly by entrapping them within its cavity through host–guest complexation. In contrast, larger molecules are detected on the outer surface of the cavity. This chapter discusses cyclodextrin-based inclusion complexes containing various chromophoric groups, such as naphthalene, ferrocene, anthracene, pyrene, and dye molecules, which function as sensors for recognizing organic molecules such as steroids, pesticides, and nitro aromatics, which will be very helpful for new generation researchers.

Keywords Chemical sensors · Chromophore · Cyclodextrins · Inclusion complex · Organic molecules

5.1 Introduction: Chemical Sensors

Chemical sensors are essential devices with a significant impact across diverse sectors including environmental monitoring, healthcare, food safety, and industrial processes. Their primary function is to identify and measure particular chemical substances, spanning from gases to liquids, thereby facilitating continuous monitoring and analysis in real time [1]. Chemical sensors employ various principles,

R. Das (✉)
Department of Chemistry, Chandidas Mahavidyalaya, Khujutipara, Birbhum, West Bengal, India
e-mail: chem.rima@gmail.com

S. Krishnamoorthy (ed.), *Cyclodextrins for Chemosensing*,
https://doi.org/10.1007/978-3-031-89299-8_5

encompassing electrochemical, optical, piezoelectric, and resistive techniques. Electrochemical sensors, for instance, detect alterations in electrical properties arising from chemical reactions at electrode interfaces. Optical sensors exploit interactions between light and matter to identify changes in absorbance, fluorescence, or refractive index prompted by target analytes. Piezoelectric sensors detect shifts in mechanical properties triggered by chemical interactions, whereas resistive sensors gauge fluctuations in electrical resistance caused by chemical adsorption or desorption [2].

In recent times, notable progress has been achieved in the technology of chemical sensors, elevating their sensitivity, selectivity, and dependability. A significant advancement involves the incorporation of nanomaterials like carbon nanotubes, graphene, and metal oxides, which provide augmented surface area and reactivity, consequently enhancing sensing capabilities [3]. Optical chemosensors, particularly, are highly intriguing and beneficial. This can be attributed to the direct observation of optical changes, such as colour, absorption, and/or emission, which occur when the host molecule detects the target guest analyte, often visible to the naked eye. In optical chemosensors, chromophores are integrated with key sensing structures to harness these optical shifts, enabling the detection of successful recognition or sensing events [4]. The microenvironment surrounding the employed chromophores induces alterations in colour and fluorescence patterns. To achieve optimal performance, researchers worldwide have modified carbon dots with various dyes in addition to fluorescent components. By employing chromophore-modified CDs, the detection of a diverse array of hydrophobic guest molecules within hydrophobic compartments has been extensively investigated [5, 6].

This chapter outlines the chromophore appended CD-based inclusion complexes with various chromophore moieties such as naphthalene, ferrocene, anthracene, pyrene, and dye molecules that function as a chemical sensor for the recognition of organic molecules.

5.2 Sensoring by Chromophore-Appended CD Inclusion Complex

5.2.1 CDs Bearing Naphthalene Moieties

The focus of major studies has centred around α-cyclodextrin and β-cyclodextrin due to their accommodating cavity sizes for numerous organic compounds. However, γ-cyclodextrin has garnered significant interest recently due to its ability to host two guests within its spacious cavity [7–11]. In the early 1980s, Ueno et al. demonstrated that attaching a naphthalene moiety to γ-cyclodextrin transforms it into an effective host for small molecules through induced-fit complexation (Fig. 5.1) [12, 13].

The preparation method involves sulfonation of γ-cyclodextrin at the 6-hydroxy position through a reaction with naphthalene-2-sulfonyl chloride in pyridine, resulting in compound (1). This sulfonated 7-γ-cyclodextrin is then converted into the

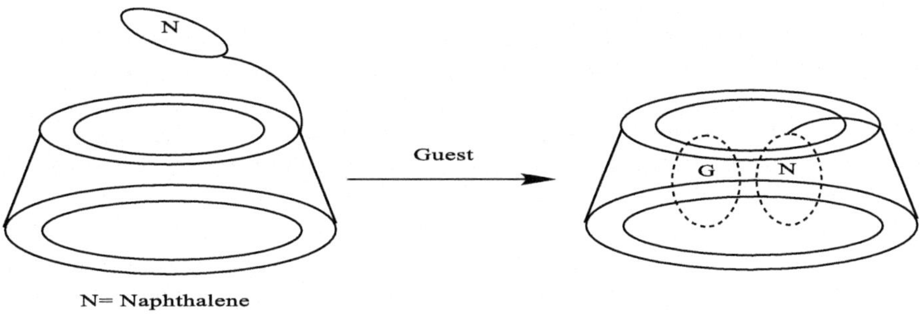

Fig. 5.1 Induced-fit type complexation with CD-bearing naphthalene

ester (2) (Fig. 5.2) through a replacement reaction with sodium 2-naphthylacetate. These γ-cyclodextrin derivatives feature an attached naphthalene moiety, serving as a probe to assess its surroundings. The complexation behaviour with organic molecules such as cyclohexanol, (+)-fenchone, and (−)-borneol were analysed through circular dichroism spectra in the absorption regions of the chromophore.

γ-Cyclodextrin derivatives with dual 2-naphthylsulfonyl groups display excimer emission, which is notably affected by guest molecules, allowing the detection of a range of neutral organic compounds. In a subsequent study, Ueno et al. synthesized four cyclodextrin inclusion complexes by reacting 2-naphthylsulfonyl chloride with γ-cyclodextrin in pyridine [14]. Among these four hosts, the third demonstrated the highest sensitivity to guest molecules, with selectivity in the following order:

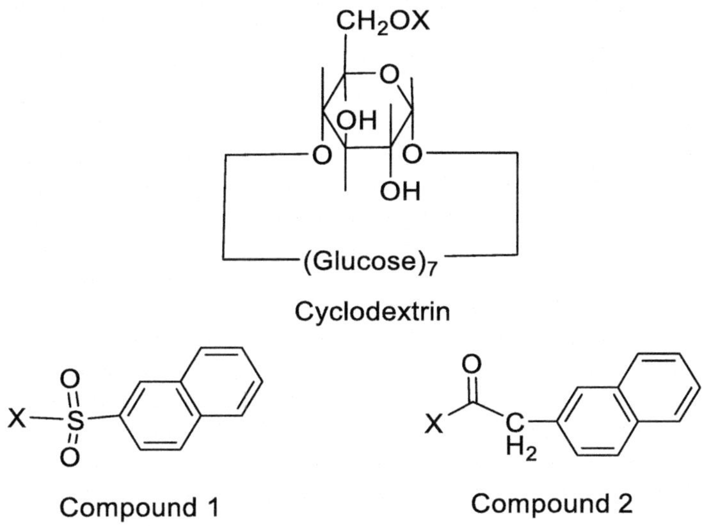

Fig. 5.2 Sulfonyl compound **1** and ester compound **2**

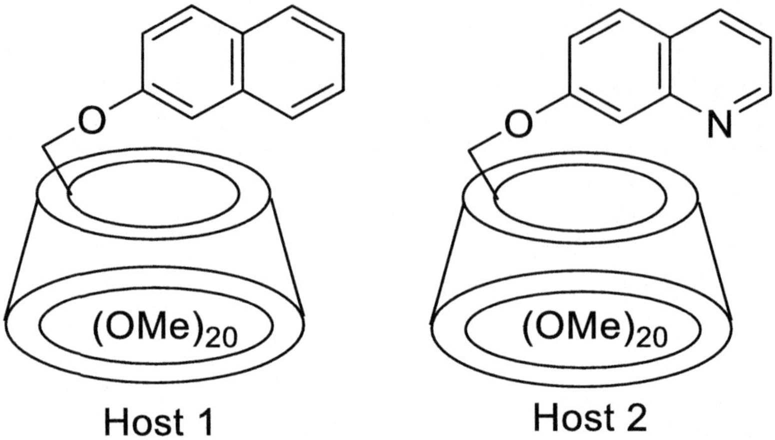

Fig. 5.3 PM-β-CD derivatives with appended naphthalene and quinoline fluorophores

cyclododecanol > adamantane-1-carboxylic acid (1-ACA) = (−)-borneol > cyclohexanol. Hosts 2 and 4 showed similar guest binding patterns. These systems can detect various organic compounds through fluorescence changes in the hosts, even if the compounds themselves are optically inactive.

Liu et al. synthesized two novel PM-β-CD derivatives (1 and 2), illustrated in (Fig. 5.3), with appended naphthalene and quinoline fluorophores. They further examined their molecular binding ability and selectivity for bile salts [15]. Shifting the chromophore groups from the central cavity to the more hydrophobic torus rim led to a significant increase in fluorescence intensity and extended fluorescent lifetimes in hosts 1 and 2 with the gradual addition of bile salts. This represents a significant departure from the molecular recognition observed with chromophore-modified β-CD species and bile salts.

5.2.2 CDs Bearing Ferrocene Moieties

Cyclodextrins have the ability to encapsulate lipophilic guest molecules within their cavities, forming a hydrophobic microenvironment in aqueous solutions. Therefore, studies on cyclodextrin inclusion phenomena are typically conducted in water-based solutions [16]. However, Ueno et al. found that ferrocene-appended β-cyclodextrin can act as a host in various organic solvents. This discovery demonstrates that cyclodextrins can participate in inclusion chemistry even without water [17]. The compound was synthesized by reacting 6-O-tosyl β-cyclodextrin with potassium ferrocene carboxylate in DMSO at 80 °C, followed by purification through recrystallization using a mixture of 1-butanol, ethanol, and water. A circular dichroism band observed within the ferrocene absorption range indicated that the ferrocene moiety was involved within the chiral cavity of the compound. The researchers examined how guest addition affected the circular dichroism peak by introducing cyclohexanol and adamantane carboxylic acid as guests, noting that the intensity of the circular dichroism signal decreased with increasing guest concentration in each solvent system.

5.2.3 CDs Bearing Anthracene Moieties

Two years after their previous report, the same research group presented another instance of Induced-Fit Types of Host–Guest Complexation involving anthracene-appended γ-cyclodextrin derivatives. The complexation behavior of these γ-cyclodextrin derivatives depends on the size of the attached aromatic moiety. When the size of the aromatic group is suitable, two host molecules can interact with each other through the aromatic moiety. Consequently, they delve into the complexation of γ-cyclodextrin, encompassing association, photodimerization, and induced-fit types of complexations of anthracene-appended γ-cyclodextrin derivatives. When anthracene guests align parallel to the axis of the cyclodextrin (CD) cavity (axial inclusion), the induced circular dichroism of the CD-anthracene complexes shows a positive signal in the wavelength regions corresponding to electronic transitions polarized along the long axis of the arenes. Similar circular dichroism changes were observed when 1-adamantanecarboxylic acid was used as the guest. With the introduction of borneol into the solution, there is a significant decrease in dichroism intensities across the entire region, and a distinct negative band emerges at 250 nm, replacing the exciton coupling band. The presence of a negative band when (−)-borneol is present suggests that the long axis of the anthracene moiety is oriented perpendicular to the axis of the beta-cyclodextrin (β-CD) cavity within the complex [18].

5.2.4 CDs Bearing Pyrene Moieties

In solution, the presence of pyrene-appended γ-cyclodextrin (γ-CD) induces changes in the fluorescence intensity of pyrene (Py) excimers and monomers due to the conversion from its dimeric form into a host–guest complex. This fluorescence alteration has been utilized for the detection of various organic compounds. The pyrene moiety is linked to the primary hydroxyl group of γ-cyclodextrin via amide (Fig. 5.4) and ester bonds. In this configuration, γ-cyclodextrin acts as a molecular flask or container, facilitating the meeting and reaction of two species, as reported by Ueno in 1989.

The binding affinity of both 1 and 2 increases as the guest molecules become larger, following the sequence: cyclohexanol < n-octanol < l-menthol < l-borneol < 1-adamantanecarboxylic acid < cyclododecanol. Importantly, host 1 consistently surpasses host 2, displaying higher host–guest association constants for all guest molecules. This advantage is particularly noticeable with larger guests such as

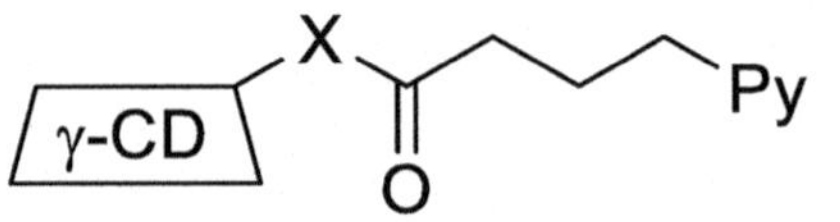

Fig. 5.4 Primary hydroxyl group of γ-cyclodextrin via amide and ester bonds

l-borneol, 1-adamantanecarboxylic acid, and cyclododecanol. Given the low likelihood of such large guests being co-included with the pyrene moiety, it is suggested that in these complexes, the pyrene moiety likely acts as a flexible cap rather than a spacer. Conversely, cyclohexanol, being the smallest guest, might share the cavity with the pyrene moiety. These systems hold promise as distinctive sensors for detecting different guest molecules by leveraging changes in the pyrene excimer-to-monomer fluorescence ratio [19]. Pyrene-γ-CD inclusion complexes demonstrate heightened sensitivity, especially towards certain steroid compounds. Among the guests utilized, which are either similar or smaller in molecular size compared to cyclohexanol, 1 exhibits the highest sensitivity. This sensitivity is notably influenced by the molecular structure of the guest, as evidenced by a group of guest substrates such as nerol, geraniol, d-fenchone, l-fenchone, d-camphor, and 1-adamantanol. The variation in sensitivity of 1 is further accentuated by changes in the intensity of pyrene excimer and monomer emissions [20].

The pioneering work by the Ueno group introduced the first host–guest sensory systems capable of detecting organic compounds, including biologically important substances, through pyrene-CD excimer fluorescence with remarkable sensitivities and selectivity [21]. In their research, the pyrene excimer emission generated by the association dimers of pyrene-appended γ-cyclodextrins, referred to as γ-PI and γ-P2, was used as a method to detect a range of organic compounds. Upon the addition of guests, the association dimers, wherein two pyrene moieties form an excimer, were transformed into 1:1 host–guest complexes. The sensitivities of γ-PI and γ-P2, assessed by the decrease in excimer emission intensity after guest addition, were influenced by the polarity, shape, and size of the guest molecules. The molecular shape plays a crucial role in the tight inclusion of guests within the cavities of γ-PI and γ-P2. Monocyclic compounds such as nerol, geraniol, and 1-menthol exhibited significantly lower sensitivities compared to the bicyclic compounds [22]. Notably, these sensitivities were notably high towards certain steroidal compounds such as chenodeoxycholic acid, ursodeoxycholic acid, and lithocholic acid. Many ketosteroids and amines led to reductions in both monomer and excimer emission intensities due to their quenching effects. However, alkaloids such as quinine and strychnine were barely detectable with these systems.

5.2.5 *CDs Bearing Dansyl Moiety*

Molecular recognition of steroidal compounds by dansyl-modified β-cyclodextrin (Fig. 5.5), leading to changes in fluorescence intensity. Nearly three decades ago, A.

Ueno observed a decrease in fluorescence intensity upon binding of guest molecules to dansyl-modified β-cyclodextrin, illustrating a monomolecular system for guest binding within the host cavity, facilitating molecular recognition [23].

Dansyl-modified β-cyclodextrins were synthesized through the reaction of dicyclohexyl carbodiimide, dansyl glycine, and 6-deoxy-6-amino-β-cyclodextrins in DMF. The fluorescence spectra of synthesized complex were analysed both in the presence and absence of 1-borneol, revealing a decrease in fluorescence intensity with increasing borneol concentration. The observed variation in fluorescence spectra, induced by the guest, suggests that the dansyl moiety relocates from the interior of the hydrophobic cavity towards the bulk water environment outside the cavity [24].

Figure 5.6 displays a range of steroids with varying sensitivities, with ursodeoxycholic acid (G) and chenodeoxycholic acid (H) demonstrating notably high sensitivity.

Conversely, deoxycholic acid (F), which differs only in the position of its OH group compared to these acids, exhibits significantly lower sensitivity. Cholic acid (I), featuring an additional OH group, is scarcely detected due to its heightened polarity. Ketosteroids like corticosterone (B) and cortisone (C), which possess two OH groups, as well as prednisolone (D) and hydrocortisone (E), with three OH groups, prove challenging to detect using this system. Progesterone (A), lacking OH groups and exhibiting greater hydrophobicity compared to ketosteroids, is detected with higher but still limited sensitivity [25].

The interaction between a guest molecule and a host cavity leads to a relocation of the attached chromophore, shifting it from within to outside the cavity. This alteration results in a significant modification of the spectrum of the chromophore-modified CDs. Such modified CDs find utility in the detection of medium to large cyclic compounds, including adamantane derivatives and steroidal compounds. Furthermore, research conducted by the Ueno group has unveiled a method for detecting linear compounds such as aliphatic alcohols, employing α-cyclodextrin (α-CD) with a suitably narrow cavity size. Adding a p-dimethylaminobenzoyl group to α-cyclodextrin yields a compound that displays dual emissions stemming from both regular planar and twisted intramolecular charge transfer excited states.

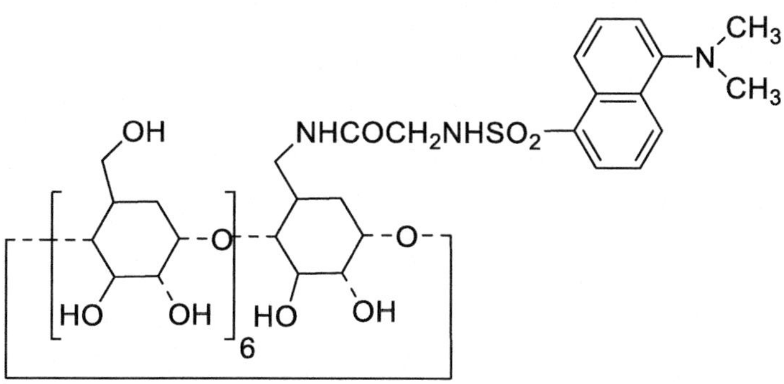

Fig. 5.5 Dansyl-modified β-cyclodextrins

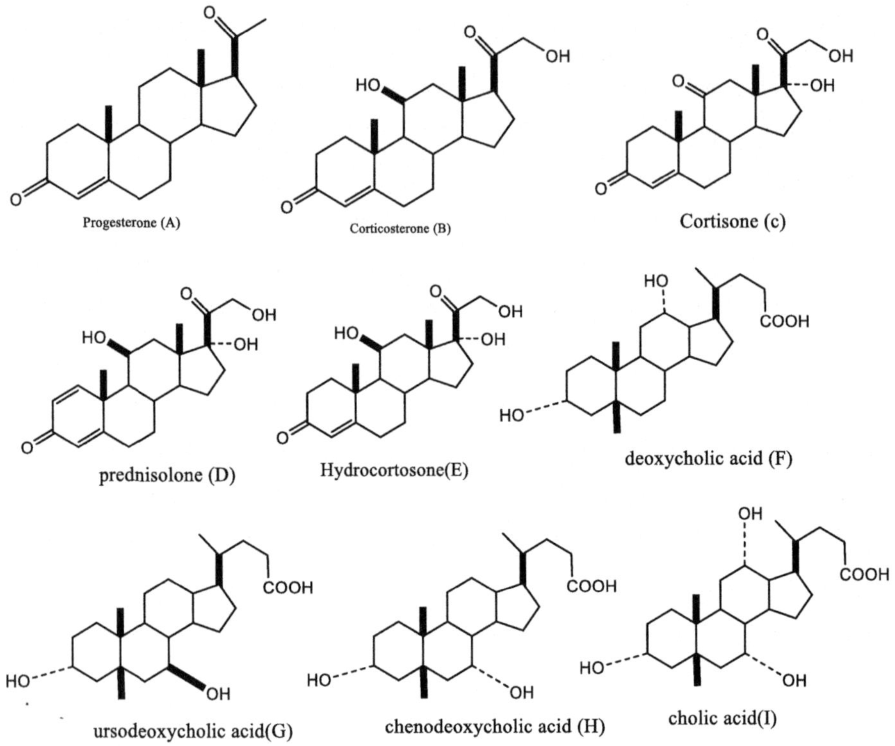

Fig. 5.6 Detection of a range of steroids with varying sensitivities

This compound functions as a fluorescent sensor, specifically detecting aliphatic compounds like n-pentanol and n-hexanol [26].

They synthesized a novel sensor, DMAB-α-CD, by covalently attaching a p-dimethyl aminobenzoyl (DMAB) moiety to α-cyclodextrin (α-CD), aimed at detecting linear compounds. The fluorescence responses to aliphatic alcohols were observed. The synthesis of DMAB-α-CD was carried out by reacting mono-6-deoxy-6-amino-α-cyclodextrin with p-dimethylaminobenzoic acid in N,N-dimethylformamide at temperatures below 0 °C, using N,N-dicyclohexylcarbodiimide (DCC) as a coupling agent. For two alcohol series, C_4H_9OH and $C_5H_{11}OH$, the sensitivity of DMAB-α-CD followed the order: normal > iso > tert. Among potential cyclodextrin oligomers, cyclodextrin dimers have garnered the most thorough investigation. These oligomers exhibit activity in regulating molecular recognition through conformational changes. The diverse range of linker moieties that connect the cyclodextrin recognition sites enables the creation of distinct specificity and functionalities [27].

Dansyl-modified CD dimer induces changes in fluorescent intensity within aqueous solutions, serving as a means for detecting steroidal compounds [22, 27].

Synthesized through the reaction of dansyl-L-aspartic acid and 6-deoxy-6-amino-beta-CD in DMF and hydroxybenzotriazole, this compound demonstrates varying responses to steroidal compounds such as ursodeoxycholic acid (UDCA), chenodeoxycholic acid (CDCA), deoxycholic acid (DCA), and cholic acid (CA) based on their concentrations. Notably, UDCA and CDCA elicit negative sensitivity values, while DCA and CA yield positive responses, with negligible sensitivity exhibited by the latter two compounds [28]. The decrease in fluorescence implies that the dansyl component, housed within one β-CD cavity of the CD dimer, transitions from the hydrophobic cavity to the bulk water environment. However, due to the asymmetry of this dimer, it raises the question of which cavity among the two CD units accommodates the dansyl moiety [29].

Kuroda et al. have introduced novel dimeric and trimeric cyclodextrins linked by biphenyl moieties, demonstrating solubility in both water and a range of organic solvents. The dimer exhibits a distinctive conformational shift, characterized by the formation of an excimer between two biphenyl moieties. It was observed that the cyclodextrin dimer can adopt at least two conformations: a stacked form and a separate form. The predominant factor influencing the balance between these conformations is expected to be their hydrophobic characteristics. The novel doubly bridged cyclodextrin oligomers were prepared by reacting per-O-methyl-6A,6D-diamino-β-cyclodextrin with 4,4′-biphenyl dicarbonyl chloride in THF, achieving yields of 30% and 9%, respectively. Following this, the molecular recognition abilities of the newly developed cyclodextrin oligomers were examined utilizing the anthracene derivative, which features two hydrophobic alkyl chains. Through spectroscopic titration, the association constants were determined to be approximately 5200, 27,000, and 8600 M^{-1}, respectively. These findings underscore the unique characteristics of the novel cyclodextrin oligomers and propose intriguing avenues for regulating their conformations [30].

5.2.6 pH Indicator Modified CDs

In qualitative chemical analysis, the detection of neutral organic molecules is usually done by the presence of a colour change indicator. Some ionophores exhibit colour change upon binding with metal or ammonium ions in solution [31–33]. Detection of organic molecules in solution becomes difficult in such cases. Fluorophore-modified CD shows certain fluorescent intensity by host–guest complexation with organic guest molecules which is discussed in the following section.

Over the past decade, significant attention has been directed towards Alizarine yellow modified CDs. The systems are capable of detecting chemical and biological compounds without necessitating any chemical modification of the analytes [34, 35]. Central to this endeavour are the absorption and fluorescence alterations exhibited by suitably designed hosts upon binding with guests [36, 37]. Ueno et al. reported Alizarin yellow modified CDs (ACD) that have been used as an effective chemo sensor for organic host molecules such as adamantane, borneol, and bile acids

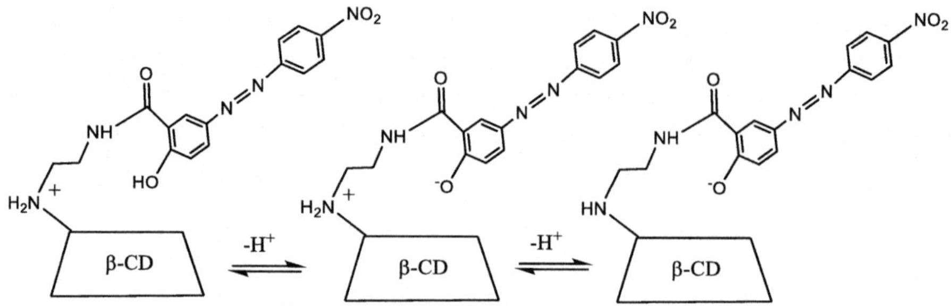

Fig. 5.7 Protonation-deprotonation equilibrium of the ACD

via absorption change indicator. Alizarin yellow acts as a pH indicator, displaying absorption peaks at 360 nm and 480 nm in neutral and alkaline conditions, respectively. pKa values were determined for ACD, and shifts in these values were observed upon the introduction of guest species. The guest-induced shift in pKa within the alkaline range suggests deprotonation of the amine group of ACD. This occurs when the alizarin yellow portion is displaced from the cyclodextrin cavity to make room for the guest, leaving it exposed to the alkaline medium.

Alizarin yellow, connected to β-cyclodextrin by means of an ethylenediamine spacer, was synthesized as a novel indicator of absorption changes for molecules. Within this compound, the alizarin yellow component is covalently bonded to β-cyclodextrin via an ethylenediamine linkage. Researchers discovered that the absorption changes responsive to guests do not stem solely from the detachment of the hydroxyl group of alizarin yellow. Rather, they result from a shift in the protonation-deprotonation equilibrium of the amine group in the ethylenediamine segment (Fig. 5.7) [38].

5.2.7 Methyl Red Modified CDs

The same group has announced the development of two variations of methyl red modified *β*-CD, serving as guest-responsive colour change indicators. These varieties, namely methyl red modified (A) and p-methyl red modified CDs (B), were synthesized to achieve colour alteration through host–guest complexation (Fig. 5.8) [39].

Their structural arrangement and binding characteristics were extensively investigated. In Complex A, the methyl red group is positioned within the β-CD cavity, aligned parallel to the CD axis, forming an intramolecular complex. In contrast, Complex B hosts the p-methyl red group in its β-CD cavity, with the orientation perpendicular to the CD axis.

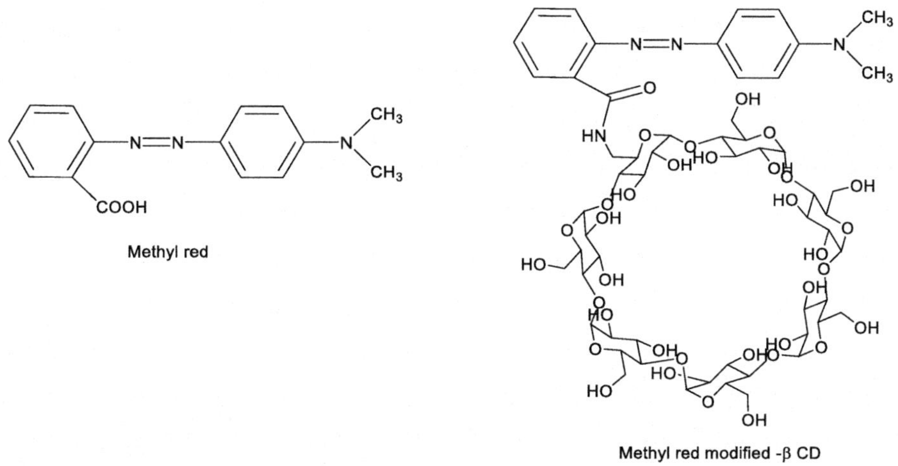

Fig. 5.8 Methyl red modified CD inclusion complex

The addition of guest species under acidic conditions, both Complex A and B transfer the dye components from inside to outside the cavities, creating a 1:1 inclusion complex with the guests. The structural changes prompted by the guests result in colour shifts from yellow to red for A and from orange to red for B. These changes are due to the transformation of the dye components from the azo form to the azonium form. In acidic conditions, the binding constants of B are higher than those of A, and A loses its binding capacity in neutral media, whereas B maintains its binding ability (Fig. 5.9). The binding capacities and responsiveness to guests significantly vary between the hosts. These indicators have been effectively utilized for sensing organic guest molecules such as adamantane derivatives [39]. In addition, the same group have tested the sensing ability of methyl red modified CD for a series of organic molecules categorised as bicyclic monoterpenes like L-Borneol, L-Camphor, D-Camphor, L-Fenchone, L-Fenchone; monocyclic like L-Menthol, D-Menthol; monoterpene like Gereneol and Nerol; cyclohexanol, cyclooctanol, 1-adamentanol, 1-adamentanecarboxylic acid [40]. This indicates that a wide range of organic molecules can be detected by this technique.

5.2.8 CDs Bearing Biquinoline System

Over the past decade, there has been a growing interest in molecular sensors utilizing the guest-induced response of fluorescent-labeled Cyclodextrins (CDs) [41]. Ueno et al. have extensively investigated a variety of monosubstituted CDs that have been modified with chromogenic groups. These groups demonstrate sensitivity to changes in the binding state of the CDs. These chromogenic aromatic groups, initially located

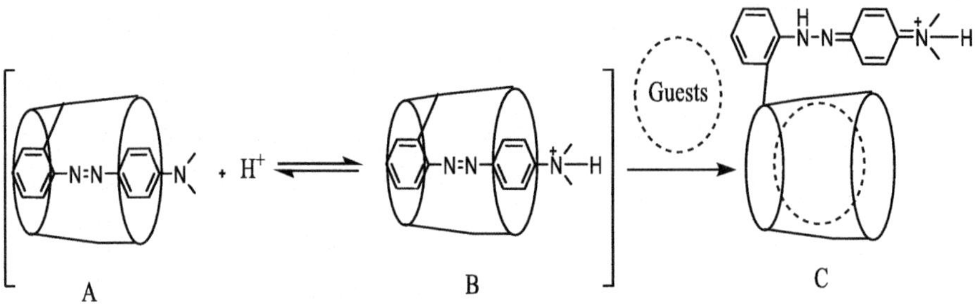

Fig. 5.9 Guest-induced conformational changes of methyl red modified CD

within the CD cavity, experience notable conformational changes upon guest inclusion, resulting in significant spectral shifts. The binding efficiency of a fluorescent-labeled CD can be reliably assessed by examining the spectral changes caused by guest inclusion [42–47]. Liu et al. were the first to introduce dimeric Cyclodextrins (CDs) connected by a chromogenic bridge or metal-coordinated complexes of such CDs as fluorescence probes capable of identifying optically inactive guest molecules [48]. In their investigation, they synthesized a range of bis(β-CD)s incorporating a fluorescent biquinolino tether, alongside the Cu II complexes of these compounds.

A set of bridged β-cyclodextrin (β-CD) dimers, featuring functional tethers of varying lengths, was synthesized with moderate yield. This synthesis involved treating 2,2′-biquinoline-4,4′-dicarboxylic dichloride with β-CD or mono[6-oligo(ethylenediamino)-6-deoxy]-β-CDs. Upon reacting with copper perchlorate, their copper (II) complexes were obtained with satisfactory yields exceeding 77%. All bis(β-CD)s serve as effective fluorescent sensors, exhibiting significant fluorescence enhancement upon the addition of optically inert steroids. The guest-induced fluorescence enhancement observed in these bis-(CD)s paves the way for novel sensor material design strategies. These investigations enhance our understanding of steroid recognition mechanisms and hold promise for the development of novel therapeutic strategies [49]. The stability constants (Ks) resulting from the complexation of steroids with cyclodextrins (CDs) are interpreted through cooperative host-tether-guest interactions. Moreover, a molecular multiple recognition mechanism is proposed for the interaction between guest steroids and CD hosts. Particularly noteworthy is the exploration of molecular recognition of biological steroid guests by bis(CD)s using a fluorescence-sensing mechanism [50].

5.2.9 CD Bearing NBD Amine Moiety

Detection of colourless neutral molecules relies on alterations in fluorescence intensity, absorption, or circular dichroism, facilitated by chemo sensors utilizing chromophore-modified cyclodextrins (CDs). A novel fluorescent chemo sensor, employing modified cyclodextrins with the NBD-amine fluorophore unit, has been

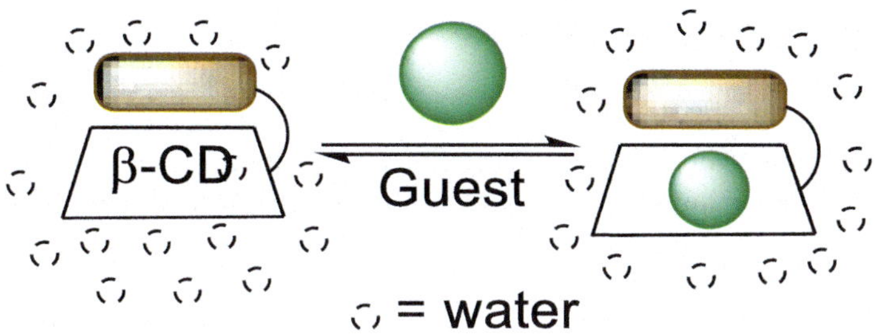

Fig. 5.10 Guest-induced conformational changes by a new type of chromophore-modified CD

developed [51]. The operational mechanism of these chemosensors is illustrated in Fig. 5.10. Although effective in molecule detection, this chemosensor system presents certain drawbacks:

1. Self-inclusion of the chromophore may impede guest accommodation.
2. Modifying the chromophore or spacer unit can influence chemo sensor selectivity, yet the impact is limited as guest selectivity primarily hinges on CD selectivity.
3. Traditional chemosensors typically exhibit decreased fluorescence intensity upon guest detection, whereas enhancing emission intensity due to guest response proves more beneficial for chemical sensing systems.

A novel approach addressing these shortcomings is depicted in (Fig. 5.10).

The crucial point to note is that Ikeda et al. reported on two new types of chemo sensors: the first, NC0-β-CD, exhibits sensitivity to adamantane and borneol derivatives while being insensitive to bile acids; the second, NC0-γCD, displays sensitivity to bile acids but not to adamantane derivatives [52]. In a separate study by Ikeda, the molecular sensing abilities and selectivities of a "turn-on" chemosensor, NBD amine-appended α-cyclodextrin (NC0αCD), were examined in relation to cyclic and acyclic alcohols [53]. This chemosensor exhibits greater sensitivity to cyclic alcohols compared to acyclic alcohols, despite having a stronger binding affinity for acyclic alcohols over cyclic ones.

5.2.10 CD Bearing Biphenyl Moiety

Lu et al. carried out the encapsulation of a long-lifetime phosphorescent Au(I)-diethynylbiphenyl emitter within the large cavity of gamma-cyclodextrin (γ-CD) to create a rotaxane. This process effectively mitigated oxygen quenching to some extent while facilitating the concurrent binding of a guest molecule of particular significance. This methodology offers a sensitive approach to explosive detection [54]. Oxygen quenching occurs through a diffusion-controlled energy transfer process. Consequently, it makes sense to encapsulate the Au(I)-ethynyl complexes within a confined space to protect them from collisions with molecular oxygen.

5.3 Conclusion and Outlook

The utilization of cyclodextrin (CD) derivatives for sensing organic molecules has become increasingly crucial, despite its relatively short history of less than 30 years. Traditionally, chemical sensors employing chromophore-modified CD have been prevalent. However, progress in fields such as supramolecular chemistry, transition metal chemistry, polymer chemistry, carbon nanomaterials, and nanotechnology has led to a shift toward creating advanced sensing systems by incorporating concepts from these areas.

Integration of CD chemistry with other disciplines has facilitated the creation of innovative chemical sensors. CD offers several advantages in organic molecule sensing. Firstly, researchers prefer using cyclodextrins (CDs) to enhance the intrinsic sensitivity of organic chemosensory molecules by encapsulating or tagging them with CDs. Secondly, the target species for detection can be directly encapsulated as guest molecules within the CD host molecule.

References

1. Giannetti A, Bocková M (2020) Optical chemosensors and biosensors. Chem 8:33. https://doi.org/10.3390/chemosensors8020033
2. Krämer J, Kang R, Grimm LM, De CL, Picchetti P, Biedermann F (2022) Molecular probes, chemosensors, and nanosensors for optical detection of biorelevant molecules and ions in aqueous media and biofluids. Chem Rev 122:3459–3636. https://doi.org/10.1021/acs.chemrev.1c00746
3. Thomas SW, Joly GD, Swager TM (2007) Chemical sensors based on amplifying fluorescent conjugated polymers. Chem Rev 107:1339–1386. https://doi.org/10.1021/cr0501339
4. Ikeda H, Murayama T, Ueno A (2005) Skeleton-selective fluorescent chemosensor based on cyclodextrin bearing a 4-amino-7-nitrobenz-2-oxa-1,3-diazole moiety. Org Biomol Chem 3:4262–4267. https://doi.org/10.1039/b508477f
5. Rather IA, Hasan A, Ali R (2022) Cyclodextrin-based sensors for the recognition of small molecules. https://doi.org/10.5772/intechopen.108500
6. Moorthy JN, Venkatesan K, Weiss RG (1992) Photodimerization of coumarins in solid cyclodextrin inclusion complexe. J Org Chem 57:3293–3297
7. Kobayashi N, Ueno A, Osa T (1981) Promotion of charge-transfer complex formation by γ-cyclodextrin. J Chem Soc Chem Commun 340
8. Kobayashi N, Saito R, Hino Y, Ueno A, Osa T (1982) Chirality of the two pyrene molecules included in γ-cyclodextrin. J Chem Soc Chem Commun 706–707
9. Emert J, Kodali D, Catena R (1981) Enhanced intramolecular excimer formation of guest molecules included in the cavity of γ-cyclodextrin. J Chem Soc Chem Commun 758–759
10. Turro NJ, Okubo T, Weed GC (1982) Enhancement of intramolecular excimer formation of 1,3-bichromophoric propanes via application of high pressure and via complexation with cyclodextrins. Protection from oxygen quenching. Photochem Photobiol 35:325
11. Kano K, Takenoshita I, Ogawa T (1982) γ-Cyclodextrin-enhanced excimer fluorescence of pyrene and effect of *n*-butyl alcohol. Chem Lett 1982:321–324
12. Ueno A, Tomita Y, Osa T (1983) Promoted binding ability of y-cyclodextrin appended by a space regulating naphthalene moiety. J Chem Soc Chem Commun 976–977
13. Ueno A, Moriwaki F, Osa T (1987) Excimer formation in inclusion complexes of modified cyclodextrins. Tetrahedron 43(7):1571–1578

14. Minato S, Osa T, Ueno A (1991) Detection of neutral organic compounds by excimer-forming bichromophoricy-cyclodextrins. J Chem Soc Chem Commun 107–108
15. Liu Y, Shi J, Guo D-S (2007) Novel permethylated *â*-cyclodextrin derivatives appended with chromophores as efficient fluorescent sensors for the molecular recognition of bile salts. J Org Chem 72:8227–8234
16. Bender ML, Komiyama M (1978) Cyclodextrin chemistry. Springer, Berlin
17. Ueno A, Moriwaki F, Matsue T, Tetsuo O (1985) Host-guest complexation of ferrocene-appended b-cyclodextrin in organic solvents Makromol. Chem Rapid Commun 6:231–233
18. Ueno A, Moriwaki F, Osa T, Hamada F, Murai K (1988) Association, photodimerization, and induced-fit types of host-guest complexation of anthracene-appendedy-cyclodextrin derivatives. J Am Chem Soc 110:4323–4328
19. Ueno A, Suzuki I, Osa T (1989) Association dimers, excimers, and inclusion complexes of pyrene-appended y-cyclodextrins. J Am Chem Soc I 11:6391–6397
20. Suzuki I, Suzuki I, Osa T (1989) Host-guest sensory sensory system for detecting a variety of organic compounds by variation in pyrene excimer and monomer fluorescence intensities. Chem Lett 1059–1062
21. Ueno A, Suzuki I, Osa T (1990) Host-guest sensory systems for detecting organic compounds by pyrene excimer fluorescence. Anal Chem 62:2461–2466
22. Ueno A, Minato S, Suzuki I, Fukushima M, Ohkubo M, Osa T, Hamada F, Murai K (1990) Host-guest sensory system of Dansyl-modified b-cyclodextrin for detecting steroidal compounds by Dansyl fluorescence. Chem Lett 605–608
23. Rettig W (1986) Charge separation in excited states of decoupled systems—TICT compounds and implications regarding the development of new laser dyes and the primary process of vision and photosynthesis. Angew Chem Int Ed Engl 25:971
24. Breslow R, Halfon S, Zhang B (1995) Molecular recognition by cyclodextrin dimers. Tetrahedron 51:377
25. Okabe Y, Yamamura H, Obe K, Ohta K, Kawai M, Fujita K (1995) Synthesis of a 'head-to-tail' type cyclodextrin dimer linked by a disulfide bridge. J Chem Soc Chem Commun 581
26. Veneam F, Rowan AE, Nolte RJM (1996) Binding of porphyrins in cyclodextrin dimers. J Am Chem Soc 257
27. Hamasaki K, Ueno A, Toda FA (1993) Fluorescent α-Cyclodextrin as a sensor for detecting aliphatic alcohols by dual fluorescence arising from normal planar and twisted intramolecular charge transfer excited states. J Chem Soc Chem Commun 331–333
28. Nakamura M, Ikeda Y, Nakamura A, Ikeda H, Ueno A, Toda F (1995) Remarkable molecular recognition of Dansyl-modified cyclodextrin dimer. Chem Lett 343–344
29. Jong de MR, Engbersen JFJ, Huskens J, Reinhoudt DN (2000) Cyclodextrin dimers as receptor molecules for steroid sensors. Chem Eur J 6(21):4034–4040
30. Sasaki K, Nagasaka M, Kuroda Y (2001) New cyclodextrin dimer and trimer: formation of biphenyl excimer and their molecular recognition. Chem Commun 2630–2631
31. Loehg HG, Voegtle F (1985) Chromo- and fluoroionophores. A new class of dye reagents. Acct Chem Res 18:65–72
32. Cram DJ, Carmack RA, Helgeson RC (1988) Host-guest complexation. 45. A highly preorganized chromogenic spherand indicator system specific for sodium and lithium ions. J Am Chem Soc 110:571–577
33. Kaneda T, Umeda S, Ishizaki Y, Kuo H, Misumi S (1989) Azophenolic acerands: amine-selective coloration and crystal structure of a piperidinium saltex. J Am Chem Soc 111:1881–1883
34. James TD, Sandanayake KRAS, Shinkai S (1995) Recognition of sugars and related compounds by "reading-out"-type interfaces. Supramol Chem 6:141–157
35. Vance DH, Czarnik AW (1994) Real-time assay of inorganic pyrophosphatase using a high-affinity chelation-enhanced fluorescence chemosensor. J Am Chem Soc 116:9397
36. Aoki I, Kawahara Y, Sakaki T, Harada T, Shinkai S (1993) Molecular design of a new fluorescent barbiturate receptor. Sensitive detection of barbiturates through solvent extraction. Bull Chem Soc Jpn 66:927–933

37. Czarnik A (ed) (1992) Fluorescent chemosensors for ion and molecule recognition. In: ACS symposium series 538. American Chemical Society, Washington, DC
38. (a) de Jong MR, Engbersen JFJ, Huskens J, Reinhoudt D (2000) Cyclodextrin dimers as receptor molecules for steroid sensors. Chem Eur J 6(21):4034–4040. https://doi.org/10.1002/1521-3765(20001103)6:21<4034::AID-CHEM4034>3.0.CO;2-3. (b) Aoyagi T, Nakamura A, Ikeda H, Ikeda T, Mihara H, Ueno A (1997) Alizarin yellow-modified-cyclodextrin as a guest-responsive absorption change sensor. Anal Chem 69:659–663
39. Kuwabara T, Nakamura A, Ueno A, Toda F (1994) Inclusion complexes and guest-induced color changes of pH-indicator-modified @-cyclodextrins. J Phys Chem 98:6297–6303
40. Ueno A, Kuwabara T, Garcia-Garibay Nakamura A, Toda F (1992) A modified cyclodextrin as a guest responsive colour-change indicator. Nature 356:136–137
41. McAlpine SR, Garcia-Garibay MA (1998) Studies of naphthyl-substituted β-cyclodextrins. Self-aggregation and inclusion of external guests. J Am Chem Soc 120:4269–4275
42. Bugler J, Engberson JFJ, Reinhoudt DN (1998) Novel water-soluble β-cyclodextrin–calix[4]arene couples as fluorescent sensor molecules for the detection of neutral analytes. J Org Chem 63:5339-5344
43. Nakamura M, Ikeda A, Ise N, Ikeda H, Toda F, Ueno A (1995) Dansyl-modified β-cyclodextrin with a monensin residue as a hydrophobic, metal responsive cap. J Chem Soc Chem Commun 721–722
44. Ikeda H, Nakamura M, Ise N, Toda F, Ueno A (1997) NMR studies of conformations of *N*-Dansyl-l-leucine-appended and *N*-Dansyl-d-leucine-appended β-cyclodextrin as fluorescent indicators for molecular recognition. J Org Chem 62:1411–1418
45. Ikeda H, Nakamura M, Ise N, Oguma N, Nakamura A, Ikeda T, Toda F, Ueno A (1996) Fluorescent cyclodextrins for molecule sensing: fluorescent properties, NMR characterization, and inclusion phenomena of *N*-dansylleucine-modified cyclodextrins. J Am Chem Soc 118:10980–10988
46. Aoyagi T, Nakamura A, Ikeda H, Ikeda T, Mihara H, Ueno A (1997) Alizarin yellow-modified β-cyclodextrin as a guest-responsive absorption change sensor. Anal Chem 69:659–663
47. Kuwabara T, Nakajima H, Nanasawa M, Ueno A (1999) Color change indicators for molecules using methyl red-modified cyclodextrins. Anal Chem 71:2844–2849
48. Liu Y, Song Y, Chen Y, Li X-Q, Ding F, Zhong R-Q (2004) Biquinolino-modified β-cyclodextrin dimers and their metal complexes as efficient fluorescent sensors for the molecular recognition of steroids. Chem Eur J 10:3685–3696. https://doi.org/10.1002/chem.200305724
49. Wallimann P, Marti T, Fürer A, Diederich F (1997) Steroids in molecular recognition. Chem Rev 97:1567–1608
50. (a) Davis AP, Wareham RS (1999) Carbohydrate recognition through noncovalent interactions: a challenge for biomimetic and supramolecular chemistry. Angew Chem Int Ed Engl 38:2978–2996. (b) Henning TP, Cunningham DC (1998) In: Ramsay G (ed) Commercial biosensors: applications to clinical, bioprocess, and environmental samples. Wiley, New York
51. Ikeda H, Murayama T, Ueno A (2005) Skeleton-selective fluorescent chemosensor based on cyclodextrinbearing a 4-amino-7-nitrobenz-2-oxa-1,3-diazole moiety. Org Biomol Chem 3:4262–4267
52. Ikeda H, Murayama T, Ueno A (2006) Bile acids-selective chemosensors based on NBD-amine-modified cyclodextrins. J Incl Phenom Macrocycl Chem 56:101–105
53. Ikeda H (2017) Fluorescent cyclodextrins as chemosensors for the detection of cyclic and acyclic alcohols. J Incl Phenom Macrocycl Chem. https://doi.org/10.1007/s10847-017-0730-8
54. Yu X, Wan S, Wu W, Yang C, Lu W (2022) Cyclodextrin-based [2] rotaxane stoppered with gold(I)–ethynyl complexation: phosphorescent sensing for nitroaromatics. Chem Commun 58:6284–6287

Dr. Rima Das received her M.Sc degree in chemistry with specialization in organic chemistry from University of Burdwan, West Bengal in 2007. Afterwards, she finished her Ph.D. from Indian Institute of Technology (IIT) Madras in 2012. Her doctoral work was based on metal mediated catalytic organic transformations. Currently she is working as an Assistant Professor in chemistry at Chandidas Mahavidyalaya affiliated by University of Burdwan, West Bengal. She has ten years of teaching and research interest. She has published 12 research papers in various international journals. Her current research interest is metal mediated catalytic transformations, nano chemistry, cyclodextrin chemistry.

Chapter 6
Intriguing Role of Cyclodextrin in Fluorescence-Based Metal Ion Sensors

Muhammed Shahsad, S. Jayaram, and John Prakash

Abstract The widespread presence of metal ions poses significant environmental and health risks, necessitating highly sensitive and selective detection methods. Fluorescence spectroscopy is a powerful tool for trace metal detection, offering sensitivity, cost-effectiveness, and quick analysis with minimal sample preparations. This chapter summarizes the role of cyclodextrin (CD)-based fluorescent probes in metal ion detection, exploring their recognition performance and mechanisms. CD is utilized to develop fluorescent metal ion sensors, enhancing selectivity through its ability to form inclusion complexes. CD-modified fluorescent nanoparticles demonstrate excellent anti-interference properties, allowing for selective metal ion detection. The Chapter also highlights recent advancements in supramolecular structures, including CD dimers and polymers, and discusses the incorporation of lipophilic polymer dots (PDs) in nanomicelles with AIE properties. Most research focuses on 'turn-off' and 'turn-on' sensors, where emission intensity is influenced by mechanisms like Photoinduced Electron Transfer (PET), Förster Resonance Energy Transfer (FRET), Aggregation-Induced Emission (AIE), and intramolecular hydrogen transfer. These developments underscore the significance of CD-based fluorescent sensors in detecting metal ions.

Keywords Cyclodextrin derivatives · Fluorescence spectroscopy · Nanomicelles · Inclusion complex · Metal ion sensing · Modified cyclodextrin

6.1 Introduction

Quantitative monitoring of metal ions in aqueous media in a sensitive manner is of great importance due to their toxicity and threat to the environment and living organisms. Increased human activities have led to the release of metal ions into water resources that has put aquatic ecosystems and the environment at risk as these are

M. Shahsad · S. Jayaram · J. Prakash (✉)
Department of Chemistry, School of Basic and Applied Sciences, Central University of Tamil Nadu, Thiruvarur, India
e-mail: johnprakash@cutn.ac.in

S. Krishnamoorthy (ed.), *Cyclodextrins for Chemosensing*,
https://doi.org/10.1007/978-3-031-89299-8_6

highly toxic, non-degradable, and have a tendency to bioaccumulate and biomagnify as a result of the food chain [1]. Metal ions, including iron, copper, magnesium, zinc, cobalt, chromium, etc. are indispensable in biological processes such as such as osmotic regulation, catalysis, metabolism, and cell signalling [2]. From microbes to mammals, Copper (Cu^{2+}) is essential for many physiological processes. It serves as an enzymatic cofactor in pathways for cellular respiration, iron oxidation, peptide amidation, neurotransmitter synthesis, and antioxidant defence. Inborn errors in copper metabolism results in Menken's disease and Wilson's disease [3]. Although copper is essential, excessive intake results in diarrhea, jaundice, hemoglobinuria, hematuria, anuria, and oliguria [4].

Zinc (Zn^{2+}) is another vital metal ion, uniquely present across all enzyme classes, typically functioning either in catalysis or in the structural stabilization of over 300 enzymes, and contributing to an even greater number of metalloproteins [5, 6]. Although zinc plays an essential biological role, toxicity is infrequently observed [7]. However, some studies have linked zinc imbalances to neurological disorders such as Alzheimer's and Parkinson's diseases [8].

Metal ions like mercury, cadmium, and lead have no biological importance, and tiny amounts of these metal ions can cause immediate poisoning, and damage to liver, kidneys, and brain. They are also reported as carcinogenic [9]. Considering the mitigation of health issues associated with elevated levels of these metal ions, the detection strategy holds utmost significance. Metal ion sensing techniques play a pivotal role in the meticulous monitoring of metal ion concentrations in environmental samples and aid in the early diagnosis and effective management of conditions arising from the dysregulation of these ions.

Notably, various techniques such as electrochemistry [10], atomic absorption spectrometry [11], and atomic emission spectrometry [12] have been widely employed for different sample types and analytical needs, offering both versatility and flexibility. Moreover, advanced functional systems, such as nanomaterials, polymers, porous materials, and biomaterials, have also been incorporated to enhance analytical performance [13]. Spectroscopy offers several advantages in detecting minute quantities of metal ions, even at trace levels, making them ideal for environmental [14] and biological analysis [2]. These techniques can identify different metal ions, providing valuable information about the specific contaminants present. Spectroscopic methods cater to other sample types and analytical needs, offering adaptability and flexibility. Spectroscopic techniques such as fluorescence spectroscopy enable analysis without altering the original sample, preserving its integrity for subsequent investigations. Fluorescence spectroscopy is a widely used spectroscopic technique well known for its sensitivity, cost-effectiveness, and quick analysis. It is a powerful tool for trace metal detection with minimal sample disturbance [15].

The development of suitable chemical sensors is essential for the detection and quantification of ions due to their inherent lack of fluorescence properties. Combining fluorescence with appropriate probes offers an effective approach for detecting metal ions [16]. Numerous types of fluorescent probes are employed for metal ion sensing, including small molecule-based probes [17], nanoprobes utilizing noble metal nanomaterials, up-conversion nanoparticles, semiconductor quantum dots, and carbon-based nanomaterials [18]. Additionally, conjugated polymers [19] and host–guest recognition systems, employing hosts such as crown ethers, cyclodextrins, calixarenes, cucurbiturils, pillararenes, and other macrocyclic hosts, are also integral to this field [20, 21].

CDs have no fluorescence characteristics, but they can be converted into spectroscopically active hosts by incorporation of a fluorophore and it is found that their spectral parameters change upon guest accommodation. CDs combined with fluorescence spectroscopy offer a powerful and versatile approach for metal ion detection due to their synergistic properties to form inclusion complexes with guest molecules [22]. Hence, it has been utilized to develop fluorescent-based metal sensors. These sensors use receptors or fluorescent dyes to target metal ions selectively. The rigid fit afforded by CD enhances the chemosensory compound's selectivity and sensing capabilities towards the target metal ion, enhancing the solubility and stability of formed complexes. The fluorescence intensity can increase (Enhance) or decrease (Quench) upon binding, depending on the CD-metal ion interaction [23, 24]. This provides a clear and sensitive signal for metal ion detection. Fluorescence quenching can occur through several mechanisms, such as energy transfer or electron transfer, offering multiple avenues for designing sensitive sensors [25]. The selective binding of CDs towards specific metal ions can be utilized in competitive binding assays to detect other metal ions in a sample.

The book chapter provides a comprehensive summary of the role played by CD-based fluorescent probes in detecting metal ions. It elucidates the recognition performance and mechanisms involved, aiming to establish a theoretical foundation for the application of CD-derivatives and inclusion complexes in the realm of fluorescent detection.

6.2 Analytical Methods for Metal Ion Detection

Heavy metal ions analysis has traditionally relied on well-established methods like electrochemical and spectroscopic techniques, including potentiometry [26] amperometry [27], voltammetry [28], coupled plasma mass spectrometry (ICP-MS) [29], atomic absorption spectroscopy [30], and atomic emission spectroscopy [31]. Electrochemical techniques are favored for their economic viability, user-friendly nature, and reliability, employing straightforward procedures for monitoring contaminated samples. Nevertheless, these methods exhibit lower sensitivity and higher limits of detection (LOD) when compared to their spectroscopic and optical techniques.

Atomic spectroscopic techniques are versatile in simultaneously determining a large range of metal ions with extremely low detection limits. However, these techniques are costly, cumbersome, and require trained personnel for technical operations. Such challenges limit their real-time applicability in environmental monitoring [32]. The exploration and advancement of sensor materials possessing heightened sensitivity, rapid response times, and selectivity for metal ion detection have become focal points, particularly in the context of sophisticated experimental setups.

Among the various sensing approaches, optical techniques have garnered great attention due to their numerous advantages, including simple and naked eye detection, real-time monitoring, cost-effectiveness, high specificity towards target analytes, rapid response times, and the use of less complex analytical equipment [33, 34]. In recent times, optical sensors have advanced and found use in a wide range of applications, including food safety [35], cancer diagnosis [36, 37], environmental monitoring [38], DNA sensing [39], and blood glucose monitoring [40].

Optical spectroscopic techniques such as UV–VIS absorption and Fluorescence spectroscopy are extensively used for metal ion sensing [41–43]. UV–VIS absorption spectroscopy is renowned for its versatility, being applicable to a large number of compounds, and its simple instrumentation setup. However, its sensitivity is comparatively lower than that of fluorescence spectroscopy. Fluorescence spectroscopy is capable of detecting analytes at concentrations as low as the nanomolar to picomolar ranges. This elevated sensitivity, coupled with its impressive combination of selectivity and rapid analytical capabilities, positions fluorescence spectroscopy as a potent tool for trace metal detection [44, 45]. Fluorescence spectroscopy with the unique attributes of CD derivatives opens a pathway to sophisticated sensors with unprecedented precision in metal ion detection. The hydrophobic cavities of CD serve as molecular pockets, influencing the microenvironment around fluorophores. Careful selection of fluorophores is the core of fluorescence-based sensing. Within CD-based sensors, the choice of fluorophores is intricately tied to the unique host–guest interactions facilitated by these derivatives [46].

6.3 Fluorescent Modifications of CDs

Novel avenues have emerged to harness the potential of fluorescence detection of diverse metal ions through modifications of native CDs. This relatively underexplored, strategic manipulation includes:

Polymers: Integration into polymeric matrices signifies a promising frontier for native CDs in fluorescence detection. Embedding CDs within polymers enhances their metal ion sensing capabilities, ensuring improved performance and adaptability [44].

Nanostructures: Exploiting the unique attributes of nanostructures extends the applicability of native CDs in fluorescence detection at the nanoscale. This research capitalizes on the inherent advantages of CDs within nanostructures, facilitating precise and efficient metal ion detection with heightened sensitivity [45]

Inclusion Complexes: Despite existing limitations, the creation of inclusion complexes remains a viable strategy for fluorescence detection. Tailoring native CDs to formulate inclusion complexes enables the specific capture and identification of targeted metal ions, contributing to the diversification of fluorescence-based detection methodologies.

Derivatization: Systematic derivatization of native CDs amplifies their fluorescence detection capabilities. This approach enhances sensitivity and selectivity, broadening the spectrum of identifiable metal ions [43].

6.4 Polymer

The limited solubility and weak recognition capacity of CDs hinder their broader application in analytical chemistry. To address these limitations, a variety of CD-containing polymers with diverse architectures have been developed, ranging from linear, grafted, and block copolymers to branched, hyperbranched, and dendritic structures. In these polymers, the CD units are either covalently incorporated into the main polymer chains or linked as side groups [47]. With numerous external interaction sites for guest inclusion, the supramolecular polymer enhances recognition capability for molecular self-assembly [48].

CD-polymers offer a versatile platform for designing materials with dual functionalities that combine detection, adsorption, and detoxification capabilities. For example, a non-porous β-CD polymer modified with an amidoxime group has been employed for highly sensitive and selective detection of uranyl U(IV) ions. The emissive properties of this tailored polymer were notably quenched upon interaction with uranyl ions through the PET mechanism. The LOD was calculated to be 10.3 nM in deionized water [49]**.** PET involves either the transfer of an electron from a receptor to the ground state of an excited fluorophore, or the transfer of excited-state electrons from the fluorophore to another ground-state species. This process typically results in the inhibition of fluorescence [50, 51].

Poly β-CDs (Fig. 6.1), a type of polymer formed through the crosslinking reaction of CDs, demonstrate excellent water solubility, robust host–guest inclusion capabilities, and high stability. A fluorescence sensor is constructed by poly β-CD, 8-hydroxy-2-aminoquinoline, and an aryl dialdehyde, allowing for effective sensing of zinc ions in pure water environment [52]. Poly(β-CD-co-citric acid) composites integrated with carbon quantum dots (QDs), exhibit dual functionality as both adsorbents and photoluminescent probes for detecting divalent metal ions such as Ni(II), Cu(II), Cd(II), and Pb(II) [53].

Fig. 6.1 Structure of poly β-CDs

6.5 Nanostructures

The integration of CDs with nanostructures presents significant potential for enhancing the efficacy and versatility of nanoscale sensors. This fusion has opened new in the area of nanoscale fluorescence detection. CDs possess abundant hydroxyl groups, which stabilize nanoparticles (NPs) while also enabling the delivery or adsorption of small molecules within the system [54].

CD-coated nanoparticles (Fig. 6.2) primarily composed of noble metals such as gold and silver have been successfully employed in various applications, including chemical sensing, drug delivery, and catalysis [55].

Gold nanoclusters with covalent attachment of β-CDs exhibit inherent fluorescence, enabling selective detection of metal ions such as Co^{2+} [56], Cu^{2+} [57], Pb^{2+}, Al^{3+}, Cu^{2+}, Fe^{3+}, Ag^{+} [58], and Hg^{2+} [59]. Additionally, an inclusion complex has been developed from γ-CD and 1,5-dihydroxynaphthalene, decorated with a silver nanoparticle fluorescent probe for the detection of Cu^{2+} and Hg^{2+} [60].

QDs are luminescent nanocrystals that are extensively researched for their exceptional optical properties and high photo-stability [61]. Mn-doped ZnS QDs were synthesized and then modified with β -CDs. The FL emission of β -CDs@QDs was quenched by Co^{2+}. The LOD was calculated to be 0.26 mmol L^{-1} [62]· Chitosan (CTS) and β-CD have been effectively combined through formaldehyde cross-linking to form a CTSCD polymer, Sulfur Quantum Dots (S QDs) incorporated into this polymer leading to the formation of S QDs/CTSCD nanocomposites. In the presence of Ag^{+} ions, these nanocomposites exhibit rapid aggregation resulting in aggregation-induced quenching (AIQ) [63]. Fluorescent Carbon Quantum Dots (CQDs) demonstrate advantages over other metallic nanoparticles for various applications due to their low toxicity, cost-effectiveness, excellent biocompatibility, and biodegradability [64]. A fluorescent probe was prepared with fluorescent CQDs, isophorone diisocyanate (IPDI) and β-CD to detect Cr(VI) ions in water and soil samples [65]. Determination of Mn^{2+} was achieved by using chitosan and β-CD functionalized N–C QDs [66]. diethylenetriamine-β-CD (3 N-β-CD)-modified CQDs were synthesized and utilized as a fluorescent probe for the selective detection of Hg^{2+} and Fe^{3+} from aqueous solution [67] Additionally, a ratiometric fluorescent sensor was constructed by coupling blue-green emitting β-CD derived carbon dots (β-CD-CDs)

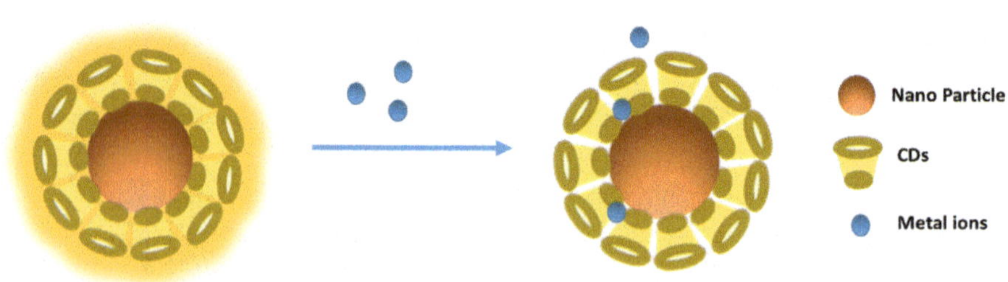

Fig. 6.2 Fluorescence quenching of CD-coated nanoparticles upon binding with metal ions

to a red-emitting CdSe/ZnS QDs through host–guest interaction. The resulting hybrid micelles were employed to detect Hg^{2+} and Cu^{2+} with the detection limits down to 1.6 and 2.74 μM respectively [68]. Polymer dots (PDs) are recognized as one of the recently explored variants of CQDs. β- CD are exploited for the implantation of lipophobic PDs and it is used for fluorescent detection of Ag^+ ions in aquatic streams [69].

6.6 Inclusion Complexes

Metal ion sensing is crucial for mitigating adverse effects caused by metals on the earth's ecosystem and living organisms. With CDs as a pivotal host molecule in chemosensor systems, understanding its enhancement of selectivity and sensitivity becomes paramount [9]. CDs behave as host molecules, forming inclusion complexes that encapsulate a diverse range of guest molecules, including fluorophores (Fig. 6.3). The creation of these complexes hinges on non-covalent interactions such as hydrogen bonding, hydrophobic interaction, and van der Waals forces. This encapsulation induces notable alterations in the physicochemical properties of the guest molecules, endowing CDs with versatility across various disciplines. In the context of β-CD's application for chemosensory purposes, it provides a precise fit for the chemosensory (guest) compound within its cavity, enhancing the selectivity and sensitivity of the target metal ion.

Natural pigments are novel guest molecules used as complexing agents for CD. Curcumin is a yellow hydrophobic polyphenol natural fluorescent pigment with outstanding photostability and low toxicity, Curcumin: β-cyclodextrin inclusion complex is utilized for sensing mercury ions, relying on an "on–off" fluorescence response and the mechanism behind it was Excited-State Intramolecular Hydrogen Transfer (ESIHT) [70].

Similarly, molecules coming under the triazine class are also used for chemosensory applications. Benzoguanamine, recognized for its potential as a cross-linking agent, is utilized in conjunction with CDs for the sensing of metal ions. The multitopic N-donor ligands which can coordinate up to five metal centers give this molecule a good complexation capacity. The 1:1 inclusion complex between CD and guest molecule senses various metal ions [71].

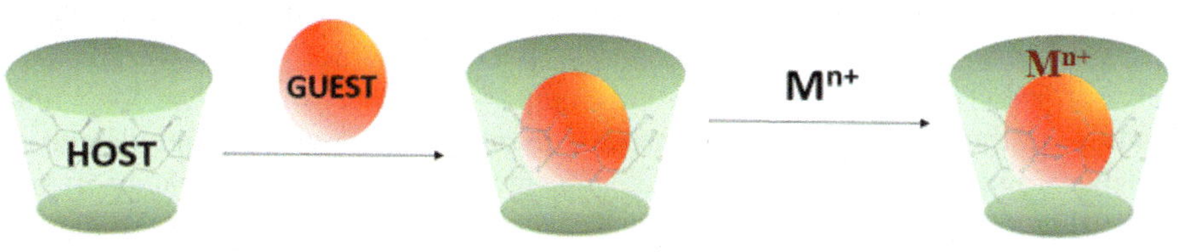

Fig. 6.3 Guest host chemistry adapted in the formation of inclusion complex and its interaction with metal ions

Notably, a dual-emission N-hydroxy 1, 8-naphthalimide (NHN) sensor has been developed for the selective determination of Bi^{3+} [72]. Additionally, a highly sensitive and selective "turn-on" fluorescence sensor for Hg^{2+} and Fe^{3+} ions has been reported, utilizing a solid inclusion complex of 3, 3′-dihydroxybenzidine and α-cyclodextrin (DHB:α-CD) within an acetonitrile–water system [70, 72, 73]. Furthermore, fluorescence quenching-based sensing of Al^{3+} has been achieved through an imidazolylpiperazine derivative integrated with a β-CD host–guest system [70, 72, 73]. In another example a host–guest system derived from β-CD and an organic Schiff base molecule designed for the selective fluorescence sensing of Al^{3+} ions in aqueous environments, where Al^{3+} binding is attributed to a fluorescence resonance energy transfer mechanism [74]. A comprehensive list of various guests and the inclusion of complex formations between CDs are listed in Table 6.1.

6.7 Derivatization

The strategic derivatization of CDs has emerged as a solid tool in advancing the field of fluorescence detection of diverse metal ions. This approach not only augments the sensing capabilities but also broadens the scope of application in metal ion detection methodologies. Specifically, derivatives of CDs exhibit enhanced capabilities in terms of sensitivity and selectivity, thereby expanding the spectrum of identifiable metal ions. Through systematic derivatization, CDs undergo modifications that amplify their fluorescence detection potential, allowing for a more nuanced and precise identification of targeted metal ions. This multifaceted strategy amalgamates scientific precision with practical utility, making derivatives of CDs indispensable tools in the ongoing quest for innovative and effective solutions in the realm of metal ion detection technologies. Derivatization of CD can be viewed in a number of ways.

Fluorophore-Appended CDs

CDs are inherently spectroscopically inert, and can be transformed into active hosts through the modification of chromophoric groups, thereby enabling their application as chemosensors. The derivatization of CDs for metal ion sensing serves several purposes, primarily addressing the challenges of poor water solubility and biocompatibility encountered with fluorophores in ion detection. Furthermore, derivatization facilitates the formation of aggregates upon metal ion coordination or prevents non-radiative deactivation pathways and other quenching mechanisms, resulting in fluorescence "turn-on" responses. For example, β-CD, when derivatized with chitosan polymer and coated with SQDs, has been employed as a sensor for Ag^{+} ions, functioning via the AIQ mechanism. In AIQ, fluorescence decreases significantly as molecules aggregate or enter a solid state. The incorporation of cross-linked polymers effectively mitigates issues related to poor biocompatibility and solubility, enhancing the sensor's performance in aqueous environments [63]. Similarly, the AIE-active CD derivative, tetraphenylethylene (TPE)-triazole-CD (Fig. 6.4a), demonstrates a highly

Table 6.1 Various inclusion complexes with CDs for sensing metal ions with their LOD and associated mechanism

Name and structure of the complex	Metal ion(s) and sensing mechanism	LOD (μM)
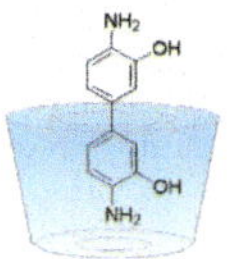 3, 3′-dihydroxybenzidine- α-CD [75]	Hg^{2+} and Fe^{3+}, *PET*	0.1 (Hg^{2+}) 0.05 (Fe^{3+})
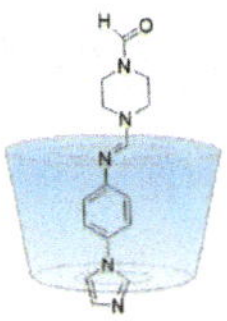 Imidazolylpiperazine derivative–β-CD [76]	Al^{3+}, *FRET*	0.06
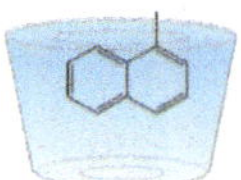 Naphthalene-methyl- β-CD [77]	Mn^{2+} and Co^{2+} *PET*	-
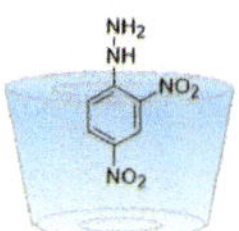 Dinitrophenylhydrazine-β-CD [78]	Ce^{4+}	2.9×10^{-4}
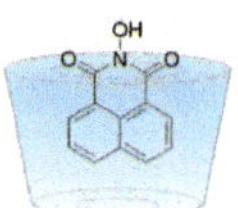 Naphthalimide-β-CD [72]	Bi^{3+} *PET*	2.78
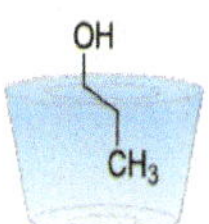 Hydroxypropyl-β-CD [79]	Hg^{2+} *PET*	2.7×10^{-4}

(continued)

Table 6.1 (continued)

Name and structure of the complex	Metal ion(s) and sensing mechanism	LOD (μM)
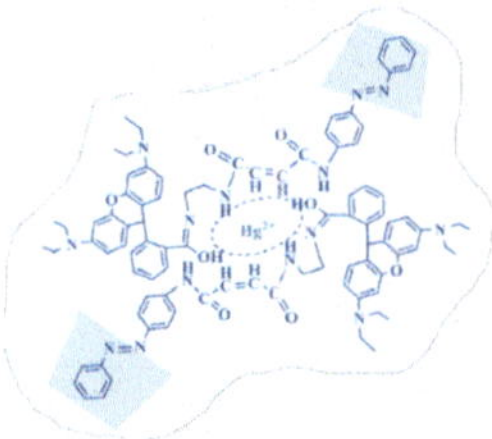 Rhodamine B, maleic anhydride, and azobenzene-β-CD [80]	Hg^{2+}	–
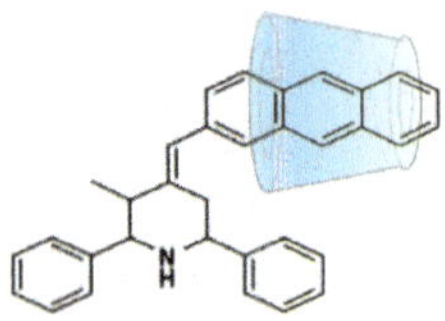 Anthracene piperidin-4-one derivative-β-CD [81]	Ca^{2+}	0.0518
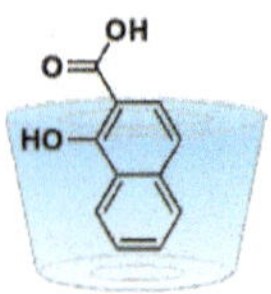 1-hydroxy-2-naphthoic acid: β-cd (1H2NA: β-cd) [82]	Cr^{3+}	30
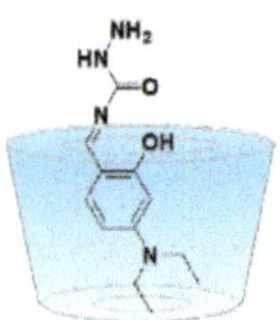 Benzaldehyde derivative: β-cd [83]	Al^{3+}	0.00029
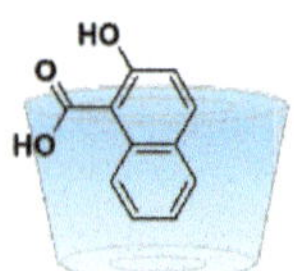 2H1NA: β-cd [84]	Ag^{+}	0.1

(continued)

Table 6.1 (continued)

Name and structure of the complex	Metal ion(s) and sensing mechanism	LOD (μM)
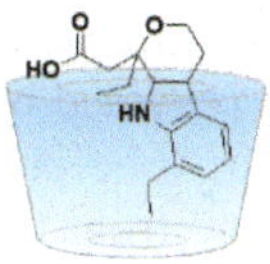 Etodolac: β-cd [85]	Ba^{2+}	24
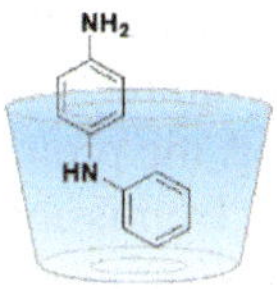 N-phenyl p-phenylenediamine: β-cd [86]	Fe^{2+}	0.01
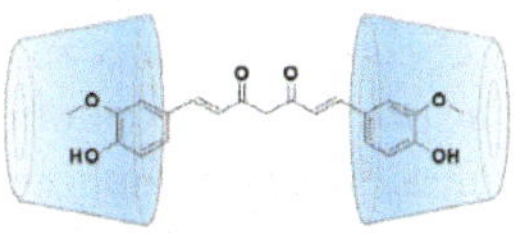 Curcumin: β-cd [70]	Hg^{2+}	5.02

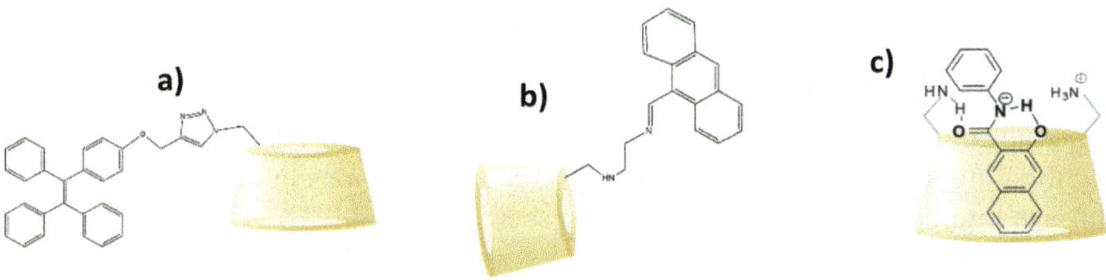

Fig. 6.4 Structure of **a** tetraphenylethylene (TPE)-triazole-CD. **b** anthracene derivative appended β-CD. **c** aminocyclodextrin:3-hydroxy-N-phenyl-2-naphthamide

selective fluorescence "turn-on" response to Cd^{2+} ions. The fluorescence enhancement is attributed to the restricted intramolecular rotation (RIR) mechanism, when coordination with Cd $^{2+}$ inhibits the non-radiative deactivation of the TPE moieties, leading to a significant increase in fluorescence intensity [87].

Fluorescent β-CD based probes containing a naphthalimide group were synthesized for the specific detection of Hg^{2+}/Hg^{+}. This work by disruption of PET and results in fluorescence enhancement [88]. Similarly, for Zn^{2+} detection, a tetraethylenepentamine-modified β-CD derivative (TEPA-β-CD-L) was developed. The detection mechanism is attributed to a chelation-induced fluorescence enhancement (CIFE) process, where Zn^{2+} coordination enhances the planarity of the TEPA-L moiety, facilitating fluorescence emission. This coordination also triggers the PET mechanism between Zn^{2+} and TEPA-β-CD-L [89].

Another example is the anthracene derivative appended β-CD (Fig. 6.4b) for the detection of Pb^{2+}. The fluorescence enhancement mechanism is attributed to photoinduced charge transfer (PCT) and the conformational restriction provided by metal ion binding. Additionally, a possible PET may occur due to the lone pair of electrons on the NH- group, with fluorescence enhancement resulting from the inhibition of PET upon metal ion binding [90].

Aminocyclodextrin as host and 3-hydroxy-N-phenyl-2-naphthamide [Fig. 6.4 c] as fluorophore moiety made for the detection of Hg^{2+} through both colorimetric and fluorescence techniques. This probe undergoes an Excited-State Intramolecular Proton Transfer (ESIPT) process in which the naphthoxide ion abstracts the proton from the –NH group resulting in a localized emission as well as ESIPT emission. The fluorescence quenching observed in this system is attributed to the disruption of the ESIPT process upon Hg^{2+} binding [91].

Using CD supramolecular systems

Supramolecular structures built from CDs are employed as chemosensors because of their distinctive ability to enclose guest compounds within spacious hydrophobic environments. Covalently connected CDs, like dimers and trimers, form hydrophobic binding cavities that efficiently capture larger guest compounds as compared to monomeric CDs. This feature has spurred exploration into sensing systems based on CD dimers and trimers. Recent research has demonstrated the potential of CD-based supramolecular structures in chemosensor design, as exemplified by the development of a water-soluble FRET-based Fe (III) sensor utilizing a β-CD/dye complex [92]. Recent advances in fluorescence sensing have showcased the potential of supramolecular architectures based on pillar[n]arenes, with a comprehensive review highlighting their synthesis, self-assembly, and applications in detecting metal ions through intelligent and controllable fluorescence sensors [93].

6.8 Conclusion

To brief it up, fluorescence spectroscopy combined with CD derivatives presents a cutting-edge approach for detecting metal ions with precision and efficiency. Smart sensors can be crafted through the careful selection of fluorophores influenced by host–guest interactions within CDs. While traditional methods like electrochemical sensors and mass spectrometry have been used for metal ion analysis, they are often costly, complex, and not ideal for real-time application. Fluorescence-based sensors, particularly those utilizing CDs, overcome these issues.

Modifications to CDs are crucial in boosting their fluorescence detection capabilities. Techniques like derivatization, incorporation into polymers, use in nanostructures, and the formation of inclusion complexes broaden the spectrum of identifiable metal ions and enhance sensitivity. CD derivatives, with their improved sensitivity and selectivity through systematic modifications, are emerging as versatile platforms for various molecular configurations. Fluorescent sensors, classified as "turn-on" or "turn-off," are effective in detecting metal ions through unique chromophore mechanisms. Furthermore, integrating CDs into polymers provides a multifunctional platform that combines detection, adsorption, and detoxification. CD nanostructures, such as gold nanoclusters, quantum dots, and nanomicelles, show exceptional sensitivity and selectivity for nanoscale fluorescence detection, particularly within live cells. Inclusion complexes formed by CDs emphasize their role as host molecules, encapsulating a wide range of guest molecules, enhancing the selectivity and sensitivity of chemosensory compounds for target metal ions. Guest molecules like natural pigments (e.g., curcumin) and molecules from the triazine class contribute to the versatility of CDs in chemosensory applications.

In summary, the diverse strategies for modifying CDs highlight their indispensable role in advancing fluorescence-based metal ion detection technologies. Ongoing developments in this field hold great promise for practical, accessible, and real-time solutions, especially in environmental and aquatic monitoring.

References

1. Ayangbenro AS, Babalola OO (2017) A new strategy for heavy metal polluted environments: a review of microbial biosorbents. Int J Environ Res Public Heal 14:94. https://doi.org/10.3390/IJERPH14010094
2. Zheng X, Cheng W, Ji C et al (2020) Detection of metal ions in biological systems: a review. Rev Anal Chem 39:231–246. https://doi.org/10.1515/revac-2020-0118
3. Karim N (2018) Copper and human health- a review. J Bahria Univ Med Dent Coll 08:117–122. https://doi.org/10.51985/JBUMDC2018046
4. Bremner I (1998) Manifestations of copper excess. Am J Clin Nutr 67:1069S-1073S. https://doi.org/10.1093/ajcn/67.5.1069S
5. Broadley MR, White PJ, Hammond JP et al (2007) Zinc in plants. New Phytol 173:677–702. https://doi.org/10.1111/j.1469-8137.2007.01996.x
6. Haase H, Rink L (2014) Zinc signals and immune function. BioFactors 40:27–40. https://doi.org/10.1002/biof.1114
7. Wani AL, Parveen N, Ansari MO et al (2017) Zinc: An element of extensive medical importance. Curr Med Res Pract 7:90–98. https://doi.org/10.1016/j.cmrp.2017.02.006
8. Xu Z, Liu X, Pan J, Spring DR (2012) Coumarin-derived transformable fluorescent sensor for Zn2+. Chem Commun 48:4764. https://doi.org/10.1039/c2cc30963g
9. Sivakumar K, Parinamachivayam G (2022) Review on β-cyclodextrin inclusion complex based chemosensors for heavy metals. J Incl Phenom Macrocycl Chem 102:603–618. https://doi.org/10.1007/S10847-022-01153-2
10. Sulthana SF, Iqbal UM, Suseela SB et al (2024) Electrochemical sensors for heavy metal ion detection in aqueous medium: a systematic review. ACS Omega 9:25493–25512. https://doi.org/10.1021/acsomega.4c00933
11. Arienzo M, Capasso R (2000) Analysis of metal cations and inorganic anions in olive oil mill waste waters by atomic absorption spectroscopy and ion chromatography. Detection of metals bound mainly to the organic polymeric fraction. J Agric Food Chem 48:1405–1410. https://doi.org/10.1021/jf990588x

12. He Q, Zhu Z, Hu S (2014) Flowing and Nonflowing Liquid Electrode Discharge Microplasma for Metal Ion Detection by Optical Emission Spectrometry. Appl Spectrosc Rev 49:249–269. https://doi.org/10.1080/05704928.2013.820195
13. Li M, Shi Q, Song N et al (2023) Current trends in the detection and removal of heavy metal ions using functional materials. Chem Soc Rev 52:5827–5860. https://doi.org/10.1039/D2CS00683A
14. Bargossi C, Fiorini MC, Montalti M et al (2000) Recent developments in transition metal ion detection by luminescent chemosensors. Coord Chem Rev 208:17–32. https://doi.org/10.1016/S0010-8545(00)00252-6
15. Miller JN (2006) Some recent advances in fluorescence spectroscopy. Anal Lett 39:851–862. https://doi.org/10.1080/00032710600605758
16. Li J, Yim D, Jang W-D, Yoon J (2017) Recent progress in the design and applications of fluorescence probes containing crown ethers. Chem Soc Rev 46:2437–2458. https://doi.org/10.1039/C6CS00619A
17. Fang Y, Dehaen W (2021) Small-molecule-based fluorescent probes for f-block metal ions: a new frontier in chemosensors. Coord Chem Rev 427:213524. https://doi.org/10.1016/j.ccr.2020.213524
18. Zhang J, Cheng F, Li J et al (2016) Fluorescent nanoprobes for sensing and imaging of metal ions: Recent advances and future perspectives. Nano Today 11:309–329. https://doi.org/10.1016/j.nantod.2016.05.010
19. Pak YL, Wang Y, Xu Q (2021) Conjugated polymer based fluorescent probes for metal ions. Coord Chem Rev 433:213745. https://doi.org/10.1016/j.ccr.2020.213745
20. Sahoo SK, Sharma D, Bera RK et al (2012) Iron(iii) selective molecular and supramolecular fluorescent probes. Chem Soc Rev 41:7195. https://doi.org/10.1039/c2cs35152h
21. Guo C, Sedgwick AC, Hirao T, Sessler JL (2021) Supramolecular fluorescent sensors: an historical overview and update. Coord Chem Rev 427:213560. https://doi.org/10.1016/j.ccr.2020.213560
22. Ueno A (1996) Review: fluorescent cyclodextrins for molecule sensing. Supramol Sci 3:31–36. https://doi.org/10.1016/0968-5677(96)00016-8
23. Ogoshi T, Harada A (2008) Chemical sensors based on cyclodextrin derivatives. Sensors 8:4961–4982. https://doi.org/10.3390/s8084961
24. Rasheed T, Bilal M, Nabeel F et al (2018) Fluorescent sensor based models for the detection of environmentally-related toxic heavy metals. Sci Total Environ 615:476–485. https://doi.org/10.1016/j.scitotenv.2017.09.126
25. Wong J, Todd M, Rutledge P (2017) Recent advances in macrocyclic fluorescent probes for ion sensing. Molecules 22:200. https://doi.org/10.3390/molecules22020200
26. Huang MR, Li XG (2022) Highly sensing and transducing materials for potentiometric ion sensors with versatile applicability. Prog Mater Sci 125:100885. https://doi.org/10.1016/J.PMATSCI.2021.100885
27. Lee HJ, Lagger G, Pereira CM et al (2009) Amperometric tape ion sensors for cadmium(II) ion analysis. Talanta 78:66–70. https://doi.org/10.1016/J.TALANTA.2008.10.059
28. Lu Y, Liang X, Niyungeko C et al (2018) A review of the identification and detection of heavy metal ions in the environment by voltammetry. Talanta 178:324–338. https://doi.org/10.1016/J.TALANTA.2017.08.033
29. Otero-Romaní J, Moreda-Piñeiro A, Bermejo-Barrera P, Martin-Esteban A (2009) Inductively coupled plasma–optical emission spectrometry/mass spectrometry for the determination of Cu, Ni, Pb and Zn in seawater after ionic imprinted polymer based solid phase extraction. Talanta 79:723–729. https://doi.org/10.1016/J.TALANTA.2009.04.066
30. Ghaedi M, Fathi MR, Shokrollahi A, Shajarat F (2006) Highly selective and sensitive preconcentration of mercury ion and determination by cold vapor atomic absorption spectroscopy. Anal Lett 39:1171–1185. https://doi.org/10.1080/00032710600622167
31. Wu J, Yu J, Li J et al (2007) Detection of metal ions by atomic emission spectroscopy from liquid-electrode discharge plasma. Spectrochim Acta Part B At Spectrosc 62:1269–1272. https://doi.org/10.1016/J.SAB.2007.10.026

32. Malik LA, Bashir A, Qureashi A, Pandith AH (2019) Detection and removal of heavy metal ions: a review. Environ Chem Lett 17:1495–1521. https://doi.org/10.1007/S10311-019-00891-Z
33. Saylan Y, Akgönüllü S, Denizli A (2020) Plasmonic sensors for monitoring biological and chemical threat agents. Biosensors 10:142. https://doi.org/10.3390/bios10100142
34. Pirzada M, Altintas Z (2020) Recent progress in optical sensors for biomedical diagnostics. Micromachines 11:356. https://doi.org/10.3390/mi11040356
35. Sun R, Li Y, Du T, Qi Y (2023) Recent advances in integrated dual-mode optical sensors for food safety detection. Trends Food Sci Technol 135:14–31. https://doi.org/10.1016/j.tifs.2023.03.013
36. Rumaling MI, Chee FP, Bade A et al (2022) Methods of optical spectroscopy in detection of virus in infected samples: a review. Heliyon 8:e10472. https://doi.org/10.1016/j.heliyon.2022.e10472
37. Maryam S, Nogueira MS, Gautam R et al (2022) Label-free optical spectroscopy for early detection of oral cancer. Diagnostics 12:2896. https://doi.org/10.3390/diagnostics12122896
38. Chapman J, Truong VK, Elbourne A et al (2020) Combining chemometrics and sensors: toward new applications in monitoring and environmental analysis. Chem Rev 120:6048–6069. https://doi.org/10.1021/acs.chemrev.9b00616
39. Kumar S, Singh R (2021) Recent optical sensing technologies for the detection of various biomolecules: review. Opt Laser Technol 134:106620. https://doi.org/10.1016/j.optlastec.2020.106620
40. Alsunaidi B, Althobaiti M, Tamal M et al (2021) A review of non-invasive optical systems for continuous blood glucose monitoring. Sensors 21:6820. https://doi.org/10.3390/s21206820
41. Yan Z, Cai Y, Zhang J, Zhao Y (2022) Fluorescent sensor arrays for metal ions detection: a review. Measurement 187:110355. https://doi.org/10.1016/j.measurement.2021.110355
42. Carter KP, Young AM, Palmer AE (2014) Fluorescent sensors for measuring metal ions in living systems. Chem Rev 114:4564–4601. https://doi.org/10.1021/cr400546e
43. Guo Y, Liu C, Ye R, Duan Q (2020) Advances on water quality detection by UV-vis spectroscopy. Appl Sci 10:6874. https://doi.org/10.3390/app10196874
44. Bridgeman J, Bieroza M, Baker A (2011) The application of fluorescence spectroscopy to organic matter characterisation in drinking water treatment. Rev Environ Sci Bio/Technol 10:277–290. https://doi.org/10.1007/s11157-011-9243-x
45. Li P, Hur J (2017) Utilization of UV-vis spectroscopy and related data analyses for dissolved organic matter (DOM) studies: a review. Crit Rev Environ Sci Technol 47:131–154. https://doi.org/10.1080/10643389.2017.1309186
46. Elbashir AA, Dsugi NFA, Mohmed TOM, Aboul-Enein HY (2014) Spectrofluorometric analytical applications of cyclodextrins. Luminescence 29:1–7. https://doi.org/10.1002/bio.2504
47. Fülöp Z, Kurkov SV, Nielsen TT et al (2012) Self-assembly of cyclodextrins: formation of cyclodextrin polymer based nanoparticles. J Drug Deliv Sci Technol 22:215–221. https://doi.org/10.1016/S1773-2247(12)50032-8
48. Liu P, Sun S, Guo X et al (2015) Competitive host-guest interaction between β-cyclodextrin polymer and pyrene-labeled probes for fluorescence analyses. Anal Chem 87:2665–2671. https://doi.org/10.1021/ac503301q
49. Sun Y, Leng R, Ma X et al (2023) Economical amidoxime-functionalized non-porous β-cyclodextrin polymer for selective detection and extraction of uranium. Chem Eng J 459:141687. https://doi.org/10.1016/j.cej.2023.141687
50. Bryan AJ, de Silva AP, De Silva SA et al (1989) Photo-induced electron transfer as a general design logic for fluorescent molecular sensors for cations. Biosensors 4:169–179. https://doi.org/10.1016/0265-928X(89)80018-5
51. Bissell RA, de Silva AP, Gunaratne HQN et al (1992) Molecular fluorescent signalling with 'fluor–spacer–receptor' systems: approaches to sensing and switching devices via supramolecular photophysics. Chem Soc Rev 21:187–195. https://doi.org/10.1039/CS9922100187

52. Zhu W, Chen Y, Liu Z et al (2023) Fabrication of supramolecular system derived from poly β-cyclodextrin coupling Guinoline Dderivative and its fluorescence sensing of zinc ion in pure water environment. J Fluoresc 33:2241–2252. https://doi.org/10.1007/s10895-023-03222-2
53. Silva EC, Gomes CG, Pina J et al (2024) Carbon quantum dots-containing poly (β-cyclodextrin) for simultaneous removal and detection of metal ions from water. Carbohydr Polym 323:121464. https://doi.org/10.1016/j.carbpol.2023.121464
54. Narayanan G, Shen J, Matai I et al (2022) Cyclodextrin-based nanostructures. Prog Mater Sci 124:100869. https://doi.org/10.1016/j.pmatsci.2021.100869
55. Cengiz B, Gevrek TN, Chambre L, Sanyal A (2023) Self-assembly of cyclodextrin-coated nanoparticles: fabrication of functional nanostructures for sensing and delivery. Molecules 28:1076. https://doi.org/10.3390/molecules28031076
56. Lakkakula JR, Divakaran D, Thakur M et al (2018) Cyclodextrin-stabilized Gold nanoclusters for bioimaging and selective label-free intracellular sensing of Co2+ ions. Sensors Actuators B Chem 262:270–281. https://doi.org/10.1016/j.snb.2018.01.219
57. Bindhu MR, Saranya P, Sheeba M et al (2021) Functionalization of gold nanoparticles by β-cyclodextrin as a probe for the detection of heavy metals in water and photocatalytic degradation of textile dye. Environ Res 201:111628. https://doi.org/10.1016/j.envres.2021.111628
58. Wang Y, Guo H, Zhang Y et al (2018) Achieving highly water-soluble and luminescent gold nanoclusters modified by β–cyclodextrin as multifunctional nanoprobe for biological applications. Dye Pigment 157:359–368. https://doi.org/10.1016/j.dyepig.2018.05.015
59. Paau MC, Lo CK, Yang X, Choi MMF (2010) Synthesis of 1.4 nm α-cyclodextrin-protected gold nanoparticles for luminescence sensing of mercury (II) with picomolar detection limit. J Phys Chem C 114:15995–16003. https://doi.org/10.1021/jp101571k
60. Ganesan V, Senguttuvan S, Narayanan V et al (2022) Silver nanoparticle decorated γ-cyclodextrin with 1,5-dihydroxy naphthalene inclusion complex; as a sensitive fluorescence probe for dual metal ion sensing employing spectrum techniques. Chem Phys Lett 796:139537. https://doi.org/10.1016/j.cplett.2022.139537
61. Pinaud F, Michalet X, Bentolila LA et al (2006) Advances in fluorescence imaging with quantum dot bio-probes. Biomaterials 27:1679–1687. https://doi.org/10.1016/j.biomaterials.2005.11.018
62. Yang S, Fan W, Wang D et al (2019) Fluorescence analysis of cobalt (II) in water with β-cyclodextrin modified Mn-doped ZnS quantum dots. Anal Methods 11:3829–3836. https://doi.org/10.1039/C9AY00709A
63. Wang S, Wang H, Yue J et al (2022) Highly sensitive and selective fluorescent sensor for Ag+ detection using β-cyclodextrin/chitosan polymer-coated S QDs based on an aggregation-induced quenching mechanism. Cellulose 29:4083–4095. https://doi.org/10.1007/s10570-022-04524-7
64. Baker SN, Baker GA (2010) Luminescent carbon nanodots: emergent nanolights. Angew Chemie Int Ed 49:6726–6744. https://doi.org/10.1002/anie.200906623
65. Wang J, Qiu F, Wu H et al (2017) Fabrication of fluorescent carbon dots-linked isophorone diisocyanate and β-cyclodextrin for detection of chromium ions. Spectrochim Acta Part A Mol Biomol Spectrosc 179:163–170. https://doi.org/10.1016/j.saa.2017.02.031
66. Wang S, Liu J, Zhao H, Zhang F (2021) Carboxymethyl chitosan crosslinked β-cyclodextrin containing hydrogen bonded N C QDs nanocomposites to design fluorescence probes for manganese ion (II) sensing. Mater Sci Eng C 119:111556. https://doi.org/10.1016/j.msec.2020.111556
67. Yang J, Yang Y, Su L et al (2023) Diethylenetriamine-β-CD-modified carbon quantum dots for selective fluorescence sensing of Hg2+ and Fe3+ and cellular imaging. Spectrochim Acta Part A Mol Biomol Spectrosc 291:122364. https://doi.org/10.1016/j.saa.2023.122364
68. Zhu S, Zhao F, Deng M et al (2019) Construction of β-cyclodextrin derived CDs-coupled block copolymer micelles loaded with CdSe/ZnS QDs via host-guest interaction for ratiometric fluorescence sensing of metal ions. Dye Pigment 168:369–380. https://doi.org/10.1016/j.dyepig.2019.04.051

69. Almahri A, Al-bonayan AM, Attar RMS et al (2023) Multifunctional lipophobic polymer dots from cyclodextrin: antimicrobial/anticancer laborers and silver ions chemo-sensor. ACS Omega 8:16956–16965. https://doi.org/10.1021/acsomega.3c00873
70. Prabu S, Mohamad S (2020) Curcumin/beta-cyclodextrin inclusion complex as a new "turn-off" fluorescent sensor system for sensitive recognition of mercury ion. J Mol Struct 1204:127528. https://doi.org/10.1016/j.molstruc.2019.127528
71. Pršir K, Matić M, Grbić M et al (2023) Naphthalimide-Piperazine derivatives as multifunctional "On" and "Off" fluorescent switches for pH, Hg2+ and Cu2+ ions. Molecules 28:1275. https://doi.org/10.3390/molecules28031275
72. Ramasamy K, Thambusamy S (2017) Dual emission and pH based naphthalimide derivative fluorescent sensor for the detection of Bi3+. Sensors Actuators B Chem 247:632–640. https://doi.org/10.1016/J.SNB.2017.03.043
73. Sivakumar K, Parinamachivayam G, Murali Krishnan M et al (2018) Preparation, characterization and molecular modeling studies of the beta-cyclodextrin inclusion complex with benzoguanamine and its analytical application as chemosensor for the selective sensing of Ce 4+. Spectrochim Acta Part A Mol Biomol Spectrosc 200:212–225. https://doi.org/10.1016/j.saa.2018.04.034
74. Liu Z, Ren K, Chen Y et al (2017) A supramolecular self-assembly host–guest system from cyclodextrin as an absolute water-soluble fluorescence sensor for aluminium ions: synthesis, characterization and sensing activity. RSC Adv 7:38160–38165. https://doi.org/10.1039/C7RA06915D
75. Maniyazagan M, Rameshwaran C, Mariadasse R et al (2017) Fluorescence sensor for Hg2+ and Fe3+ ions using 3,3′–Dihydroxybenzidine:α–Cyclodextrin supramolecular complex: characterization, in-silico and cell imaging study. Sens Actuators B Chem 242:1227–1238. https://doi.org/10.1016/j.snb.2016.09.093
76. Kaliyamoorthi K, Maniraj S, Govindaraj TS et al (2020) Unusual fluorescence quenching-based Al3+ sensing by an Imidazolylpiperazine derivative. β-Cyclodextrin encapsulation-assisted augmented sensing. J Fluoresc 30:445–453. https://doi.org/10.1007/s10895-020-02511-4
77. Ay U, Sarlı SE (2018) Investigation by fluorescence technique of the quenching effect of Co2+ and Mn2+ transition metals, on naphthalene-methyl-beta-cyclodextrin host-guest inclusion complex. J Fluoresc 28:1371–1378. https://doi.org/10.1007/s10895-018-2301-9
78. Sivakumar K, Komathi V, Krishnan MM (2018) Dinitrophenylhydrazine: β-cyclodextrin inclusion complex as a novel fluorescent chemosensor probe for Ce4+. Res Chem Intermed 44:5301–5327. https://doi.org/10.1007/s11164-018-3424-y
79. Abbo H, Cherian AR et al (2022) An efficient inclusion complex based fluorescent sensor for mercury (II) and its application in live-cell imaging. J Fluoresc 32:1109–1124. https://doi.org/10.1007/s10895-022-02931-4
80. Shi D, Zeng J, Ye J et al (2017) Detection and removal of metal ion based on a fluorescent sensor composed of a photoresponsive rhodamine derivative by host-gust interaction. J Chin Chem Soc 64:986–992. https://doi.org/10.1002/jccs.201600842
81. Sumithra M, Sivaraj R, Selvan GT et al (2017) Ca 2+ ion sensing by a piperidin-4-one derivative and the effect of β-cyclodextrin complexation on the sensing. J Lumin 185:205–211. https://doi.org/10.1016/j.jlumin.2017.01.021
82. Sivakumar K, Bhakyajothi V, Parameswari M et al (2013) Spectral studies on the supramolecular assembly of 1H2NA: β-CD complex and its analytical application as chemosensor for the selective sensing of Cr 3+. Polycycl Aromat Compd 33:221–235. https://doi.org/10.1080/10406638.2013.781044
83. Liu ZC, Zhu WP, Chen YH et al (2015) Water-soluble host–guest system from β-cyclodextrin as a fluorescent sensor for aluminium ions: synthesis and sensing studies. Dalt Trans 44:16528–16533. https://doi.org/10.1039/C5DT02220G
84. Sivakumar K, Ragi TR (2016) 2H1NA/β-cyclodextrin inclusion complex: preparation, characterization and chemosensory application for detecting Ag +. J Macromol Sci Part A 53:677–686. https://doi.org/10.1080/10601325.2016.1224625

85. Sivakumar K, Parameswari M, Stalin T (2016) Etodolac:β-cyclodextrin inclusion complex as a novel fluorescent chemosensor probe for Ba 2+. J Carbohydr Chem 35:118–130. https://doi.org/10.1080/07328303.2016.1149713
86. Prema D, Thamaraiselvi S, Yamuna R (2020) Encapsulation of N-phenyl p-phenylenediamine into β-CD: spectral, molecular modelling studies and sensor application for detecting Fe2+ ion. J Mol Liq 319:113990. https://doi.org/10.1016/j.molliq.2020.113990
87. Serhan M, Sprowls M, Jackemeyer D, et al (2019) Total iron measurement in human serum with a smartphone. AIChE Annu Meet Conf Proc 2019-Nov: https://doi.org/10.1039/x0xx00000x
88. Liu K, Marin L, Cheng X (2023) Water-soluble β-cyclodextrin based turn-on amplifying fluorescent probes for sensitive and selective detection of Hg2+/Hg+ ions. Sens Actuators B Chem 377:133060. https://doi.org/10.1016/j.snb.2022.133060
89. Liu Z, Yang W, Li Y et al (2015) A facile synthesis of a highly water-soluble and selective fluorescent sensor towards zinc ions derived from β-cyclodextrin based on an unexpected sensing process. RSC Adv 5:100482–100487. https://doi.org/10.1039/C5RA21038K
90. Antony EJ, Raj M, Paulpandi RQ et al (2015) A Highly selective fluorescent sensor for Pb2+ based on a modified β-cyclodextrin. J Fluoresc 25:1031–1036. https://doi.org/10.1007/s10895-015-1588-z
91. Kanagaraj K, Bavanidevi K, Chow TJ, Pitchumani K (2014) Selective "turn-off" fluorescent sensing of mercury ions using aminocyclodextrin:3-hydroxy-N-phenyl-2-naphthamide complex in aqueous solution. RSC Adv 4:11714. https://doi.org/10.1039/c3ra45837g
92. Xu M, Wu S, Zeng F, Yu C (2010) Cyclodextrin supramolecular complex as a water-soluble ratiometric sensor for ferric ion sensing. Langmuir 26:4529–4534. https://doi.org/10.1021/la9033244
93. Li X, Jin Y, Zhu N et al (2024) Recent developments of fluorescence sensors constructed from Pillar[n]arene-based supramolecular architectures containing metal coordination sites. Sensors 24:1530. https://doi.org/10.3390/s24051530

Muhammed Shahsad a research scholar at the Central University of Tamil Nadu, completed his Integrated M.Sc. in Chemistry. His research focuses on combining White Light Excitation Fluorescence (WLEF) techniques with machine learning for environmental applications. He works on modifying fibre optics to improve fluorescence measurements, addressing spectral overlap in multicomponent systems using chemometrics, and adding dimensions to WLEF spectra to eliminate matrix effects.

S. Jayaram completed his Integrated Master's in Chemistry at the Central University of Tamil Nadu, focusing on chemical reactions of atomically precise nanomaterials and clean water technologies. Under the supervision of Dr. John Prakash, he developed fluorescence lifetime-based techniques for rapid data acquisition. He is currently a Research Assistant at IIT Madras, working on Nano engineered materials to create innovative solutions for clean water, aiming to address environmental challenges through material science and spectroscopy.

Dr. John Prakash is an Assistant Professor at the Central University of Tamil Nadu. He earned his PhD from the Indian Institute of Technology, Madras, and was awarded the prestigious Eli Lilly Outstanding Thesis Award in 2013. His research interests encompass analytical fluorimetry, the development of modular spectroscopic instrumentation for environmental and biologically relevant systems, and the application of chemometric tools to spectroscopic data.

Chapter 7
Cyclodextrins as Sensors in Metals Determination: An Exploration Through Fluorometry

Suvarna Yenduri and K. Naga Prashant

Abstract Sensing of heavy metals is one of the major thrust areas of research. Monitoring and control of heavy metal levels are essential to safeguard our biological and ecosystems. At the same time, the use of simple, sensitive, and economical sensors is also required for the immediate detection of heavy metals. Chemosensors can satisfy these requirements and can give rapid results. Cyclodextrins are one of the widely used chemosensors and have revolutionized supramolecular chemistry. Cyclodextrins (CDs) can form inclusion complexes with several chemical and inorganic compounds, including metal ions. These materials can selectively bind and detect metals, making them excellent for this. The application of cyclodextrins as chemosensors for fluorescent metal detection is a fascinating and innovative analytical chemistry approach. This chapter discusses the role of cyclodextrins as chemosensors in the detection of metal ions by fluorescence and a detailed discussion on probes used for fluorescent detection. Fluorescent methods can detect and measure metal ion concentration with respect to changes in fluorescent intensities. Fluorophores are used to quantify metal ion concentrations by affecting fluorescence when they bind to fluorophores. These changes may be detected and evaluated to determine sample metal ion concentration. The use of cyclodextrins as chemosensors in fluorometric methods has several benefits. These include high selectivity, sensitivity, and metal detection in complex sample matrices. The approach suggested is adaptable and has been used in many disciplines. These methods detect metal ions in samples sensitively and reliably. Environmental monitoring, industrial operations, pharmaceutical analysis, and biological research need metal ion concentration detection and quantification. Analytical chemistry is exploring cyclodextrins as fluorometric metal chemosensors. This method for metal ion detection is promising since it is accurate and effective. This approach has wide-ranging applications in science and practice.

Keywords Cyclodextrins · Fluorophore · Sensor · Probe metals · Chemosensing

---

Y. Suvarna (✉) · K. Naga Prashant
Department of Pharmaceutical Chemistry and Analysis, Sri Adichunchanagiri College of Pharmacy, Adichunchanagiri University, Mandya, Karnataka, India
e-mail: suvarna.ph52@gmail.com

S. Krishnamoorthy (ed.), *Cyclodextrins for Chemosensing*,
https://doi.org/10.1007/978-3-031-89299-8_7

7.1 Introduction

Heavy metals are the elements which are present in nature have high atomic weight and are five times denser than water. Because of their wide applications in many fields like industrial, domestic, agricultural, medicinal, and technological fields, they are distributed in the environment in a wide range which causes adverse effects on the environment and human health [1]. Heavy metals are the most common environmental hazards due to their toxicity, tendency to linger in the atmosphere, and bioaccumulative nature in humans. Some heavy metals are synthesized by humans but most heavy metals are found in nature. The two most distinctive qualities of heavy metals are their large atomic mass and toxicity to living organisms. Most heavy metals have the potential to kill people and damage the environment and climate. Heavy metals may become exceedingly dangerous when they interact with other environmental elements like soil, water, or air. Through the food chain, people and other living creatures may be exposed to heavy metals [2].

Over the past century, there has been a noticeable increase in human exposure to heavy metals due to industrial activities. There are several heavy metals that can be harmful to humans, including arsenic, mercury, lead, chromium, and cadmium [3]. The high rate of exposure to these toxic metals created research interest to study the toxic effects of these metals on human health. The challenges posed due to the environmental contamination by heavy metals have become more pronounced as there is widespread use of distinct metals in industry and everyday human activities. Toxic metals are primarily derived from various environmental sources, including organic wastes, transportation, electricity generation, trash burning, and industrial effluents. Depending on whether they are particulate or gaseous, the wind can carry them many miles from the sources. The surface of waterways or the land eventually receives the discharge of metallic contaminants from the atmosphere. In light of this, air pollution contributes to environmental degradation [4].

As heavy metals are enduring in the environment, they get accumulated in plants and soil. Plant physiological processes, including photosynthesis, gas exchange, and nutrient absorption, are disrupted by heavy metals, leading to decreased plant growth, dry matter accumulation, and yield. Additionally, plant's antioxidant levels are disrupted by heavy metals, which lowers the produce's nutritional value. Dietary absorption of heavy metals through plant sources imposes chronic adverse effects on human health [5]. In recent decades, several analytical techniques have been utilized to identify trace heavy metals like colorimetry, atomic absorption spectroscopy, inductively coupled plasma, chromatographic techniques, and electrochemical analysis which led to the usage of toxic chemicals, sophisticated instruments and also consuming more time. Consequently, it is significant to develop innovative sensors for the swift, reliable, cost-effective, eco-friendly, sensitive, and convenient analysis of heavy metals [6].

In view of these, chemosensors are effective tools for detecting heavy metals as they offer several advantages such as portability, simple, economic, and also biodegradability. As heavy metals are colorless, they can be detected by monitoring variations in fluorescence intensity or color variations in chemosensors [7].

7.2 Cyclodextrins as Traditional Chemical Sensors

Chemosensors, also known as molecular sensors, are supramolecular-based devices which can convert chemical information into a signal which can be useful for analytical study. The data can encompass a wide range of information, starting from determining the analyte concentration to a detailed interpretation of the structure [8].

7.3 Features of Cyclodextrins

Cyclodextrins are cyclic oligosaccharides characterized by hydrophilic exterior and hydrophobic core cavities facilitating the formation of low polar inclusion complexes. Cyclodextrins exist as α, β, and γ isomers which are having their significant and unique properties. Among them, β-cyclodextrin is recognized for its superior applicability due to its ease of complexation and low cost. Enhancement of overall solubility and stability was observed when cyclodextrins formed an inclusion complex with water-insoluble or poorly water-soluble analytes [9–11].

7.4 Fluorescent Chemosensors

Fluorescent chemosensors are molecules that change fluorescence in response to their environment or chemical recognition events. Quantifying molecules, emission spectra, and fluorescent intensity changes with fluorescent sensors is typical. Decay life after interacting with analytes like cations, anions, and neutral molecules in photophysical signaling mechanisms like Electronic Energy Transfer (EET), Intramolecular Charge Transfer (ICT), Photo-induced Electron Transfer, Fluorescence Resonance Energy Transfer (FRET), and also non-covalent excimer/exciplexes formation [12]. The ability of the microenvironment in which chromophores interact to modify both color and fluorescence patterns has driven scientists around the world to functionalize CDs with fluorescent components and different dyes for better results [13].

7.5 Cyclodextrins as Fluorophores

Exploration into metal ion detection is the most important step in the process of minimizing the adverse effects that metals have on living organisms and the environment. Among the 27 β-CD and chemosensor systems that have been described, the role of β-CD, which is the host molecule that can often be utilized for metal ion sensing, is taken into consideration and explored. Among the most significant are naturally occurring CDs, because of their remarkable water solubility, non-toxic nature, biocompatibility, availability, low cost, easy to use, and biocompatible nature [14].

7.6 Turn On-Turn Off Mechanism

Several research organizations worldwide have manufactured a multitude of CDs with chromophore affixed and have also successfully completed their sensing operations. However, *Ueno* et al. have made the most groundbreaking contributions to this subject; they have reported numerous CD-based fluorescent chemical sensors by utilizing a flexible linker to install various fluorophores (dansyl, pyrene, anthracene, etc.) in CDs. This stiff spacer has prevented the creation of any self-inclusion complexes. As a result, the fluorophore is exposed to a hydrophilic environment, which causes the CDs' fluorescence to quench. Thus, when a hydrophobic guest is added, it becomes included in the hydrophobic cavity of CDs, which places the attached fluorophore in a more hydrophobic environment than when the CD is in its free state. Therefore, the "turn-on" fluorescence reaction is caused by an increase in fluorescence. Whereas "turn-off' fluorescence" results in a decrease in fluorescence due to the displacement of the fluorophore from inside to outside the cavity (Fig. 7.1) [14].

7.7 Factors Affecting Fluorescence Intensity

Fluorescence chemosensing of metal ions has emerged as a prominent area of research for the detection of metals in biological and environmental samples [15]. Through a changed fluorescence response, the CD molecule can provide the guest molecule it encapsulates with increased metal-detecting capabilities. To achieve metal ion binding, chelating metal atoms should be present outside the cavity [16]. Chemosensors based on cyclodextrin can withstand variations in pH, temperature, and salinity levels and are thus useful for their application in fluorescence studies [17]. Chemosensors utilizing chromophore-modified CDs have the ability to detect colorless neutral substances by observing changes in fluorescence, absorption, or circular dichroism intensity. The image provided showcases the functioning of these

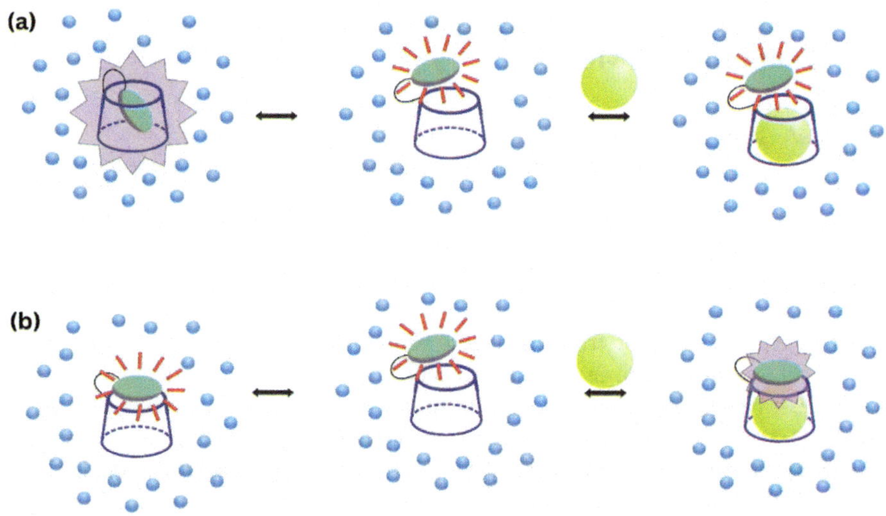

Fig. 7.1 Illustration showing the fluorescent CD-based chemosensors in turn-off (**a**) and turn-on (**b**) modes. (Adapted from Ikeda, H. 2018)

chemosensors. Several conformations can be observed in aqueous solutions when the CDs are modified with chromophores. The simplified two-state model provides an explanation for the conformation equilibrium [18].

In most cases, the "self-inclusion state" constitutes the principal conformation. In order to include guests in the CD cavity, the chromophore displaces itself outside the cavity and thereby undergoing induced fit conformational change. When the concentration of guests increases, the "non-self-inclusion state" also increases. When the CD cavity includes fluorophore, the hydrophobic environment makes the molecule to fluoresce. The displacement of the fluorophore from the hydrophobic cavity resulted in a decrease in fluorescent intensity due to the accommodation of water in the cavity. Variability in fluorescence intensity is dictated by the chemosensor's affinity for a visitor. Self-inclusion of chromophores may impede the visitor's ability to be accommodated. For example, dansyl-D-leucine-modified β-CD is twice stable in its self-inclusion state as là-L-leucine-modified β-CD, whereas the former possesses a binding ability that is approximately half that of the latter [19].

The fluorescence intensity of substituted CD usually diminishes with the inclusion of guest molecules due to competitive inclusion. This is because of the displacement of fluorescent from the cavity to a more hydrophilic environment by the guest's preferential attachment to the cyclodextrin cavity, dimming the substituent's fluorescence [20].

7.8 Cyclodextrins as Chemosensors for Fluorometric Estimation of Metals

α-Cyclodextrins as Chemosensors

Yue Zhang et al. developed a novel colorimetric and fluorometric chemosensor for the determination of Cu^{2+} which relies on guest–host interaction. It produced Rhodamine 6G modified with azobenzene and $Fe_3O_4{\cdot}SiO_2$ was modified with α-cyclodextrin to form the inclusion of complex magnetic nanoparticles (IFIC MNPs). Cu^{2+} ions can be used to produce selective "turn-on" type fluorescence improvements and also an apparent change of color from light yellow to pink in the UV-visible spectra and colorless to yellow-green in the fluorescence study. Exhibiting a detection limit of 2.5×10^{-7} M, the IFIC MNPs result in high sensitivity and selectivity to Cu^{2+} in $MeCN/H_2O = 1{:}1$. Due to their strong magnetism, the magnetic nanoparticles of IFIC can be easily separated and recycled. Furthermore, the suggested method is an easy and effective method as it relies on grafting reaction and self-assembly process, and also the recycling of nanoparticles is easy and productive. It proved to be an effective stand-in for developing high-performance, highly sensitive, and selective fluorescence sensing materials for the detection of copper ions in aqueous solutions [21].

Maniyazagan M. et al. in their recent study introduced a solid inclusion complex of 3,3′-dihydrobenzene: α-Cyclodextrin (DHB: α-CD) in the ACN-water system. This complex is being hailed as a remarkable fluorescent sensor, capable of detecting Hg^{2+} and Fe^{3+} ions with high selectivity and sensitivity. The DHB: α-CD inclusion complex was developed by kneading and co-precipitation techniques. The complex was subsequently characterized by FTIR, XRD, SEM, and ^{1}H NMR studies. Interpretation of the fluorescence spectrum revealed the complex formed distinct fluorescence properties on exposure to Fe^{3+} and Hg^{2+} ions in the ACN-water system. When Hg^{2+} and Fe^{3+} are introduced, the fluorescence of the host–guest system experiences a significant increase. The strong binding constants ($k = 3.1626 \times 10^4\ M^{-1}$, $1.8225 \times 10^4\ M^{-1}$) between DHB: α–CD and the metal ions (Hg^{2+}, Fe^{3+}) contribute to their effective coordination with the fluorophore. In addition, despite the interference from various metal ions, in-depth studies revealed the highly selective capacity of the sensor to Hg^{2+} and Fe^{3+} ions. Based on molecular docking studies, it has been observed that the docked structure of DHB: α–CD complies with the obtained results. On the basis of molecular dynamic simulations, it is evident that the molecules exhibit promising structural and energetic interactions, aligning with the experimental findings. Further research utilizing bioimaging has revealed that the DHB: α–CD has the ability to identify Hg^{2+} and Fe^{3+} ions in HepG2 cells in a laboratory setting. The author concludes that in acetonitrile–water media, the DHB: α–CD inclusion complex showed both selectivity and sensitivity for mercuric and ferric ions than other heavy metals. The presence of Hg^{2+} and Fe^{3+} ions resulted in the host–guest system displaying remarkable fluorescence imaging properties, providing evidence of the probe's ability to permeate cells. The probe is capable of functioning as a sensor for determining Hg^{2+} and Fe^{3+} in live cells because of its non-toxic nature [22].

***β*-Cyclodextrins as Chemosensors**

Ying-Ming Zhang et al. established a water-soluble fluorescent for Cd^{2+} sensors using a quinolinotriazole-β-CD and its adamantane carboxylic acid (AdCA) complex as an effective choice. They utilized click chemistry to generate a novel derivative named 1, which contained 8-hydroxyquinolino and triazole groups. The resulting complex demonstrated favorable water solubility to a maximum of 0.03 mol/l and showed an effective switch-on fluorescence response selectively to Cd^{2+} under physiological conditions. Cooperative coordination of Cd^{2+} with the triazole moiety and the 8-hydroxyquinolino moiety, which was omitted from the β-CD cavity, was identified as a fundamental and essential element of the fluorescence sensing process, according to research on the recognition mechanism. Significantly, spectrophotometric analyses also revealed that the 1/AdCA system improved fluorescence sensitivity to Cd^{2+} through a cyclodextrin/substrate/Cd^{2+} triple binding mode subsequent to 1-adamantanecarboxylic acid sodium salt inclusion complexation. Furthermore, the author emphasized the criticality of considering the pH effect when selecting the optimal sensing condition for fluorescent probes employing the electron donor/acceptor, as their performance in detecting metal ions is exceedingly sensitive to pH levels. An emission peak was observed at 485 nm at a pH of 1.13. The acidic environment resulted in a slight hypsochromic shift towards 472 nm on reducing the concentration, and also there is the appearance of a shoulder peak at 416 nm. Alkaline conditions, on the other hand, did not demonstrate a significant distinction. The fluorescence sensing investigations of 1 towards Cd^{2+} were conducted in HEPES buffer solution. The intensity of 1's fluorescence emission at 418 nm increased progressively as Cd^{2+} was added in a sequential fashion to a solution containing 1. Prior to coordination with Cd^{2+}, 1 exhibited a diffuse fluorescence; however, coordination with Cd^{2+} resulted in an increase in fluorescence [23].

Yu Liu et al. devised a practical and exceptionally efficient sensing method by fluorescence for determining mercuric ions (Hg^{2+}) in thin films and aqueous solutions. The quinolino CD MQAS-β-CD was synthesized. MQAS is chemically N-(2-methyl-8-aminoquinolyl)-p-aminobenzene sulfonamide. Subsequent analysis indicates that Zn^{2+} forms a 2:1 stable stoichiometric complex with water. Notable characteristics of the quinolinocyclodextrin/Zn^{2+} composite gave a distinct fluorescence response to Hg^{2+} than other metal ions and also made it a simple method for determination in aqueous solution and PVA-based thin film. The aforementioned finding would render the Zn^{2+} complex an exceptionally functional and pragmatic fluorescence sensor in the context of Hg^{2+} detection. Moreover, the author deduces that a water-soluble quinolinocyclodextrin/Zn^{2+} alloy was successfully synthesized in the laboratory for sensing Hg^{2+} by fluorescence. Transparent fluorescence responses were observed on the application of this sensor for sensing Hg^{2+} present in aqueous solutions and thin organic films; these responses are readily discernible through fluorescence spectroscopy and visual inspection. Moreover, the fluorescence sensing system exhibited negligible observable effects from a variety of cations. By virtue of its user-friendly

Hg^{2+} detection method, favorable water solubility, and higher sensitivity towards Hg^{2+} than other similar cations, the complex formed with this sensor can be potentially utilized in environmental domains to satisfy the detection requirements of Hg^{2+} assays [24].

Narayanan V. et al. investigated the spectral properties of a synthetic compound known as symmetric syringaldehyde di-Schiff azine. It was found that these compounds formed an inclusion complex with β-CD which acts as a fluorescent probe. This probe demonstrated high selectivity and sensitivity in detecting various metal ions, including transition metals and alkali ions. The findings highlight the potential of organic fluorescence probes in sensor applications. A Schiff-based symmetrical azine derivative has been successfully synthesized. This derivative, known as SyA, is a profitable compound with promising potential. It is derived from 4,4′((1E,1′E)-hydrazine-1,2-diylidenebis(methanylylidene)) *bis*2,6-methoxyphenyl dihydrazone, or syringaldehyde azine. The synthesis process was straightforward and yielded positive results. Rapid and simultaneous determination of both Cu^{2+} and Pb^{2+} can be done under optimized conditions in an acetonitrile medium by using Chelation-Enhanced Fluorescence Mechanism (CHEF) and Ligand to Metal Charge Transfer (LMCT) approaches. The mode of interaction between the metals and SyA was also studied using Benesi-Hildebrand and ^1H NMR spectroscopy, later, β-CD: SyA inclusion complex was developed to enhance solubility and sensing. It resulted in the enhancement of properties as well as led to the determination of a new ion Al^{3+} (due to the interaction of OH group of β-CD with metal ions) of β-CD with metal ions) without causing any effect on Cu^{2+} and Pb^{2+} detection. The absorbance and fluorescence titration profiles were used for calculating LOD and binding constants of SyA and β-CD: SyA for Cu^{2+} and Pb^{2+}. The assessment of antibacterial activity and biological activities was conducted with a focus on practical applications [25].

Li Q. et al. devised a technique that utilizes β-CD/AD adamantane host–guest self-assembly interactions to create a new versatile fluorescent chemosensor. This chemosensor exhibits exceptional selectivity with a distinct "off–on" response and can be easily recovered and reused. The $Fe_3O_4@SiO_2$ inclusion complex magnetic nanoparticles (FFIC MNPs) are chemically modified using adamantane and cyclodextrin to exhibit a specific amplification of green fluorescence upon exposure to Zn^{2+}. These nanoparticles show a LOD of 4.5×10^{-7} mol/L in methyl cyanide and water mixture of 1:4 v/v. Upon addition of Zn^{2+}, the spirolactam of the fluorophore undergoes ring opening and it can be reversible on removal of the Zn complex. Each gram of nanoparticle can absorb 6.1 mg of Zn. Due to the fluorescence quenching when exposed to ultraviolet (UV) light, the modified chemosensor FFICMNP-Zn exhibits an LOD of 7.7×10^{-7} mol/L and is capable of reacting to CN. FFIC MNPs show the advantage of reusability and recyclability in water-based systems because of their magnetic properties and the ability for repeated usage. The authors efficiently absorbed the Zn complex with an excess of CN, and then reused the residual MNPs to detect Zn^{2+} on four separate occasions. The identical procedure also operates in the other direction. In order to ensure that the probes remain effective in detecting Zn^{2+} and CN^- ions for at least seven cycles, if the fluorophore component becomes non-functional, the inactive fluorescent molecules can be removed from the host MNP and can be replaced with new fluorescent small guest molecules. It was concluded that the described method can allow for easy and high yield of recycling of both nanoparticles and inclusion complex due to its self-grafting reaction as well as self-assembly technique. Moreover, this method offers a very favorable option for the creation of efficient magnetic sensing materials that can be used for the detection and separation of Zn^{2+} and CN^- ions from aqueous solutions. The flour-Ad/Fe_3O_4@SiO_2-β-cyclodextrin inclusion complex magnetic nanoparticles (FFIC MNPs) have excellent adsorption capacity for Zn^{2+} ions and may efficiently and rapidly extract zinc ions from an aqueous solution using a readily available magnet. Furthermore, the FFIC MNPs-Zn that have been swapped may be isolated and gathered following their reaction with CN^- [26].

Wang J. et al. described a technique for identifying copper ions using a novel fluorescent probe called L. This probe was developed by combining β-cyclodextrin (β-CD) with thiosemicarbazone moieties. A reduction in fluorescence was observed for probe L in the tris–HCl buffer of pH 7.4, however, this was only observed when Cu^{2+} and other interfering ions were present. After being complexed in a 1:1 ratio with Cu^{2+}, the LOD of the probe was found to be 1.37 μM. The obtained results showed probe L as a rapid chemosensor for the determination of Cu^{2+} in both environmental and biological samples. The procedure incorporates the "Jobs plot" mechanism, which elucidates the potential for cooperation between S and N atoms of thiosemicarbazone moiety of probe L and Cu^{2+} ion during the binding process. A color change of light yellow to colorless attributes to the formation of probe L-Cu^{2+} coordination complex. The occurrence of fluorescence emission "turn-off" was also shown due to the formation of the probe L-Cu^{2+}. The hypothesized mechanism suggests that the interaction between Cu^{2+} and probe L is responsible for the sensor's "turn-off" function [27].

CD-37, a fluorescent sensor was developed by Liu et al. through the utilization of a triazole crosslinker to connect 8-hydroxyquinoline to β-cyclodextrin, which functions as the fluorophore. Intramolecular photoinduced electron transfer (PET) from the N atom of triazole to the 8-hydroxyquinoline results in a negligible amount of fluorescence in the solution of the β-cyclodextrin derivative. In contrast, enhancement in fluorescence was obtained when Cd^{2+} was added to the solution and obstructed PET. Addition of a co-chelator, i.e., sodium adamantine carboxylate results in a higher binding constant value for the complex (Cd-37-Cd^{2+}) from 2.10 ± 0.23 to $3.38 \pm 0.230.09$ logarithmically, while the emission wavelength remains unchanged. The graphic of the work confirmed that the binding ratio was 1:1. Cd-38, an additional 8-hydroxyquinoline-based reagent, was developed through the utilization of 2,8-dithia-5-aza-2,6-pyridinophane as the ionophore. An elevated CHEF response is observed in the chemosensor when exposed to aqueous Cd^{2+}, sodium dodecyl sulphate, liposomes, and living cells. Job's plot provides confirmation of 1:1 stoichiometry. XRD analysis confirms the formation of the Cd-38-Cd^{2+} complex. Confocal microscopy provides evidence that Cd-38 a potential imaging reagent for intracellular Cd^{2+}. Cd-39 and Cd-40, two sensors that utilized the CHEF mechanism to select Cd^{2+} and Zn^{2+} with exceptional specificity, exhibited a "turn-on" reaction. An "off" reaction, was also resulted in the addition of cupric and mercuric ions via the CHEQ pathway.

Job's plotting validates the stoichiometry of 1:1 binding. XRD analysis confirmed the formation of Cd-39-Cd^{2+} and Cd-40-Cd^{2+} complexes [28].

O. Sayo et al. reviewed a method that describes the detection of copper, mercury, and zinc has recently implemented modified CDs that are capable of combining with various heavy metals. A linear response was observed at a 1:1 molar ratio when a modified CD with a dansyldiethylenetriamine metal-binding group (6-deoxy-6-N(N′-dansyldiethylenetriamino)-CD) was fluorescent probe for detection of Cu^{2+}. When comparing Cu (II) to Fe (II), Co (II), Ni (II), and Zn (II) ions, it is observed that the sensors exhibit an exceptionally high degree of selectivity. Additionally, Hg (II) and Zn (II) ion determination has been reported using a fundamental fluorometric method that uses CD/porphyrin supramolecular inclusion complex in aqueous solution. The aforementioned methods rely on the quenching effect shown by the entry of metal ions into the inclusion complex. These sensitive sensors have the capability to detect Hg (II) at a concentration range of 5×10^{-8} to 2×10^{-5} M, with a LOD of 2.0×10^{-9} M. The sensor can detect Zn^{2+} ions in an aqueous solution at a range of concentration from 5.0×10^{-7} to 2.5×10^{-4} M [29].

Ajath I. A. et al. performed their studies on the synthesis of a new combination of 7-aminoflavone-modified β-CD with a hydrazine carbothioamide linker. This new combination was interpreted for its structure through elemental analysis, NMR, and ESI-MS studies. The study aimed to find the ability of this new combination for its selectivity and fluorescence ability for the detection of L-histidine and cupric ions. As suggested by the nuclear overhauser effect The presence of a phenyl ring within the cyclodextrin cavity is shown by spectroscopy and emission spectra. The hydrazine carbothioamide linker functions as the metal-chelating moiety. The CD that has been treated with flavone exhibits preferential binding with Cu (II) in preference to other metal ions. The compound becomes non-fluorescent upon binding with Cu (II). The inclusion of histidine, along with other amino acids, leads to the restoration of the fluorescence enhancement. This occurs on displacement of cupric ions by l-histidine from the complex thereby releasing free flavone followed by restoration of fluorescence. This new probe can be successfully applied for the selective determination of cupric ions in the case of heavy metals and L-histidine in the case of amino acids. The author explains that the reported results of the study may be explained by the interaction between the metal and the fluorophore, which impacts the fluorescence. The binding of metal ions to the chelator indirectly enhances metal-receptor contact and thus decreases the interaction between fluorophore and metal ions. The inquiry primarily focuses on understanding the recognition mechanism, with experiments being conducted to gather information on the process of fluorescence activation, specifically regarding the role of the cyclodextrin cavity in forming complexes [30].

Sareen D. et al. conducted a review and reported that 1,8-Dihydroxyanthraquinone (1,8-DHAQ) has the ability for complexation with cations making it a versatile chromophore and strong chelating agent. 1,8-DHAQ 16 acquired water solubility by forming a tight inclusion complex with β-CD in an aqueous solution. The 16: β-CD supramolecular complex was found to readily bind Cu^{2+} in the presence of ammonia. This led to the formation of intricate compound 17 that causes Chelation-Enhanced Quenching (CHEQ). This enhanced its potential use for detecting Cu^{2+} ions in actual samples, since the identification of Cu^{2+} was also confirmed in samples of tea, tap water, and contaminated river water. 1,4-DHAQ 18 doped cellulose microporous nanofiber film is an effective fluorescent probe for detecting Cu^{2+} and Cr^{3+} ions at very low concentrations in the nanomolar range. The phenolate complex formed between 1,4-DHAQ 18 and Cu^{2+} is the principle underlie for Cu^{2+} detection. The presence of Cr^{3+} was further identified by utilizing the 1,4-DHAQ and Cu^{2+} co-doped film. The two sensing events exhibited high sensitivity and demonstrated alterations in emission and visual color variations [31].

In order to measure trace amounts of lead ions in an aqueous solution fluorometrically, Ying Zhang and co-authors developed a method based on the inclusion complex formed between β-CD and α, β, γ, δ-tetrakis(3,5-dibromo-2-hydroxylphenyl)porphyrin (TDBHPP). A significant increase in porphyrin fluorescence intensity was observed in an aqueous solution containing β-CD and TDBHPP. Moreover, once TDBHPP's fluorescence was quenched at its maximum emission wavelength, the enhanced fluorescence brought by β-CD was vulnerable to metal ions. At room temperature, the fluorescence rapidly and nearly entirely decreased when lead ions interacted with TDBHPP. The ligand's shielding from the microenvironment and the β-CD medium's organizing power allowed for higher selectivity, detection limit, and sensitivity. This made it possible to determine lead concentrations ranging from 2.8×10^{-7} to 7.4×10^{-5} mol/L in comparison to findings without β-CD. According to the author's findings, β-CD greatly increased the fluorescence of TDBHPP in water. Moreover, it was found that the quenching in fluorescence intensity was due to lead ions and this quenching effect is higher at ambient temperature. Consequently, a simple-sensitive fluorometric method was developed for the determination of Pb^{2+} ions [32].

Curcumin/β-CD inclusion complex is a novel "turn-off" fluorescence sensor method suggested by Samikannu Prabu ab and Sharifah Mohamad for sensitive identification of Hg^{2+} ions. Using this technique, the formation of a supramolecular complex between β-CD/probe LC and curcumin was confirmed by means of AAS and AES spectroscopy. The binding properties of probe LC with Hg^{2+} ions in water were studied for the first time using AAS and AES. The application of β-CD for the selective identification of dangerous Hg^{2+} ions has been developed effectively. A simple probe LC shows a noticeable change in absorbance and fluorescence emission spectrum (turn-off fluorescence) behavior when toxic Hg^{2+} ions are present. Two phenolic and one aliphatic hydroxy group are present in curcumin. There was also beta-cyclodextrin and toxic Hg^{2+} ions, which led to the formation of a cationic complex. Furthermore, due to the preferential adsorption of the Hg^{2+} ion, curcumin's aliphatic hydroxy group deprotonation resulted in spectrum and color changes from yellow to colorlessness and strong fluorescent to weak fluorescent, which were visible to the unaided eye. There were no noticeable changes in color or spectrum when other metal cations such as [Ag^{+}, K^{+}, Na^{+}, Cs^{+}, Ba^{2+}, Fe^{2+}, Mg^{2+}, Pb^{2+}, Mn^{2+}, Ni^{2+}, Cd^{2+}, Co^{2+}, Cr^{3+}, Sn^{2+}, and Zn^{2+}] were introduced. The dimming of emission spectra caused by Hg^{2+} and the blue shift in absorption spectra indicated a strong potential for a strong interaction between Hg^{2+} ions and probe LC. The concentration-dependent fluorescence response is well described by the conventional Stern–Volmer model. Compared to earlier findings, this approach is significantly more readily biodegradable of probe LC, selective, sensitive, and inexpensive [33].

Maniyazagan M. et al. have made β-CD-encapsulated rhodamine derivative core shell microspheres to enhance aqueous solubility and biocompatibility. The microspheres were developed by mixing the solutions of β-CD and rhodamine-B hydrazide and stirred for 6 h, followed by autoclaving for 24 h at 180 °C and centrifugation to obtain the precipitate. The developed microspheres demonstrated intense and stable fluorescence with Au^{+3} ions in aqueous solution. The fluorescent sensing of Au^{+3} in viable cells was successfully done with a simple, eco-friendly approach with triangular and hexagonal gold microplates. The interference by none of the other related ions for ring-opening reaction that leads to fluorescence was also observed. This chemosensor has overcome the disadvantages shown by the previously developed chemosensors like low aqueous solubility and sensitivity [10].

Golfamsadat et al. developed a turn-off fluorescent β-CD chemosensor using novel thienothiophene-based polymers as probes for the detection of heavy metals. Suzki coupling method was used for the synthesis of probes and their characterization and properties and was analyzed using suitable analytical approaches. Among the developed probes, G2-CD probe inclusion with β-CD showed promising sensitivity and had selectivity towards Hg^{2+} ions by avoiding the interference caused by other ions. The chemosensors act by turn-off fluorescence mechanism and have a sensitivity of less than 1 ppm. This chemosensor can be further extended for studying the ions that have a role in toxicological and biological aspects [34].

Manuel et al. designed a new Cr (VI) selective electrode using a modified polymeric β-CD membrane with sulfur donors. In the study, the glassy carbon electrode was modified by adding carbon quantum dots and β-CDs which increase electrical properties and thus lead to the formation of a polymeric membrane and again modified by adding sulfur donor groups which can act as probes. This resulted in a chemosensor with high selectivity towards Cr (IV) ions. ANOVA study was performed to optimize the electrode. It was found that the sensor at pH 5 with 50 cycles with carbon quantum dots was found to give better results. The limit of detection of the optimized chemosensor was found to be 5.2×10^{-7} mol L^{-1}. It was found to have excellent selectivity towards chromium ions without getting any interference effect. It carried the added advantages of minimum cost, ease of handling, construction, rapid response and portability [35].

γ-Cyclodextrins as Chemosensors

For the detection of Pb^{2+}, Hayashita et al. (2003) synthesized Podand fluoroionophore (Pb-7), an additional probe, from γ-cyclodextrin. When γ-CD is absent, there is no discernible fluorescence. In contrast, there was a discernible change in the fluorescence's intensity when γ-CD (12 mM) was present. This fluorometric shift forms a molecule with γ-CD that has an enhanced quantum yield, indicating the aqueous solubility of Pb-7. Pb^{2+} is present, which indicates that the broad emission band observed in the larger wavelength range (471 nm) deepened and that the emission of pyrene monomers (370–410 nm) decreased. The association constant with a binding stoichiometry of 2:1 was determined to be $1.17 \pm 0.75 \times 109\ M^{-2}$. Moreover, the absorption PD-1/γ-CD complex with Pb^{2+} shows relatively less resolution and intensity than without Pb^{2+}. This unique property is due to the interaction of two pyrenyl moieties in the ground state. According to the authors' design idea, the findings unambiguously demonstrate that Pb^{2+} binding causes PD-1 to dimerize inside γ-CD [36].

Vignesh Kumar et al. developed an excellent sensitive and selective chemosensor for the detection of Cu^{2+} using azine-based probe with γ-cyclodextrin. The developed sensor acts by fluorescent quenching of the probe which acts as a Schiff's base. The study has overcome the challenge of poor solubility azine-based probes by the inclusion of 2-hydroxybenzaldehyde azine (2-HBA) in γ-CD which enhanced its solubility. The inclusion complex emits fluorescence. The synthesized γ-CD/2-HBA sensor shows selectivity and sensitivity towards copper ions and the limit of detection was found to be 1.53 nM [37].

Cyclodextrins and the fluorophores used for preparing modified CDs to act as chemosensors for the fluorometric determination of metals are listed in Table 7.1.

Table 7.1 Cyclodextrins and their fluorophores in modified cyclodextrins for metal ion detection

S. No	Cyclodextrin	Fluorophores	Metal ions detected
1	α-CD	Azobenzene modified rhodamine 6G	Cu^{2+}
2	α-CD	3,3′-Dihydroxy benzene	Fe^{3+}, Hg^{2+}
3	β-cyclodextrin	Dansyldiethylenetriamine	Cu^{2+}, Fe^{2+}, Co^{2+}, Ni^{2+}, Zn^{2+}, Hg^{2+}
4	β-cyclodextrin	Dihydroxyanthraquine	Cu^{2+}
5	β-cyclodextrin	Curcumin	Hg^{2+}
6	β-cyclodextrin	Thienothiphene polymers	Hg^{2+}
7	β-cyclodextrin	Carbon quantum dots + sulfur donor groups	Cr^{4+}
8	β-cyclodextrin	7-Aminoflavone + Hydrazinecarbothioamide linker	Cu^{2+}
9	β-cyclodextrin	α, β, γ, δ-tetrakis(3,5-dibromo-2-hydroxylphenyl)porphyrin	Pb^{2+}
10	β-cyclodextrin	$Fe_3O_4@SiO_2$	Zn^{2+}
11	β-cyclodextrin	Quinolinotriazole + Adamantane carboxylic acid	Cd^{2+}
12	β-cyclodextrin	N-(2-methyl-8-aminoquinolyl)-p-aminobenzene sulfonamide	Hg^{2+}
13	β-cyclodextrin	8-hydrozy quinoline	Cd^{2+}
14	β-cyclodextrin	Thiosemicarbazone moieties	Cu^{2+}
15	β-cyclodextrin	Rhodamine-B-hydrazide	Au^{3+}
16	β-cyclodextrin	Syringaldehyde di-Schiff azine	Cu^{2+}, Pb^{2+}
17	γ cyclodextrin	2-Hydroxy benzaldehyde azine	Cu^{2+}
18	γ cyclodextrin	Podand fluoroionophore	Pb^{2+}

7.9 Conclusion

This book chapter on the role of fluorescent cyclodextrins in metal sensing explicitly demonstrates that the analytical chemistry field prioritizes the advancement of systems that can perform real-time analysis. This interest is justified as it addresses the environmental concern and is simultaneously applicable in several other fields also Cyclodextrins have proven to be highly effective chemosensors for the detection of heavy metals. Their unique molecular structure allows them to form inclusion complexes with metal ions, leading to enhanced selectivity and sensitivity in detection. This makes cyclodextrins even better at selectively binding to heavy metals by adding specific ligands to them. This has opened up new avenues for the development of efficient and reliable detection methods for heavy metal pollutants in various environmental samples. Moreover, the use of cyclodextrins as chemosensors offers several advantages, such as low cost, simplicity of synthesis, and potential for miniaturization. The turn-on and turn-off fluorescent properties of CDs on the formation of inclusion complexes with heavy metals, the stability of inclusion complexes, biocompatibility, and non-toxic nature make it a primary choice of fluorescent chemosensor for researchers. In spite of these advantages, some of the limitations of cyclodextrins acting as chemosensors need to be addressed. It includes instability at extreme pH and temperatures and interference by other ions and similar ions. However, further research is necessary to enhance the performance of cyclodextrins as chemosensors, particularly in terms of improving detection limits and real-time monitoring. Nonetheless, the application of cyclodextrins in heavy metal detection holds great promise for addressing the growing concerns of heavy metal contamination in our environment and ensuring the safety of human health. Cyclodextrins have future directions towards designing novel cyclodextrin derivatives and also investigating the guest host interaction through computational studies and also development of cyclodextrins as nanosensors.

References

1. Tchounwou PB, Yedjou CG, Patlolla AK, Sutton DJ, (2012) Heavy metal toxicity and the environment. 133–164
2. Mitra S, Chakraborty AJ, Tareq AM, Bin ET, Nainu F, Khusro A, Idris AM, Khandaker MU, Osman H, Alhumaydhi FA, Simal-Gandara J (2022) Impact of heavy metals on the environment and human health: novel therapeutic insights to counter the toxicity. J King Saud Univ Sci 34:101865. https://doi.org/10.1016/j.jksus.2022.101865
3. Balali-Mood M, Naseri K, Tahergorabi Z, Khazdair MR, Sadeghi M (2021) Toxic mechanisms of five heavy metals: mercury, lead, chromium, cadmium, and arsenic. Front Pharmacol 12. https://doi.org/10.3389/fphar.2021.643972
4. Mahurpawar M (2015) Effects of heavy metals on human health. Int J Res Granthaalayah 3:1–7. https://doi.org/10.29121/granthaalayah.v3.i9SE.2015.3282
5. Sharma RK, Agrawal M (2005) Biological effects of heavy metals: an overview. J Environ Biol 26:301–313

6. Xu X, Yang S, Wang Y, Qian K (2022) Nanomaterial-based sensors and strategies for heavy metal ion detection. Green Anal Chem 2:100020. https://doi.org/10.1016/j.greeac.2022.100020
7. Ikeda H (2020) Diversely responsive turn-on fluorescent cyclodextrin chemosensors: guest selectivities and mechanism insights. Chemosensors 8:48. https://doi.org/10.3390/chemosensors8030048
8. Hulanicki A, Glab S, Ingman F (1991) Chemical sensors: definitions and classification. Pure Appl Chem 63:1247–1250. https://doi.org/10.1351/pac199163091247
9. Shanmuga Priya A, Sivakamavalli J, Vaseeharan B, Stalin T (2013) Improvement on dissolution rate of inclusion complex of Rifabutin drug with β-cyclodextrin. Int J Biol Macromol 62:472–480. https://doi.org/10.1016/j.ijbiomac.2013.09.006
10. Munisamy M, Rathinam B, Shanmugasundaram E, Ganesan V, Narayanan V, Balakrishnan SB, Kaliyamoorthy S, Thambusamy S (2023) β-cyclodextrin-encapsulated rhodamine derivatives core-shell microspheres—based fluorescent sensor for Au^{3+} and template for generating microplates of gold. Micromachines (Basel) 14:1443. https://doi.org/10.3390/mi14071443
11. Loftsson T, Duchene D (2007) Cyclodextrins and their pharmaceutical applications. Int J Pharm 329:1–11. https://doi.org/10.1016/j.ijpharm.2006.10.044
12. Dongare P, Kolekar G, Gore A (2021) A Systematic review on fluorescent and colorimetric chemosensors for ion recognition. Curr Chin Sci 1:494–512. https://doi.org/10.2174/2210298101666210831162500
13. Yu X, Wan S, Wu W, Yang C, Lu W (2022) γ-Cyclodextrin-based [2]rotaxane stoppered with gold(I)–ethynyl complexation: phosphorescent sensing for nitroaromatics. Chem Commun 58:6284–6287. https://doi.org/10.1039/D2CC02256G
14. Sivakumar K, Parinamachivayam G (2022) Review on β-cyclodextrin inclusion complex based chemosensors for heavy metals. J Incl Phenom Macrocycl Chem 102:603–618. https://doi.org/10.1007/s10847-022-01153-2
15. Goswami P, Das DK (2012) A new highly sensitive and selective fluorescent cadmium sensor. J Fluoresc 22:391–395. https://doi.org/10.1007/s10895-011-0971-7
16. Poomalai S, Govindaraj TS, Soundrapandian S, Paulraj MS, Enoch IVMV (2018) A new fluorescent chemosensor for cadmium(II) based on a pyrene-appended piperidone derivative and its β-cyclodextrin complex. Luminescence 33:538–544. https://doi.org/10.1002/bio.3443
17. DiScenza D, Culton E, Verderame M, Lynch J, Serio N, Levine M (2017) Towards rational chemosensor design through improved understanding of experimental parameter variation and tolerance in cyclodextrin-promoted fluorescence detection. Chemosensors 5:34. https://doi.org/10.3390/chemosensors5040034
18. Aoyagi T, Ikeda H, Ueno A (1999) Synthesis and molecular recognition abilities of modified γ-cyclodextrins bearing two pyrene moieties. In: Proceedings of the ninth international symposium on cyclodextrins. Springer Netherlands, Dordrecht, pp 137–140
19. Ikeda H, Murayama T, Ueno A (2005) Skeleton-selective fluorescent chemosensor based on cyclodextrin bearing a 4-amino-7-nitrobenz-2-oxa-1,3-diazole moiety. Org Biomol Chem 3:4262. https://doi.org/10.1039/b508477f
20. Liu Y, Shi J, Guo D-S (2007) Novel permethylated β-cyclodextrin derivatives appended with chromophores as efficient fluorescent sensors for the molecular recognition of bile salts. J Org Chem 72:8227–8234. https://doi.org/10.1021/jo071131m
21. Zhang Y, Li Q, Guo J, Li Y, Yang Q, Du J (2015) A novel magnetic fluorescent chemosensor for Cu^{2+} based on self-assembled systems of azobenzene and α-cyclodextrin. RSC Adv 5:66674–66680. https://doi.org/10.1039/C5RA11251F
22. Maniyazagan M, Rameshwaran C, Mariadasse R, Jeyakanthan J, Premkumar K, Stalin T (2017) Fluorescence sensor for Hg^{2+} and Fe^{3+} ions using 3,3′–dihydroxybenzidine:α–cyclodextrin supramolecular complex: characterization, in-silico and cell imaging study. Sens Actuators B Chem 242:1227–1238. https://doi.org/10.1016/j.snb.2016.09.093
23. Zhang Y-M, Chen Y, Li Z-Q, Li N, Liu Y (2010) Quinolinotriazole-β-cyclodextrin and its adamantanecarboxylic acid complex as efficient water-soluble fluorescent Cd^{2+} sensors. Bioorg Med Chem 18:1415–1420. https://doi.org/10.1016/j.bmc.2010.01.024

24. Liu Y, Yu M, Chen Y, Zhang N (2009) Convenient and highly effective fluorescence sensing for Hg^{2+} in aqueous solution and thin film. Bioorg Med Chem 17:3887–3891. https://doi.org/10.1016/j.bmc.2009.04.018
25. Narayanan V, Ganesan V, Shanmugasundram E, Durganandini S, Vellaisamy K, Amirthalingam H, Rajamohan R, Thambusamy S (2023) Spectral studies of synthetic symmetric syringaldehyde di-Schiff azine with β-cyclodextrin inclusion complexes as a fluorescent probe for the detection of metal ions. J Photochem Photobiol A Chem 445:115069. https://doi.org/10.1016/j.jphotochem.2023.115069
26. Li Q, Zhang Y, Jin Y, Yang Q, Du J, Li Y (2015) Fluorescent magnetic nanosensors for Zn^{2+} and CN^- in aqueous solution prepared from adamantane-modified fluorescein and β-cyclodextrin-modified Fe_3O_4 @SiO_2 via host–guest interactions. RSC Adv 5:68815–68821. https://doi.org/10.1039/C5RA12258A
27. Wang J, Yang X-Q, Li N, Wang L-L, Xu X-Y, Zhang C (2023) A cyclodextrin-based turn-off fluorescent probe for naked-eye detection of copper ions in aqueous solution. Spectrochim Acta A Mol Biomol Spectrosc 287:122069. https://doi.org/10.1016/j.saa.2022.122069
28. Rasheed T, Bilal M, Nabeel F, Iqbal HMN, Li C, Zhou Y (2018) Fluorescent sensor based models for the detection of environmentally-related toxic heavy metals. Sci Total Environ 615:476–485. https://doi.org/10.1016/j.scitotenv.2017.09.126
29. Fakayode SO, Lowry M, Fletcher KA, Huang X, Powe AM, Warner IM (2007) Cyclodextrins host-guest chemistry in analytical and environmental chemistry. Curr Anal Chem 3:171–181. https://doi.org/10.2174/157341107781023811
30. Azath IA, Pitchumani K (2013) Flavone modified-β-cyclodextrin as a highly selective and efficient fluorescent chemosensor for Cu^{2+} ions and l-histidine. Sens Actuators B Chem 188:59–64. https://doi.org/10.1016/j.snb.2013.07.009
31. Sareen D, Kaur P, Singh K (2014) Strategies in detection of metal ions using dyes. Coord Chem Rev 265:125–154. https://doi.org/10.1016/j.ccr.2014.01.015
32. Zhang Y, Xiang W, Yang R, Liu F, Li K (2005) Highly selective sensing of lead ion based on α-, β-, γ-, and δ-tetrakis(3,5-dibromo-2-hydroxylphenyl)porphyrin/β-CD inclusion complex. J Photochem Photobiol A Chem 173:264–270. https://doi.org/10.1016/j.jphotochem.2005.04.005
33. Prabu S, Mohamad S (2020) Curcumin/beta-cyclodextrin inclusion complex as a new "turn-off" fluorescent sensor system for sensitive recognition of mercury ion. J Mol Struct 1204:127528. https://doi.org/10.1016/j.molstruc.2019.127528
34. Hoda G, Isci R, Faraji S, Ozturk T (2024) β-cyclodextrin and calix[4]arene-functionalized thieno[3,2-b]thiophene based polymers as fluorescent ion sensors. Opt Laser Technol 177:111135. https://doi.org/10.1016/j.optlastec.2024.111135
35. Ochoa-Pérez M, Juárez-Gómez J, Meneses-Aparicio S, Guzmán-Hernández DS, Lara V, Rojas-Hernández A, Ramírez-Silva MT (2024) Design and optimization of a Cr(VI)-selective electrode based on a polymeric ß-cyclodextrin membrane modified with sulfur donor groups. Sens Actuators Rep 8:100226. https://doi.org/10.1016/j.snr.2024.100226
36. Hayashita T, Qing D, Minagawa M, Lee JC, Ku CH, Teramae N (2003) Highly selective recognition of lead ion in water by a podand fluoroionophore/γ-cyclodextrin complex sensor. Chem Commun 2160–2161. https://doi.org/10.1039/B305758E
37. Ganesan V, Govindasamy C, Shanmugasundram E, Vellaisamy K, Krishnan NK, Mani MK, Kaliyamoorthy S, Liu B-T, Thambusamy S (2024) 2-Hydroxybenzaldehyde azine and γ-cyclodextrin based colorimetric/fluorescent probe for highly sensitive and selective detection of Cu^{2+}. ChemPhotoChem 8:e202400170. https://doi.org/10.1002/cptc.202400170

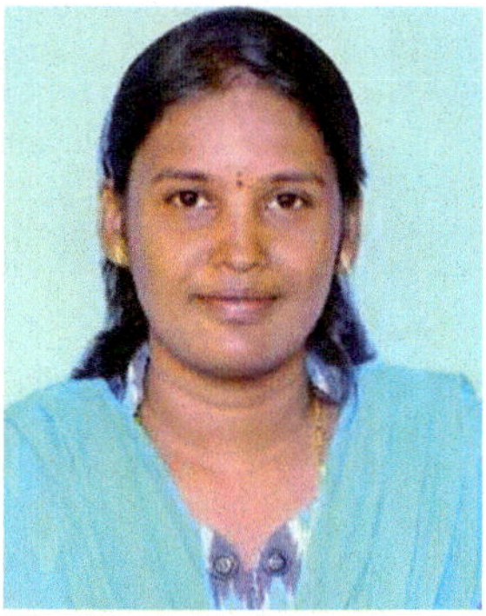

Dr. Suvarna Yenduri M.Pharm, Ph.D., working as Associate professor in department of Pharmaceutical Analysis, Sri Adichunchanagirl College of Pharmacy, Adichunchanagirl University, B.G. Nagar, Karnataka. Received her doctoral degree form Acharya Nagarjuna University and has a total of 10 years of teaching and research experience at various reputed organizations. Has 1 Indian potent and 20 peer reviewed articles in National and International journals. She is a life member at various professional bodies like IPA, APTI.

Dr. Naga Prashant Koppuravuri is a distinguished academician and researcher specializing in pharmaceutical sciences. With over 14 years of teaching and research experience, Dr. Koppuravuri has expertise in drug design and discovery, natural products, and analytical chemistry. He has published over 28 research articles in reputed international and national journals, presented his work at national and international conferences, and holds 3 Indian patents. He is a Fellow of the Association of General Education (FAGE) and has received several awards and honours for his contributions to the field of pharmaceutical sciences. Currently, Dr. Koppuravuri serves as an Associate Professor in Pharmaceutical Chemistry at Sri Adichunchanagiri College of Pharmacy, Karnataka, and actively participates in various research projects and collaborations with leading academic institutions.

Chapter 8
Fluorescence Sensing of Metal Ions Using Cyclodextrins

Nithya Mohan, Sanu K. Anand, and S. S. Sreejith

Abstract This chapter explores the principles and innovations in using modified cyclodextrins for fluorescence sensing, focusing on their role in metal detection which find applications in environmental, analytical, and biomedical fields. Cyclodextrins improve sensor performance by forming host–guest complexes and nano assemblies, enhancing solubility, selectivity, and tunability. These structures, including polymer dots and micelles, are sensitive to factors like pH and exhibit unique properties when binding with various metals. Cyclodextrins combined with quantum dots also facilitate metal ion adsorption and fluorescence detection. The chapter offers a comprehensive overview of cyclodextrins' potential in fluorescence-based metal sensing.

8.1 Introduction

Cyclodextrins (CD) are one of the naturally occurring and water-soluble macro molecules having molecular weight greater than conventional organic molecules and smaller than enzymes [1–3]. The naturally occurring cyclodextrin derivatives were first discovered in 1891 but the structure elucidations were done only after seventy-nine years in the 1970s. The cyclic arrangement of these species results in a hydrophilic outer surface and a hydrophobic central cavity [4–10].

N. Mohan (✉)
Department of Basic Science and Humanities, SCMS School of Engineering and Technology, Karukutty, Kerala, India
e-mail: nithyanedumpilly@gmail.com

S. K. Anand
Department of Applied Chemistry, Cochin University of Science and Technology, Kochi, Kerala, India

S. S. Sreejith
Department of Chemical Oceanography, Cochin University of Science and Technology, Kochi, Kerala, India

S. Krishnamoorthy (ed.), *Cyclodextrins for Chemosensing*,
https://doi.org/10.1007/978-3-031-89299-8_8

The distinctive structural characteristics of cyclodextrins confer them with remarkable properties that find applications in various industries like pharmaceuticals, food, cosmetics, and agriculture [11–29]. Their hydrophobic cavities can encapsulate and form inclusion complexes with a wide range of guest molecules, leading to improvements in the solubility, stability, and bioavailability of these guest compounds [30–32]. The inclusion of complex formation between cyclodextrins and guest molecules is driven by non-covalent interactions, including van der Waals forces, hydrogen bonding, and hydrophobic interactions. This reversible encapsulation mechanism allows for the targeted and controlled release of the guest molecules, adding a level of precision and efficiency to various applications.

Detection of metal ions is of great significance in a variety of fields like environmental, biological, and medicinal contexts. Metal ions especially heavy metals are non-biodegradable and cause a severe threat to the environment and living systems [33, 34]. The design and development of rapid, accurate, and cost-effective methods for detecting these metals is of paramount importance in minimizing or preventing exposure to them. The successful development of methodologies for the discerning detection of metal ions holds profound implications, extending from environmental monitoring to the frontiers of advanced medical diagnostics. Among the available methods, fluorescent-based sensors hold considerable appeal, not only due to their ability to quantitatively sense metal ions through a swift fluorescent response but also because this approach is rapid, nondestructive, and facilitates direct visual perception [35–39].

8.2 Overview of Fluorescence Sensing

There are three important mechanisms which explain the fluorescent recognition of the analyte (here it is metal ion) (i) chemosensors or binding site-signaling subunit approach, (ii) displacement approach, and (iii) chemodosimetric approach [40–43]. These classifications are based on either the mode of interaction between the analyte and the receptor unit or with respect to the arrangement of the binding site with the signaling subunit. Among the three different mechanisms, the chemosensor approach is widely seen in the fluorescence-based metal sensing of cyclodextrins [44, 45]. In the chemosensor approach, the binding and the signaling units are connected through a covalent bond, and the ground or excited state properties of the signaling unit may change with the approach of an analyte. Numerous photophysical mechanistic studies on fluorescent-based sensors have been reported thus far. Among them, key investigations include charge transfer (CT), photoinduced electron transfer (PET), Förster resonance energy transfer (FRET), excimer formation, and the recently explored phenomena of aggregation-induced emission (AIE) or aggregation-caused quenching (ACQ) [46–49].

The chapter provides a comprehensive insight into the intricate world of fluorescence sensing, highlighting the contributions of cyclodextrins in the detection and recognition of diverse metal ions. At the intersection of cyclodextrins, fluorescence, and metal ion sensing, we anticipate not only unveiling novel insights but also contributing to the ongoing discourse within this dynamic and interdisciplinary research domain. The chapter discusses the fluorescent recognition of metal ions in two distinct aspects, i.e.,

(i) cyclodextrin-based inclusion compounds as probes;
(ii) cyclodextrin-based nanostructures as probes.

8.3 Cyclodextrin-Based Inclusion Complexes as Sensors

The convergence of supramolecular chemistry and fluorescence technology has given rise to a particularly intriguing class of compounds—fluorescent cyclodextrin inclusion compounds [50–53] that not only showcase remarkable attributes in their own right but also hold substantial promise in the realm of metal ion sensing.

8.4 Mechanism of Inclusion Complex Formation

Cyclodextrins, cyclic oligosaccharides featuring a unique toroidal structure, have become exemplars of molecular encapsulation due to their ability to host guest molecules within their hydrophobic cavities (Fig. 8.1) [30, 54–56]. The donut-shaped molecule has a hydrophilic outer surface and possesses a hydrophobic axial opening which is capable of encapsulating guest molecules having appropriate geometries [57]. This is the basis of sensing via the formation of an inclusion compound. One or more guest molecules may become entrapped within the CD cavity, with the most frequently observed and simplest ratio between the CD and metal ion being 1:1. This structural hallmark underscores the versatility of cyclodextrins in molecular recognition, a facet that has been explored extensively across various scientific disciplines [58–61].

The amalgamation of cyclodextrins and fluorescence technology has resulted in the design of fluorescent cyclodextrin inclusion compounds for the selective and sensitive detection of metal ions [62–65]. This is of paramount importance due to the diverse roles that metal ions play in biological [66–69], environmental [70], and industrial processes [71, 72].

The most widely seen interaction of CDs for the fluorescent sensing of metal ions is the formation of inclusion compound. CDs act as hosts and the inner cavities encapsulate the metal ion by means of non-covalent interactions like hydrogen bonding, hydrophobic and van der Waals forces [31]. In the fluorescent sensing process, upon chelation of metal ions, the fluorescent properties of the host get altered without any significant alteration in the host framework. The change in emission characteristics of the host with or without blue/red shift happens on chelation [73]. Now, let's explore

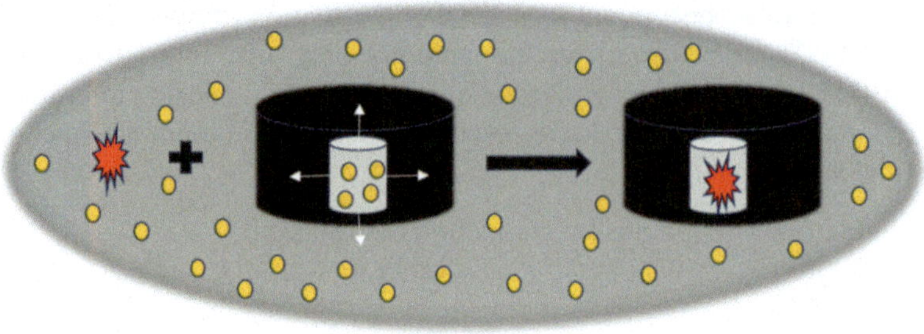

Fig. 8.1 Schematic representation of CD inclusion complex formation. The star shape is the guest molecule; the small circles represent the solvent molecules

prominent works that utilize cyclodextrin-based inclusion compounds as fluorescent sensors for metal ion detection.

8.5 Types of Cyclodextrins-Based Probes

The intriguing proposition of multimetallic sensors, with the ability for nuanced differential sensing achieved through structural tuning, adds another dimension to the realm of sensing technologies. A realization of this conceptual framework can be seen using a cyclodextrin-based inclusion compound with symmetric syringaldehyde di-Schiff azine as a fluorescent probe [74]. Here the initially prepared Schiff-based symmetrical azine derivative (SyA) is able to rapidly detect dual metal ions Cu^{2+} and Pb^{2+} via Chelation Enhanced Fluorescence Mechanism (CHEF) and Ligand to Metal Charge Transfer (LMCT) approach in optimized physiological conditions in acetonitrile solvent. Then the inclusion compound of the aforementioned compound was prepared by treating SyA with β-cyclodextrin which forms a 1:1 complex enhancing the water solubility and the development of fluorescence. This inclusion complex was able to detect Al^{3+} ion with a limit of detection of 3.2×10^{-6} M. Now changing the cyclodextrin unit into a γ-cyclodextrin and SyA with 4,4′-((1E,1′ E)-hydrazine-1,2-diylidenebis (methanylylidene)) diphenol resulted in a water-soluble inclusion compound [75]. As opposed to the earlier one, the emission intensity of the probe was quenched in the presence of Al^{3+}. On carrying out the DFT studies, it was found out the proximity of Al^{3+} and the core functional group in the sensor such as –OH and $C = N$ is the reason for the sensing and this interaction leads to the change in frontier molecular orbitals. This change is evidenced from the UV–Visible absorption spectrum as the vanishing of the absorption band at 333 nm and the blue-shift of the band at 283 nm; while in the case of fluorescence emission spectrum both quenching and shifting occur.

The medium or solvent used in the detection process also has a profound influence on sensing behavior as illustrated by the selective metal ion sensing and tuning using an antipyrine Derivative (**1**) in Aqueous and β-Cyclodextrin Media [76]. In an aqueous solution, the antipyrine exhibits enhanced fluorescence for sensing vanadyl ions, while in a cyclodextrin medium, it demonstrates sensitivity to aluminum ions. This antipyrine derivative functions as a sensor for VO^{2+} ions in water (1% DMF) and as a sensor for Al^{3+} ions in an aqueous β-CD environment. The Molecular encapsulation of the sensor influences the metal chelation center of the ligand (Fig. 8.2). During the complex formation of antipyrine derivative with β-CD, the β-CD molecule masks the $-COCH_2CO-$ moiety of the derivative, resulting in the carbonyls being unable to engage in the chelation of metal ions. The introduction of Al^{3+} significantly increased the fluorescence intensity of the sensor by nearly 25 times compared to that of the free probe. It is apparent that the encapsulating β-CD molecule shields one of the chelating atoms of the antipyrine derivative, preventing it from binding with VO^{2+}. This, in turn, allows the remaining portion of the chelating center to effectively bind with the Al^{3+} ion.

Further tailoring of the properties of the probe can be done by optimizing the probe moiety and the cyclodextrin inclusion system [77]. This optimization can provide valuable insights into the sensing mechanism as evidenced by the piperidone-based Al^{3+} ion sensor which has anthracene as the fluorescent signaling unit. This sensor shows signal enhancement in an aqueous medium via the excimer formation with Al^{3+}. But on reaction with β-cyclodextrin, it forms a 1:2 inclusion compound which quenches the excimer fluorescence. While the sensing capability is compromised with the formation of the inclusion complex, it has concurrently unveiled a pathway for confirming the generation of an excimer. This occurrence is attributed to the

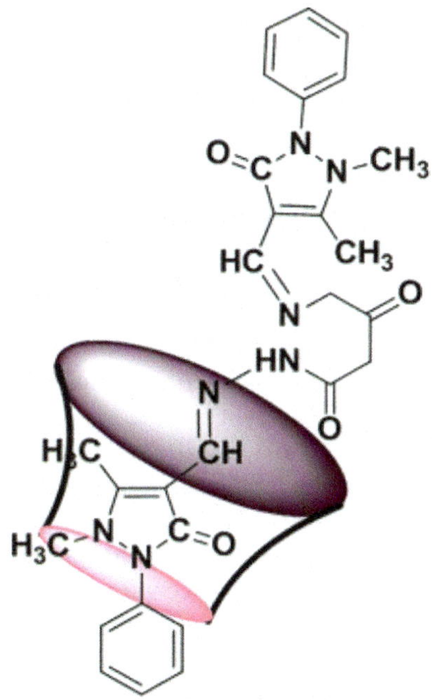

Fig. 8.2 Schematic representation of the complex formed between **1** and β-CD

persistence of the anthracene moiety, proving crucial in the sensing of Al^{3+} ions. On careful analysis, it has been found that the reason for this phenomenon is due to the disruption of $\pi\cdots\pi$ interactions between the anthracene rings. Refinement of the sensor to a simplified system (4-(diethylamino)-2-hydroxy-benzaldehyde and carbohydrazide) yielded a blue fluorescent probe, unveiling itself as a water-soluble β-cyclodextrin derived host–guest system. This transformed entity serves as a highly responsive and selective fluorescent chemosensor specifically designed for the detection of Al^{3+} ions [78]. The host–guest system exhibited exceptional bioactivity, sensing Al^{3+} in onion epidermal cells via fluorescence imaging property.

A further change in solvent to an acetonitrile water system and the replacement of non-cyclodextrin component with 3,3′–dihydroxybenzidine, and the cyclodextrin type with α–cyclodextrin resulted in the formation of a durable solid inclusion compound, marked by the specific pairing of 3,3′–dihydroxybenzidine and α–cyclodextrin, denoted as (DHB:α–CD) [79]. The synthesized compound utilizes a sophisticated integration of co-precipitation and kneading methods, thus serving as a dual-metal fluorescent "turn-on" sensor. This intricate process involves a precise 1:1 ratio between 3,3′–dihydroxybenzidine (DHB) and alpha cyclodextrin (α–CD). The synergy between the co-precipitation and kneading techniques contributes to the compound's unique properties, positioning it as a promising advancement in the realm of fluorescent sensors. This sensor (DHB:α–CD inclusion compound) exhibits a weak fluorescence in the range from 350 to 450 nm. The addition of Hg^{2+} red shifts the peak and simultaneously enhances the fluorescence. But in the case of the addition of Fe^{3+} the fluorescence enhancement is followed by a blue shift. Further studies on fluorescent titration show that these metal ions form a 1:1 complex with DHB: α–CD. It was also proved that the sensor interacts with the cytoplasmic Hg^{2+} and Fe^{3+} and produces green fluorescence. In a similar way fluorescent enhancement was achieved via binding to Ca^{2+} using another 1:1 inclusion compound synthesized using anthracene piperidin-4-one derivative and β-cyclodextrin (PP: β-CD) [80]. While the absorption spectra of the inclusion compound remain largely unchanged upon the introduction of various metal ions, the fluorescent studies reveal a noteworthy augmentation in the emission, particularly with the addition of Ca^{2+} ions. This phenomenon can be attributed to the interaction taking place between the PP and β-CD within the inclusion compound. On step-wise addition of β-CD onto PP, it was found that there is a hyperchromic shift in the absorption spectrum which indicates the interaction of fluorophore and the host molecule (β-CD). At the same time a fluorescent enhancement is seen during the host: guest compound formation along with a blue shift of about 3 nm. This is due to the dislodging of PP molecules from the aqueous media to the apolar cavity of β-CD. The encapsulation of the PP molecule by β-CD is limited to anthracene moiety alone and imino nitrogen at position 4 is exposed for complexation with Ca^{2+} ion. Thus, the inclusion phenomenon enhances the fluorescence sensing of PP and aids in the facile detection of the metal ion in β-CD media rather than in aqueous media. A very similar fluorescent enhancement was observed for another sensor based on pyrene-appended piperidone derivative [81] in β-CD media but this time the sensor was selective towards Cd^{2+} rather than Ca^{2+} ion. The inclusion compound proves instrumental in the determination of Cd^{2+} in water samples collected from the environment. The analytical procedure involves treating the collected water samples (25 ml) with concentrated nitric acid (6.5 ml), followed by heating. Upon evaporation, the sample is dried to the residue, which is then reconstituted in distilled water and filtered.

Notably, the absorption spectrum exhibits hypochromic shifts in the bands upon the addition of various metal ions, with the exception of Cd^{2+}. In the presence of Cd^{2+}, a distinctive hyperchromic effect is observed, setting it apart among the array of metal ions and significantly influencing the absorption spectrum of the inclusion compound. The fluorescence studies mirror this distinction, with the addition of Cd^{2+} ions showcasing a pronounced enhancement in fluorescence, surpassing the effects observed with other tested metal ions. This underscores the unique and heightened sensitivity of the inclusion compound, particularly in response to Cd^{2+} ions, making it a valuable tool in modulating both absorption and fluorescence spectra for precise environmental analysis.

The development of environmentally benign sensors for copper metal ions is of utmost importance, given their integral role in numerous biological processes, including various redox reactions and enzyme functions. This remarkable achievement is realized through the incorporation of the quercetin–2-hydroxypropyl-β-cyclodextrin inclusion system (Q-CD) in the presence of Cu^{2+} ions [82]. The fluorescent emission of quercetin (Q) was found to enhance in the presence of 2-hydroxypropyl-β-cyclodextrin (CD). Here the presence of Cu^{2+} ions quenches the aforementioned luminescence which results in the selective detection of this metal ion among others. This is due to the intramolecular charge transfer (ICT) which takes place in the Q-CD-Cu(II) complex. The shift in the absorption spectra (379 nm was shifted to 430 nm) indicates that the B-ring cinnamoyl system interacts with the Cu^{2+} ion, and the complexation process is first initiated by the sites involving 3-hydroxy and 4-carbonyl groups (Fig. 8.3). Further, the real-time application of the sensing was demonstrated via the detection of Cu^{2+} ion in vegetables, fruits, rivers, and lakes. In a similar fashion, an inclusion compound synthesized using 4-((benzo[d]thiazol-2-ylthio)methyl)-5,7-dihydroxy-2H-chromen-2-one (BDC) and β-cyclodextrin exhibited a turn-off fluorescence in presence of copper ions [83]. In this context also, the phenomenon of intramolecular charge transfer (ICT) is the driving force behind the quenching observed. DFT calculations show that this transfer occurs from thiazolidine unit to the coumarin unit. This is due to the fact that on inclusion, the thiazolidine group penetrates inside the primary side of the β-CD cavity and the coumarin group of BDC stays outside. This enables the binding of the Cu^{2+} ion with the sulfur atom in the 2-position of thiazolidine ring, thiazolidine nitrogen, the 5-hydroxyl group of coumarin moiety, and a primary hydroxyl group of β–Cyclodextrin which results in enhanced fluorescence quenching.

Further cyclodextrin moieties are also used for sensor stabilization. For instance, a naturally occurring and unmodified simple chemosensor is found to be stabilized by the incorporation of β-cyclodextrin (β-CD) thus forming an inclusion compound. This sensor (2-hydroxy-1,4-naphthoquinone (2HNQ)), demonstrates efficacy in both colorimetric detection of Hg^{2+} and fluorescent (on–off) detection of pH [84]. The strategic incorporation of probe is instrumental in solubilizing and stabilizing the

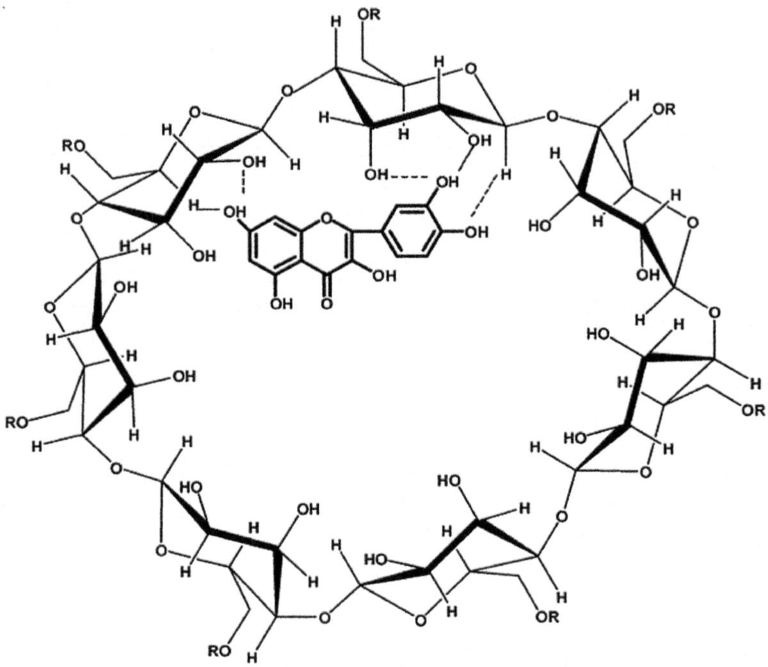

Fig. 8.3 The most probable structure of the Q-CD-Cu(II) complex

sensor in aqueous solutions, significantly expanding its multifaceted applicability. The inclusion compound of the sensor with β-CD exhibited excellent sensing properties at around biological pH. Spectral studies show that the interaction occurs between OH and CaO of 2HNQ and the Hg^{2+} ion.

Ratiometric sensors based on various emission pathways can be developed using cyclodextrin inclusion systems. For instance, a FRET-based cyclodextrin supramolecular–Rhodamine dye complex system [63] is developed which functions as a ratiometric fluorescent probe towards Fe^{3+} detection in aqueous medium. The dye system consists of an adamantyl-linked spirolactam rhodamine (AD-SRhB) and the β-cyclodextrin system is a dansyl-linked β-cyclodextrin (β-CD-DNS) one. The emission of the dansyl moiety matches the absorption of the cyclodextrin system and acts as the donor while the rhodamine system acts as the energy acceptor. The donor–acceptor distance is within the Föster distance which facilitates the ratiometric determination of ferric ion. This ideal distance also limits other undesirable interactions like π-π interactions between donor and acceptor.

8.6 Cyclodextrin-Based Nanocompounds as Sensors

A breakthrough in the development of nanomaterials has influenced researchers for the development of fluorescent nanosensors for various applications. These nanomaterials easily overcome many of the shortcomings of conventional fluorescent tags such as photobleaching, low stability, narrow absorption, broad emission, etc. [85–88]. A plethora of nanomaterials that include metal nanoparticles, metal nanoclusters, semiconductor quantum dots, carbon dots, polymer dots, their hybrid forms, etc. with submicron-sized dimensions and excellent physicochemical properties have introduced new horizons to fluorescence sensing applications [87–90]. Functionalization of these nanomaterials tunes their luminescence characteristics, makes them more water soluble, and prevents aggregation, which endows them for better sensor performances [91]. Moreover, certain capping agents can specifically interact with analytes of interest, resulting its selective and specific determination [91]. As a ligand, cyclodextrin (CD) is very promising as it is a water-soluble, commercially available, non-toxic, cost-effective, and easily functionalized macrocyclic host [92–95].

As mentioned before, CDs possess a hydrophilic exterior cavity and a hydrophobic interior cavity [96]. This hydrophobic inner cavity is capable of forming stable complexes via the host–guest interaction with its solution partners [94]. Also, it can easily functionalize other materials through the hydrophilic interaction, hence the physical as well as chemical properties of this functionalized materials will be greatly improved [94, 96, 97]. Functionalization of CDs with nanomaterials reduces their surface defects thereby exhibiting higher photostability and displaying superior sensor performances [98].

8.7 Cyclodextrin-Nanocarbon-Based Sensors

The unique optoelectronic properties of carbon-based nanomaterials have inspired researchers for hybridizing them with CDs for the development of various chemical sensors [94]. As an emerging material, the use of graphene and its derivatives such as graphene oxide, reduced graphene oxide (RGO), and graphene quantum dots have been encouraged for various applications. They are amenable to conjugating with CDs via the covalent interaction between the -OH group of CD and -COOH group of GO (Fig. 8.4), thereby extending the performances of both materials [99, 100]. Because of the high surface-to-volume ratio and numerous oxygen functionalities, this nanocomposite is widely employed as an adsorbent for the removal of toxic metal ions such as Pb^{2+}, $UO_2{}^{2+}$, Ni^{2+}, and Cd^{2+}in water samples [99, 101–103]. On the other hand, Reduced GO (RGO) is reported to be an efficient carrier of CD molecules, and the CD-RGO complex was utilized for the efficient removal of Cu^{2+} ions from wastewater, where a square planar geometry is formed between the Cu^{2+} ion and oxygen groups CD-RGO system [104]. Moreover, the incorporation of CDs into nanocarbons such as fullerenes makes them water soluble, enabling them suitable

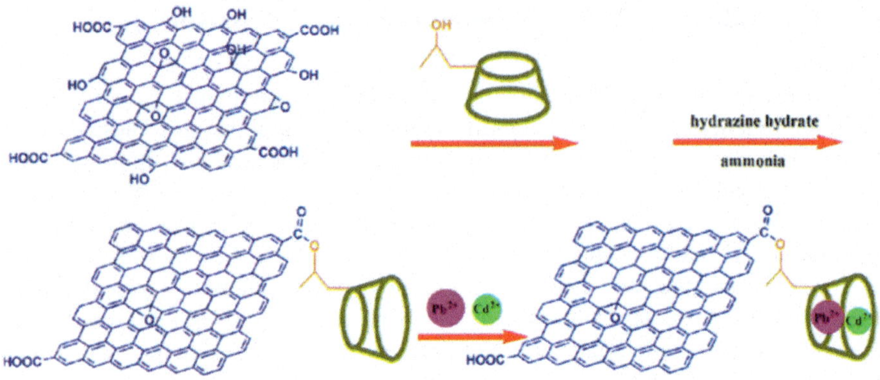

Fig. 8.4 The binding interaction between reduced GO and CD

for sensing applications [105]. There are reports for the formation of fullerene-CD conjugates via the covalent binding with CD amines or CD azides which results in enhanced water solubility and biocompatibility [105, 106]. Molecularly imprinted Polymers (MIPs) are synthetic polymers having tailor-made binding sites which are complementary to the size, shape and chemical functionalities of the target analyte [107–110]. The well-defined host–guest characteristics of CDs can be effectively used for MIP fabrication. For instance, Sedghi et al. designed a MIP sensor based on β-CD and GO for diphenylamine determination, where a stable β-CD-analyte complex is formed at the GO surface [111]. The same strategy can be adopted for the development of specific sensors for metal ion detection.

Functionalized Carbon nanotubes (CNTs) and CDs form well-known nanocomposites due to their outstanding properties such as excellent chemical and thermal stability, mechanical strength, and conductivity [112]. Murthy et al. made use of amino-functionalized CDs and carboxyl-functionalized CNTs for the generation of β-CD-CNT nanoconjugates through the specific amide linkage [113]. The resulting nanocomposite exhibited excellent water solubility and provided active sites for the interaction of analyte molecules [113].

Carbon quantum dots (CQDs) are good choice of fluorescence probes due to the availability of cost-effective raw materials, convenient synthetic procedure, strong fluorescence, and excellent photostability [98]. The low toxicity as well as biocompatibility of CQDs replace other semiconductor quantum dots in real-time applications [98]. CQDs can easily assembled with β-CD without losing its luminescence characteristics [98]. These functionalized CQDs were reported to be used for the detection as well as adsorption of hazardous metal ions such as Cd, Cu, Pb, and Ni in aqueous medium [98]. A highly photostable β-CD capped CQDs were reported with a high quantum yield of 67.2% where the trap states of CQDs were occupied by the CD molecules via amide bond and have been utilized for the fluorescence detection of Hg^{2+} and Fe^{3+} ions [114]. Moreover, the fluorescence intensity was stable between the pH range of 4–10 with a high salt tolerance. The sensor was highly selective for Hg^{2+} and Fe^{3+} ions and determination was attained in the micromolar range with a low detection limit [114]. Fluorescence sensing can also be achieved by the formation of stable chelate complexes with metal ions and probes.[116] In one of the reports, fluorescent CQDs capped with β-CD and isophorone diisocyanate formed a chelate complex with Cr (VI) ions via its –COONH– group, resulting in fluorescence quenching [115]. They have determined the toxic Cr ions in the micromolar range and also demonstrated its real-time applicability in water and soil samples [115].

Polymer dots (PDs) belong to the class of CQDs which are synthesized from carbon-containing polymeric compounds like CDs [116]. PDs are known to be characterized by spherical morphology with particle size less than 10 nm with improved optical properties. CDs became a good choice for their nucleation because the resulting PDs possess the molecular recognition ability of CDs as well as the optical properties of QDs [116]. These nanostructured PDs were effectively exploited for the sensitive determination of various metal ions. The decorated surface groups decide the fluorescence sensitivity of PDs to their target metal ions. Reports suggest that oxygen, sulfur, and nitrogen-decorated PDs efficiently monitor Ag^{+}, Cu^{2+}, and Fe^{3+} ions, respectively, in which fluorescence intensity is quenched in the presence of these metal ions [116–118].

8.8 Cyclodextrin-Metal Nanoparticles-Based Sensors

Metal nanoparticles (MNPs) find tremendous interest as chemical sensors mainly due to their size-dependent optical properties [119]. Synthesis of stable nanoparticles is the key factor for the development of ultrasensitive sensors for their targets. Functionalization results in the synthesis of smaller, stable, and conductive nanoparticles with sharp surface plasmon resonance peaks [120]. Hence, considerable attention has been made to the modification of these colloidal particles with monolayers of organic substrates like CDs. The well-defined molecular recognition ability of CDs to its solution partners helps in the development of efficient fluorescence sensors [121, 122]. The synthetic route involves the reduction of metal ions with suitable reducing agents followed by the addition of CD and in many cases CD itself acts as both reductant and stabilizer [120, 123]. As CD exists in their α, β, and γ structure, the β form is widely utilized due to its structural peculiarities and cost-effective nature [124]. β-CD acts as both reducing as well as stabilizing agents during the synthesis of MNPs [124]. The 21 –OH groups of β-CD undergo deprotonation in a basic medium and which in turn is responsible for the reduction of metal ions [124]. Nanoparticles of gold, platinum and silver with surface-attached CDs were reported, where the targets were determined specifically due to the pre-designed host characteristic of CDs [121]. Bindhu et al. used monodispersed β-CD capped AuNPs for the detection of Cu^{2+} ions [120]. Experimental results demonstrated that Cu^{2+} is co-ordinating with CD molecules thereby losing the stability of the system and inducing aggregation of AuNPs [120]. Similarly, the chelating affinity of β-CD towards Pb^{2+} was used for its determination in β-CD-AuNPs system, resulting in the aggregation of AuNPs [125].

In addition, the large surface area of this compound resulted in antibacterial action, where β-CD acts as the host for bacterial action [120]. As stated before, supramolecular chemistry was well utilized for the synthesis of β-CD-piperidine-4-one loaded AuNPS, where the fluorenyl ring is encapsulated by CD molecules [126]. Chelation-enhanced fluorescence is well executed for the sensing of Cd^{2+} ions in which the detection limit went down to the 10^{-8} M range [126].

AgNPs are always a hotspot in fluorescence sensors owing to their ultrasensitive determination ability for their targets. Water soluble α-CD capped AgNPs were synthesized using $NaBH_4$ as a reducing agent, in which the hydrophobic interaction with α-CD makes the AgNPs stable. γ-CD decorated AgNPs adopt a one-pot synthesis, where γ-CD acts as both a reducing and stabilizing agent in a basic medium [127]. The sensor has demonstrated good selectivity and sensitivity towards Hg^{2+} ions based on the redox reaction between Ag(0) and Hg (II) [127]. In addition, visual detection of metal ions was also reported, where its characteristic SPR absorption will be decreased or shifted. For instance, Rajamanikandan et.al. proposed the colorimetric detection of Hg^{2+} ions, where the SPR peak at 400 nm is decreased in the presence of Hg^{2+} ions along with a noticeable blue shift [124]. Under the optimal analytical conditions, the sensor-enabled the determination of Hg^{2+} ions in a wide linear range with a nanomolar detection limit [124]. Similarly, sodium dodecyl sulphonate-β-CD-AgNPS conjugate was probed for the naked eye detection of Cr^{2+}, where a color change of yellow to orange is observed [119]. Here, sodium dodecyl sulphonate- β-CD inclusion complex was used as the capping agent, and at pH 13, deprotonation of CD takes place resulting in the formation of CD-Cr^{2+} complex [119]. Also, CD-based magnetic nanoparticles were effectively used for the removal of heavy metals such as Pb^{2+}, Cd^{2+}, and Ni^{2+} based on the adsorption process via the electrostatic force of attraction between the -COO^- group of the nanoadsorbent and metal ions [128].

A nanomolar detection of highly toxic, Cr^{6+} was performed using bimetallic NPs of Au and Fe [129]. They have a large surface area and synergistic properties of their individual counterpart compared to monometallic forms, which ultimately helps to monitor the metal ions in lower detection limits. In this synthetic procedure, CD molecules not only replace the toxic reducing agents like $NaBH_4$, to reduce the metal ion to a metallic state but also act as stabilizing agents, to make it more stable, soluble, and reactive [129]. The strong interaction between the Cr^{6+} and –OH groups of β-CD, weakens the bonding between β-CD and Au-Fe NPs, resulting in the aggregation of NPs (Fig. 8.5) [129].

Metal nanoclusters (MNCs) belong to the class of MNPs, consisting of a few to hundred atoms where the particle size is even reduced to 2 nm [130, 131]. This ultra-small size of MNCs is comparable with the Fermi wavelength of electrons and exhibits molecule-like properties, thereby showing unique optical, electrical, and chemical features [130, 132]. CDs can easily conjugate with metal nanoclusters to form nanoassembles thereby increasing the catalytic efficiency and sensitivity of the fluorescence sensors [133]. The fluorescence intensity, quantum yield, stability, water solubility, large stoke shift, etc. have greatly enhanced in CD conjugated CuNCs compared to free CuNCs [133]. This probe was effectively utilized for

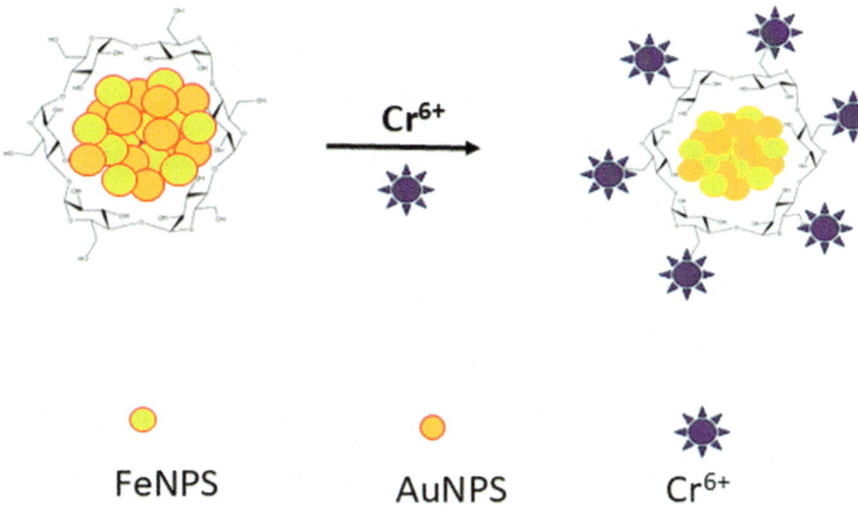

Fig. 8.5 Mechanism of the β-CD/Au-FeNPs probe for the sensing of Cr^{6+} ions

the fluorescence as well as colorimetric determination of Fe^{3+} ions in the micromolar range [133]. The specific thiol-Cu^{2+} ion interaction was employed by Zhong and his co-workers for the synthesis of monodispersed, blue-emitting 6-SH-β-CD protected CuNCs.

Metal oxide nanoparticles (MONPs) are another important class of nanostructured materials due to its enzyme-like catalytic efficiency [134, 135]. Their increased surface reactivity is due to the presence of a higher percentage of atoms at its surface. As discussed before, surface functionalization greatly influences physicochemical properties and their sensitivity. For the synthesis of water-soluble MONPs, surface functionalization with molecules like CDs is vital because its water solubility is a major concern for different applications [136]. The hydrophilic exterior of CD makes the trapped MONPs water soluble thus increasing the scope of applications [136]. Their antimicrobial studies are widely exploited especially for iron oxide NPs [136]. Like other nanosystems, the synthesis of CD-MONPs adopts a one-pot synthesis in the presence of suitable reducing agents like $NaBH_4$. The photochemical as well as thermal stability of β-CD-Fe_2O_3 NPs has been studied and reported to be an outstanding fluorescence probe for photochemical reactions, mainly as an adsorbent for the removal of environmental contaminants from water samples [137, 138]. In another report, crystalline, monodispersed β-CD conjugated GaO(OH) NPs displayed excellent biological applications compared to free GaO(OH) NPs [139]. Also, TiO_2 NPs are well-studied photocatalysts [140]. CD-capped TiO_2 was used for the photocatalytic reduction of the toxic Cr (VI) ions. Highly stable CD-conjugated MONPs such as Fe_3O_4 NPs, Co_3O_4 NPs, CeO_2, and NPs have been reported for several biomedical analyses [134, 141, 142].

8.9 Cyclodextrin-Quantum Dots Based Sensors

Semiconductor quantum dot (QDs) is an active area of research in nanotechnology and have been greatly explored in various applications due to their well-established quantum confinement property and tunable emission [143]. Broad absorption, narrow emission, high quantum yield, high photostability, etc. make them excellent for optical applications [123, 144]. Capping of QDs with organic molecules reduces their trap state emission and provides desired surface groups, making them an efficient probe in fluorescence sensing [145]. In most of the QD-based systems, the fluorescence of the QDs was quenched in the presence of metal ions. This could be due to several reasons such as, (a) metal ions can form complexes with the adsorbed anions from the buffer solution in the cavities of CD (b) metal ions can directly bind with –OH groups of CD, (c) specific electron transfer between QDs and metal ions, which in turn results in the development of fluorescence turn-off sensors [146]. Efficient FRET-based CD-QDs systems with several organic molecules were reported in which the detection went down to lower ranges [147]. The chemistry behind the functionalized CDs and QDs is well-defined. For instance, thiol-cadmium interaction has been utilized for the synthesis of thiolated β-CD capped CdSe-CdS QDs, resulting in stable and soluble compounds [145]. Their superior photophysical properties can be utilized for investigating their interactions with chemical or biological molecules in aqueous solvents [145]. Similarly, the water solubility of QDs increases with carboxyl functionalized QDs [148]. For instance, highly dispersed β-CD-L-arginine stabilized CdSe, CdSe/CdS, and CdSe/ZnSe are synthesized, where –COOH group of β-CD-L-arginine is bonded to the metal ion which is confirmed using FTIR spectroscopy [148]. Also, stable QDs can be prepared by ligand exchange mechanism [146]. For example, Jin and co-workers successfully replaced oleic acid from the CdSe surface for the synthesis of β-CD-CdSe QDs and used for the sensing of Ag^{+}, Hg^{2+}, and Co^{2+} via the fluorescence quenching mechanism [146].

Polymers of chitosan and β-CD modified Sulphur quantum dots (SQDs) have rectified many of the limitations such as poor water solubility and biocompatibility of SQDs which have been utilized for the effective quantification of Ag^{+} ions [149]. The nanocomposite aggregated in the presence of Ag^{+} ions via the specific S–Ag^{+}-S interaction resulting in an aggregation-induced fluorescence quenching [149]. Similarly, Wang et al. made use of aggregation-induced fluorescence quenching for the nanomolar detection of Mn^{2+} ions using a polymer-carbon dot nanocomposite between carboxymethyl chitosan β-CD and nitrogen-doped carbon dot [150]. This capped polymer shell provides active groups to the CD surface and quickly aggregates in the presence of the target Mn^{2+} via the specific electrostatic force of attraction [150]. A ratiometric fluorescence sensor has been reported for the selective recognition of Hg^{2+} and Cu^{2+} based on a nanocomposite of β-CD based CQDs and a polymeric micelle encapsulated CdSe/ZnS QDS [91]. The host–guest chemistry between β-CD and azo groups of polymeric micelle results in a stable conjugated system with blue, green, and red emissions [91]. The presence of Hg^{2+} and Cu^{2+} selectively quenches the fluorescence intensity of red emission due to the excited state electron transfer from the electron-rich QD moieties to the electron-deficient Hg^{2+} and Cu^{2+}.

8.10 Conclusions

In conclusion, the utilization of cyclodextrin-based fluorescent sensors, particularly in the form of nanoclusters and inclusion compounds, has demonstrated remarkable efficacy in the realm of metal ion detection. The unique structural characteristics of cyclodextrins, such as their ability to form inclusion complexes and nanoclusters, contribute significantly to the enhanced sensitivity and selectivity observed in these sensors. The incorporation of cyclodextrin nanoclusters in fluorescent sensing platforms has shown great promise in improving the overall performance of metal ion detection. The synergistic effects arising from the cooperative interactions within the nanoclusters and the encapsulation of fluorophores lead to amplified signal responses, making these sensors highly sensitive to trace amounts of target metal ions. Moreover, the modular nature of cyclodextrin derivatives allows for tailoring the nanocluster properties, offering a versatile approach to address specific detection requirements.

Similarly, the design of inclusion compounds involving cyclodextrins has provided a robust framework for developing fluorescent sensors with high metal ion selectivity. The host–guest interactions within the inclusion complexes not only enhance the stability and solubility of the fluorescent probes but also contribute to the selectivity of the sensor towards specific metal ions. This design strategy has proven effective across various metal ions, underscoring the versatility and applicability of cyclodextrin-based inclusion compounds.

Looking ahead, further advancements in cyclodextrin-based fluorescent sensors for metal ion detection could explore additional modifications of cyclodextrin structures, enabling fine-tuning of sensor properties. Additionally, efforts to enhance the biocompatibility of these sensors could pave the way for applications in biological and medical contexts. The continued exploration and refinement of cyclodextrin-based fluorescent sensors, particularly in the form of nanoclusters and inclusion compounds, hold considerable promise for addressing the evolving challenges in metal ion detection across diverse fields, ranging from environmental monitoring to biomedical diagnostics.

References

1. Szejtli J (1998) Introduction and general overview of cyclodextrin chemistry. Chem Rev 98:1743–1754
2. Saenger W (1980) Cyclodextrin inclusion compounds in research and industry. Angew Chem, Int Ed Engl 19:344–362
3. Bender ML, Komiyama M (2012) Cyclodextrin chemistry. Springer Science & Business Media

4. Egyed O (1990) Spectroscopy studies on β-cyclodextrin. Vib Spectrosc 1:225–227
5. Flourié B, Molis C, Achour L, Dupas H, Hatat C, Rambaud JC (1993) Fate of beta-cyclodextrin in the human intesntine. J Nutr 123:676–680
6. Bezerra FM, Lis MJ, Firmino HB, Dias da Silva JG, Curto Valle RCS, Borges Valle JA, Scacchetti FAP, Tessaro AL (2020) The Role of beta-cyclodextrin in the textile industry-review. Molecules 25:3624
7. Saokham P, Loftsson T (2017) γ-Cyclodextrin. Int J Pharm 516:278–292
8. Li Z, Wang M, Wang F, Gu Z, Du G, Wu J, Chen J (2007) Gamma-cyclodextrin-a review on enzymatic production and applications. Appl Microbiol Biotechnol 77:245–255
9. Domiński A, Konieczny T, Kurcok P (2019) α-cyclodextrin based polypseudorotaxane hydrogels. Materials 13:133
10. van der Veen BA, Uitdehaag JC, Penninga D, van Alebeek G-JW, Smith LM, Dijkstra BW, Dijkhuizen L (2000) Rotational design of cyclodextrin glycosyltransferase from bacillus circulans strain 251 to increase alpha-cyclodextrinproduction. J Mol Biol 296:1027–1038
11. Szejtli J (1992) The properties and potential uses of cyclodextrin derivatives. J Incl Phenom Mol Recognit Chem 14:25–36
12. Del Valle EM (2004) Cyclodextrin and their uses. Process Biochem 39:1033–1046
13. Davis ME, Brewster ME (2004) Cyclodextrin based pharmaceutics: past, present and future. Nat Rev Drug Discovery 3:1023–1035
14. Khan NA, Durakshan M (2013) Cyclodextrin: An Overview. Int J Bioassays 2:858–865
15. Loftsson T, Jarho P, Másson M, Järvinen T (2005) Cyclodextrins in drug delivery. Expert Opin Drug Deliv 2:335–351
16. Otero-Espinar F, Torres-Labandeira J, Alvarez-Lorenzo C, Blanco-Méndez J (2010) Cyclodextrins in drug delivery systemes. J Drug Deliv Sci Technol 20:289–301
17. Bonnet V, Gervaise C, Djedaïni-Pilard F, Furlan A, Sarazin C (2015) Cyclodextrin nanoassemblies: a promising tool for drug delivery. Drug Discov Today 20:1120–1126
18. Liu Y, Chen Y, Gao X, Fu J, Hu L (2022) Application of cyclodextrin in food industry. Crit Rev Food Sci Nutr 62:2627–2640
19. Szente L, Szejtli J (2004) Cyclodextrins as food ingredients. Trends Food Sci Technol 15:137–142
20. Matencio A, Navarro-Orcajada S, Garcia-Carmona F, López-Nicolás JM (2020) Applications of cyclodextrins in food science. A Rev Trends Food Sci Technol 104:132–143
21. Astray G, Gonzalez-Barreiro C, Mejuto JC, Rial-Otero R, Simal-Gandara J (2009) A review on the use of cyclodextrins in foods. Food Hydrocoll 23:1631–1640
22. Crini G, Fenyvesi É, Szente L (2021) Outstanding contribution of Professor József Szejtli to cyclodextrin applications in foods, cosmetics, drugs, chromatography and biotechnology: a review. Environ Chem Lett 19:2619–2641
23. Buschmann H-J, Schollmeyer E (2002) Applications of cyclodextrins in cosmetic products: A review. J Cosmet Sci 53:185–192
24. Szejtli J (1982) Cyclodextrins in food, cosmetics and toiletries. Starch 34:379–385
25. Lee J, Park S, Lee J, Park S (2019) Review of the usability of cyclodextrin as a cosmetic ingredient. Asian J Beauty Cosmetol 17:545–553
26. Morin-Crini N, Fourmentin S, Fenyvesi É, Lichtfouse E, Torri G, Fourmentin M, Crini G (2021) 130 years of cyclodextrin discovery for health, food, agriculture, and the industry: a review. Environ Chem Lett 19:2581–2617
27. Arockia Jency D, Umadevi M, Sathe G (2015) SERS detection of polychlorinated biphenyls using β-cyclodextrin functionalized gold nanoparticles on agriculture land soil. J Raman Spectrosc 46:377–383
28. Liu B, Duan J, Zhang Y, Wang R, Zhao L, Gao S, Ye F, Fu Y (2024) Preparation and characterization of hexaconazole/hydroxypropyl-gamma-cyclodextrin inclusion complex nanofibers for sustainable agriculture: Improved physicochemical properties and antifungal activity of hexaconazole. J Mol Struct 1299:137195
29. Decool G, Kfoury M, Paitel L, Sardo A, Fourmentin S (2024) Cyclodextrins as molecular carriers for biopesticides: a review. Environ Chem Lett 22:321–353

30. Sandilya AA, Natarajan U, Priya MH (2020) Molecular view into the cyclodextrin cavity: structure and hydration. ACS Omega 5:25655–25667
31. Ross PD, Rekharsky MV (1996) Thermodynamics of hydrogen bond and hydrophobic interactions in cyclodextrin complexes. Biophys J 71:2144–2154
32. Szente L, Fenyvesi É (2017) Cyclodextrin-lipid complexes: cavity size matters. Struct Chem 28:479–492
33. Xu X, Yang S, Wang Y, Qian K (2022) Nanomaterial-based sensors and strategies for heavy metal ion detection. Green Anal Chem 2:100020
34. Kumar A, Arya K, Mehra S, Kumar A, Mehta SK, Kataria R (2024) Detection and sorption of heavy metal ions in aqueous media using Zn-based luminescent metal-organic framework. Sep Purif Technol 333:125875
35. Czarnik AW (1993) Fluorescent chemosensors for ion and molecule recognition. American Chemical Society
36. Thomas SW, Joly GD, Swager TM (2007) Chemical sensors based on amplifying fluorescent conjugated polymers. Chem Rev 107:1339–1386
37. Valeur B, Leray I (2000) Coord Chem Rev 205:3–40
38. Gunnlaugsson T, Glynn M, Tocci GM, Kruger PE, Pfeffer FM (2006) Coord Chem Rev 250:3094–3117
39. Jiang P, Guo Z (2004) Design principles of fluorescent molecular sensors for cation recognition. Coord Chem Rev 248:205–229
40. Martínez-Máñez R, Sancen6n F (2006) Chemodosimeters and 3D inorganic functionalised hosts for the fluoro-chromogenic sensing of anions. Coord Chem Rev 250:3081–3093
41. Moragues ME, Martínez-Máñez R, Sancen6n F (2011) Chromogenic and fluorogenic chemosensors and reagents for anions. A comprehensive review of the year 2009. Chem Soc Rev 40:2593–2643
42. Silva AP, NimaláGunaratne H, MarkáLynch P, Glenn E, SamankumaraáSandanayake K (1992) Molecular fluorescent signalling with 'fluor–spacer–receptor' systems: approaches to sensing and switching devices *via* supramolecular photophysics. Chem Soc Rev 21:187–195
43. Wiskur SL, Ait-Haddou H, Lavigne JJ, Anslyn EV (2001) Teaching old indicators new tricks. Acc Chem Res 34:963–972
44. Kalchenko OI, Solovyov AV, Cherenok SA, Starodub NF, Kalchenko VI (1999). J Incl Phenom Macrocycl Chem
45. Lamb JD (2002) J Incl Phenom Macrocycl Chem. Springer
46. Wong JK-H, Todd MH, Rutledge PJ (2017) Recent advances in macrocyclic fluorescent probes for ion sensing. Molecules 22:200
47. Formica M, Fusi V, Giorgi L, Micheloni M (2012) New fluorescent chemosensors for metal ions in solution. Coord Chem Rev 256:170–192
48. Schäferling M (2012) The art of fluorescence imaging with chemical sensors. Angew Chem Int Ed 51:3532–3554
49. Wu J, Liu W, Ge J, Zhang H, Wang P (2011) New sensing mechanisms for design of fluorescent chemosensors emerging in recent years. Chem Soc Rev 40:3483–3495
50. Baker GA, Crane NJ, Mayrhofer RC, Betts TA (2002) Cyclodextrin inclusion complexes with a solvatochromic fluorescent probe. J Chem Educ 79:1261
51. Al-Soufi W, Reija B, Novo M, Felekyan S, Kühnemuth R, Seidel CA (2005) Fluorescence correlation spectroscopy, a tool to investigate supramolecular dynamics: inclusion complexes of pyronines with cyclodextrin. J Am Chem Soc 127:8775–8784
52. Nau WM, Zhang X (1999) An exceedingly long-lived fluorescent state as a distinct structural and dynamic probe for supramolecular association: an exploratory study of host−guest complexation by cyclodextrins. J Am Chem Soc 121:8022–8032
53. Cid-Samamed A, Rakmai J, Mejuto JC, Simal-Gandara J, Astray G (2022) Cyclodextrins inclusion complex: preparation methods, analytical techniques and food industry applications. Food Chem 384:132467
54. Harrison JC, Eftink MR (1982) Cyclodextrin–adamantanecarboxylate inclusion complexes: a model system for the hydrophobic effect. Biopolym Original Res Biomol 21:1153–1166

55. Hunt MA, Tonelli AE, Balik CM (2007) Effect of guest hydrophobicity on water sorption behavior of oligomer/α-cyclodextrin inclusion complexes. J Phys Chem B 111:3853–3858
56. Chen J, Lu W-L, Gu W, Lu S-S, Chen Z-P, Cai B-C, Yang X-X (2014) Drug-in-cyclodextrin-in-liposomes: a promising delivery system for hydrophobic drugs. Expert Opin Drug Deliv 11:565–577
57. Szejtli J (2004) Past, present and future of cyclodextrin research. Pure Appl Chem 76:1825–1845
58. Zhang Y-M, Xu Q-Y, Liu Y (2019) Molecular recognition and biological application of modified β-cyclodextrins. Sci China Chem 62:549–560
59. Liu Y, Han B-H, Zhang H-Y (2004) Spectroscopic studies on molecular recognition of modified cyclodextrins. Curr Org Chem 8:35–46
60. Tomatsu I, Hashidzume A, Harada A (2005) Photoresponsive hydrogel system using molecular recognition of α-cyclodextrin. Macromolecules 38:5223–5227
61. Liu Y, Han B-H, Qi A-D, Chen R-T (1997) Molecular recognition study of a supramolecular system. Bioorg Chem 25:155–162
62. Prabu S, Mohamad S (2020) Curcumin/beta-cyclodextrin inclusion complex as a new "turn-off" fluorescent sensor system for sensitive recognition of mercury ion. J Mol Struct 1204:127528
63. Xu M, Wu S, Zeng F, Yu C (2010) Cyclodextrin supramolecular complex as a water-soluble ratiometric sensor for ferric ion sensing. Langmuir 26:4529–4534
64. Yamauchi A, Hayashita T, Kato A, Nishizawa S, Watanabe M, Teramae N (2000) Selective potassium ion recognition by benzo-15-crown-5 fluoroionophore/γ-cyclodextrin complex sensors in water. Anal Chem 72:5841–5846
65. Corradini R, Dossena A, Galaverna G, Marchelli R, Panagia A, Sartor G (1997) Fluorescent chemosensor for organic guests and copper (II) ion based on dansyldiethylenetriamine-modified β-cyclodextrin. J Org Chem 62:6283–6289
66. Sigel A, Sigel H (1998) Metal ions in biological systems. CRC Press
67. Anastassopoulou J, Theophanides T (1995) The role of metal ions in biological systems and medicine, bioinorganic chemistry: an inorganic perspective of life. Springer, pp 209–218
68. Zheng X, Cheng W, Ji C, Zhang J, Yin M (2020) Detection of metal ions in biological systems: a review. Rev Anal Chem 39:231–246
69. Sigel A, Sigel H (1996) Metal Ions in biological systems: interactions of metal ions with nucleotides: nucleic acids, and their constituents, vol 32. CRC Press
70. Jorgensen S, Jensen A (1984) Processes of metal ions in the environment. In: Metal ions in biological systems, pp 61–103
71. Kołodyńska D (2013) Application of a new generation of complexing agents in removal of heavy metal ions from different wastes. Environ Sci Pollut Res 20:5939–5949
72. Li Z, Chen J, Guo H, Fan X, Wen Z, Yeh MH, Yu C, Cao X, Wang ZL (2016) Triboelectrification-enabled self-powered detection and removal of heavy metal ions in wastewater. Adv Mater 28:2983–2991
73. Szejtli J, Szejtli J (1988) Cyclodextrin inclusion complexes. In: Cyclodextrin technology, pp 79–185
74. Narayanan V, Ganesan V, Shanmugasundram E, Durganandini S, Vellaisamy K, Amirthalingam H, Rajamohan R, Thambusamy S (2023) Spectral studies of synthetic symmetric syringaldehyde di-Schiff azine with β-cyclodextrin inclusion complexes as a fluorescent probe for the detection of metal ions. J Photochem Photobiol A 445:115069
75. Ganesan V, Mani MK, Narayanan V, Shanmugasundram E, Vellaisamy K, Baskaralingam V, Jeyaraj J, Veerakanellore GB, Rajamohan R, Thambusamy S (2023) Synthesis, characterization of 4,4′-((1E,1′E)-hydrazine-1,2-diylidenebis(methanylylidene)) diphenol and the inclusion complex with γ-cyclodextrin as a fluorescent probe for detection of Al^{3+}. J Photochem Photobiol A 442:114814
76. Selvan GT, Poomalai S, Ramasamy S, Selvakumar PM, Muthu Vijayan Enoch IV, Lanas SG, Melchior A (2018) Differential metal ion sensing by an antipyrine derivative in aqueous and β-cyclodextrin media: selectivity tuning by β-cyclodextrin. Anal Chem 90:13607–13615

77. Poomalai S, Soundrapandian S, Ramasamy S, Paulraj MS, Enoch IV (2020) Fluorescence sensing of Al^{3+} by an anthracene appended piperidone. Hindrance in excimer formation offered by β-cyclodextrin complexation, Chemical Physics Letters 751:137551
78. Liu Z, Zhu W, Chen Y, Li Y, Ding Y, Yang W, Li K (2015) Water-soluble host–guest system from β-cyclodextrin as a fluorescent sensor for aluminium ions: synthesis and sensing studies. Dalton Trans 44:16528–16533
79. Maniyazagan M, Rameshwaran C, Mariadasse R, Jeyakanthan J, Premkumar K, Stalin T (2017) Fluorescence sensor for Hg^{2+} and Fe^{3+} ions using 3,3′–dihydroxybenzidine:α–cyclodextrin supramolecular complex: characterization, in-silico and cell imaging study. Sens Actuators B Chem 242:1227–1238
80. Sumithra M, Sivaraj R, Selvan GT, Selvakumar PM, Enoch IV (2017) Ca^{2+} ion sensing by a piperidin-4-one derivative and the effect of β-cyclodextrin complexation on the sensing. J Lumin 185:205–211
81. Poomalai S, Govindaraj TS, Soundrapandian S, Paulraj MS, Enoch IVMV (2018) A new fluorescent chemosensor for cadmium(II) based on a pyrene-appended piperidone derivative and its β-cyclodextrin complex. Luminescence 33:538–544
82. Yang S, Jiang W, Tang Y, Xu L, Gao B, Xu H (2018) Sensitive fluorescent assay for copper(II) determination in aqueous solution using quercetin–cyclodextrin inclusion. RSC Adv 8:37828–37834
83. Corradini R, Dossena A, Galaverna G, Marchelli R, Panagia A, Sartor G (1997) Fluorescent chemosensor for organic guests and copper (II) ion based on dansyldiethylenetriamine-modified β-cyclodextrin. J Org Chem 62:6283–6289
84. Kavitha R, Stalin T (2014) A highly selective chemosensor for colorimetric detection of Hg^{2+} and fluorescence detection of pH changes in aqueous solution. J Lumin 149:12–18
85. Singh H, Bamrah A, Bhardwaj SK, Deep A, Khatri M, Kim K-H, Bhardwaj N (2021) Nanomaterial-based fluorescent sensors for the detection of lead ions. J Hazard Mater 407:124379
86. Sharma A, Majdinasab M, Khan R, Li Z, Hayat A, Marty JL (2021) Nanomaterials in fluorescence-based biosensors: defining key roles. Nano-Struct Nano-Objects 27:100774
87. Kumar ASK, Tseng W-L (2020) Perspective on recent developments of near infrared-emitting gold nanoclusters: applications in sensing and bio-imaging. Anal Methods 12:1809–1826
88. Kang X, Zhu M (2019) Tailoring the photoluminescence of atomically precise nanoclusters. Chem Soc Rev 48:2422–2457
89. Coronado EA, Encina ER, Stefani FD (2011) Optical properties of metallic nanoparticles: manipulating light, heat and forces at the nanoscale. Nanoscale 3:4042–4059
90. Wang H, He Y (2017) Recent advances in silicon nanomaterial-based fluorescent sensors. Sensors 17:268
91. Zhu S, Zhao F, Deng M, Zhang T, Lü C (2019) Construction of β-cyclodextrin derived CDs-coupled block copolymer micelles loaded with CdSe/ZnS QDs via host-guest interaction for ratiometric fluorescence sensing of metal ions. Dyes Pigm 168:369–380
92. Zhang NMY, Qi M, Wang Z, Wang Z, Chen M, Li K, Shum P, Wei L (2019) One-step synthesis of cyclodextrin-capped gold nanoparticles for ultra-sensitive and highly-integrated plasmonic biosensors. Sens Actuators B Chem 286:429–436
93. Wang Y, Han Y, Tan X, Dai Y, Xia F, Zhang X (2021) Cyclodextrin capped gold nanoparticles (AuNP@CDs): from synthesis to applications. J Mater Chem B 9:2584–2593
94. Ogoshi T, Harada A (2008) Chemical sensors based on cyclodextrin derivatives. Sensors 8:4961–4982
95. Nazerdeylami S, Ghasemi JB, Amiri A, Ziarani GM, Badiei A (2020) Fluorescence turn off-on probe (β-cyclodextrin-hydroxyquinoline) for monitoring of Cd2+ ions and tetracycline. Methods Appl Fluoresc 8:025009
96. Gao L, He C (2021) Application of nanomaterials decorated with cyclodextrins as sensing elements for environment analysis. Environ Sci Pollut Res 28:59499–59518
97. Lin Z-Y, Kuo Y-C, Chang C-J, Lin Y-S, Chiu T-C, Hu C-C (2018) Highly sensitive sensing of hydroquinone and catechol based on β-cyclodextrin-modified carbon dots. RSC Adv 8:19381–19388

98. Silva EC, Gomes CG, Pina J, Pereira RF, Murtinho D, Fajardo AR, Valente AJ (2024) Carbon quantum dots-containing poly(β-cyclodextrin) for simultaneous removal and detection of metal ions from water. Carbohyd Polym 323:121464
99. Xiong T, Zou C, Lin S (2023) β-Cyclodextrin magnetic graphene oxide (β-CD@MGO) for efficient removal of Ni(II) and phenol: adsorption performance, reusability and adsorption mechanism studies. Diam Relat Mater 139:110269
100. Villalobos E, Marco JF, Yáñez C (2023) Reduced Graphene Oxide as a Platform for the Immobilization of Amino-Cyclodextrins. Micromachines 14:746
101. Einafshar E, Khodadadipoor Z, Nejabat M, Ramezani M (2021) Synthesis, characterization and application of α, β, and γ cyclodextrin-conjugated graphene oxide for removing cadmium ions from aqueous media. J Polym Environ 29:3161–3173
102. Javanmirpourshirzadi Z, Zamani Hargalani F, Robati M, Taghavi L (2023) Removal of lead (II) from aqueous solutions using β-cyclodextrin functionalized magnetic graphene oxide nanocomposite, performance, and optimization with response surface methodology. J Nanopart Res 25:131
103. Kocot K, Pytlakowska K, Talik E, Krafft C, Sitko R (2022) Sensitive determination of uranium using β-cyclodextrin modified graphene oxide and X-ray fluorescence techniques: EDXRF and TXRF. Talanta 246:123501
104. Cheng M, Yao C, Su Y, Liu J, Xu L, Bu J, Wang H, Hou S (2022) Cyclodextrin modified graphene membrane for highly selective adsorption of organic dyes and copper (II) ions. Eng Sci 18:299–307
105. Yuan D-Q, Koga K, Kourogi Y, Fujita K (2001) Synthesis of fullerene–cyclodextrin conjugates. Tetrahedron Lett 42:6727–6729
106. Samal S, Geckeler KE (2000) Cyclodextrin–fullerenes: a new class of water-soluble fullerenes. Chem Commun 1101–1102
107. Saylan Y, Denizli A (2020) Advances in molecularly imprinted systems: materials, characterization methods and analytical applications. Curr Anal Chem 16:196–207
108. Saylan Y, Akgönüllü S, Yavuz H, Ünal S, Denizli A (2019) Molecularly imprinted polymer based sensors for medical applications. Sensors 19:1279
109. Rebelo P, Costa-Rama E, Seguro I, Pacheco JG, Nouws HP, Cordeiro MND, Delerue-Matos C (2021) Molecularly imprinted polymer-based electrochemical sensors for environmental analysis. Biosens Bioelectron 172:112719
110. Akgönüllü S, Kılıç S, Esen C, Denizli A (2023) Molecularly imprinted polymer-based sensors for protein detection. Polymers 15:629
111. Sedghi R, Heidari B, Yassari M (2017) Novel molecularly imprinted polymer based on β-cyclodextrin@graphene oxide: synthesis and application for selective diphenylamine determination. J Colloid Interface Sci 503:47–56
112. Salem MA, Khan AM, Manea YK, Saleh HA, Ahmad M (2022) Carbon nanotubes decorated with coordination polymers for fluorescence detection of heavy-metal ions and nitroaromatic chemicals. ACS Omega 8:1220–1231
113. Bhoi VI, Imae T, Ujihara M, Murthy C (2013) Surface immobilization of carbon nanotubes by β-cyclodextrins and their inclusion ability. J Nanosci Nanotechnol 13:2604–2612
114. Yang J, Yang Y, Su L, Tao X, Zhang J, Chen Y, Yang L (2023) Diethylenetriamine-β-CD-modified carbon quantum dots for selective fluorescence sensing of Hg2+ and Fe3+ and cellular imaging. Spectrochim Acta Part A Mol Biomol Spectrosc 291:122364
115. Wang J, Qiu F, Wu H, Li X, Zhang T, Niu X, Yang D, Pan J, Xu J (2017) Fabrication of fluorescent carbon dots-linked isophorone diisocyanate and β-cyclodextrin for detection of chromium ions. Spectrochim Acta Part A Mol Biomol Spectrosc 179:163–170
116. Almahri A, Al-Bonayan AM, Attar RM, Karkashan A, Abbas B, Al-Qahtani SD, El-Metwaly NM (2023) Multifunctional lipophobic polymer dots from cyclodextrin: antimicrobial/anticancer laborers and silver ions chemo-sensor. ACS Omega 8:16956–16965
117. Wu H, Jiang J, Gu X, Tong C (2017) Nitrogen and sulfur co-doped carbon quantum dots for highly selective and sensitive fluorescent detection of Fe(III) ions and L-cysteine. Microchim Acta 184:2291–2298

118. Zhao L, Li H, Xu Y, Liu H, Zhou T, Huang N, Li Y, Ding L (2018) Selective detection of copper ion in complex real samples based on nitrogen-doped carbon quantum dots. Anal Bioanal Chem 410:4301–4309
119. Akhondi M, Jamalizadeh E (2020) Selective colorimetric detection of Mn2+ and Cr2+ ions using silver nanoparticles modified with sodium dodecyl sulfonate and β-cyclodextrin. Acta Chim Slov 67:537–550
120. Bindhu M, Saranya P, Sheeba M, Vijilvani C, Rejiniemon T, Al-Mohaimeed AM, Abdel-Gawwad MR, Elshikh MS (2021) Functionalization of gold nanoparticles by β-cyclodextrin as a probe for the detection of heavy metals in water and photocatalytic degradation of textile dye. Environ Res 201:111628
121. Liu J, Alvarez J, Ong W, Kaifer AE (2001) Network aggregates formed by C60 and gold nanoparticles capped with γ-cyclodextrin hosts. Nano Lett 1:57–60
122. Liu J, Ong W, Román E, Lynn MJ, Kaifer AE (2000) Cyclodextrin-modified gold nanospheres. Langmuir 16:3000–3002
123. Prochowicz D, Kornowicz A, Lewiński J (2017) Interactions of native cyclodextrins with metal ions and inorganic nanoparticles: fertile landscape for chemistry and materials science. Chem Rev 117(22):13461–13501
124. Rajamanikandan R, Ilanchelian M (2018) β-cyclodextrin functionalised silver nanoparticles as a duel colorimetric probe for ultrasensitive detection of Hg2+ and S2− ions in environmental water samples. Mater Today Commun 15:61–69
125. Aswathy B, Avadhani G, Suji S, Sony G (2012) Synthesis of β-cyclodextrin functionalized gold nanoparticles for the selective detection of Pb2+ ions from aqueous solution. Front Mater Sci 6:168–175
126. Sumathi P, Suganthi S, Sivaraj R, Selvan GT, Selvakumar PM, Enoch IV (2019) Novel supramolecular β-cyclodextrin-piperidin-4-one complex assembled on gold nanoparticles. Selective detection of Cd^{2+} ions. Microchem J 150:104066
127. Ganesan V, Senguttuvan S, Narayanan V, Shanmugasundaran E, Vellaisamy K, Varadharajan R, Duraisamy R, Govindasamy M, Thambusamy S (2022) Silver nanoparticle decorated γ-cyclodextrin with 1,5-dihydroxy naphthalene inclusion complex; as a sensitive fluorescence probe for dual metal ion sensing employing spectrum techniques. Chem Phys Lett 796:139537
128. Badruddoza AZM, Shawon ZBZ, Tay WJD, Hidajat K, Uddin MS (2013) Fe3O4/cyclodextrin polymer nanocomposites for selective heavy metals removal from industrial wastewater. Carbohyd Polym 91:322–332
129. Amanulla B, Subbu HKR, Ramaraj SK (2018) A sonochemical synthesis of cyclodextrin functionalized Au-FeNPs for colorimetric detection of Cr^{6+} in different industrial waste water. Ultrason Sonochem 42:747–753
130. Zhang L, Wang E (2014) Metal nanoclusters: new fluorescent probes for sensors and bioimaging. Nano Today 9:132–157
131. Qian S, Wang Z, Zuo Z, Wang X, Wang Q, Yuan X (2022) Engineering luminescent metal nanoclusters for sensing applications. Coord Chem Rev 451:214268
132. Ou G, Zhao J, Chen P, Xiong C, Dong F, Li B, Feng X (2018) Fabrication and application of noble metal nanoclusters as optical sensors for toxic metal ions. Anal Bioanal Chem 410:2485–2498
133. Mohandoss S, Palanisamy S, You S, Lee YR (2022) Synthesis of cyclodextrin functionalized photoluminescent metal nanoclusters for chemoselective Fe^{3+} ion detection in aqueous medium and its applications of paper sensors and cell imaging. J Mol Liq 356:118999
134. Grygorova G, Seminko V, Maksimchuk P, Hubenko K, Kavok N, Demchenko L, Yefimova S (2024) Stable nanometer-size beta-cyclodextrin–CeO2-x colloidal nanoparticles with high free radical scavenging activity. J Mol Liq 124091
135. Liu B, Liu J (2019) Sensors and biosensors based on metal oxide nanomaterials. TrAC Trends Anal Chem 121:115690
136. Philip S, Kuriakose S (2018) Studies on the antibacterial activity of water-soluble iron oxide nanoparticle β-cyclodextrin aggregates against selected human pathogenic bacteria. Nano-Struct Nano-Objects 16:347–353

137. Ghazimokri HS, Aghaie H, Monajjemi M, Gholami MR (2022) Removal of methylene blue dye from aqueous solutions using carboxymethyl-β-cyclodextrin-Fe3O4 nanocomposite: thermodynamics and kinetics of adsorption process. Russ J Phys Chem A 96:371–380
138. Chen D, Shen Y, Wang S, Chen X, Cao X, Wang Z, Li Y (2021) Efficient removal of various coexisting organic pollutants in water based on β-cyclodextrin polymer modified flower-like Fe3O4 particles. J Colloid Interface Sci 589:217–228
139. Ganguly BN, Verma V, Chatterjee D, Satpati B, Debnath S, Saha P (2016) Study of gallium oxide nanoparticles conjugated with β-cyclodextrin: an application to combat cancer. ACS Appl Mater Interfaces 8:17127–17137
140. Sang L, Zhao Y, Burda C (2014) TiO2 nanoparticles as functional building blocks. Chem Rev 114:9283–9318
141. Almutairi TM, Al-Rasheed HH, Alaqil ZM, Hajri AK, Elsayed NH (2023) Green synthesis of magnetic supramolecules β-cyclodextrin/iron oxide nanoparticles for photocatalytic and antibacterial applications. ACS Omega 8:32067–32077
142. Lu W, Zhang J, Li N, You Z, Feng Z, Natarajan V, Chen J, Zhan J (2020) Co3O4@β-cyclodextrin with synergistic peroxidase-mimicking performance as a signal magnification approach for colorimetric determination of ascorbic acid. Sens Actuators B Chem 303:127106
143. Frasco MF, Chaniotakis N (2009) Semiconductor quantum dots in chemical sensors and biosensors. Sensors 9:7266–7286
144. Wu D, Chen Z, Huang G, Liu X (2014) ZnSe quantum dots based fluorescence sensors for Cu^{2+} ions. Sens Actuators A 205:72–78
145. Palaniappan K, Xue C, Arumugam G, Hackney SA, Liu J (2006) Water-soluble, cyclodextrin-modified CdSe−CdS core−shell structured quantum dots. Chem Mater 18:1275–1280
146. Shang ZB, Hu S, Wang Y, Jin WJ (2011) Interaction of β-cyclodextrin-capped CdSe quantum dots with inorganic anions and cations. Luminescence 26:585–591
147. Freeman R, Finder T, Bahshi L, Willner I (2009) β-Cyclodextrin-modified CdSe/ZnS quantum dots for sensing and chiroselective analysis. Nano Lett 9:2073–2076
148. Zhao M-X, Xia Q, Feng X-D, Zhu X-H, Mao Z-W, Ji L-N, Wang K (2010) Synthesis, biocompatibility and cell labeling of l-arginine-functional β-cyclodextrin-modified quantum dot probes. Biomaterials 31:4401–4408
149. Wang S, Wang H, Yue J, Lu Y, Jia T, Chen J (2022) Highly sensitive and selective fluorescent sensor for Ag+ detection using β-cyclodextrin/chitosan polymer-coated S QDs based on an aggregation-induced quenching mechanism. Cellulose 29:4083–4095
150. Wang S, Liu J, Zhao H, Zhang F (2021) Carboxymethyl chitosan crosslinked β-cyclodextrin containing hydrogen bonded Nsingle bondC QDs nanocomposites to design fluorescence probes for manganese ion (II) sensing. Mater Sci Eng C 119:111556

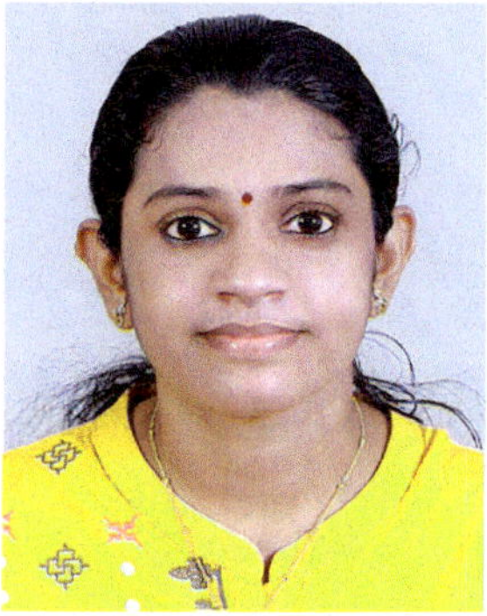

Dr. Nithya Mohan took her postgraduate degree in general chemistry from Union Christian College, Aluva, India. She did her Ph.D in inorganic chemistry from the department of Applied Chemistry, Cochin University of Science and Technology, India in 2019. Following the completion of her doctoral studies, Dr. Mohan dedicated an additional two years to postdoctoral research at the same department. In the year 2022, she joined the department of Basic Science and Humanities at SCMS School of Engineering and Technology, Kerala, where she assumed the role of Assistant Professor in Chemistry. Dr. Mohan's academic pursuits are complemented by her role as a key member of the research committee at the institute and her scholarly contributions reflect in her impressive publication records. To date, she has authored over 20 reputable international journals and contributed to the collective knowledge through the publication of 2 book chapters.

Dr. Sanu K. Anand is a chemistry professional having seven years of research experience in Analytical Chemistry. She earned her Ph.D in Analytical Chemistry from Cochin University of Science and Technology (CUSAT) in 2021. Presently she is working as a postdoctoral fellow at Department of Applied Chemistry, CUSAT. She has research experience at Institute of Animal Reproduction and Food Research of the Polish Academy of Sciences, Poland. She has authored or co-authored 15 international publications including 2 book chapters.

Dr. S. S. Sreejith earned his Ph.D in inorganic chemistry from Cochin University of Science and Technology in 2017. Following this, he served as a DST-SERB National postdoctoral fellow at IISER Kolkata (2017–2019). Notably, he served as Ecosystems editor for Elsevier, focusing on inorganic journals (2021–2022). With an h index of 18, Dr. Sreejith has authored or co-authored 33 international publications and 2 book chapters, establishing himself as a distinguished figure in the field.

Chapter 9
Electrochemical Detection of Metals Using Cyclodextrins

Mani Arivazhagan and Jaroon Jakmunee

Abstract An immense challenge in trace amounts of heavy metal ions involves the identification of electrochemical techniques and devices that possess low detection thresholds, user-friendly, and enable expedited analyses. Efforts to mitigate heavy metal pollution include implementing stricter regulations on industrial discharge, improving waste management practices, and remediation efforts to clean up contaminated sites. Additionally, there's a growing emphasis on green technologies and sustainable practices to minimize heavy metal ion pollution at its source. The fact that heavy metals like lead (Pb), cadmium (Cd), and mercury (Hg) are not biodegradable gives rise to significant human health concerns. Therefore, there is a need for the development of cost-effective, sensitive, and rapid detection methods for these metals. This chapter presents a major contribution of the electrochemical detection of metals using cyclodextrins-based electrochemical sensing platforms in the previous eras. The cyclodextrin-based electrochemical sensors are introduced after a brief review of the general characteristics of cyclodextrins and their capacity to form inclusion complexes. This comprises cyclodextrins combined with a variety of nanostructured materials (carbon nanotubes, aptamers, and enzymes, conducting polymers, reduced graphene oxide, and electropolymerized cyclodextrin films). The use of these substances in biosensors, chiral recognition agents, and electrochemical environmental pollutant detection of heavy metal ions are discussed. This chapter particularly emphasizes the advantages and limitations of these approaches to the last Water Frame Directive devoted to the Environmental Quality Standards for heavy metals. Recent trends in trace metal speciation as well as the realm of electrochemical sensors and electroanalytical techniques over the previous decades, are discussed.

M. Arivazhagan · J. Jakmunee (✉)
Research Laboratory for Analytical Instrument and Electrochemistry Innovation, Department of Chemistry, Faculty of Science, Chiang Mai University, Chiang Mai, Thailand
e-mail: jaroon.jakmunee@cmu.ac.th

J. Jakmunee
Research Laboratory on Advanced Materials for Sensor and Biosensor Innovation, Materials Science Research Center, Chiang Mai University, Chiang Mai, Thailand

Center of Excellence for Innovation in Chemistry, Faculty of Science, Chiang Mai University, Chiang Mai, Thailand

S. Krishnamoorthy (ed.), *Cyclodextrins for Chemosensing*,
https://doi.org/10.1007/978-3-031-89299-8_9

Keywords Cyclodextrins · Electrochemical sensors · Heavy metal ions (HMIs) · Inclusion complex · Nanocomposites

9.1 Introduction: Significance and Challenges in Detecting Heavy Metal Ions (HMIs)

Heavy metal ions (HMIs) are inflicting immediate harmful effects on human health and it dangerous to the existence of the ecosystem [1–3]. The most important research for addressing the detrimental impact that heavy metallic materials have on the earth's ecosystem and creatures that main focus on heavy metal ion sensing. Researchers have increasingly employed diverse nanocomposite materials in recent years to sense the Heavy Metal Ions (HMIs) by using a range of electroanalytical techniques [4–6]. The existence of the heavy metal ions present in different water resources on the surface of the earth is nearly inevitable due to industrialization. Additionally, their composition varies depending on the level of industrialization in each specific modern locality. It is important to manage exposure to these metals to minimize potential adverse effects on health, when they accumulate in the body over time, often through exposure to contaminated air, water, soil, or food. Heavy metal poisoning can lead to a various complicated health issue, which include neurological impairments, organ damage, developmental problems in children, and even cancer in some cases. Hence, it's crucial to monitor and limit exposure to heavy metal ions to protect public health. The most well-known heavy metals are iron (Fe^{2+}), zinc (Zn^{2+}), arsenic (As^{3+}), copper (Cu^{2+}), nickel (Ni^{2+}), cadmium (Cd^{2+}), mercury (Hg^{2+}), lead (Pb^{2+}), etc. [7]. These metals can be discovered naturally and they cannot be degraded or destroyed. While certain heavy metal ions are necessary for the human body to maintain organ functions effectively, excessive concentrations of these ions can be harmful to human health [8–10]. On the other hand, even in minute amounts, trace-level heavy metal contaminants such as lead (Pb^{2+}), mercury (Hg^{2+}), arsenic (As^{2+}), and cadmium (Cd^{2+}) are hazardous in the environment in the natural world. The human body needs small quantities of several heavy metals, including copper, zinc, and selenium, to maintain a healthy metabolism. Higher concentrations, though, could result in poisoning. Water resources of rivers, lakes, ponds, streams, and oceans are examples of surface water in the worldwide.

One of the most essential elements for life on Earth is water, and contamination of the water supply frequently results in a variety of illnesses in human beings. The health effects of heavy metal ions pollutions are severe worldwide. Numerous techniques have been created to find out HMIs. For instance, human beings are affected majorly by numerous health issues, including cancer, liver damage, cardiovascular impacts, kidney damage, and neurological disorders, which can result from exposure to heavy metal ions from the atmosphere. Nonetheless, the majority of these procedures whether they involve finding pollutants in the environment or cancer cells within the human body should be done in an aquatic setting. The most commonly used instrumental techniques are surface-enhanced Raman scattering (SERS), atomic absorption spectroscopy (AAS), X-ray absorption spectroscopy, and inductively coupled plasma atomic emission spectroscopy (ICP-MS). Ongoing research aims to create detection technologies that are affordable, portable, and easy to use in order to overcome these constraints [11]. Among the different analytical techniques, electrochemical sensors are intriguing instruments because of their remarkable sensitivity, exceptional accuracy, ease of use, and reasonable selectivity [12]. Overall, electroanalytical approaches play a vital role in the detection and determination of heavy metal ions due to their combination of speed, simplicity, cost-effectiveness, and analytical performance. These methods continue to be valuable tools in environmental monitoring, industrial quality control, and public health surveillance efforts [13, 14]. Overall, the combination of high sensitivity, ease of miniaturization, long-term durability, and practical compatibility makes electrochemical sensing platforms well-suited for a wider range of applications, We may anticipate more advancements in technology as well as the incorporation of detection techniques into other fields to meet new opportunities and challenges [15–17].

9.2 Merits of Electrochemical Sensors

Electrochemical sensors are incredibly useful tools that can be used a greater variety of industries, including the food storage sector, tracking the environmental pollutant, and performing the biomedical diagnostic devices. They may provide the benefits of a straightforward design, repeatable and reversible measurements, and precise and accurate analytical data [16, 18–20]. Because they are affordable and relatively simple in contrast to other methods, electrochemical sensors have seen significant development in the past 10 years, particularly in the pharmaceutical, environmental, and medical fields. Most recently, cyclodextrins and their nanocomposites have been used for the majority of detection transducers in electrochemical sensing platforms [21–23]. On the other hand, in current techniques, the active compounds play a different role. Here, they are involved in the electrochemical activation of biomolecules. This activation process often involves particle recognition, where the active compounds selectively bind to specific biomolecules or analytes of interest. Once bound, these compounds can promote the preconcentration of the analyte on the electrode surface, enhancing the sensitivity of the electrochemical measurement. Overall, the ability of these macrocyclic compounds to interact with a wide range of molecules and form host–guest complexes opens up possibilities for various applications in drug delivery, sensing, catalysis, and other areas of chemistry and biotechnology [24, 25]. Many electrochemical studies on cyclodextrins have been conducted; these studies have primarily examined the interactions between cyclodextrins in both solid and solution states in the past few eras. As a result, the process of creating CDs-modified electrodes would pave the way for the development of novel electrochemical sensors and increase their application in drug, environmental, and physiological analysis [26, 27].

9.3 Importance of Cyclodextrins and Their Nanocomposites

The development of nanocomposite materials relies on interdisciplinary research spanning materials science, chemistry, physics, and engineering is essential to building nanostructured materials. As intellectuals transport out their investigations through novel synthetic methods and tailor the properties of nanocomposites, their potential for technological innovation remains vast [28–30]. When nanoscale components are integrated into a host matrix, their special qualities provide unmatched benefits like increased electrical conductivity, mechanical strength, thermal stability, and customized functionalities. Nanostructured materials with cyclodextrin-based electrodes perform extraordinary sensitivity, selectivity, and responsiveness have made them essential for use in sensors. These materials have special qualities that can be adapted to particular sensing by integrating nanostructured materials into a host matrix. Improved surface-to-volume ratios provided by nanocomposite sensors allow for more effective interaction with analytes and higher sensitivity, even at low concentrations, and even a fast response time of three seconds. The detection of heavy metal ions for various nanostructures with cyclodextrin-based materials can be summarized in Table 9.1.

The tunable properties of nanomaterials enable the customization of sensor responses to various stimuli, offering unparalleled flexibility and versatility for detecting a wider linear range of targets with higher sensitivity and selectivity. This capability makes nanomaterials-based electrochemical sensors valuable tools for

Table 9.1 A survey of the nanostructured materials-based sensing platforms for detecting heavy metal ions in the environment

No.	Materials	Analyte	LOD (nM)	Linear range (μM)	References
1	Bi/HP-β-CD-rGO/Nafion	Pb^{2+}	0.09	0.001–0.09	[31]
2	β-CD/NH_2-rGO/	Cu^{2+}	2.8	0.05–1.0	[32]
3	mp-g-C_3N_4/β-CD	TNT	299.38	1–100	[33]
4	MXene/CNHs/βCD/MOF	CBZ	1.0	0.003–10.0	[34]
5	NH_2-rGO/β-CD	Cu^{2+}	9.0	0.05–1.0	[32]
6	BiNPs@$Ti_3C_2T_x$/GCE	Pb^{2+}	10.8	0.06–0.6	[35]
		Cd^{2+}	12.4	0.08–0.8	[35]
7	PANI-Ti_3C_2/GCE	Hg^{2+}	17.0	0.1–20.0	[36]
8	H-C_3N_4/$Ti_3C_2T_x$/GCE	Cd^{2+}	1.0	0.5–1.5	[37]
		Pb^{2+}	600.0	0.5–1.5	[37]
9	$Nb_4C_3T_x$/GCE	Pb^{2+}	12.0	0.025–0.5	[38]
10	Fe-MOF/MXene-1.5/GCE	As^{3+}	0.58	0.01–0.1	[39]

β-CD—Beta-cyclodextrins, rGO—reduced graphene oxide; CNHs—carbon nano horn; BiNPs—bismuth nanoparticles; GCE—glassy carbon electrode; TNT—trinitro toluene; CBZ—carbendazim; C_3N_4—graphitic carbon nitrite; MOF—metal organic framework applications of environmental pollutant monitoring, healthcare diagnostics, food safety, and beyond. Moreover, their ability to work with a range of fabrication processes enables the development of flexible, wearable, and small sensor devices. Due to their synergistic effect on detecting heavy metal ions with improved sensing properties, cyclodextrin-nanocomposite-based sensors have an immense attraction in recent advancements and future perspectives in the field of electrochemical sensing platforms. β-CD is a cyclic polymer with a numerous range of constituents and cheap production cost. Its excellent applications include biodegradability, environmental protection, and enhanced solubility of functional nanostructured materials. It possesses the features of an enzyme model due to its distinct hydrophobic inner cavities and hydrophilic exterior surface structure, and as such, it has been thoroughly studied in the domains of food science, medicine, analysis, catalysis, and the chemical industry. In addition, some functional groups can be added and further good qualities can be obtained by chemically altering the hydroxyl groups on the surface, which expands the field of application. For instance, amphiphilic cyclodextrins can form monolayers, micelles, or thermotropic liquid crystals when they are grafted onto a cyclic oligosaccharide core along with one or more hydrophobic groups. Glycosylated amphiphilic cyclodextrins with varying molecular structures exhibit aggregation characteristics that are contingent upon the hydrophobic-hydrophilic balance, or the length of the alkyl chain. Shorter hydrophobic chain cyclodextrins have a tendency to form micelles.

Combining the CDs with graphene is one of the more intriguing combinations, and this strategy has garnered a lot of attention, especially in the past 10 years. Guo et al. originally showed in 2010 that the combination of CDs and graphene produced a synergistic effect that allowed for the remarkable electrochemical detection of drugs and biomolecules. Graphene has been widely used in the development of electrochemical sensors because of its many beneficial properties [27]. Its higher surface-active sites, excellent electrical conductivity, excellent stability, good thermal and mechanical qualities, and ease of synthesis make it an appealing material for CD supports free of metal. While pristine graphene may be challenging to produce and work with, graphene oxide and its composites offer a versatile and promising alternative for various applications, particularly in the field of electrochemical sensors. Because of their hydrophobic interior and hydrophilic exterior, cyclodextrins have a unique structure that gives them useful properties in a variety of fields, from environmental science to pharmaceuticals. Their broad application and continuous interest in research are due in part to their capacity to improve solubility, aid in molecular recognition, and enrich target compounds (Fig. 9.1). As a result, CDs-based nanostructured materials have garnered substantial interest in the field of electrochemical sensing.

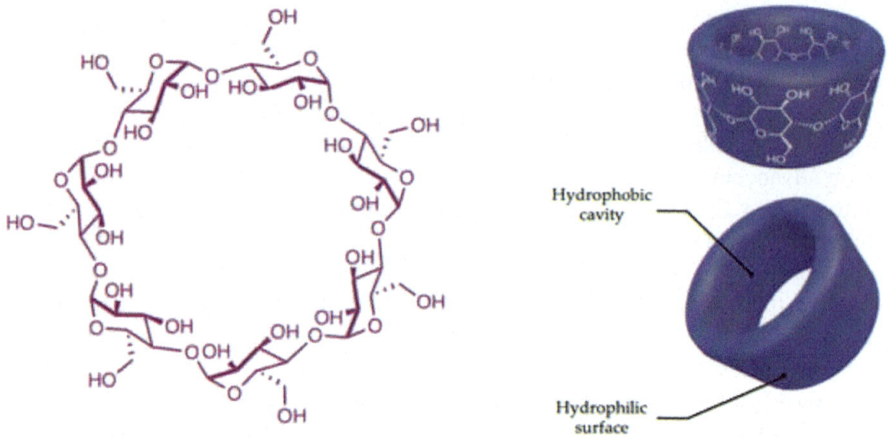

(a) (b)

Fig. 9.1 Pictorial diagram of (**a**) chemical structure and (**b**) three-dimensional structure of β-CD [40]

9.4 The Mechanism of Host–Guest Inclusion in β-CDs

The host–guest inclusion mechanism in cyclodextrins (CDs) refers to the process by which cyclodextrins, which are cyclic oligosaccharides with a hydrophilic exterior and a hydrophobic interior, form non-covalent complexes with guest molecules. This inclusion happens due to the unique structure of CDs and the ability to accommodate a variety of guest molecules inside their hydrophobic cavity shown in Fig. 9.2. Cyclodextrins are composed of glucose units linked by α-(1 → 4) glycosidic bonds, forming a toroidal or donut-like shape. The outer surface is hydrophilic (due to hydroxyl groups), while the inner cavity is relatively hydrophobic [40]. The hydrophobic interior of the cyclodextrin molecule acts as a host, while a guest molecule with appropriate size and hydrophobic character is attracted to the cavity. The guest molecule fits into the cyclodextrin cavity primarily through weak van der Waals forces and in some cases, the guest molecule can enhance stability due to hydrogen bonding [41]. The host–guest inclusion complex is highly versatile, and the ability to encapsulate various molecules makes cyclodextrins valuable in electrochemical sensing capability. Because of their varied cavity sizes and adaptable interaction mechanisms, cyclodextrins can encapsulate guest molecules of different sizes and stoichiometries. Cyclodextrins have the ability to form inclusion complexes based on the chemical properties and molecular dimensions of the guest.

A class of cyclic oligosaccharides known as cyclodextrins (CDs) has drawn a lot of interest from a variety of industries, including material science, food technology, and pharmaceuticals. Their distinct chemical structure and adaptable characteristics render them perfect for the creation of inclusion complexes, also referred to as cyclodextrin complexes (Fig. 9.2). The stability, solubility, and bioavailability of

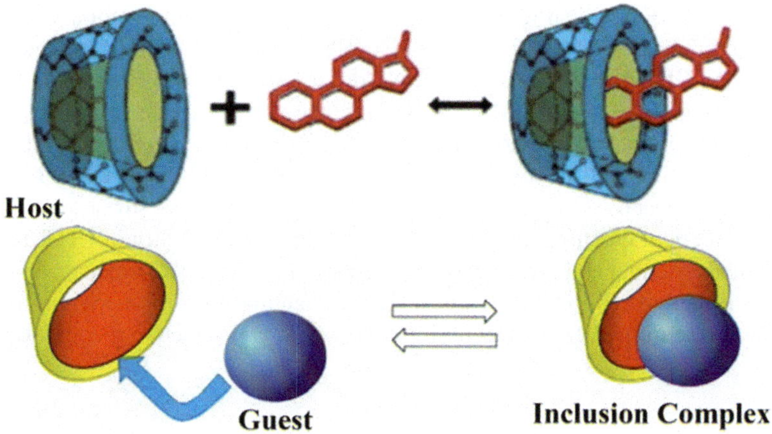

Fig. 9.2 Schematic diagram for guest encapsulated host to form an inclusion complex [41]

guest molecules are all improved by these complexes. We will examine the methods used to make cyclodextrin complexes and investigate their various uses in this blog post. A guest molecule is encapsulated inside the hydrophobic cavity of a cyclodextrin molecule to form a cyclodextrin complex. A medicine, perfume, dye, or any other hydrophobic or poorly water-soluble substance can be the guest molecule. This encapsulation improves the guest molecule's physicochemical properties, making it more useful for a variety of applications.

9.5 Cyclodextrin and Its Composites-Based HMIs Detection

The combination of appropriate recognition elements with cyclodextrins' ability to form stable host–guest inclusion complexes offers a versatile platform to create highly sensitive and selective electrochemical detectors for sensing HMIs contaminated in the environment. This approach holds great potential for addressing the challenges associated with heavy metal pollution and ensuring environmental monitoring and safety [42]. Screen-printed carbon electrodes (SPCEs) have been modified through the use of host–guest interactions between CNT-CD and nanostructured materials to achieve high sensitivity and selectivity in electrochemical sensing platforms. Later novel testing strips that are produced and highly helpful for electrochemical analysis in many fields (food, clinical, and environmental) due to their large-scale production, affordability, availability, and capacity to work with microsensors. For instance, Jin Zhao et al. developed the RGO-CD-MWCNT-POM tetra component hybrid structure offers a versatile and sensitive platform for electrochemical detection of dopamine (DA), uric acid (UA), and ascorbic acid (AA), with prospective applications in biomedical diagnostics, food safety and environmental monitoring. Initially, graphene oxide (GO), β-CD, and MWCNT were actively assembled to build up the RGO-CD-MWCNT-POM. Afterward, GO was chemically reduced with the help of polyoxometalate (POM) [43]. Hongyuan et al. introduced a novel electrochemical detector for the determination of CBZ, fabricated using β-cyclodextrin (β-CD) functionalized carbon nanosheets@carbon nanotubes (CNS@CNT). This innovative approach combined the larger electrochemical active surface area of CNS with the excellent electrical conductivity and high ionic mobility of CNT, resulting in significantly enhanced electrocatalytic and rapid sensing performance. Moreover, β-CD exhibited exceptional host–guest supramolecular uniform recognition ability, thereby augmenting CBZ's capacity for enrichment and selective recognition [1].

9.6 Lead Ion Detection

When lead (Pb^{2+}) is present in sources and drinking water for an extended period of time, it can cause effects such as behavioral disorders and neurobehavioural symptoms are created in children, elevated blood pressure, renal failure in adults, and brain, lung, or kidney cancer. For instance, M. J. Deen et al., A straightforward, easy, and inexpensive way to modify multiwalled carbon nanotubes (MWCNT) incorporated with β-cyclodextrin (βCD) as a mediator by using a physical (Phys) and covalent method through Steglich esterification (SE) methodology is used for the sensing of lead (Pb^{2+}) in the water sample. The Pb^{2+} is monitored based on the quantity of β-CD present on the MWCNT-β-CD electrodes, determining both the outermost layer and the physical/chemical adherence of β-CD to MWCNT. Chemically induced MWCNT-βCD functions as a sensitive electrode, as depicted in Fig. 9.3, exhibiting a higher sensitivity of 38.6 nA/ppb and a lower limit of detection (LOD) of 2.3 ppb. Conversely, physically induced MWCNT-βCD modified electrodes demonstrate a high sensitivity of 98 nA/ppb with a lower LOD of 0.9 ppb. The modified electrodes are highlighted in their robustness and cost-effectiveness, making them suitable for various applications that require reliable and durable sensing technology. The suggested sensors present a promising technological advancement for the creation of an extremely sensitive and reasonably priced electrochemical sensing system for determining the lead content in drinking water samples [1]. Additionally, the MWCNT-βCD-based modified electrodes demonstrated strong lead (Pb^{2+}) sensing ability and high selectivity even in the existence of higher concentrations of some potential interferences such as Zn^{2+} and Cd^{2+}. Based on the multiple reuses of the electrodes and the cost of the materials, the cost-effective benefits of the MWCNT-βCD/SPE were calculated. The fabricated sensor could be employed in point-of-care, low-cost sensing systems to monitor the quality of drinking water.

Lead is typically absorbed by the human body, as well as by animals, plants, and other living things. Once inside biological systems, lead ions interact with biological macromolecular affinity sites to cause acute or long-term toxic effects on various organs of the organism. In addition, Guy B. P. Ngassa et al., developed by utilizing

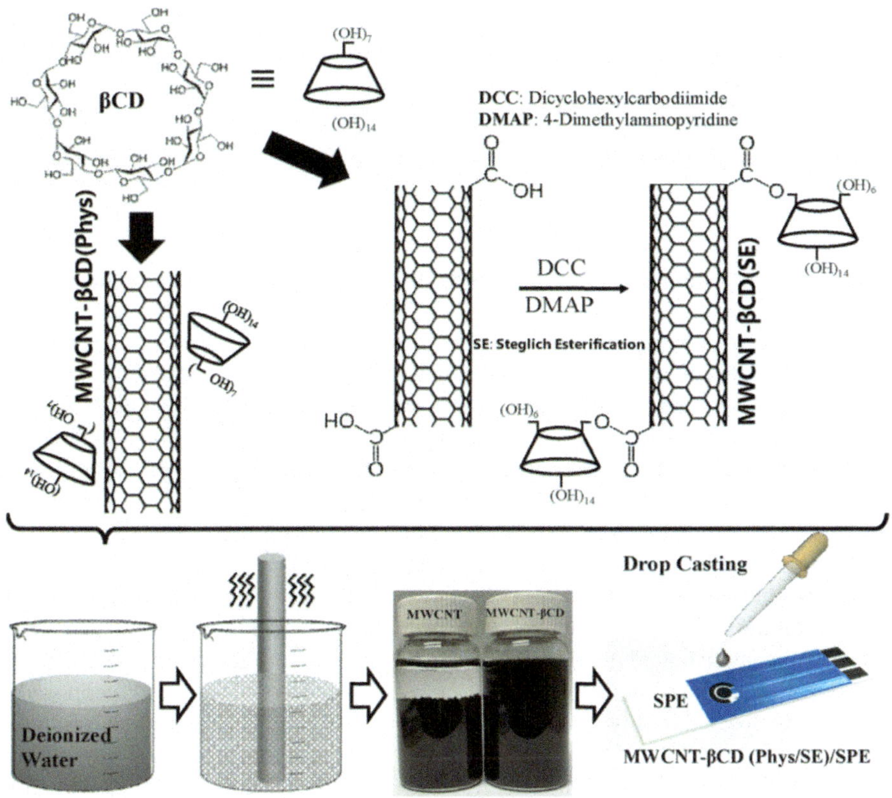

Fig. 9.3 Pictorial representation of the MWCNT-βCD(Phys/SE) modified screen-printed carbon electrode fabrication processes for the sensing of lead ions [1]

citric acid as a cross-linker provides a sustainable, versatile, and effective way to enhance the properties of various polymers to facilitate the β-CD polymerization on the surface of natural hydroxyapatite, a composite material was utilized as the active electrode material for Pb (II) detection. A new and highly sensitive Pb^{2+} detection sensor was suggested, which relies on a glassy carbon electrode modified with a thin layer of hydroxyapatite/β-CD nanocomposite materials. This study showed that hydroxyapatite can be used to prepare analytical devices at a low cost, which could have used in environmental protection and pollution control [44].

9.7 Cadmium Ion Detection

Cadmium (Cd) contamination is indeed a significant environmental issue, as it poses a serious threat to both ecosystems and human health. Food crops grown in contaminated soil or irrigated with Cd-contaminated water can accumulate dangerous levels of this heavy metal, eventually leading to human exposure through the food chain. Cd is highly toxic, even in small amounts, and long-term exposure can lead to health issues such as kidney damage, bone demineralization, and increased risks of cancer. For instance, Gang Liu et al. developed a novel glassy carbon electrode that incorporated a bismuth film matrix and a graphene-β-cyclodextrin-Nafion nanocomposite (GR-β-CD-Nafion-GCE) for the detection of trace levels of cadmium using square wave anodic stripping voltammetry (SWASV) techniques (Fig. 9.4). The surface altered electrode's morphological traits and electrochemical characteristics were assessed using scanning electron microscopy (SEM), chronoamperometry (CA), cyclic voltammetry (CV), and SWASV [27]. It was discovered that the GR incorporated β-CD-Nafion modified GCE demonstrated exceptional enhancing the catalytic effects for the detection of cadmium ions, which significantly facilitated the fast electron transport and increased the electrode's active surface area. The anodic stripping currents, under optimal conditions, showed good linear ranges from 800–2000 $\mu g\ L^{-1}$ and 1–500 $\mu g\ L^{-1}$, with a lower detection limit of 0.3 μg/L (S/N = 3). Additionally, the sensor was used to detect cadmium ions selectively in soil extracts, showing good stability and high reusability in the process of sensing cadmium ions.

These days, the environment contains HMIs like Cd^{2+} and Pb^{2+} that are not biodegradable and can build up in the body, harming the immune, reproductive, and central nervous systems (CNS). Therefore, the development of straightforward, quick, and sensitive analytical techniques is of paramount importance for the accurate quantification of them in the environment. Hongmei Jiang et al. proposed a straightforward and highly sensitive and selective electrochemical sensor capable of simultaneous detection of Cd^{2+} and Pb^{2+}. The sensing platform utilized functionalized α-cyclodextrin (α-CD)-graphene hybrids (AuNPs-CD-GS) integrated with gold nanoparticles (Fig. 9.5). Employing hydrazine and α-CD as reducing agents to reduce graphene oxide and gold ions, a straightforward one-pot synthesis was employed to produce the dispersed AuNPs-CD-GS. The stripping peak currents, under ideal

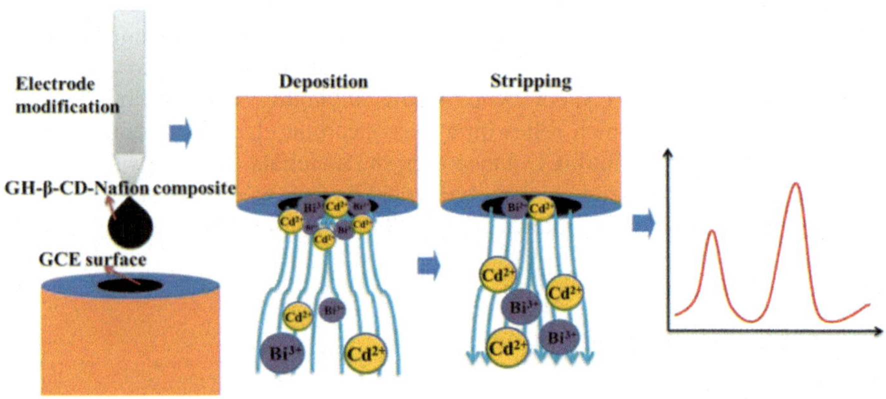

Fig. 9.4 Pictorial representation of the development of GR-β-CD-Nafion modified GCE for the detection of Cd^{2+} [27]

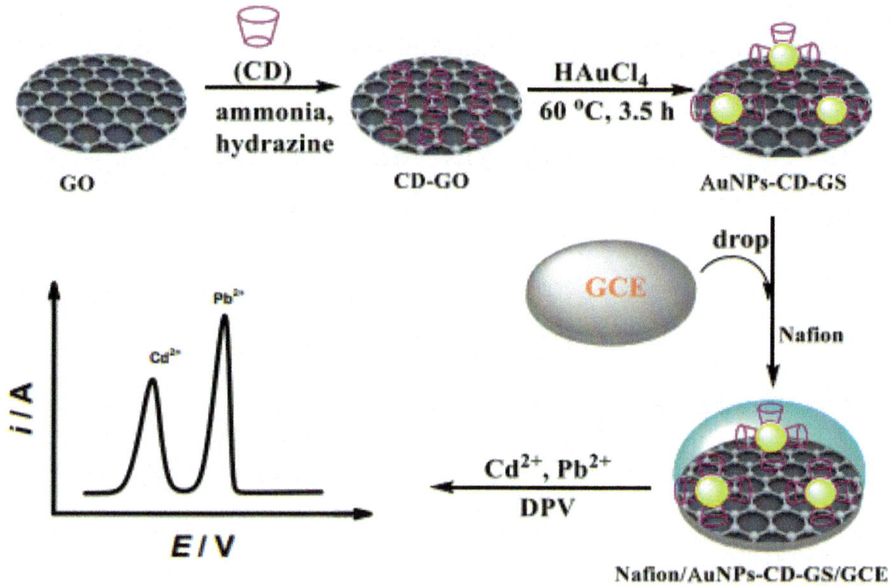

Fig. 9.5 Schematic representation of fabrications of the AuNPs-CD-GS electrodes for the simultaneous detection of Cd^{2+} and Pb^{2+} [2]

circumstances, have a linear relationship with the Cd^{2+} and Pb^{2+} concentrations within the 40–1200 mg L^{-1} range. In real water samples, simultaneous and sensitive detection of Cd^{2+} and Pb^{2+} was also carried out, with satisfactory results.

Due to their detrimental effects on the environment and human health, toxic heavy metals have become a significant concern. Water pollution stemming from these metals, such as lead (Pb^{2+}), cadmium (Cd^{2+}), and nickel (Ni^{2+}), has attracted considerable attention. Even at relatively low concentrations, these heavy metal ions pose hazards and are carcinogenic. Because they are not biodegradable, heavy metals can build up within living things. There are serious hazards to the environment and the general public's health when heavy metals and other pollutants are released from industrial processes. For these risks to be reduced and a safer environment to be ensured, more efficient regulation, better industrial practices, and increased public awareness are essential. For instance, Abu Zayed et al. developed Fe_3O_4 nanoparticles modified with carboxymethyl-cyclodextrin (CM-CD) polymer were created to specifically remove Pb^{2+}, Cd^{2+}, and Ni^{2+} ions from water. The cyclodextrin polymer/ Fe_3O_4 nanocomposites, according to our outcomes, can be employed as a recyclable absorbing material for the quick, simple, and effective detoxification of metallic ions from wastewater management. Various electrochemical sensing electrodes based on cyclodextrin and nanostructured materials for lead and cadmium ions detection are summarized in Table 9.2.

Table 9.2 Comparison of lead and cadmium ions detection results using electrochemical sensors based on cyclodextrin and various nanostructured materials

No.	Electrode	Analyte	LOD	Linear range	References
1	Bi/HP-β-CD-RGO/ Nafion/GCE	Pb^{2+}	0.09 nM L^{-1}	1.0–90 nM L^{-1}	[31]
		Cd^{2+}	0.07 nM L^{-1}	5.0–90 nM L^{-1}	
2	PCys/Pβ-CD/GCE	Pb^{2+}	0.53 nM	1.60–700 nM	[45]
		Cd^{2+}	1.13 nM	3.42–700 nM	
3	PET-TeMs	Pb^{2+}	–	0.01–4.25 mg/L	[46]
		Cd^{2+}	–	0.05–4.25 mg/L	
4	rGO/CeO_2/FTO nanoribbons	Pb^{2+}	0.04 μg L^{-1}	0.2–500 μg L^{-1}	[7]
		Cd^{2+}	0.35 μg L^{-1}	1–300 μg L^{-1}	
5	$CuZrO_3$ nanocomposites	Pb^{2+}	0.1 μg L^{-1}		[47]
		Cd^{2+}	0.5 μg L^{-1}	1–100 μg L^{-1}	
6	Bismuth modified (3-mercaptopropyl) trimethoxy silane (MPTMS)	Pb^{2+}	1.3 μg L^{-1}	0.5–80 μg L^{-1}	[48]
		Cd^{2+}	0.37 μg L^{-1}	0.56–11.24 μg L^{-1}	
7	Multiwalled CNTs	Pb^{2+}	6.6 μg L^{-1}	–	[49]
		Cd^{2+}	8.4 μg L^{-1}	–	
8	rGO	Pb^{2+}	0.2 μg L^{-1}	20–200 μg L^{-1}	[50]
		Cd^{2+}	0.6 μg L^{-1}	20–200 μg L^{-1}	

Bi—bismuth film; GCE—glassy carbon electrode; Cys: L-cysteine; RGO—reduced graphene oxide; BCD—beta cyclodextrin; TeMs—track-etched membranes; CeO_2—cerium oxide

9.8 Mercury Ion Detection

Among the most hazardous metal elements, mercury has a high capacity for bioaccumulation. As our society continues to grow, a number of environmental issues are also appearing more frequently. Due to the fact that heavy metal ions significantly affect human health and the environment, there has been a lot of interest in the precise and credible sensing of these HMIs. Furthermore, heavy metal ions in industrial wastewater are the primary source of environmental pollution [21]. In particular, mercury (Hg^{2+}) is regarded as one of the most hazardous elements that can readily enter both the environment and the human body. For instance, Claudia et al., described the electrochemical detection of Hg^{2+} in an aqueous solution employing a nanoassembly based on ferrocenylcarnosine (FcCAR) and cationic carboxylextrin carbon nanotubes (CNT-CDs), as shown in Fig. 9.6. CV and electrochemical impedance spectroscopy were employed to evaluate the electrochemical catalytic response of the CNT-CD/FcCAR nanoassembled electrodes. Chronoamperometry and differential pulsed voltammetry were utilized to evaluate the electroanalytical performance of the Hg^{2+} sensor. With a heightened sensitivity of 0.12 μA/nM, a lower detection limit of 0.50 nM, and a quantification limit of 1 nM, the oxidation peak

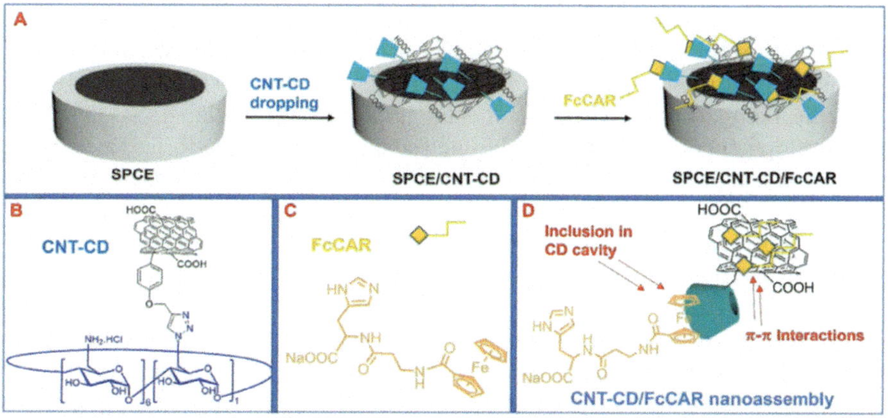

Fig. 9.6 Pictorial representation of development of SPCE modified with CNT-βCD-FcCAR electrodes (**a**), Chemical bonding structures of cationic CNT-CD (**b**) and FcCAR (**c**); CNT-CD/FcCAR nanoassembled electrodes with molecular interaction sites between CNT-CD and FcCAR [21]

current (i_{pa}) exhibited a wider linear concentration dependence within the ranges from 1 to 100 nM.

To manage the effects of mercury on the environment, it is crucial to continuously monitor the levels of the metal in the soil, water, and air. Mercury (Hg) is a naturally occurring element found in soil, rocks, water, and air. Mercury exposure can have severe neurological effects, especially in young children. It can impair cognitive function, motor skills, and cause developmental issues. For instance, Shen-Ming Chen et al., fabricated by utilizing a nanocomposite incorporated screen-printed carbon electrode (SPCE) decorated with polypyrrole (PPy) and β-cyclodextrin (GR-CD), proposed an electrochemical sensor that is both disposable and sensitive for detecting mercury in different water samples. Using $FeCl_3$, the PPy monomer in the GR-CD solution was chemically oxidized to create the GR-CD/PPy nanocomposite electrode materials. In comparison to bare, GR, GR-CD, and PPy-modified SPCEs, the GR-CD/PPy composite-modified SPCE has a higher sensitivity to Hg^{2+} when measured using differential pulse voltammetry (DPV) [51, 52].

9.9 Summary and Outlook

This book chapter discusses the advancement of electrochemical sensing platforms based on cyclodextrins and their nanocomposites based active materials to enrich the sensing property of HMIs detection. In addition, this chapter addresses the theoretical foundation and progressive advancements in electrochemical measurement in cyclodextrin-based nanostructured compounds, achieving detection limits as low as μM/nM. This multifunctional cyclodextrin-based nanocomposite was initially utilized in an electrochemical detector for the identification of metal ions. It has since been utilized with the goal of identifying pollutants from the environment of toxic heavy metal ions and other analytical uses. Cyclodextrin-based sensors have shown tremendous promise for a range of analytical applications, such as food safety, biomedical diagnostics, and environmental monitoring, by utilizing these qualities. The development of sensitive, selective, and adaptable sensor platforms for detecting a variety of analytes can be further advanced by ongoing investigations in this field of study. The field of cyclodextrin-based electrochemical detectors has actually consolidated its position in the fields of pharmaceutical science, biochemistry, supramolecular chemistry, polymer chemistry, and nanotechnologies. By modified cyclodextrin-based chemistry with fundamental chemistry principles, a variety of novel chemical sensors possessing excellent sensitivity, outstanding performance, and broad adaptability have been created for analytical pollutants of a common fascination. According to the authors, this chapter will provide readers with fresh perspectives on cyclodextrin-based electrochemical sensors and could help them gain a deeper comprehension of cyclodextrins and the materials that make up their nanocomposite components for electrochemical monitoring, which is used to detect metal ions in the real word application. Because cyclodextrins (CD) can form host–guest complexes with metal ions, they have shown promise as electrochemical sensors for the detection of heavy metal ions. While cyclodextrin-based electrochemical sensors offer advantages like simplicity and the ability to form host–guest complexes with metal ions, their limitations in selectivity, stability, sensitivity, and real-world applicability remain significant challenges. Improving functionalization techniques, enhancing binding affinities, and developing strategies to overcome interference are crucial areas of ongoing research.

Acknowledgements This work was supported by a discovery grant from the PMU-B, NSRF, via the Program Management Unit for Human Resources and Institutional Development, Research, and Innovation [Grant Number B13F660056], and Fundamental Fund 2024, Chiang Mai University. Chiang Mai University and the Center of Excellence for Innovation in Chemistry are gratefully acknowledged for partial support.

References

1. Alam AU, Howlader MMR, Hu N-X, Deen MJ (2019) Electrochemical sensing of lead in drinking water using β-cyclodextrin-modified MWCNTs. Sens Actuators B Chem 296:126632. https://doi.org/10.1016/j.snb.2019.126632
2. Qin X, Tang D, Zhang Y, Cheng Y, He F, Su Z, Jiang H (2020) An electrochemical sensor for simultaneous stripping determination of Cd(II) and Pb(II) based on gold nanoparticles functionalized β-cyclodextrin-graphene hybrids. Int J Electrochem Sci 15:1517–1528. https://doi.org/10.20964/2020.02.59
3. Themsirimongkon S, Waenkaew P, Fang L, Jakmunee J, Kuntalue B, Kuimalee S, Saipanya S (2024) Development of a Pd-xCNT-Ni-yCB composite catalyst to enhance the oxidation of various alcohols in alkali mediums. Biomass Bioenergy 181:107032. https://doi.org/10.1016/j.biombioe.2023.107032

4. Zhang S, Zhao W, Zeng J, He Z, Wang X, Zhu Z, Hu R, Liu C, Wang Q (2023) Wearable non-invasive glucose sensors based on metallic nanomaterials. Mater Today Bio 20:100638. https://doi.org/10.1016/j.mtbio.2023.100638
5. Redhwan TZ, Ali Y, Howlader MMR, Haddara YM (2023) Electrochemical sensing of lead in drinking water using copper foil bonded with polymer. Sensors 23. https://doi.org/10.3390/s23031424
6. Yaiwong P, Iamsawat K, Wiratchan S, Jumpathong W, Semakul N, Bamrungsap S, Jakmunee J, Ounnunkad K (2024) A toluidine blue/porous organic polymer/2D MoSe2 nanocomposite as an electrochemical signaling platform for a sensitive label-free aflatoxin B1 bioassay in some crops. Food Chem 439:138147. https://doi.org/10.1016/j.foodchem.2023.138147
7. Zhao N, Ren L, Du G, Liu J, You X (2022) Determination of heavy metals in water using an FTO electrode modified with CeO_2/rGO nanoribbons prepared by an electrochemical method. RSC Adv 12:21851–21858. https://doi.org/10.1039/d2ra03153a
8. Ashok A, Nguyen TK, Barton M, Leitch M, Masud MK, Park H, Truong TA, Kaneti YV, Ta HT, Li X, Liang K, Do TN, Wang CH, Nguyen NT, Yamauchi Y, Phan HP (2023) Flexible nanoarchitectonics for biosensing and physiological monitoring applications. Small 19. https://doi.org/10.1002/smll.202204946
9. Khaleque MA, Hossain MI, Ali MR, Bacchu MS, Saad Aly MA, Khan MZH (2023) Nanostructured wearable electrochemical and biosensor towards healthcare management: a review. RSC Adv 13:22973–22997. https://doi.org/10.1039/d3ra03440b
10. Anuthum S, Wiratchan S, Semakul N, Jakmunee J, Ounnunkad K (2024) Signaling redox probe/DNA aptamer complexes on a new POP/2D WSe2 composite-based immunosensor towards the simultaneous detection of three-protein overexpression as an alternative severe SARS-COV-2 infection diagnosis. Sens Actuators B Chem 404:135196. https://doi.org/10.1016/j.snb.2023.135196
11. Arivazhagan M, Shankar A, Maduraiveeran G (2020) Hollow sphere nickel sulfide nanostructures–based enzyme mimic electrochemical sensor platform for lactic acid in human urine. Microchim Acta 187. https://doi.org/10.1007/s00604-020-04431-3
12. Arivazhagan M, Santhosh YM, Maduraiveeran G (2021) Non-enzymatic glucose detection based on nis nanoclusters@nis nanosphere in human serum and urine. Micromachines 12. https://doi.org/10.3390/mi12040403
13. Ko M, Mendecki L, Eagleton AM, Durbin CG, Stolz RM, Meng Z, Mirica KA (2020) Employing conductive metal-organic frameworks for voltammetric detection of neurochemicals. J Am Chem Soc 142:11717–11733. https://doi.org/10.1021/jacs.9b13402
14. Zhang L, Ye C, Li X, Ding Y, Liang H, Zhao G, Wang Y (2018) A CuNi/C nanosheet array based on a metal–organic framework derivate as a supersensitive non-enzymatic glucose sensor. Nano-Micro Lett 10. https://doi.org/10.1007/s40820-017-0178-9
15. Bao C, Lu Y, Liu J, Gao Y, Lu L, Liu S (2022) Sensitive detection of Cu^{2+}
16. Feng Z, Lim HN, Ibrahim I, Endot NA, Malek EA, Gowthaman NSK (2024) Enhanced simultaneous voltammetric detection of dopamine and uric acid using Au@Ni-metal–organic framework-modified electrode. Appl Organometall Chem 1–16. https://doi.org/10.1002/aoc.7350
17. Arivazhagan M, Maduraiveeran G (2020) Ultra-fine nickel sulfide nanoclusters @ nickel sulfide microsphere as enzyme-free electrode materials for sensitive detection of lactic acid. J Electroanal Chem 874:114465. https://doi.org/10.1016/j.jelechem.2020.114465
18. Zhang X, Sun R (2023) Construction of an electrochemical sensor for detection of nitrite by gold nanoparticles immobilized on biochar. Int J Electrochem Sci 18:100219. https://doi.org/10.1016/J.IJOES.2023.100219
19. Krishnan SK, Prokhorov E, Bahena D, Esparza R, Meyyappan M (2017) Chitosan-covered Pd@Pt core-shell nanocubes for direct electron transfer in electrochemical enzymatic glucose biosensor. ACS Omega 2:1896–1904. https://doi.org/10.1021/acsomega.7b00060
20. Fan B, Spindler BD, Zhao W, Chan H, Wang Z, Kim M, Chipangura Y, Bühlmann P, Stein A (2023) Comparison of Copper(II) oxide nanostructures with different morphologies for nonenzymatic glucose sensing. ACS Appl Nano Mater 6:1475–1486. https://doi.org/10.1021/acsanm.2c05433

21. Abate C, Neri G, Scala A, Mineo PG, Fazio E, Mazzaglia A, Fragoso A, Giuffrè O, Foti C, Piperno A (2023) Screen-printed carbon electrodes with cationic cyclodextrin carbon nanotubes and ferrocenyl-carnosine for electrochemical sensing of Hg(II). ACS Appl Nano Mater 6:17187–17195. https://doi.org/10.1021/acsanm.3c03480
22. Zhu G, Yi Y, Chen J (2016) Recent advances for cyclodextrin-based materials in electrochemical sensing. TrAC—Trends Anal Chem 80:232–241. https://doi.org/10.1016/j.trac.2016.03.022
23. Arivazhagan M, Kannan P, Maduraiveeran G (2022) Gold nanoclusters dispersed on gold dendrite-based carbon fibre microelectrodes for the sensitive detection of nitric oxide in human serum. Biosensors 12. https://doi.org/10.3390/bios12121128
24. Niu X, Mo Z, Yang X, Sun M, Zhao P, Li Z, Ouyang M (2018) Advances in the use of functional composites of β-cyclodextrin. Microchimica Acta 185
25. Arivazhagan M, Maduraiveeran G (2022) Gold-dispersed hierarchical flower-like copper oxide microelectrodes for the sensitive detection of glucose and lactic acid in human serum and urine. Biomater Sci 10:4538–4548. https://doi.org/10.1039/d2bm00527a
26. Sumathi P, Suganthi S, Sivaraj R, Selvan GT, Selvakumar PM, Enoch IVMV (2019) Novel supramolecular β-cyclodextrin-piperidin-4-one complex assembled on gold nanoparticles. Selective detection of Cd^{2+} ions. Microchem J 150:104066. https://doi.org/10.1016/j.microc.2019.104066
27. Zhao G, Wang H, Liu G (2016) Electrochemical determination of trace cadmium in soil by a bismuth film/graphene-β-cyclodextrin-nafion composite modified electrode. Int J Electrochem Sci 11:1840–1851. https://doi.org/10.1016/s1452-3981(23)16065-2
28. Arivazhagan M, Maduraiveeran G (2023) Hierarchical gold dispersed nickel oxide nanodendrites microarrays as a potential platform for the sensitive electrochemical detection of glucose and lactate in human serum and urine. Mater Chem Phys 295:127084. https://doi.org/10.1016/j.matchemphys.2022.127084
29. Luo L, Xu QQ, Chen KK, Liu ZG, Guo Z (2023) Carbon spheres decorated with $Fe_2O_3/Fe_3C/Fe$ composite nanoparticles for highly sensitive and selective electrochemical determination of nitrite (NO^{2-}). Sens Actuators B Chem 382:133488. https://doi.org/10.1016/j.snb.2023.133488
30. Arivazhagan M, Kannan P, Maduraiveeran G (2023) Nanostructured transition metal sulfide-based glucose and lactic acid electrochemical sensors for clinical applications. Curr Top Med Chem 23:284–294. https://doi.org/10.2174/1568026623666221205093154
31. Lv M, Wang X, Li J, Yang X, Zhang C, Yang J, Hu H (2013) Cyclodextrin-reduced graphene oxide hybrid nanosheets for the simultaneous determination of lead(II) and cadmium(II) using square wave anodic stripping voltammetry. Electrochim Acta 108:412–420. https://doi.org/10.1016/j.electacta.2013.06.099
32. Huang S, Lu S, Huang C, Sheng J, Su W, Zhang L, Xiao Q (2015) Sensitive and selective stripping voltammetric determination of copper(II) using a glassy carbon electrode modified with amino-reduced graphene oxide and β-cyclodextrin. Microchim Acta 182:2529–2539. https://doi.org/10.1007/s00604-015-1627-0
33. Wang Z-W, Liu H-J, Li C-Y, Chen X, Weerasooriya R, Wei J, Lv J, Lv P, Wu Y-C (2020) Mesoporous g-C3N4/β-CD nanocomposites modified glassy carbon electrode for electrochemical determination of 2,4,6-trinitrotoluene. Talanta 208:120410. https://doi.org/10.1016/j.talanta.2019.120410
34. Tu X, Gao F, Ma X, Zou J, Yu Y, Li M, Qu F, Huang X, Lu L (2020) Mxene/carbon nanohorn/β-cyclodextrin-metal-organic frameworks as high-performance electrochemical sensing platform for sensitive detection of carbendazim pesticide. J Hazard Mater 396:122776. https://doi.org/10.1016/j.jhazmat.2020.122776
35. He Y, Ma L, Zhou L, Liu G, Jiang Y, Gao J (2020) Preparation and application of bismuth/MXene nano-composite as electrochemical sensor for heavy metal ions detection. Nanomaterials 10:866. https://doi.org/10.3390/nano10050866
36. Cheng H, Yang J (2020) Preparation of Ti_3C_2-PANI composite as sensor for electrochemical determination of mercury ions in water. Int J Electrochem Sci 15:2295–2306. https://doi.org/10.20964/2020.03.24

37. Lv X, Pei F, Feng S, Wu Y, Chen S-M, Hao Q, Lei W (2020) Facile synthesis of protonated carbon nitride/$Ti_3C_2T_x$ nanocomposite for simultaneous detection of Pb^{2+} and Cd^{2+}. J Electrochem Soc 167:067509. https://doi.org/10.1149/1945-7111/ab7e22
38. Rasheed PA, Pandey RP, Gomez T, Naguib M, Mahmoud KA (2020) Large interlayer spacing $Nb_4C_3T_x$ (MXene) promotes the ultrasensitive electrochemical detection of Pb^{2+} on glassy carbon electrodes. RSC Adv 10:24697–24704. https://doi.org/10.1039/D0RA04377J
39. Xiao P, Zhu G, Shang X, Hu B, Zhang B, Tang Z, Yang J, Liu J (2022) An Fe-MOF/MXene-based ultra-sensitive electrochemical sensor for arsenic(III) measurement. J Electroanal Chem 916:116382. https://doi.org/10.1016/j.jelechem.2022.116382
40. Kravtsova AA, Skuredina AA, Malyshev AS, Le-Deygen IM, Kudryashova EV, Budynina EM (2023) The solubility studies and the complexation mechanism investigations of biologically active Spiro[cyclopropane-1,3′-oxindoles] with β-cyclodextrins. Pharmaceutics 15. https://doi.org/10.3390/pharmaceutics15010228
41. Pandey A (2021) Cyclodextrin-based nanoparticles for pharmaceutical applications: a review. Environ Chem Lett 19:4297–4310. https://doi.org/10.1007/s10311-021-01275-y
42. Ji J, Qu L, Wang Z, Li G, Feng W, Yang G (2022) A facile electrochemical chiral sensor for tryptophan enantiomers based on multiwalled carbon nanotube/hydroxypropyl-β-cyclodextrin functionalized carboxymethyl cellulose. Microchem J 175:107133. https://doi.org/10.1016/j.microc.2021.107133
43. Ma C, Xu P, Chen H, Cui J, Guo M, Zhao J (2022) An electrochemical sensor based on reduced graphene oxide/β-cyclodextrin/multiwall carbon nanotubes/polyoxometalate tetracomponent hybrid: simultaneous determination of ascorbic acid, dopamine and uric acid. Microchem J 180:107533. https://doi.org/10.1016/j.microc.2022.107533
44. Tchoffo R, Ngassa GBP, Doungmo G, Kamdem AT, Tonlé IK, Ngameni E (2022) Surface functionalization of natural hydroxyapatite by polymerization of β-cyclodextrin: application as electrode material for the electrochemical detection of Pb(II). Environ Sci Pollut Res 29:222–235. https://doi.org/10.1007/s11356-021-15578-8
45. Tang LW, Alias Y, Woi PM (2024) Simultaneous quantification of cadmium and lead ions using β-cyclodextrin functionalized poly(amino acid) electrochemical sensor. J Electroanal Chem 967:118427. https://doi.org/10.1016/j.jelechem.2024.118427
46. Zhumanazar N, Korolkov IV, Yeszhanov AB, Shlimas DI, Zdorovets MV (2023) Electrochemical detection of lead and cadmium ions in water by sensors based on modified track-etched membranes. Sens Actuators A 354:114094. https://doi.org/10.1016/j.sna.2022.114094
47. Lu Y, Liang X, Xu J, Zhao Z, Tian G (2018) Synthesis of $CuZrO_3$ nanocomposites/graphene and their application in modified electrodes for the co-detection of trace Pb(II) and Cd(II). Sens Actuators B Chem 273:1146–1155. https://doi.org/10.1016/j.snb.2018.06.104
48. Dimovasilis PA, Prodromidis MI (2013) Bismuth-dispersed xerogel-based composite films for trace Pb(II) and Cd(II) voltammetric determination. Anal Chim Acta 769:49–55. https://doi.org/10.1016/j.aca.2013.01.040
49. Tarley CRT, Santos VS, Baêta BEL, Pereira AC, Kubota LT (2009) Simultaneous determination of zinc, cadmium and lead in environmental water samples by potentiometric stripping analysis (PSA) using multiwalled carbon nanotube electrode. J Hazard Mater 169:256–262. https://doi.org/10.1016/j.jhazmat.2009.03.077
50. Xuan X, Park JY (2018) A miniaturized and flexible cadmium and lead ion detection sensor based on micro-patterned reduced graphene oxide/carbon nanotube/bismuth composite electrodes. Sens Actuators B Chem 255:1220–1227. https://doi.org/10.1016/j.snb.2017.08.046
51. Badruddoza AZM, Shawon ZBZ, Tay WJD, Hidajat K, Uddin MS (2013) Fe_3O_4/cyclodextrin polymer nanocomposites for selective heavy metals removal from industrial wastewater. Carbohyd Polym 91:322–332. https://doi.org/10.1016/j.carbpol.2012.08.030
52. Palanisamy S, Thangavelu K, Chen SM, Velusamy V, Chang MH, Chen TW, Al-Hemaid FMA, Ali MA, Ramaraj SK (2017) Synthesis and characterization of polypyrrole decorated graphene/β-cyclodextrin composite for low level electrochemical detection of mercury (II) in water. Sens Actuators B Chem 243:888–894. https://doi.org/10.1016/j.snb.2016.12.068

Dr. Mani Arivazhagan is a Post doctoral researcher (Faculty of Science) in the Department of Chemistry at Chiang Mai University, Chiang Mai, Thailand. He has more than 8 years of experience in teaching and research. He has published 15 research papers in the reputed journals and 1 Indian patent grant. His h-index is 9, and i10 index is 8. His current research interest includes the synthesis of nanomaterials, flexible electrodes and applications related to the electrochemical sensors, Immunosensors, and biosensors.

Dr. Jaroon Jakmunee is an Associate Professor in the Department of Chemistry at Chiang Mai University, Chiang Mai, Thailand. He has more than 30 years of experience in teaching and research. He has published more than 250 research papers. His h-index is 42, and i10 index is 124. He has received funding from various Thailand funding agencies, like PMU-B, NRCT, NIA. His current research interest is developing application materials using Flexible SPCE electrodes in biosensors and Environmental applications, Flow analysis and Microfluidics.

Chapter 10
Electrochemical Detection of Pollutants and Biomolecules Using Cyclodextrins

Vinitha Packirisamy, Ariyamuthu Rajabhuvaneswari, and M. V. Arularasu

Abstract Cyclodextrins (CDs) are a type of oligosaccharide. The existence of glycosidic oxygen bridges and hydrogen atoms from OH groups show activity in the inner cavity. The hydrophobicity of CD facilitates the selective binding of a variety of guest molecules, which leads to the formation of host–guest inclusion complexes or supramolecular complexes. In addition to that, CDs can serve as functional polymers to increase the dispersibility of practical materials in solvents due to their hydrophilic surface. These fascinating characteristics of CDs led to their substantial and exciting potential uses in sensing various pollutants and biologically significant molecules. Electrochemistry-based sensors offer a sensitive and selective determination of analytes among various sensing techniques due to their unique electron transfer mechanism. Electrochemical sensors provide an intriguing way of analyzing a sample concentration due to the direct conversion of an electrochemical reaction to an electronic signal. CDs-based functional materials such as CDs/metal nanomaterials, CDs/graphene, CDs/conducting polymers, and CDs/carbon nanomaterials are enormously employed as electrode material that promotes the excellent sensing of precise quantification of analytes with their large surface area and high conductivity nature. In this chapter, the most standard electrochemical techniques, such as cyclic voltammetry, linear sweep voltammetry, amperometry, differential pulse voltammetry, and square wave voltammetry techniques adopted to detect the concentration of organic and inorganic pollutants, and biomolecules by using CDs-based functional materials have been discussed. Cyclodextrin-modified electrochemical sensors allow for the selective detection of both environmental contaminants and essential biomolecules with increased precision. This novel technique brings up new possibilities for real-time environmental monitoring, wearable biomedical diagnostics, and multi-analyte sensing technologies.

V. Packirisamy (✉) · M. V. Arularasu
Centre for Global Health Research, Saveetha Medical College and Hospital, Saveetha Institute of Medical and Technical Sciences, Saveetha University, Thandalam, Chennai, India
e-mail: vinithaprs18@gmail.com

A. Rajabhuvaneswari
Department of Chemistry, Karpaga Vinayaga College of Engineering and Technology, Padalam, India

S. Krishnamoorthy (ed.), *Cyclodextrins for Chemosensing*,
https://doi.org/10.1007/978-3-031-89299-8_10

Keywords Cyclodextrin-based functional materials · Electrochemical sensor · Pollutants · Biomolecules · Amperometry · Voltammetry

10.1 Introduction

The inclusion of complex formations between the guest molecules and cyclodextrin often leads to an increase in absorbance and fluorescence intensity. The cyclodextrins, specifically 6α-, 7β-, and 8γ-cyclodextrin are capable of forming macrocycles by connecting doughnut-shaped oligosaccharides through α-1, 4-glycosidic linkages. Additionally, it can form an inclusion complex with hydrophobic guest molecules with the appropriate size and shape through its external hydrophilic surface [1]. β-cyclodextrin can generate uniform dispersed solutions in water, allowing it to interact with various chemical compounds and pharmaceutical drugs by host–guest interactions [2]. Cyclodextrins are widely recognized for their supramolecular structures, which makes them valuable building blocks for a variety of applications. However, their versatility extends beyond this, finding applications in many sectors such as analytical chemistry, agriculture, medicine, cosmetics, and more [3, 4].

Absorption-based technology in sensing has attracted significant attention as a rapid and effective method for detecting chemical and bio-species through its user-friendly nature, versatility, high selectivity, and sensitivity [5, 6]. There has been a growing interest in supramolecular chemistry for sensing heavy metal cations due to their crucial roles in biological, pharmacological, and environmental processes. Enhanced, very sharp, and exceptionally responsive sensors have the potential to be utilized in various applications such as environmental analysis [7], wearable bio/chemical sensors [8], and sensors designed for monitoring municipal air quality [9]. Since the discovery of carbon nanotubes and graphene, there has been significant interest in exploring various two-dimensional, layered, and nonlayered materials due to their distinct features [10, 11]. Electrochemical measurements have numerous advantages, such as affordability, exceptional selectivity and sensitivity, minimal environmental impact, and real-time operation. These benefits have made them highly relevant and influential in molecular recognition [12, 13]. The design of β-CD based electrochemical sensors can be accomplished through various techniques, such as cyclodextrin-modified screen-printed electrodes [14], polymer films [15], Langmuir–Blodgett self-assembly [16], and nanomaterials (multiwalled carbon nanotubes, graphene, etc.) [17, 18]. Molecular recognition systems (MRS) encompass a diverse range of substances, including enzymes, proteins, aptamers (single-stranded DNA or RNA), as well as supramolecular entities like cyclodextrins possess excellent sensing performance [19, 20]. In this chapter, the design and construction of CD-based electrochemical sensors for the detection of pollutants and biomolecules were discussed briefly.

10.2 Design of Cyclodextrin-Based Materials for Sensor Applications

The inclusion of nanomaterials (e.g., AuNPs, CNTs, graphene) into CD-based sensors markedly improves their electrochemical characteristics. The enhanced surface area and conductivity of nanomaterials enhance the selective binding characteristics of CDs, leading to increased sensitivity and expedited reaction times. Cyclodextrins are frequently employed to alter electrode surfaces in electrochemical sensors, improving the affinity for target molecules. These alterations enhance their detection sensitivity, especially in systems intended for identifying harmful contaminants or biomolecules in trace quantities.

10.3 Cyclodextrins/Graphene Oxide

CD-graphene oxide integration has grown in popularity in the past decade. Shao et al. showed that the combination of CDs and graphene oxide (GO) enabled the detection of biomolecules and drugs. Graphene oxide's unique features make it ideal for electrochemical sensors [21]. Pure graphene is difficult to manufacture and sustain, but GO can be synthesized using the modified Hummers method [22]. Chemical or electrochemical reduction can yield reduced GO (rGO), which improves conductivity. High conductivity and electron transport make rGO ideal for electrochemical sensors. The functional groups of CDs allow covalent attachment to graphene sheets [23]. The hydroxyl (-OH) groups on cyclodextrins (CDs) and the oxygen-containing groups on graphene oxide (GO) or reduced graphene oxide (rGO) sheets form hydrogen bonds, increasing CD's bonding ability [24]. CD-treated rGO sheets had less aggregation and a comparable high surface area to well-dispersed sheets and decreased restacking [25]. The fabrication of rGO/CD-based non-covalent sensors is easy and requires few chemical modifications. After dispersing graphene oxide (GO) in a solution, CDs are mixed with it to generate stable aqueous dispersions [11, 26, 27]. Thus, reducing agents like hydrazine or the more environmentally friendly ascorbic acid can chemically decrease GO/CD and generate rGO/CD [28, 29]. Other modified cyclodextrins, including 2, 6-dimethyl-β-CD and mercapto-β-CD, have been utilized alongside SH-β-CD [30]. Noncovalent π-π interactions have been used to immobilize dopamine-β-cyclodextrin and pyrene-β-cyclodextrin onto reduced graphene oxide (rGO) for efficient electrochemical sensors [31, 32].

A straightforward wet chemical strategy for making β-CD/rGO hybrid nanosheets was initially presented by Guo et al. [11]. The process for producing β-CD/rGO involves combining graphene oxide and β-CD as shown in Fig. 10.1. The nanohybrids show excellent stability and dispersibility in various kinds of solvents. Eight probe molecules (dopamine, uric acid, norepinephrine, tyrosine, tryptophan, acetaminophen, rutin, and thioridazine) were electrochemically sensed using the β-CD/rGO. The synergistic interactions between rGO and β-CD molecules, β-CD/

rGO hybrid nanosheets could exhibit extremely high supramolecular recognition as well as a significantly stronger electrochemical response toward analytes than unmodified rGO and CNT [11].

Polymers and linker molecules modify or functionalize rGO sheets to make them appropriate for sensing. A substance that links the rGO and CD layers is 3, 4, 9, 10-perylene tetracarboxylic acid [33], and poly(diallyl dimethyl ammonium chloride), which can be used to produce well-functionalized supramolecular structures [25]. Increased electron transfer efficiency has been achieved with cytochrome c, Prussian blue-chitosan functionalized reduced graphene oxide (rGO), and carbon dots (CDs) [34, 35]. The aforementioned methods have demonstrated that combining CD and rGO with metal nanoparticles can enable highly effective electron transport. In addition, they have been widely employed in the electrochemical sensing of various substances and offer a more cost-effective alternative to bulk versions [36].

One example is using a straightforward one-pot synthesis method to reduce $HAuCl_4$ in situ and produce reduced graphene oxide (rGO) with AuNPs and β-CD [37]. The analysis revealed that the β-CD served a dual role as both a dispersion agent and a reducing agent. This facilitated the conversion of GO to rGO and $AuCl_4^-$ to gold nanoparticles.

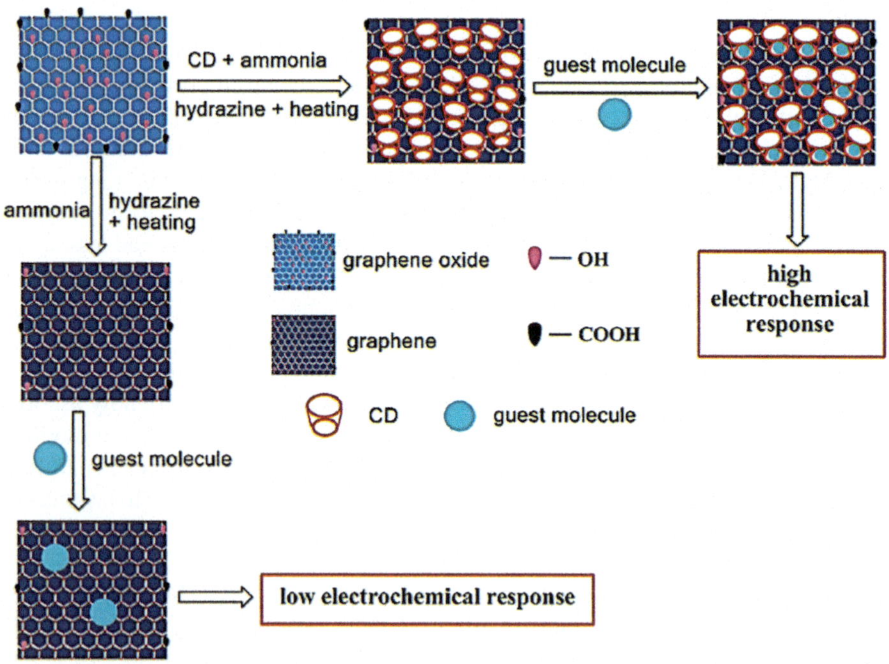

Fig. 10.1 Schematic representation of the procedure for making CD/graphene hybrid nanosheets and GNs, and electrochemical sensing of guest molecules. Reproduced from Guo et al. [11] 2010 with permission from American Chemical Society, 2010

CDs are also modified with various antibodies, enzymes, and aptamers to detect critical chemicals, such as hydrogen peroxide, and more complex structures [38]. rGO, MWCNTs, and NPs are typically used to modify CDs to produce biosensors. Graphitic carbon and MXenes are two examples of developing materials that are being incorporated into CD-based sensor designs. While many of these sensors incorporate MWCNTs, SWCNTs, and other metal nanoparticles are also being adopted to generate excellent sensors.

10.4 Cyclodextrins/Carbon Nanotubes

During the past ten years, substantial attention has been directed toward combining carbon nanotubes (CNTs) and cyclodextrins (CDs) due to extensive research proving the beneficial effects of integrating these two substances. In a recent study [39, 40], it was demonstrated that the improved electron-transfer processes that occur when the exceptional qualities of carbon nanotubes (CNTs) are combined with the capacity of cyclodextrins (CDs) to form supramolecular inclusion complexes are a result of the combination of these two properties.

Carbon materials can improve the performance of supporting materials in various processes, including electrochemistry, chemiluminescence, and catalysis while preserving the advantageous properties of these materials. Consequently, the utilization of functionalized carbon nanomaterials has the potential to considerably improve the efficiency of bioassays and assessments of the environment [28, 41]. CDs-modified with various carbon nanostructures to take advantage of their detection capacity and enhance the efficiency of electron transfer movement. These customized CDs have been utilized as sensors to conduct environmental studies.

To integrate carbon nanotubes (CNTs) and carbon dots (CDs), the process of wet chemical synthesis, analogous to the method utilized in producing graphene sheets, is used successfully [42]. It is because of the excellent dispersion of carbon nanotubes (CNTs) in a solution of cyclodextrin (CD). The treatment of carbon nanotubes (CNTs) with nitric acid, results in the creation of -COOH groups [43]. Incorporating -COOH groups into this process results in a further dispersion of the carbon nanotubes (CNTs), which in turn damages the environment due to the rapid release of these functionalized CNTs into water [44]. It is possible to find the CDs in the middle of the MWCNTs. According to Alam et al. [17] the insertion of MWCNTs coated on glassy carbon electrodes resulted in the distribution of the particles that displayed a consistent voltammogram [45, 46]. In addition, functionalized carbon dots (CDs) have been integrated with carbon nanotubes (CNTs) [47], and CNTs/CDs have undergone additional modifications using metal nanoparticles such as gold [48, 49], copper [50], cobalt [51], silver [38], and platinum [52] used for sensing biomolecules and pollutants.

10.5 Cyclodextrins/Metal or Carbon Organic Frameworks

Metal–organic frameworks (MOFs) and carbon-organic frameworks (COFs) are examples of porous materials. These materials are made up of organic ligands that are coupled with a large number of metal ions or clusters. In the 1990s, several pioneering groups led by Ferey, Moore, Yaghi, Kitagawa, and Robson [53], were responsible for discovering and developing metal–organic frameworks (MOFs). Depending on the framework, nanoporous materials can display one-, two, or three-dimensional structures. Various metallic species, including p-block elements, actinides, lanthanides, and alkaline earth metals, can be incorporated into the structures with appropriate proportions. In addition, they have been manufactured by combining various organic linkers, including a wide range of functional groups, including carboxylate, phosphonate, sulfonate, pyridyl, imidazole, and isolate among others [54]. These materials can obtain surface areas ranging from one thousand to ten thousand square meters per gram, significantly higher than the surface areas of other materials with small holes or voids. Synthetic methods such as hydrothermal or solvothermal procedures, vapor diffusion, or microwave irradiation have been utilized to include CDs in MOFs based on alkaline or alkaline earth metals as inorganic equivalents [18]. CDs can be inserted into MOFs without further functionalization activities, which is a substantial advantage independent of the technology used for sensing.

10.6 Cyclodextrins/Conducting Polymers

Electrochemical sensors that contain immobilized nanoparticles (NPs) have been achieved by utilizing several different polymer matrices. Polythiophene derivatives, such as poly (3,4-ethylenedioxythiophene) (PEDOT), polyaniline (PANI), and polypyrrole (PPy), are conducting polymers that have been the subject of substantial research [15, 55]. The capacity of these polymeric materials to provide a conducting framework, possess ion exchange characteristics, and have a very porous surface area that can be precisely regulated by adjusting the rate of electropolymerization. Electropolymerizing pyrrole with a dopant at potentials ranging from approximately 0.6 to 0.8 V versus SCE on various conducting substrates can produce an adhesive conducting film [56, 57]. The utilization of cyclodextrins in various conducting polymers has demonstrated that the CDs significantly improve the detection of the molecules that are the focus of the investigation. The films immobilized with CDs displayed the highest level of performance among all of these combinations.

10.7 Electrochemical Sensors for the Detection of Pollutants

B-cyclodextrin (β-CD) detects harmful contaminants by building inclusion complexes with hydrophobic organic pollutants such pesticides, PAHs, and hormone-disrupting compounds. The hydrophobic cavity of β-CD encapsulates pollutants, thereby concentrating them on the sensor surface and enhancing the electrochemical signal during detection [58, 59]. The selective binding enhances the sensitivity and specificity of electrochemical sensors, facilitating the sensing of pollutants at low concentrations within complex environmental samples.

10.8 Cyclic and Linear Sweep Voltammetry

CV demonstrates significant versatility, offering essential insights into the redox behavior of analytes and the processes occurring at electrodes. This method facilitates the examination of reaction mechanisms and kinetics while providing a wide range for detecting and examining oxidation and reduction peaks. Sinniah et al. reported the sensing of bisphenol A by magnetic nanoparticles (Fe_3O_4) coated with CDs functionalized-ionic liquid through cyclic voltammetry studies. This sensor material recognizes bisphenol A at the lower electrochemical potential of 0.5 V [60]. Xu et al. reported hydroxypropyl(HP)-β-CD modified graphene nanosheets for the detection of nitrophenol through cyclic voltammetric studies with a detection limit of 1×10^8 M [30]. Ali et al. reported the bisphenol A determination by using MWCNT with β-CD coated on screen printed carbon electrode through linear sweep voltammetric measurement. Two protons and two electrons were involved in a diffusion-controlled oxidation process on the electrochemical detection of BPA (shown in Fig. 10.2). The sensor displayed a two-step linear response at optimal conditions, from 125 nM to 2 μM and 2 to 30 μM with LOD of 13.76 nM [61]. Glassy carbon electrode (GCE) modified SWCNT coated with β-CD was used to construct a sensitive electrochemical sensor for methyl parathion (MP). The sensor demonstrated sensitive MP detection, exhibiting a low limit of detection at 7.23 nM. The integration of β-CD and SCN markedly enhanced the electrochemical properties of the sensor [62].

10.9 Amperometric Sensor

Abnormal levels of uric acid(UA) in biological fluids lead to several diseases. Carboxylic-acid-functionalized MWCNT/ β-CD nanocomposite deposited on GCE and coated with nafion and hydrothane polyurethane adopted as an enzyme-free electrochemical sensor for the determination of UA through amperometric titration method. In this method, the changes in current have been recorded at various time

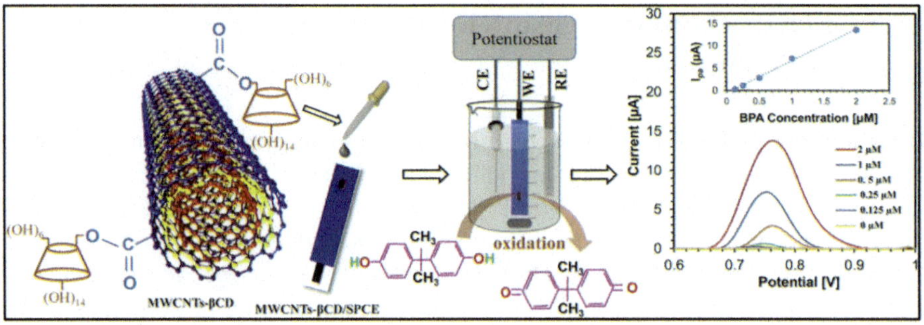

Fig. 10.2 Fabrication of MWCNT-βCD coated SPCE for the detection of bisphenol A concentration through linear sweep voltammetric measurement. Reproduced from Ali et al. [61] 2020 with permission from Elsevier, 2020

intervals by the gradual addition of UA. This sensor material is sensitive to a UA detection range of 100–700 μM and selective to UA in the presence of various interferents [63]. Fei et al. employed a thiolated β-CD functionalized AuNPs modified electrode for the amperometric detection of quercetin. The research suggested that the inclusion complexation of β-CD markedly improved the sensor's sensitivity, enabling the detection of quercetin at concentrations as low as 285 pM [64].

10.10 Differential Pulse Voltammetry

On the other hand, by delivering a sequence of voltage pulses overlaid on a linear sweep, DPV provides more sensitivity than CV. This method minimizes background noise and enhances signal resolution, thereby facilitating the detection of trace levels of analytes, such as pollutants and biomolecules.1-Chloro-4-nitrobenzene (CNB) is an organic pollutant that has been used in daily activities that contain hazardous substances that cause respiratory problems. Kingsford et al. prepared carbon nanohorns and β-CD nanohybrids for the sensitive detection of CNB with a detection limit of 9.0 nM. The sensing parameters can be optimized through sensitive electrochemical techniques, i.e., DPV which offers better resolution and great current sensitivity. Through this technique, pH, time, and concentration electrode material were optimized. This sensor shows excellent redox activity towards CNB with a detection range of 0.05–1.0 μM [65]. Mercury (II) is one of the most toxic elements that accumulation leads to severe disorders and diseases. Graphene- β-CD /PPy composite modified screen-printed carbon electrode shows a potential sensing activity towards Hg(II) ions with a lower detection limit of 0.47 nM/L. For Hg(II), the graphene-CD/PPy modified SPCE exhibits a distinct and precise DPV response, with the anodic peak appearing at 0.22 V. The lone pair of nitrogen atoms in the PPy acts as a binding site for Hg(II) and facilitates the sensing of Hg as shown in Fig. 10.3

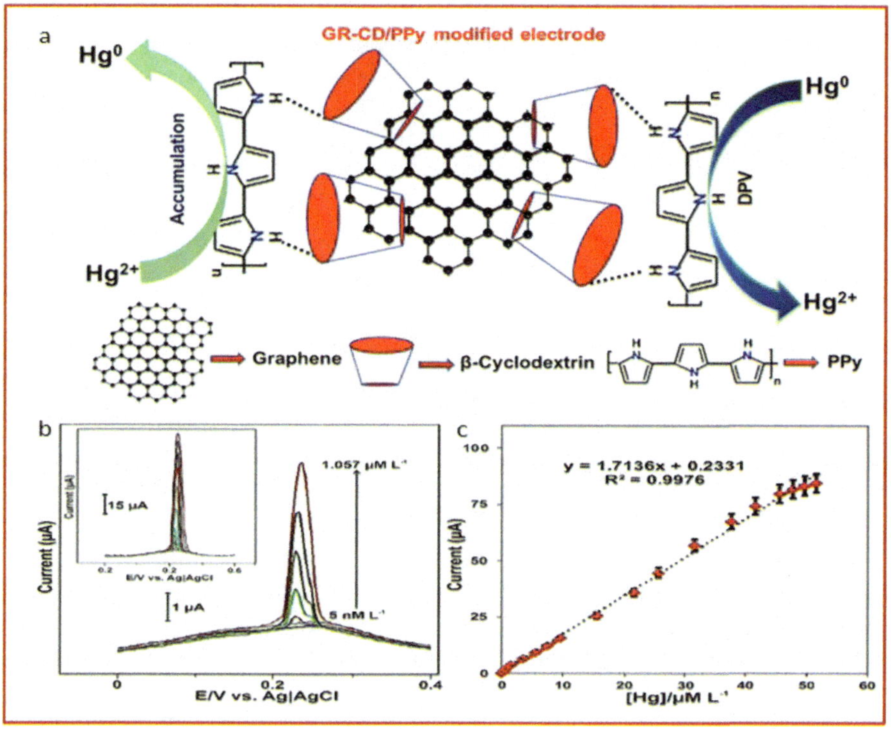

Fig. 10.3 The plausible electrochemical mechanism for mercury sensing at GR-CD/PPy composite. Reproduced from Palanisamy et al. [66] 2017 with permission from Elsevier, 2017

[66]. Chen et al. developed a poly (β-CD)/carbon quantum dots-based sensor material for the simultaneous determination of UA, dopamine, and tryptophan [42].

10.11 Square Wave Voltammetry

Square wave Voltammetry (SWV) is an electrochemical method that utilizes a square-shaped potential waveform applied to the electrode. This method alternates between forward and reverse potentials during a scan, facilitating rapid and sensitive analyte detection. SWV provides high sensitivity and low detection limits by effectively reducing background currents and improving the signal-to-noise (S/N) ratio. This renders it suitable for the detection of trace levels of toxic pollutants. Furthermore, SWV facilitates swift screening of pollutants through its rapid data acquisition capabilities. Xie et al. developed a sensor for the detection of dichlorophenol which causes severe threat to humans. A composite of β-CDs with boron-doped graphene aerogels (BGAs/CDs) offers high-performance electrochemical sensing of dichlorophenol (Dcp) in nM ranges with a LOD of 0.33 nM. Graphene aerogels possess higher

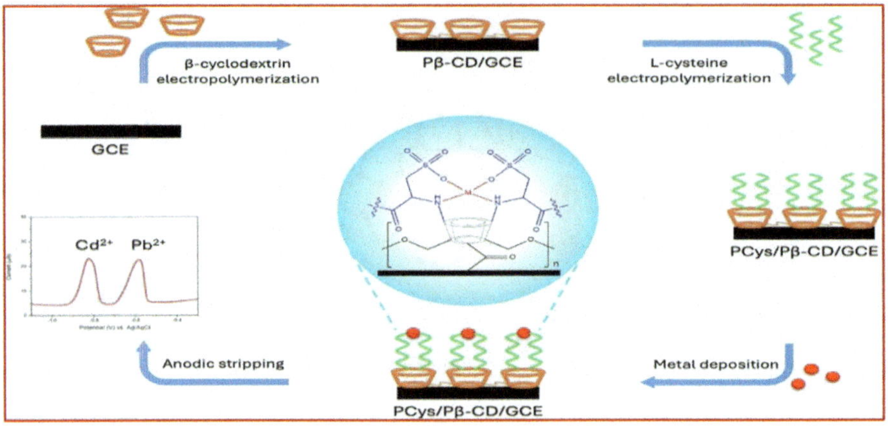

Fig. 10.4 Electropolymerization of poly cysteine/β-CD on glassy carbon electrode for the detection of Pb(II) and Cd(II) ions through square wave anodic stripping voltammetric analysis. Reproduced from Tang et al. [68] 2024 with permission from Elsevier, 2024

surface area and conductivity, which are essential for rapid electron transfer, boron doping promotes the active sites for Dcp and β-CDs capture. Dcp through host–guest recognition. The combination of these materials into a composite facilitates the enhanced sensing of Dcp [67]. Fenitrothion(FNT) is one of the major components in pesticides commonly used to increase crop yield but causes adverse effects on the nervous system in living beings. Wang et al. reported the electro-polymerization of β-CDs on a glassy carbon electrode for the sensing of FNT with a detection limit of 6 ng/ml. The sensing ability of this material towards FNT was demonstrated with real samples like water and vegetables, which gave satisfactory results [55]. The electrode was generated utilizing an environmentally friendly method, incorporating β-CD and l-cysteine through electropolymerization on the surface. The sensor exhibited high selectivity, sensitivity, stability, reproducibility, and a low LOD across a broad linear range (shown in Fig. 10.4). The sensor developed using stripping voltammetric analysis demonstrated sensitivities of 346.26 $\mu A \mu M^{-1}$ cm^{-2} and 250.54 $\mu A \mu M^{-1}$ cm^{-2}, with LODs of 1.13 nM and 0.53 nM for Cd^{2+} and Pb^{2+}, respectively [68].

10.12 Electrochemiluminescence

Electrochemiluminescence (ECL) is a method that integrates electrochemical reactions with luminescence for the detection of analytes. An electrochemical process in ECL initiates a light-emitting reaction upon the excitation of specific luminophores at the electrode surface. The emitted light's intensity is directly proportional to the target analyte concentration, facilitating quantitative detection. Over the past few decades, electrochemiluminescence (ECL), which is the production of an optical signal initiated by an electrochemical reaction, has received significant attention due to its sensitive, straightforward optical setup, low background signal, and excellent temporal and spatial control. As a result, it has grown to be a significant and valuable detection method in biosensors. To develop a sensitive sensor probe for the determination of Hg(II) concentration, Cheng et al. fabricated the Fc-Probe/$Ru(bpy)_3{}^{2+}$/Cyclodextrin-AuNPs/Nafion/GCE sensor, which shows a detection limit of 0.1 nM. The ECL measurement was carried out using a triangular potential scan with a scan rate of 100 mV/s. The quantification of Hg(II) ions is done through increasing ECL intensity [69]. Table. 10.1 provides details on the electrochemical sensing of various pollutants by using CD-based sensor probes and adopting different electrochemical methods. Most of the sensors listed in the table are performed as a good sensor probe to detect the various pollutants and analytes within their permissible concentration. These materials show higher sensitivity and selectivity towards the trace concentration of specific toxins. The sensing performance of these materials also tested with real-time samples to check their activity, shows satisfactory results.

10.13 Summary

This chapter reveals the supramolecular properties of cyclodextrins, which are used in the fabrication of various efficient sensor probes with sensitive materials like graphene, carbon nanotube, MXene, conducting polymers, metal nanoparticles, etc., that possess higher surface areas to facilitate the determination of the concentration of pollutants and materials threatening the environment. Furthermore, these sensors have been used in difficult settings. Biological and environmental media are particularly challenging because of complex samples and interfering chemicals.

According to the study, CDs superior supramolecular recognition properties can be exploited to create electrochemical sensors that detect several molecules. These sensors have uses in biology, ecology, and chemistry. Additionally, these sensors have been used in harsh environments. Biological and ecological media present unique challenges due to interfering chemicals and sample complexity. Electropolymerization leads to the covalent bonding of CDs, which promotes the stability of sensor materials.

Although CDs are biocompatible, electron conductivity additives like rGO and CNTs can have an environmental impact if discharged into the environment. Reducing the release of CNTs and GO into the atmosphere is challenging due to their high water dispersion. CD supramolecules must properly bind and trap these additions to improve electron transmission efficiency. CDs have great potential for sensor technology, chemical identification, and immobilizing enzymes and aptamers in hydrophobic compartments. The capacity to discern chirality is promising for developing chiral medicinal compounds. New functionalized CDs must be created using more straightforward and more efficient synthetic processes to boost yields and unleash the potential for CD supramolecular applications.

Table 10.1 Electrochemical sensing of pollutants by CD-based composites

Sensor materials	Electrochemical method	Detected substance	Limit of detection	References
NH_2-rGO/β-CD	SWV	Cu^{2+}	0.009 μM	[70]
β-CD–Au@graphene nanohybrids	DPV	Hydroquinone, p-nitrophenol	0.009 μM	[71]
Au@Silicon carbide-SH/ NH_2-β-CD	DPV	Nitrophenol isomers	0.1 nM	[72]
Bi/Hydroxypropyl-β-CD-rGO/ Nafion	SWV	Pb^{2+} and Cd^{2+}	0.09 nM (Pb^{2+}) 0.07 nM (Cd^{2+})	[73]
Hydroxypropyl-β-CD-rGO	CV	o-nitrophenol	0.01 μM	[30]
SH-β-CD/AuNP/SWCNTs	DPV	Methyl parathion	0.1 nM	[47]
Hydroxypropyl-β-CD/Graphene nanoribbons	DPV	p-aminophenol; Rhodamine B	0.0008 μM; 0.001 μM	[74]
NH_2-β-CD/Graphene Sheet	CV	Dopamine	0.43 μM	[75]
β-CD/Single wall CNT	CV	Methyl parathion	7.23 nM	[62]
β-CD functionalized poly (amino acid)	SWASV	Cd^{2+} and Pb^{2+}	1.13 nM (Cd^{2+}); 0.53 nM (Pb^{2+})	[68]
SnO_2-β-CD-GO modified GCE	SWV	α-lipoic acid	1.42 nM	[76]
SPCE/CNT-CD/FcCAR	CV	Hg^{2+}	0.5 nM	[77]
CuO nanoleaf/β-CD-rGO	DPV	2-chlorophenol (2-CP); 2, 4–Dichlorophenol (2, 4–DCP)	0.22 nM (2-CP); 0.52 nM (2,4-DCP)	[78]
RGO-CD-AgNPs-polyoxometalate	DPV	Uric acid, L-tyrosine	0.11 μM (UA); 0.23 μM (L-Tyr)	[79]
Polycaprolactone/PPy/β-CD	SWV	Dinotefuran	0.05 μM	[80]

10.13.1 Future Perspectives

CD-based electrochemical sensors have a lot of appealing qualities and good molecular recognition features, but there are still a lot of issues that need to be resolved before these sensors can move from the proof-of-concept to the commercialization stage in the biomedical, environmental, and healthcare industries. Even though it's a frequently used and straightforward procedure, immobilizing CDs by mixing and spreading them onto a transducer surface causes the CD to come away from the electrode surface gradually. Design of electrode material to be performed to attain high sensitivity to detect lower concentration, selective to distinguish target analyte, rapid response to change along with analyte concentration, consistent performance, trace analysis, cost-effectiveness, and ease of fabrication. The electrode materials to be chosen to achieve rapid electron transfer kinetics to generate a signal quickly, and low impedance to improve accuracy and stability. To fabricate a good sensor, more robust approaches are needed to achieve greater loadings and more robust integration of the CDs within the sensor unit, which is essential to designing a commercial sensor.

References

1. Zarrabi A, Shokrgozar MA, Vossoughi M, Farokhi M (2014) In vitro biocompatibility evaluations of hyperbranched polyglycerol hybrid nanostructure as a candidate for nanomedicine applications. J Mater Sci Mater Med 25:499–506
2. Kavitha R, Stalin T (2014) A highly selective chemosensor for colorimetric detection of Hg 2+ and fluorescence detection of pH changes in aqueous solution. J Lumin 149:12–18
3. Prabu S, Sivakumar K, Swaminathan M, Rajamohan R (2015) Preparation and characterization of host-guest system between inosine and β-cyclodextrin through inclusion mode. Spectrochim Acta Part A Mol Biomol Spectrosc 147:151–157
4. Prabu S, Swaminathan M, Sivakumar K, Rajamohan R (2015) Preparation, characterization and molecular modeling studies of the inclusion complex of Caffeine with Beta-cyclodextrin. J Mol Struct 1099:616–624
5. Steinegger A, Wolfbeis OS, Borisov SM (2020) Optical sensing and imaging of pH values: spectroscopies, materials, and applications. Chem Rev 120:12357–12489
6. Meher N, Barman D, Parui R, Iyer PK (2022) Recent development of the fluorescence-based detection of volatile organic compounds: a mechanistic overview. J Mater Chem C 10:10224–10254
7. Tajik S, Beitollahi H, Garkani Nejad F et al (2021) Performance of metal-organic frameworks in the electrochemical sensing of environmental pollutants. J Mater Chem A 9:8195–8220
8. Fonsaca JES, Moreira MP, Farooq S, et al (2022) Surface plasmon resonance platforms for chemical and bio-sensing. In: Encyclopedia of sensors and biosensors, 1st edn, vol 1–4, pp 316–353
9. Rizwan K, Bilal M, Iqbal HMN (2022) Role of nanomaterials in sensing air pollutants. In: Hybrid and combined processes for air pollution control: methodologies, mechanisms and effect of key parameters, pp 1–17
10. Alam AU, Howlader MMR, Hu NX, Deen MJ (2019) Electrochemical sensing of lead in drinking water using B-cyclodextrin-modified MWCNTs. Sens Actuators B Chem 296:126632
11. Guo Y, Guo S, Ren J et al (2010) Cyclodextrin functionalized graphene nanosheets with high supramolecular recognition capability: synthesis and host-guest inclusion for enhanced electrochemical performance. ACS Nano 4:4001–4010
12. Zhang J, Tan W, Tao Y et al (2018) A novel electrochemical chiral interface based on sandwich-structured molecularly imprinted SiO2/AuNPs/SiO2 for enantioselective recognition of cysteine isomers. Electrochem Commun 86:57–62
13. Fan L, Hao Q, Kan X (2018) Three-dimensional graphite paper based imprinted electrochemical sensor for tertiary butylhydroquinone selective recognition and sensitive detection. Sens Actuators B Chem 256:520–527
14. Palanisamy S, Thirumalraj B, Chen SM (2016) A novel amperometric nitrite sensor based on screen printed carbon electrode modified with graphite/β-cyclodextrin composite. J Electroanal Chem 760:97–104
15. Jiang Z, Li G, Zhang M (2016) Electrochemical sensor based on electro-polymerization of β-cyclodextrin and reduced-graphene oxide on glassy carbon electrode for determination of gatifloxacin. Sens Actuators B Chem 228:59–65

16. Gao Y, Jiao T, Xing R et al (2017) Construction and self-assembly of beta-cyclodextrin derivative composite Langmuir films: host-guest reaction and nanostructures. Colloids Surfaces A Physicochem Eng Asp 533:68–75
17. Alam AU, Qin Y, Howlader MMR et al (2018) Electrochemical sensing of acetaminophen using multi-walled carbon nanotube and B-cyclodextrin. Sens Actuators B Chem 254:896–909
18. Liu Z, Xue Q, Guo Y (2017) Sensitive electrochemical detection of rutin and isoquercitrin based on SH-β-cyclodextrin functionalized graphene-palladium nanoparticles. Biosens Bioelectron 89:444–452
19. Saylan Y, Erdem Ö, Inci F, Denizli A (2020) Advances in biomimetic systems for molecular recognition and biosensing. Biomimetics 5:20
20. Song N, Lou XY, Ma L et al (2019) Supramolecular nanotheranostics based on pillarenes. Theranostics 9:3075
21. Shao Y, Wang J, Wu H et al (2010) Graphene based electrochemical sensors and biosensors: a review. Electroanalysis 22:1027–1036
22. Eda G, Chhowalla M (2010) Chemically derived graphene oxide: towards large-area thin-film electronics and optoelectronics. Adv Mater 22:2392–2415
23. Lenik J (2016) Cyclodextrins based electrochemical sensors for biomedical and pharmaceutical analysis. Curr Med Chem 24:2359–2391
24. Zhong Y, He Y, Ge Y, Song G (2017) β-Cyclodextrin protected Cu nanoclusters as a novel fluorescence sensor for graphene oxide in environmental water samples. Luminescence 32:596–601
25. Fu L, Lai G, Yu A (2015) Preparation of β-cyclodextrin functionalized reduced graphene oxide: application for electrochemical determination of paracetamol. RSC Adv 5:76973–76978
26. Upadhyay SS, Kalambate PK, Srivastava AK (2019) Enantioselective biomimetic sensor for discrimination of R and S-clopidogrel promoted by β-cyclodextrin complexes employing graphene and platinum nanoparticle modified carbon paste electrode. J Electroanal Chem 840:305–312
27. Yakout AA, El-Hady DA (2016) A combination of β-cyclodextrin functionalized magnetic graphene oxide nanoparticles with β-cyclodextrin-based sensor for highly sensitive and selective voltammetric determination of tetracycline and doxycycline in milk samples. RSC Adv 6:41675–41686
28. Guo Y, Guo S, Li J et al (2011) Cyclodextrin-graphene hybrid nanosheets as enhanced sensing platform for ultrasensitive determination of carbendazim. Talanta 84:60–64
29. Lu D, Lin S, Wang L et al (2012) Synthesis of cyclodextrin-reduced graphene oxide hybrid nanosheets for sensitivity enhanced electrochemical determination of diethylstilbestrol. Electrochim Acta 85:131–138
30. Xu C, Wang J, Wan L et al (2011) Microwave-assisted covalent modification of graphene nanosheets with hydroxypropyl-β-cyclodextrin and its electrochemical detection of phenolic organic pollutants. J Mater Chem 21:10463–10471
31. Chekin F, Mishyn V, Barras A et al (2019) Dopamine-functionalized cyclodextrins: modification of reduced graphene oxide based electrodes and sensing of folic acid in human serum. Anal Bioanal Chem 411:5149–5157
32. Sun Q, Wang L, Qi L et al (2013) Enhanced electrochemical determination of trinitrophenol based on pyrenecyclodextrin functionalized reduced graphene oxide. Sens Lett 11:2227–2232
33. Niu X, Yang X, Mo Z et al (2019) Perylene-functionalized graphene sheets modified with β-cyclodextrin for the voltammetric discrimination of phenylalanine enantiomers. Bioelectrochemistry 129:189–198
34. Zhao H, Ji X, Wang B et al (2015) An ultra-sensitive acetylcholinesterase biosensor based on reduced graphene oxide-Au nanoparticles-β-cyclodextrin/Prussian blue-chitosan nanocomposites for organophosphorus pesticides detection. Biosens Bioelectron 65:23–30
35. Bin GC, Guo CC, Jiang D et al (2014) Graphene-cyclodextrin-cytochrome c layered assembly with improved electron transfer rate and high supramolecular recognition capability. Mater Sci Eng C 39:281–287

36. Breslin CB, Branagan D, Garry LM (2019) Electrochemical detection of Cr(VI) with carbon nanotubes decorated with gold nanoparticles. J Appl Electrochem 49:195–205
37. Tian X, Cheng C, Yuan H et al (2012) Simultaneous determination of L-ascorbic acid, dopamine and uric acid with gold nanoparticles-β-cyclodextrin-graphene-modified electrode by square wave voltammetry. Talanta 93:79–85
38. Gao J, Ma H, Lv X et al (2015) A novel electrochemical immunosensor using β-cyclodextrins functionalized silver supported adamantine-modified glucose oxidase as labels for ultrasensitive detection of alpha-fetoprotein. Anal Chim Acta 893:49–56
39. Alam AU, Qin Y, Catalano M et al (2018) Tailoring MWCNTs and β-cyclodextrin for sensitive detection of acetaminophen and estrogen. ACS Appl Mater Interfaces 10:21411–21427
40. Kor K, Zarei K (2013) β-Cyclodextrin incorporated carbon nanotube paste electrode as electrochemical sensor for nifedipine. Electroanalysis 25:1497–1504
41. Zhu G, Yi Y, Chen J (2016) Recent advances for cyclodextrin-based materials in electrochemical sensing. TrAC-Trends Anal Chem 80:232–241
42. Chen J, He P, Bai H et al (2017) Poly(B-cyclodextrin)/carbon quantum dots modified glassy carbon electrode: preparation, characterization and simultaneous electrochemical determination of dopamine, uric acid and tryptophan. Sens Actuators B Chem 252:9–16
43. Tofighy MA, Mohammadi T (2016) Copper ions removal from aqueous solutions using acid-chitosan functionalized carbon nanotubes sheets. Desalin Water Treat 57:15384–15396
44. Petersen EJ, Zhang L, Mattison NT et al (2011) Potential release pathways, environmental fate, and ecological risks of carbon nanotubes. Environ Sci Technol 45:9837–9856
45. Yu X, Chen Y, Chang L et al (2013) β-cyclodextrin non-covalently modified ionic liquid-based carbon paste electrode as a novel voltammetric sensor for specific detection of bisphenol A. Sens Actuators B Chem 186:648–656
46. Abbaspour A, Noori A (2011) A cyclodextrin host-guest recognition approach to an electrochemical sensor for simultaneous quantification of serotonin and dopamine. Biosens Bioelectron 26:4674–4680
47. Fu XC, Zhang J, Tao YY et al (2015) Three-dimensional mono-6-thio-β-cyclodextrin covalently functionalized gold nanoparticle/single-wall carbon nanotube hybrids for highly sensitive and selective electrochemical determination of methyl parathion. Electrochim Acta 153:12–18
48. Lian W, Huang J, Yu J et al (2012) A molecularly imprinted sensor based on β-cyclodextrin incorporated multiwalled carbon nanotube and gold nanoparticles-polyamide amine dendrimer nanocomposites combining with water-soluble chitosan derivative for the detection of chlortetracycline. Food Control 26:620–627
49. Kan X, Zhang T, Zhong M, Lu X (2016) CD/AuNPs/MWCNTs based electrochemical sensor for quercetin dual-signal detection. Biosens Bioelectron 77:638–643
50. Bathinapatla A, Kanchi S, Singh P et al (2015) Fabrication of copper nanoparticles decorated multiwalled carbon nanotubes as a high performance electrochemical sensor for the detection of neotame. Biosens Bioelectron 67:200–207
51. Hu Y, Zhang Z, Zhang H et al (2012) A sensitive and selective sensor-coated molecularly imprinted sol-gel film incorporating β-cyclodextrin-multi-walled carbon nanotubes and cobalt nanoparticles-chitosan for oxacillin determination. Surf Interface Anal 44:334–341
52. Yuan MM, Zou J, Huang ZN et al (2020) PtNPs-GNPs-MWCNTs-β-CD nanocomposite modified glassy carbon electrode for sensitive electrochemical detection of folic acid. Anal Bioanal Chem 412:2551–2564
53. Liu Z, Zhang A, Guo Y, Dong C (2014) Electrochemical sensor for ultrasensitive determination of isoquercitrin and baicalin based on DM-β-cyclodextrin functionalized graphene nanosheets. Biosens Bioelectron 58:242–248
54. Zhou Z, Gu C, Chen C et al (2019) An ultrasensitive electrochemical sensor for quercetin based on 1-pyrenebutyrate functionalized reduced oxide graphene /mercapto-B-cyclodextrin / Au nanoparticles composite film. Sens Actuators B Chem 288:88–95
55. Wang R, Wang S, Qin C et al (2023) An Electrochemical sensor based on electropolymerization of β-cyclodextrin on glassy carbon electrode for the determination of fenitrothion. Sensors 23:435

56. Carragher U, Breslin CB (2018) Polypyrrole doped with dodecylbenzene sulfonate as a protective coating for copper. Electrochim Acta 291:362–372
57. Carragher U, Branagan D, Breslin CB (2019) The influence of carbon nanotubes on the protective properties of polypyrrole formed at copper. Materials (Basel) 12:2587
58. Tian B, Hua S, Tian Y, Liu J (2021) Cyclodextrin-based adsorbents for the removal of pollutants from wastewater: a review. Environ Sci Pollut Res 28:1317–1340
59. Rizzi V, Gubitosa J, Signorile R et al (2021) Cyclodextrin nanosponges as adsorbent material to remove hazardous pollutants from water: The case of ciprofloxacin. Chem Eng J 411:128514
60. Sinniah S, Mohamad S, Manan NSA (2015) Magnetite nanoparticles coated with β-cyclodextrin functionalized-ionic liquid: synthesis and its preliminary investigation as a new sensing material. Appl Surf Sci 357:543–550
61. Ali MY, Alam AU, Howlader MMR (2020) Fabrication of highly sensitive bisphenol a electrochemical sensor amplified with chemically modified multiwall carbon nanotubes and β-cyclodextrin. Sens Actuators B Chem 320
62. Guo M, Han J, Mishchenko Y et al (2024) Fabrication of methyl parathion electrochemical sensor based on β-cyclodextrin decorated single-wall carbon nanotubes. Mater Res Innov 28:154–160
63. Wayu MB, Schwarzmann MA, Gillespie SD, Leopold MC (2017) Enzyme-free uric acid electrochemical sensors using β-cyclodextrin-modified carboxylic acid-functionalized carbon nanotubes. J Mater Sci 52:6050–6062
64. Zhou Z, Zhao P, Wang C et al (2020) Ultra-sensitive amperometric determination of quercetin by using a glassy carbon electrode modified with a nanocomposite prepared from aminated graphene quantum dots, thiolated β-cyclodextrin and gold nanoparticles. Microchim Acta 187:1–9
65. Kingsford OJ, Qian J, Zhang D et al (2018) Electrochemical sensing for 1-chloro-4-nitrobenzene based on β-cyclodextrin/carbon nanohorn nanohybrids. Anal Methods 10:5372–5379
66. Palanisamy S, Thangavelu K, Chen SM et al (2017) Synthesis and characterization of polypyrrole decorated graphene/β-cyclodextrin composite for low level electrochemical detection of mercury (II) in water. Sens Actuators B Chem 243:888–894
67. Xie Y, Tu X, Ma X et al (2019) High-performance voltammetric sensor for dichlorophenol based on β-cyclodextrin functionalized boron-doped graphene composite aerogels. Nanotechnology 30:185502
68. Tang LW, Alias Y, Woi PM (2024) Simultaneous quantification of cadmium and lead ions using β-cyclodextrin functionalized poly(amino acid) electrochemical sensor. J Electroanal Chem 967:118427
69. Cheng L, Wei BG, He LL et al (2017) "Off-On"switching electrochemiluminescence biosensor for mercury(II) detection based on molecular recognition technology. Anal Biochem 518:46–52
70. Huang S, Lu S, Huang C et al (2015) Sensitive and selective stripping voltammetric determination of copper(II) using a glassy carbon electrode modified with amino-reduced graphene oxide and β-cyclodextrin. Microchim Acta 182:2529–2539
71. Yang L, Zhao H, Li Y, Li CP (2015) Electrochemical simultaneous determination of hydroquinone and p-nitrophenol based on host-guest molecular recognition capability of dual β-cyclodextrin functionalized Au@graphene nanohybrids. Sens Actuators B Chem 207:1–8
72. Wu S, Fan S, Tan S et al (2018) A new strategy for the sensitive electrochemical determination of nitrophenol isomers using β-cyclodextrin derivative-functionalized silicon carbide. RSC Adv 8:775–784
73. Lv M, Wang X, Li J et al (2013) Cyclodextrin-reduced graphene oxide hybrid nanosheets for the simultaneous determination of lead(II) and cadmium(II) using square wave anodic stripping voltammetry. Electrochim Acta 108:412–420
74. Zhu G, Yi Y, Liu Z et al (2016) Highly sensitive electrochemical sensing based on 2-hydroxypropyl-β-cyclodextrin-functionalized graphene nanoribbons. Electrochem Commun 66:10–15

75. Tan L, Wang G, Chen N et al (2015) Layer-by-layer assembled multilayers of graphene/mono-(6-amino-6-deoxy)-β-cyclodextrin for detection of dopamine. Chin J Chem 33:185–191
76. Agadi NP, Korgaonkar K, Mathad AS, Seetharamappa J (2023) Green synthesis of SnO2-β-cyclodextrin graphene oxide composite for electrochemical sensing of an antioxidant drug, alpha lipoic acid. Ionics (Kiel) 29:1593–1603
77. Abate C, Neri G, Scala A et al (2023) Screen-Printed carbon electrodes with cationic cyclodextrin carbon nanotubes and ferrocenyl-carnosine for electrochemical sensing of Hg(II). ACS Appl Nano Mater 6:17187–17195
78. Akond US, Mahanta A, Jasimuddin S (2022) CuO nanoleaf and β-cyclodextrin functionalized reduced graphene oxide: a highly selective and sensitive electrochemical sensor for the simultaneous detection of 2-chlorophenol and 2, 4-dichlorophenol. Anal Methods 15:436–444
79. Chai X, Li Y, Ma C et al (2023) A voltammetric sensor based on a reduced graphene oxide/β-cyclodextrin/silver nanoparticle/polyoxometalate nanocomposite for detecting uric acid and tyrosine. Anal Methods 15:2528–2535
80. Mei X, Wang X, Huang W et al (2024) A novel polycaprolactone/polypyrrole/β-cyclodextrin electrochemical flexible sensor for dinotefuran pesticide detection. Food Chem 434:137194

Dr. Vinitha Packirisamy is a senior lecturer at the Centre for Global Health Research, Saveetha Medical College and Hospital (Deemed to be University), Saveetha Institute of Medical and Technical Sciences, Chennai, Tamil Nadu, India. She has completed her Ph.D. at the University of Madras. She has received funding from DST-INSPIRE to pursue her doctorate. Her current research interest is designing novel nanomaterials for sensing, energy storage, and biomedical applications.

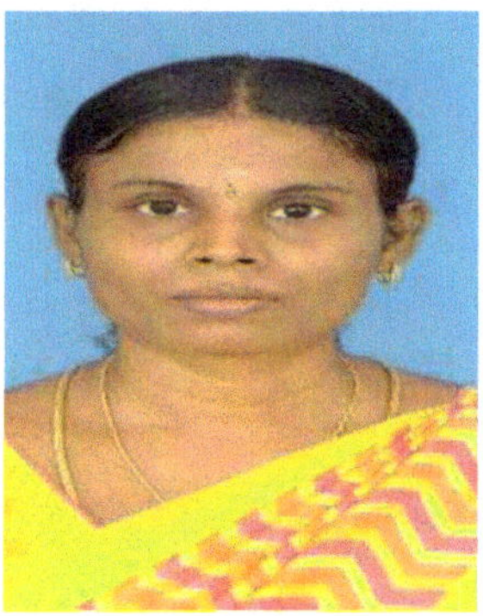

Dr. A. Rajabhuvaneswari is a distinguished educator and researcher with over 15 years of experience in higher education. Currently, she serves as a professor of chemistry at Karpaga Vinayaga College of Engineering and Technology in Chengalpattu, Chennai. She is an expert in chemistry and applied chemistry. Her studies and publications have made substantial contributions to the academic world, and she holds a Ph.D. degree in Chemistry with a specialization in Polymer Chemistry. She has authored fifteen publications. Her research interests encompass a diverse array of themes, including environmental chemistry, nutrition, and food preservation. Her roles as a reviewer for international publications, participation in the board of studies, and contributions as an examiner and question paper setter for multiple universities have led to the recognition and acknowledgment of her expertise.

Dr. M. V. Arularasu Assistant Professor at the Centre for Global Health Research, Saveetha Medical College and Hospitals, SIMATS. He obtained his Ph.D degree from the University of Madras, Chennai, India. During his Ph.D had worked as a Project fellow funded by Tamil Nadu State Council of Science and Technology funded project. Completed his Postdoctoral scientist at Chang Gung University, Taiwan (ROC). His research contributes significantly to the fields of Energy, Environmental and Biomedical research.

Chapter 11
Chemosensing of Small Molecules Using Cyclodextrins

Most. israt Jahan, Israt Jahan Heme, and Md. Abu Bin Hasan Susan

Abstract Cyclodextrins (CDs), cyclic oligosaccharides composed of glucose units, have emerged as versatile molecular hosts in the development of sensors for the detection of small molecules. Due to their exceptional biocompatibility, water solubility, non-toxicity, economical accessibility, and facile functionalization, CDs serve as pivotal building blocks for the creation of highly effective supramolecular architectures with fundamental significance. CDs have gained significant attention within the scientific community, particularly for their role in the development of chemical sensors based on the molecular recognition phenomenon. Given the remarkable sensing capabilities of CDs, this contribution aims to present recent advancements in the detection of various biologically significant small molecules using CDs, employing colorimetric, fluorescence, electrochemical, and potentiometric responses. By elucidating the distinct optical features associated with these sensing modalities, this chapter provides a comprehensive understanding of the intricate interactions between CDs and target molecules. This exploration not only showcases the versatility of CD-based sensors but also underscores their potential impact on various fields, including health care, environmental monitoring, and beyond. Finally, an insightful perspective on future directions and advancements has been offered, thereby contributing to the evolving landscape of supramolecular sensors, with a particular focus on the unique attributes of CD-based sensing technologies.

Keywords Chemosensor · Cyclodextrins · Small molecules · Sensing molecules · Colorimetric sensor · Electrochemical sensor

---

Most. Jahan · I. J. Heme · Md. A. B. H. Susan (✉)
Department of Chemistry, University of Dhaka, Dhaka, Bangladesh
e-mail: susan@du.ac.bd

Md. A. B. H. Susan
Dhaka University Nanotechnology Center, University of Dhaka, Dhaka, Bangladesh

S. Krishnamoorthy (ed.), *Cyclodextrins for Chemosensing*,
https://doi.org/10.1007/978-3-031-89299-8_11

11.1 Introduction

In the world of chemo- and biosensor development research, the goal is to measure small compounds using surface-bound receptor molecules; requiring effective detection and surface modification. Chemosensors offer flexibility, ease of use, quick response, and simplicity, relying on supramolecular chemistry to detect small molecules or ions. A typical chemosensor includes a transducer and signal receiver, where the transducer converts chemical information into an analytical signal [1]. One effective approach for small molecule sensing is the use of supramolecular structures that form inclusion complexes. Cyclodextrins (CDs), known for creating cavities that attract small molecules, have gained particular attention for sensor development [2]. A chemosensor is a molecular device that facilitates direct identification without the need for intricate devices or systems by detecting an analyte and producing a measurable signal. Chemosensors use three main methods to interact with analytes: displacement, binding-site-signaling, and chemo-dosimetric, each of which changes the way the analyte looks or behaves optically. They are classified as optical, luminescent, or electrochemical based on their signal type. These sensors detect changes in electronic properties through colorimetric, luminescent, or electrochemical reactions. They offer a simple, sensitive, and selective method for detecting neutral molecules or ions without complex equipment, providing both quantitative and qualitative results [3].

A fundamental component of supramolecular chemistry is the emphasis on host–guest interaction, which is non-covalent and crucial for the identification of guest entities, especially through non-covalent supramolecular interactions [4]. CDs are valued for their biocompatibility, water solubility, non-toxicity, affordability, and easy functionalization. They are versatile in chemical sensing, allowing precise detection of analytes, and are also used in molecular detection, drug delivery, and sensing applications. The distinctive benefits of CDs are originated by D-glucopyranoside units connected by α-1, 4-glycosidic linkages [5]. The most common CDs are α-CD, β-CD, and γ-CD. They have diameters of 0.5 nm, 0.6 nm, and 0.8 nm with 6, 7, and 8 glucose subunits, respectively. The inside cavity of CDs is hydrophobic because of the orientation of glycoside linkages, whereas the hydroxyl groups on the borders of CDs give them hydrophilic properties. Because of this unique feature, CDs can combine with hydrophobic guest molecules to create inclusion complexes [6]. Crucially, to improve the recognition capabilities of CD-based supramolecular hosts toward analytes, specific strategies for the easy functionalization of CD scaffolds have been developed on a regular basis over the years [7]. The field of chemical sensors, in particular, CD-based sensors has developed quickly and is solidifying its hold on a variety of facets of our daily lives while also bringing about a revolution in a wide range of scientific and technological fields.

The theoretical foundations and recent advancements of CD-based chemical sensors for small-molecule sensing have been emphasized in this chapter. Also, it has been discussed how small molecules are targeted by CD-based sensors, spanning from volatile organic compounds and environmental pollutants to biologically relevant analytes such as nucleotides and drugs. The fascinating world of chemosensing has been highlighted with a focus on the utilization of CDs as molecular receptors for detecting small molecules, ranging from the modification of structures of CDs to the integration of responsive materials and signal transduction mechanisms.

11.2 Chemosensing: CDs as a Chemosensor

A chemosensor is any molecule that changes its optical signal when interacting with an analyte. This change occurs due to the optical activity of the sensor. In optical sensors, auxochromes, chromophores, or fluorophores are crucial for analyte detection. The sensing unit recognizes the analyte, triggering the signaling unit to produce a detectable signal. Thus, the signaling unit and the sensing unit are the two fundamental functional components of any molecular sensor architecture. Consequently, it is evident that the signaling element generates a detectable signal due to its selectivity and binding, whereas the recognition element affects the binding strength and the selectivity of a chemosensor toward an analyte. Typically, a connector or spacer connects the recognition and signaling moieties either directly or indirectly [8].

In order to construct inclusion complexes, CDs are utilized as the host molecule, encasing several kinds of guest molecules, including fatty acids, aldehydes, ketones, amines, organic acids, and aromatic and aliphatic alcohols. Non-covalent interactions are the binding force of attraction that results in the development of the inclusion complex between the CDs and the guest molecules [9]. CDs offer a large number of applications in numerous domains because the encapsulation of guest molecules into the cavity of CDs affects the physicochemical properties of the guest molecules. When CDs are used in chemosensory applications, they encapsulate the chemosensory component within the cavity and provide a tight fit. The selectivity of the chemosensory compound and sensitivity towards the target analyte are improved by the tight fit. When a guest accommodation is provided for a convulsion induced by fluorophore-modified CDs, the conformational change moves the fluorophore from the inside of the cavity of the CD to the outside. This movement results in a shift in the fluorescence intensity, as seen in Fig. 11.1. Because of the hydrophobic environment inside the CD cavity, a fluorescent CD shows intense fluorescence when it is in the self-inclusion condition. Its fluorescence intensity is reduced when the fluorophore is sealed off from the bulk water inside the cavity [10].

CDs with various materials allow their immobilization on sensor surfaces. This technique creates a CD array that captures specific analytes, enhancing sensitivity and selectivity. The non-covalent binding between immobilized CDs and analytes is strong enough to keep the analytes near the electrode but weak enough to allow electron transfer for reduction and oxidation reactions. This process generates a signal with varying current flow based on analyte concentration [11]. The creation of inclusion complexes between CDs and the target substance allows for selective and sensitive detection, which can occasionally be aided by an accumulation phase before

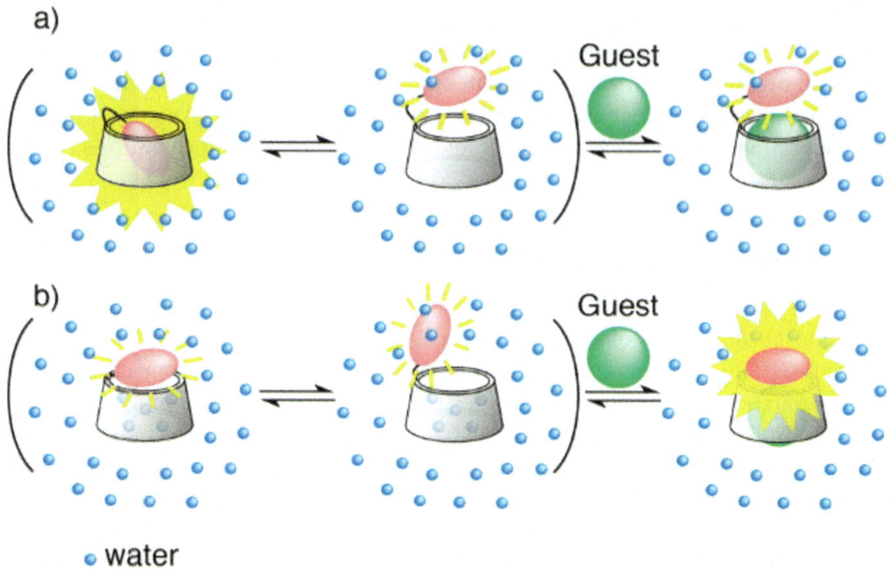

Fig. 11.1 Chemosensors that can be turned off and on are **a** an equilibrium model for changes in conformation caused by guests in aqueous solution in a type of traditional chromophore-modified CDs; **b** an equilibrium model for conformational changes caused by guests in a new kind of chromophore-modified CDs (Reproduced with permission from Ikeda, Hiroshi, Chemosensors 8.3, 48) [10]

sensing events and usually results in a detectable current flow. Figure 11.2 provides an illustration of this procedure [12]. Accumulation periods prior to sensing events allow target molecules to bind within CD cavities, enhancing detection sensitivity.

Competitive host–guest recognition, where an indicator molecule competes with the target analyte for binding to the CD, can also be employed. This displacement results in a measurable change in signal and facilitates quantification of the target analyte. For example, the electrochemically inert compound cinchonine is detected by displacement of 1,4-hydroquinone (HQ) by cinchonine, which results in a drop in current and allows cinchonine measurement. As the schematic in Fig. 11.2b illustrates, higher currents are produced by longer accumulation times. These accumulation periods are usually brief, lasting only a few seconds. Alternatively, the conflict between a signal molecule and the analyte of interest for the same host can be employed for competitive host–guest recognition [13]. As seen in Fig. 11.2a, the target molecule in this instance travels to the electrode surface and shifts the indicator molecule, causing a noticeable change in the signal. In this case, the indicator molecule is initially linked to the CD. An immobilized CD and this approach have been used to detect cinchonine, an electrochemically neutral molecule. As the HQ probe molecule oxidizes and forms an inclusion complex in the hydrophobic cavity of CD, a current signal is produced. However, the current decreases as cinchonine

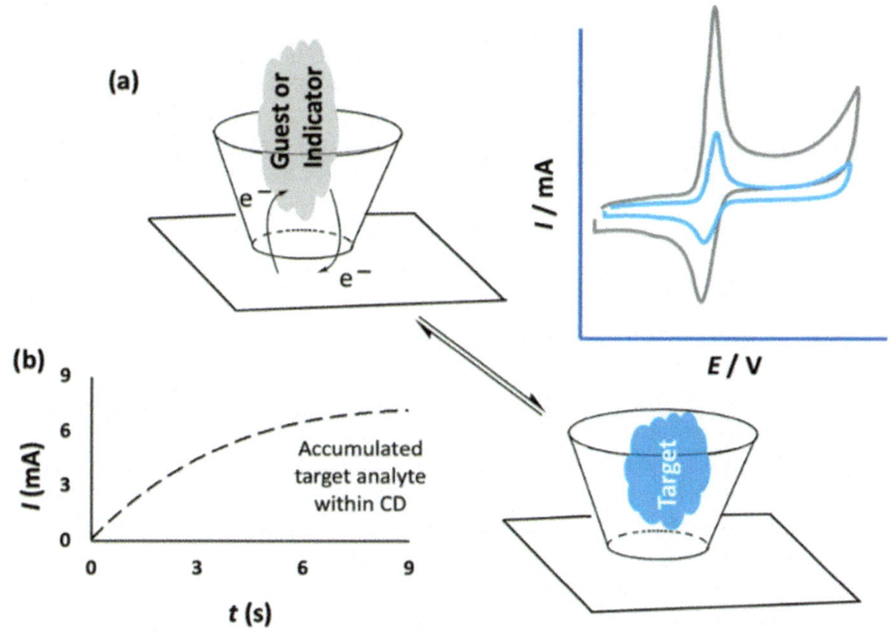

Fig. 11.2 **a** Transfer of electrons from an inserted electroactive molecule to a nonelectroactive target, together with the resulting decrease in the current that is measured; **b** the current that increases after an accumulation time (Reproduced with permission from Healy, B., Yu, et al. Materials, 14(7), 1668) [12]

displaces the HQ molecules, and this change in current can be utilized to measure the cinchonine concentration [14].

11.3 Some Derivatives of CDs to Enhance Chemosensing Properties

Derivatization modifies CD structures to enhance selectivity for specific analytes by altering their functional groups, improving binding affinity for certain molecules while reducing interactions with others. The color and fluorescence of chromophores depend on their environment, allowing the identification of guest molecules in CD cavities by adding dyes or fluorescent chemicals. This modification converts spectroscopically inactive CDs into active substances, enabling the detection of inactive organic compounds through observable spectroscopic changes. From this point, a dansyl-modified γ-CD system has been used to detect several organic guests [15]. In this system, the dansyl group acts as a chromophore since its fluorescence is highly sensitive towards the polar environment, which is the desired property for a chemosensor with greater sensitivity. Also, the larger cavity size of γ-CD enables the dansyl-modified γ-CD to detect cyclododecanol, 1-adamatanecaboxylic acid, borneol, cyclohexanol, etc. successfully [16–19]. Another study has reported modified *β* and *γ*-CDs, where fluorescamine has been added as it can produce pyrrolinone [20]. Several biologically active compounds, including vasopressin, angiotensin, oxytocin, and insulin yield highly fluorescent derivatives with fluorescamine. Therefore, modification of CDs with fluorescamine can act as a good receptor. In this system, the pyrrolinone moiety acts as a hydrophobic cap that enhances the guest-binding ability of the host and as a fluorescent probe, as it is a highly fluorescent substance. This modified sensor can detect *p*-aminobenzoic acid, 1-adamantane carboxylic acid, phthalic acid, etc. [21]. Alizarin yellow-modified *β*-CD (ACD) has been used to detect menthol, camphor, cyclohexanol, nerol, etc. Here, alizarin yellow is linked to *β*-CD by a spacer of ethylenediamine. Alizarin yellow changes its color from yellow to red when the solution becomes basic. This color change occurs due to the dissociation of the phenolic hydroxyl group that produces a phenolate form [22]. However, the absorption change of ACD occurs from the deprotonation-protonation equilibrium shift caused by the amine group of the ethylenediamine part [23]. Another sensor based on the oligomers made of per-*O*-methyl-6A, 6D-diamino-*β*-CD has been investigated as oligomers can provide an improved binding site which enables them to detect organic molecules with greater specificity [24]. CDs can create inclusion compounds when they are in the form of polymers. Polymeric films with CD (CDP) are attracting much attention as they exhibit good reproducibility, high stability, and an increased number of active sites [25, 26]. They can be prepared on the surface of a carbon or metal electrode by a poly-condensation mechanism. CDPs are a characteristic medium for the redox mediators and the immobilization of enzymes and this property makes CDP a promising candidate for implying as the amperometry biosensors, where the mediator (i.e., ferrocene, HQ) is immobilized in the hydrophobic cavity of CD and an enzyme is directly linked to the CDP by covalent bond providing transfer of charge from and to the enzyme by diffusion process [27]. Another immobilizing matrix, multi-walled carbon nanotubes (MWCNTs) have gained much attention as they exhibit significantly larger surface area, high chemical stability, enhanced electrical conduction, and comparatively less interaction with chemicals, particularly in solution. They are combined with *β*-CDs or functionalized CD nano-composites (*β*-CD-MWCNTs), which can act as a better electrochemical sensor when assembled on gold, glassy carbon, etc. [28]. In comparison with carbon nanotubes (CNTs), graphene is a stronger, more feasible substance produced from cheap graphite. Graphene sheets are two-dimensional, carbon substances and show some properties of the broad potential window, high conductivity, and a biocompatible nanomaterial that shows better conductive assistance for the deposition of some electrocatalytic nanoparticles by creating CD/graphene/NPs films [29]. An electrochemical sensor based on functionalized MWCNTs/*β*-CD nanocomposite (FMWCNTs/*β*-CD) has been designed for sensing some anti-inflammatory agents (for example, meloxicam, tenoxicam, piroxicam, etc.). FMWCNTs/*β*-CD composite modified some of the screen-printed sensors, (2-fluorophenyl 2-nitrophenyl ether and hyamine) and showed satisfactory electroanalytical performances. Because of the accommodation of CNT and CNTs/*β*-CD composite within the matrix of the selected electrode, selectivity and sensitivity increase, as CNT composite can significantly enhance the lifetime and potential stability of the designed sensors [30]. Gold nanoparticles (GNPs) coated with CD (for example, β-CD-GNP nanocomposites) have been used as sensors since GNPs have unique properties of high catalytic efficiency, immobilization of biological molecules, and efficient conducting interfaces [31–33]. Another GNPs-based modified electrode with high selectivity, efficiency, and enhanced electrocatalysis by using carboxymethyl-β-CDs as modification agent via electropolymerization and co-electrodeposition process, has been used to detect glucose molecules [34]. Zinc oxide nanorods (ZNRs) functionalized with poly(vinyl chloride) (PVC) in combination with β-CD as an ionophore, can detect the biological molecules, by a potentiometric method, with high selectivity and no considerable interference has been observed in the presence of some interfering molecules such as ascorbic acid, glucose, and urea, etc. In this work, ZNRs have been synthesized on an Au-coated glass substrate and then β-CD and potassium tetrakis (4-chlorophenyl) borate have been immobilized on the ZNR surface. The possible reason for the exhibited phenomena may be the strong electronic coupling between the Zn^{2+} ion and hydroxyl group and the pore diameter of β-CD [35]. A low-cost electrochemical sensor with greater sensitivity has been designed for detecting (+)-Catechin (CAT) in tea, commercial cocoa as well as in urine media. CAT is widely known as an antioxidant and its applications in the quantization of reducing equivalents in biological fluids and foods, hence it becomes crucial to detect CAT from food and beverage samples. Here, a modified electrode by incorporating hydroxypropyl-β-CD on carbon paste has been constructed in a fast and simple process [36]. The fluorescence of a fluorescent polymer, poly(phenylene ethynylene) employing β-CD shows observable changes upon the addition of aliphatic 1-adamantane carboxylic acid. The polymers containing π-conjugated systems are promising candidates for use in light-emitting diodes, electroluminescence, and chemosensors [37]. Here, the fluorescence changes take place in an aqueous solution and the selected polymer backbone is a π-conjugated system, surrounded by water-soluble, biocompatible CD moieties, making it a characteristic molecular sensor in the field of drug delivery systems [38]. In summary, it can be concluded that the derivatization of CDs enhances their chemosensing properties by fine-tuning binding affinities, improving selectivity, sensitivity, stability, and solubility. This optimization boosts the performance and utility of CD-based sensors in various analytical applications.

11.4 Chemosensing of Small Molecules: Water Pollutants

CDs can form inclusion complexes through the accommodation of several water pollutants in their central hydrophobic cavities in an aqueous solution. They can detect organic pollutants (phenols, dyes), pharmaceutical residues, pesticides, heavy metals (Pb^{2+}, Hg^{2+}), organic pollutants (PAHs), and nitroaromatic compounds by forming host–guest complexes in their hydrophobic cavity. As a commonly used explosive, 2,4,6-trinitrotoluene (TNT) can leave environmental residues that might cause serious health problems, such as anemia and damaged liver function, among other things [39]. The fluorescent technique, among other approaches, has garnered a lot of interest in the identification of TNT solutions especially due to its advantages in terms of simplicity, sensitivity, and improved signal output [40]. Assays that use the fluorescence resonance energy transfer (FRET) approach have a significant increase in the sensitivity for the detection of TNT [41]. Generally, if the emission band of a fluorescent material overlaps the absorption band of an element, the FRET will result in the superquenching of donor fluorescence due to polar-polar interactions at a close distance [42]. A novel fluorescent sensor based on β-CD has been employed in the identification of TNT by FRET, which is water-soluble and acts as a support carrier for the acceptor and donor complex. A fluorescein 5(6)-isothiocyanate (FITC), acts as a probe dye that is covalently bounded with the per-6-NH_2-β-CD, which acts as the detection vehicle for detecting TNT molecules shown in Fig. 11.3. Usually, the main objective of the FRET-based fluorescence sensing approach is to choose a suitable support carrier that will minimize the distance between donor and receptor. CDs are ideal for use as the carrier for a FRET-based sensor.

A TNT-binding moiety and a fluorescent moiety make up these two moieties of the FRET system. The fluorescence moiety, FITC, was applied to the rim of a single per-6-NH_2-β-CD molecule. The remaining amino groups of per-6-NH_2-β-CD were used as the selective identification of TNT and binding portion because of their higher TNT selectivity [43]. The fluorescence spectrum of the FITC moiety closely mirrors the spectrum of absorption of TNT-amine complexes because of the FRET between the energy acceptor and donor. This allows the fluorescence quenching of FITC to be observed on the platform of CDs. Applying a CD molecule as a carrier, the fluorescence sensor can greatly reduce the distance between the acceptor and donor. In addition, this method increases the energy transfer efficiency, which raises the sensitivity of the identification of a target matter. Some other small molecules

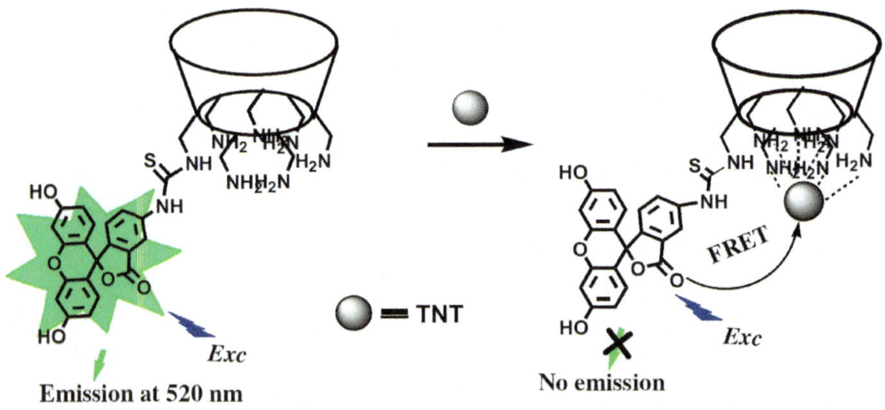

Fig. 11.3 FRET mechanism for detecting TNT by FTIC-(NH_2)-CD as a fluorescent sensor (Reproduced with permission from Feng, L., Tong, C., et al. Journal of Luminescence, 146, 502–507) [43]

are halomethanes that act as water pollutants. These are usually detected by gas chromatography, but it is now possible to detect them by "turn on" fluorescent chemosensing of modified CDs. The fluorophore stays at the entrance of the CD cavity and the addition of a guest increases the intensity of fluorescence. Here, the accommodation of the guest is not inhibited by the fluorophore. It is anticipated that this "turn-on" process will enable the construction of an excellent α-CD chemosensor, which will detect halomethanes. NBD-amine-appended α-CD (NCOαCD) showed a good binding behavior as a "turn-on" chemosensor [44]. The addition of CCl_4 decreases the absorption intensity of NCOαCD due to the hydrophobicity near the 4-nitro-2,1,3-benzoxadiazol-7-amine (NBD-amine) moiety of NCOαCD and the fluorescence intensity of NCOαCD increases upon the addition of CCl_4. NBD amine-appended α-CD is a fluorescent chemosensor, effective for halomethanes detection and exhibited selectivity towards perhalogenated methanes, as they show characteristic multisite interactions with CD molecules. Triclosan (TCS), an antimicrobial agent, has been reported as a toxic and hazardous chemical not only for humans but also for aquatic organisms [45]. It is a water pollutant that exhibits hydrophobicity and has an anionic site, allowing it to be detected at low levels using a potentiometric sensor. The fabrication of potentiometric sensors with polymeric membranes can be performed based on the selective separation of the chemical of interest between the hydrophobic ion-selective membrane and the sample solution. 2-Hydroxy-propyl β-CD shows higher water solubility and higher binding affinity to TCS [46]. Therefore, it was added to the inner filling solution as a complexing agent to optimize the electrochemical sensor, which enables it to improve the selective and sensitive determination of the contaminants, TCS with a low limit of detection values [47]. Paracetamol, a common pain reliever can be excreted which is more prevalent in source water systems due to its increasing usage and disposal in manufacturing industry effluents [48]. The host–guest interaction property of CD can be combined with the high redox capability and large surface area of MWCNTs [49]. Since this approach combines CD with unmodified MWCNTs, no chemical transformation is necessary. The hydrophilicity of CD resulted in better dispersion of MWCNTs [50]. The electrochemical interaction of acetaminophen with the sensor was controlled by an equal quantity of protons and electrons in a surface-limited process. Used extensively in the synthesis of epoxy resins and polycarbonates for use in packaging, beverage cans, and food storage, bisphenol A (BPA) is an endocrine disruptive molecule also known by its chemical name, 2,2-bis (4-hydroxyphenyl propane) [51]. BPA is capable of binding to estrogen receptors by phenol groups in its structure, which is comparable to that of endocrine hormones like diethylstilbestrol and estradiol [52]. This can have detrimental impacts on aquatic and human creatures by leaching into the water system. Graphene oxide and β-CD functionalized MWCNTs are used in this low-cost, straightforward, and sensitive electrochemical sensor to detect BPA at low levels in water. The high surface area of graphene oxide, CNTs, and the host–guest interaction feature of β-CD are combined in this electrochemical sensor electrode device. Here, an oxidation method with diffusion control was used to identify BPA electrochemically. Water contaminants called chlorophenols are commonly found in water and are known to have detrimental effects on the kidneys, liver, reproductive system, and development. For these reasons, there is growing concern about the establishment of sensors that can detect these kinds of pollutants. β-CD polymer conjugated with azo dye has been employed as a fluorescence method sensor to identify chlorophenols [53]. The fluorescence of the polymer might be extinguished by the chlorophenols. In comparison to phenol and 4-chlorophenol, the polymer exhibited higher sensitivity to 2,4-dichlorophenol. As the concentration of pollutants rises, the fluorescence intensity falls, suggesting a reaction between the contaminant and the CD. Pollutant concentration increases cause the CD cavities to become more saturated, which prevents further noticeable changes in fluorescence. The fluorescence intensity of CDs reduces due to comparable inclusion, which causes the guest to exclude the fluorescent moiety from the hydrophobic interior of CDs and move it outside the cavity into an aqueous environment [54].

More quenching of fluorescence has been observed with phenol and 4-chlorophenol because the 2,4-dichlorophenol guest replaces the majority of the azo dye moiety. This suggests a significant interaction between the 2,4-dichlorophenol guest complex and the CD cavity. Also, the use of pesticides enables the effective elimination of undesirable insects and microbes. Unfortunately, most of these chemicals are persistent in the environment, thereby increasing their potential danger to humans by causing bioaccumulation in streams and soils. For monitoring the purity of drinking water, β-CDs have been modified with naphthol to develop an innovative chemosensor [55]. Their anion species originate from the activated naphthol unit and show a broad band of fluorescence with a peak. The guest-induced migration of the naphthol molecule from the interior to the exterior of the CD cavity may be connected to these fluorescence variations. They showed high sensitivity to various alkyl halides, including carbon tetrachloride, and odorants like 2-methylisoborneol and geosmin. These findings show that it will be helpful in ensuring that drinking water quality is being monitored [56].

11.5 Chemosensing of Small Molecules: Food and Environmental Toxicants

The use of CDs in sensors has allowed for the sensitive detection of organic pollutants, which helps early warning systems lessen the impact of impending calamities. Predictive models can forecast environmental pollution similarly to early-warning systems for natural disasters. Various physicochemical and biological indicators can measure emissions and detect low quantities of pollutants, allowing for proactive disaster prevention. An early-warning network with sensitive external environmental sensors can provide real-time alerts for emerging risks to air, water, and soil. It's crucial to position these monitors near emission sources or sensitive areas, like major cities. CDs play a key role in enhancing the sensitivity and selectivity of sensors for identifying environmental contaminants in gas and water.

CDs can be altered to increase detection specificity or improve adhesion to sensor surfaces. The cavity of CDs is coated with signal-emitting moieties that create self-inclusion complexes, hence improving the specificity of detection. The signal changes when competitive environmental contaminants that have a higher affinity for the cavity of CDs are present. For example, a CD derivative modified with fluorescein was created to detect pentachlorophenol with high sensitivity [57]. The harmful nitroarene chemicals trinitrophenol (TNP) and nitrobenzene are widely used in the chemical manufacture of colors, insecticides, and additives for incendiary devices. There have been worries about human consumption and polluting the environment as well as their unpredictable fate and mobility in aquatic settings [58]. Using a one-pot technique, a *β*-CD based porous fluorescence-based polymer (FL-PFP) has been created, displaying unique photophysical and textural properties that were responsive to nitroarene guest binding. *β*-CD inclusion sites and stiff subunits within the polymer framework are responsible for the rigidity of the FL-PFP and mesoporous character. This is supported by data from gas adsorption studies and the development of stable host–guest complexes with nitroarenes in aqueous settings. Time-resolved fluorescence decay lifetime analysis revealed the involvement of two binding sites in the development of a simple fluorescence "turn-off" technique for the sensitive detection of TNP in aqueous solutions. The presence of these two binding sites in FL-PFP allows for regulated TNP adsorption and detection, as shown by the results of both static and dynamic fluorescence quenching [59]. Further investigation is being conducted to understand the mechanism underlying this phenomenon by means of synthetic alterations, with the goal of improving molecular identification and facilitating the selective detection of guest systems that share structural similarities [60]. With continued efforts to progress sensor materials to enhance molecular identification of pollutants in the environment in field-based applications, this work is noteworthy in the creation of engineered materials and the demonstration of an "on–off" fluorescence detection technique for model nitroarene compounds [61]. Persistent organic pollutants represent a significant contemporary environmental challenge due to their inherent resistance to degradation, capacity to bioaccumulate in tissues of organisms, and profound adverse effects on both human health and the environment. Among the foremost risks to consumer health, are food contaminants such as xenobiotics (e.g., phytoalexins, xenoestrogens, mycotoxins) and residues (e.g., phenyls, aromatic hydrocarbons, polychlorinated, dioxins, pesticides) are of particular concern for the food industry [62].

It is commonly known that CDs increase the fluorescence of toxins. As a result, several significant mycotoxins have been taken into consideration, including *α* and *β* zearalenone, ochratoxin, and aflatoxin. Furthermore, resveratrol a well-known bioactive substance with significant antioxidant properties in food and frequent use in nutraceuticals, has been given consideration [63]. Every one of these substances exhibits inherent fluorescence and has been examined earlier for its capacity to create complexes with *β*-CD. The fluorescence signal can be either enhanced or quenched by *β*-CD, depending on the direction of insertion. There are two possible ways in which the binding/quenching process imposed by *β*-CD can function. Aflatoxin B1 and ochratoxin, have been taken as two opposing cases, as an example [64]. While ochratoxin reveals the fluorescent group to water and β-CD may contain it, protection against water quenching is not guaranteed in the latter scenario. In the former scenario, the fluorescent group is fully integrated within the CD cavity, shielding it from water quenching. To protect the fluorescent side, the receptor must occasionally be functionalized. However, rigorous simulation analysis is needed to determine which CD modifications are most advantageous.

Chemosensors incorporating CDs are employed for quality control in the food industry. They can detect contaminants, adulterants, and indicators of food spoilage, such as volatile organic compounds produced during microbial contamination or decomposition processes. These sensors quickly and sensitively detect allergens, aiding in labeling and ensuring safety for allergic consumers. They identify vitamins, antioxidants, and beneficial compounds for nutritional labeling and assess food quality. Additionally, they detect pesticide residues and environmental contaminants, ensuring compliance with regulations and consumer safety. An increased fluorescence emission is produced when a fluorescent substance is incorporated by the CD cavity and this phenomenon is inspired to detect pesticide molecules by fluorescent chemosensing by CD. A modified β-CD (pyridinoindolizin-β-CD) has been used as a sensor to detect some pesticides. Also, the findings revealed that the inclusion phenomenon of guests in the modified β-CD cavity is what causes the quenching of fluorescence and indicated that the complexations are thermodynamically beneficial [65].

11.6 Chemosensing of Small Molecules: Pharmaceuticals

Chemosensing of small drug molecules using CDs is a crucial research area with applications in drug development, delivery, and pharmacology. CDs are ideal for detecting, encapsulating, and modulating drug molecules due to their selective molecular recognition based on size, shape, and functional groups. CD-based chemosensors allow real-time drug concentration monitoring in biological fluids, enabling personalized dosing and continuous monitoring through implantable or wearable devices. Advancements in this technology promise to improve drug efficacy, safety, and patient outcomes. CDs and their derivatives are increasingly used to create customized voltammetric and potentiometric electrodes. Their ability to encapsulate drug molecules through host–guest interactions enhances the sensitivity of these electrodes, especially when combined with carbon nanomaterials or CD nanocomposites. These modified electrodes offer lower detection limits and faster response times but may have a shorter operational lifespan due to fewer components leaking into solutions.

About the sensors, the addition of CDs has the impact of enhancing both the surface analyte accumulation and the kinetics of the electrode process. As a consequence, it helps to raise the sensitivity of the electrode. The improved electrodes are highly reproducible, stable, and selective. The identification of chiral enantiomers by means of CD-based electrochemical sensors is of tremendous importance to the medical and pharmaceutical industries [66]. The surface of the glassy carbon electrode (GCE) was modified using copper ions, MWCNTs, hydroxypropyl β-CD mounted cellulose, and tryptophan (D-Trp/L-Trp) chiral enantiomers to construct a sensitive electrochemical sensor [67]. The built electrochemical sensor is thought to have a greater affinity for L-Trp than for D-Trp, as depicted in Fig. 11.4. The amount of D-Trp in the racemic mixture has also been effectively monitored using the devised electrochemical sensor. Many CD-based potentiometric sensors that operate via electrochemical processes have been successfully created over the last few years. Potentiometric sensors based on functionalized β-CDs can measure naproxen and ketoprofen, two valuable medicinal drugs. The host functionalized β-CD cavity contains a partially or fully enclosed guest naproxen molecule [68].

On the other hand, potentiometric sensors for psychotic drugs, such as diazepam and midazolam, have also been made using functionalized β-CDs [69]. Butylcholine, acetylthiocholine, and acetylmethylcholine are acetylcholine derivatives that have been identified by creating carbon paste electrodes (CPE) using β-CD contained PVC [70]. β-CDs-graphene oxide-SO_3H composite modified GCE can adsorb lisinopril and increase its electron transfer and thus supplies greater electroactive surface area and lower charge transfer resistance compared with GCE. The electrochemical process exhibits the characteristics of an adsorption-controlled system and is irreversible. Over a broad range, there is a linear relationship between the peak current and lisinopril concentration. This might be used to measure lisinopril in medications as anodic peak sensitivity at the electrode increases with lisinopril use [71].

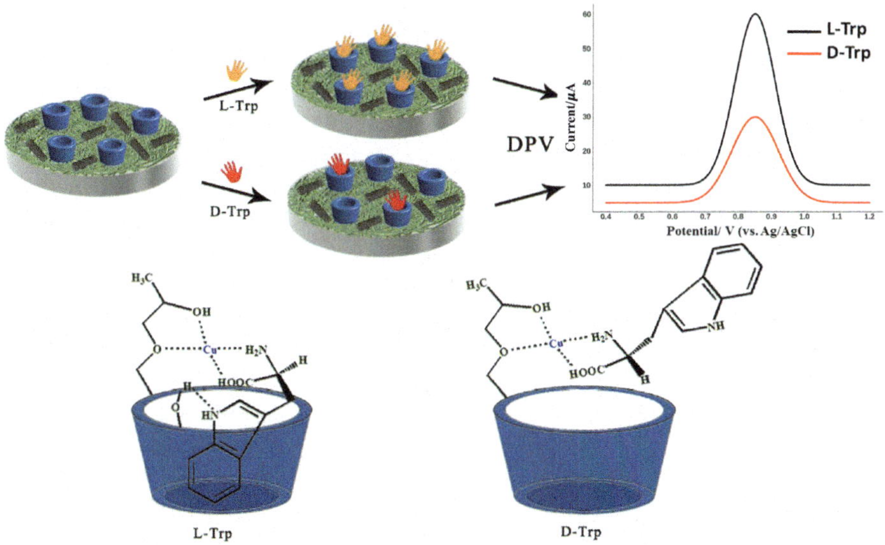

Fig. 11.4 Electrochemical identification of chiral enantiomers of Trp using an electrochemical sensor based on hydroxypropyl β-CD (Reproduced with permission from Ji, J., et al. Microchemical Journal, 175, 107133) [67]

A novel paracetamol electrochemical sensor based on β-CD functionalized reduced graphene oxide (β-CD/RGO) modified electrode has been demonstrated. A straightforward wet chemical process was used to create this composite. Since the sensor combined the benefits of both β-CD and RGO materials, the electrochemical behavior of paracetamol was significantly improved. The produced electrode, in particular, displayed a low detection limit and a short detection range. The suggested paracetamol sensor showed a strong anti-interference characteristic [72].

11.7 Chemosensing of Biological Small Molecules

CDs can selectively bind small molecules, making them effective sensing elements in detection systems. Researchers have created CD-based sensors to detect environmental pollutants, pharmaceuticals, and biomolecules like glucose and cholesterol. These sensors detect changes in the properties of inclusion complexes when target molecules bind, utilizing spectroscopic, electrochemical, or other analytical techniques. Additionally, modified CDs can specifically bind aromatic amino acids (e.g., tryptophan or tyrosine) for protein detection. They can also function as chemosensors for enzyme activity by forming inclusion complexes with substrates or inhibitors. Furthermore, CDs can detect nucleic acids through interactions with nucleobases or nucleotide analogs. Functionalized CDs have been designed to target cell surface carbohydrates, glycoproteins, or lipid rafts, allowing for the detection of specific cell types or signaling events. Moreover, CD-based sensors have been explored for monitoring receptor-ligand interactions or studying cell membrane dynamics. The GCEs have been modified to generate β-CD/MWNT-modified electrodes by preparing and depositing a new nanocomposite of β-CDs and MWCNTs. A new method for the simultaneous measurement of guanine (G), adenine (A), and thymine (T) is provided by these innovative modified electrodes.

In the analyte mixture, well-separated voltammetric peaks are found between G and A, A and T. The selective deposition of β-CDs as a filter membrane on the nanocomposite surface of the electrode may result in the superior electrochemical performance of relevant bases. This straightforward and trustworthy method, which uses an electrochemical methodology at CDs-MWNTs/GCE, could be used to construct a sensitive voltammetric biosensor that can measure guanine, adenine, and thymine simultaneously in a biological process [73]. Numerous adenosine nucleotide selective fluorescent or colorimetric chemosensing devices have been developed. The majority of previously documented receptors are, in fact, primarily designed to target the negatively charged phosphate group through interactions with a cationic unit or metal attached to a fluorophore. Furthermore, most of these sensors are used in organic or aqueous-organic environments; only a few systems can work in complicated aqueous fluids, but they are not suitable for real-world use [74]. The selective ability of nucleotides over analogous inorganic phosphate equivalents, such as phosphate, diphosphate, and triphosphate, with comparable anionic charge densities, must also be considered. While developing an aqueous medium nucleotide sensor system.

A large body of research has documented many recognition interactions and selective binding between positively charged CDs [75]. However, despite their extreme adaptability, these systems have not yet produced selective chemosensors for use in real time. A guest-targeting azo/coumarin-modified CDs-copper complex that may be used as a probe to separate adenosine tri-phosphate (ATP) from adenosine mono-phosphate (AMP) and adenosine di-phosphate (ADP) was recently shown by Hayashita and coworkers [76].

Few instances of specific nucleotide detectors are currently known. Thus, an intriguing objective for many researchers in the fields of biological and supramolecular chemistry is the fabrication and construction of a solitary supramolecular sensor that can discriminate between nucleotides with identical structures in terms of their differences in overall anionic charge concentrations. There has been an increasing interest in the design and implementation of specialized identification and sensing systems for physiologically important anionic species. A novel supramolecular sensing probe based on bis-quinoline-based CDs receptors was developed to enable extremely sensitive nucleotide discrimination. Because of the different spatial configurations of guests after binding, this receptor or probe demonstrated a very noticeable absorption and emission differentiation following interaction with nucleotide homologs of AMP, ADP, and ATP. This allowed for a colorimetric distinction of nucleotides that could be seen with the naked eye. Through a combination of theoretical calculations and several spectroscopic experiments, a clear-cut distinguishing mechanism was explained. The prepared receptor can be used for real-time monitoring of the enzymes-catalyzed phosphorylation/dephosphorylation processes. This has led to the demonstration of an unprecedented visualizable strategy for biological process monitoring in real-time via colorimetric and fluorescent changes, as shown in Fig. 11.5, as well as for selectively differentiating structurally similar nucleotides [77]. As noted earlier, β-CD as an ionophore combined with ZNRs functionalized with PVC can be employed as a direct physical adsorbent for the detection of dopamine. Since dopamine is polar, it causes a shift in charge distribution on the ZNR surface through the creation of metal–ligand bonds, which produce a detectable electrical signal. The ZNRs produced showed good crystal quality and were oriented properly. The sensor device exhibits adequate sensitivity and consistent potential response for dopamine detection across a broad concentration range. The electrode has outstanding reproducibility and a reasonable reaction time. This may help with routine analysis in laboratories looking into catechol amine neuropharmacology [35]. Furthermore, a dopamine electrochemical sensor system integrates MWCNTs as an amplifier of electron transport and β-CDs as a molecular receptor. The molecular host–guest recognition-based sensor of this kind for dopamine measurement has a high electrochemical response [78].

In the presence of ascorbic acid, uric acid was quantified using data that could be obtained using a CPE that had been physically changed by a β-CD film (CPE/β-CD). Enhancing the selectivity of detection of uric acid was the formation of an inclusion complex between β-CD and uric acid. Given its sensitivity and specificity in detecting uric acid in complex samples such as biological fluids, the results suggest that the CPE/β-CD is a promising choice. Since CDs can form particular inclusion complexes

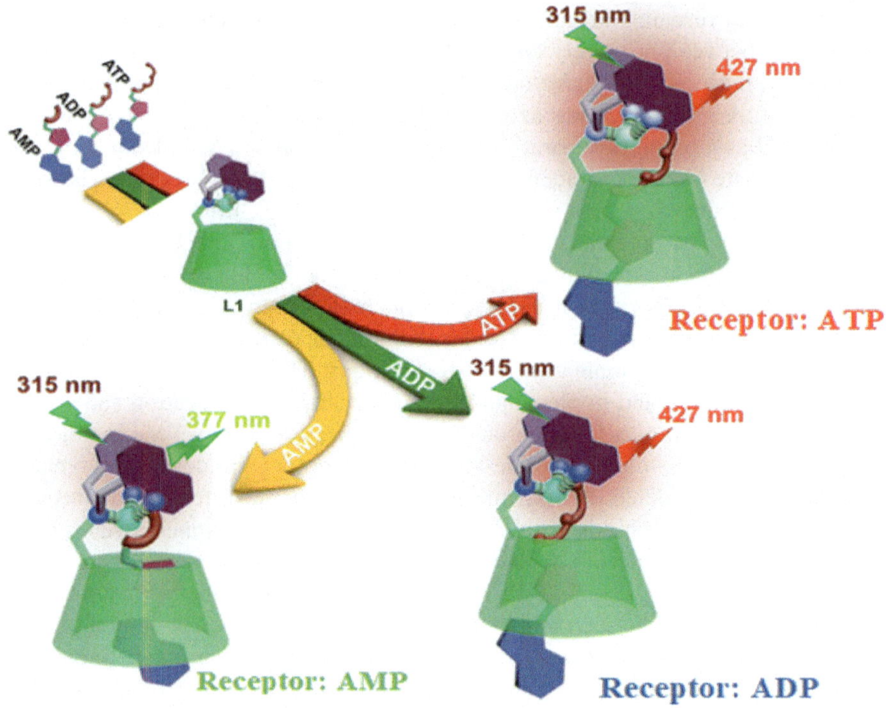

Fig. 11.5 Bis-quinoline-based CDs receptors to have sensing and identifying binding patterns with AMP, ADP, and ATP in water (Reproduced with permission from Kanagaraj, K., et al. Science, 23(3)) [77]

with uric acid in the presence of ascorbic acid, the oxidative splitting process between uric acid and ascorbic acid is dependent on these properties. When applied to actual samples (urine), the results of amperometric methods closely matched those of the commercial colorimetric approach. The following benefits come with using CPE/β-CD for uric acid estimation in the presence of ascorbic acid: no prior treatment is required for the saliva and urine samples, direct electrochemical dating is used to detect the concentration of uric acid, and safe, non-toxic materials are used to make the electrodes. The use of the CPE/β-CD as an electrochemical sensor for the identification of uric acid in real biological samples, including saliva and urine, has been both feasible and practicable [79]. An innovative enzymeless biosensor for the quantitative measurement of creatinine has been developed using β-CD embedded poly-3,4-ethylenedioxythiophene (PEDOT) modified GCE. To elucidate the precise interactions between creatinine and the PEDOT film in combination with β-CD, electrochemical impedance measurement was employed. Regular variations in the electrode potential versus creatinine concentration suggested the creation of a complex between β-CD and creatinine. With a decreased detection limit and a quick response time, the biosensor electrode demonstrated a linear shift in potential during the potentiometric evaluation. The molecular recognition connections between β-CD and creatinine are considered to be mediated by weak non-covalent connections between the amide hydrogen of creatinine and the glucopyranose oxygen atom in the β-CD.

A prospective outlook for the results of the studies on the interference brought about by urea, uric acid, ascorbic acid, glucose, and ammonium ions is given [80]. CDs offer a versatile platform for the chemosensing of biological molecules, with applications ranging from protein and enzyme detection to nucleic acid sensing and cell surface receptor recognition. By designing CD-based sensors with specific recognition elements and detection modalities, it is possible to achieve specific and accurate detection of an extensive variety of biological targets, with potential applications in biomedical research, diagnostics, and therapeutics.

11.8 Future Perspective and Challenges

CD-based chemosensing has emerged as a promising approach for the detection and quantification of small molecules in various fields, including food science, environmental monitoring, pharmaceuticals, and clinical diagnostics. One of the key goals in CD-based chemosensing is to improve the selectivity and sensitivity of detection towards target analytes. Future research efforts may focus on the design and synthesis of novel CD derivatives or functionalization strategies to achieve enhanced molecular recognition properties and lower detection limits. The development of multifunctional chemosensors capable of detecting multiple analytes simultaneously or performing additional tasks, such as signal amplification or imaging, represents an exciting direction for future research. Integration of CDs with other sensing elements or nanomaterials could lead to the generation of versatile and highly efficient chemosensing platforms. CD-based chemosensors hold great potential for biomedical applications, including disease diagnosis, drug monitoring, and therapeutics. Future research may explore the use of CDs as targeting ligands for selective recognition of biomarkers or as carriers for controlled release of therapeutic agents. Also, it may play a significant role in monitoring pollutants, toxins, and contaminants in air, water, and soil, contributing to environmental protection and public health. The successful commercialization and scalability of CD-based chemosensing technologies require addressing challenges related to production costs, scalability of synthesis methods, and regulatory compliance. Overcoming these challenges is crucial for translating research findings into practical applications and market adoption. The development of new chemical sensors has been made possible by combining the chemistry of CDs with other chemistries and concepts. As other research fields advance, new CD chemical sensors with high incremental values, such as high sensitivity, broad adaptability, and high functionality should be developed. Research has concentrated on creating complex sensing systems by combining concepts from various fields, such as supramolecular chemistry, nanotechnology, biochemistry, polymer chemistry, and inorganic chemistry, and, in light of advancements and discoveries in these areas.

11.9 Conclusions

In conclusion, this chapter highlights the significant advancements and promising prospects in the field of molecular sensing facilitated by CDs. The fundamental principles underlying CD-based chemosensing, including molecular recognition, host–guest interactions, and signal transduction mechanisms have been explored. CD-based chemosensors offer several distinct advantages, including high selectivity, sensitivity, and versatility, making them valuable tools for the detection and quantification of small molecules in various applications ranging from food science to biomedical diagnostics. Advances in molecular design, sensor fabrication, and signal amplification strategies hold the potential to further enhance the performance and applicability of CD-based chemosensors. The exploration of new applications, such as real-time monitoring, biomedical diagnostics, and environmental sensing, presents exciting avenues for future research and innovation. They have specific qualities and attributes that are acquired through the sensing event of different guest molecules using optical and electrochemical signatures on CD-based sensors. This chapter underscored the pivotal role of CDs in the development of advanced chemosensing technologies and emphasized the importance of ongoing research and collaboration in harnessing the full potential of CD-based chemosensors for addressing pressing societal challenges and improving human health and well-being.

References

1. Hulanicki A, Glab S, Ingman F (1991) Chemical sensors: definitions and classification. Pure Appl Chem 63:1247–1250. https://doi.org/10.1351/pac199163091247
2. Janshoff A, Steinem C, Michalke A, Henke C, Galla HJ (2000) Monofunctionalized β-cyclodextrins as sensor elements for the detection of small molecules. Sens Actuators B Chem 70:243–253. https://doi.org/10.1016/S0925-4005(00)00576-1
3. Hazra A, Mondal A, Paul S, Bej S, Mondal U, Nag S, Banerjee P (2022) In: Feliu Spain JM (ed) Chemosensing technology for rapid detection of emerging contaminants. Elsevier, Spain, pp 407–464
4. Rather IA, Wagay SA, Ali R (2020) Emergence of anion-π interactions: the land of opportunity in supramolecular chemistry and beyond. Coord Chem Rev 415:213327. https://doi.org/10.1016/j.ccr.2020.213327
5. Jansook P, Ogawa N, Loftsson T (2018) CDs: structure, physicochemical properties and pharmaceutical applications. Int J Pharm 535:272–284. https://doi.org/10.1016/j.ijpharm.2017.11.018
6. Van der Veen BA, Uitdehaag JCM, Penninga D, Van Alebeek G, Smith LM, Dijkstra BW, Dijkhuizen L (2000) Rational design of cyclodextrin glycosyltransferase from Bacillus circulans strain 251 to increase alpha-cyclodextrin production. J Mol Biol 296:1027–1038. https://doi.org/10.1006/jmbi.2000.3528

7. Khan AR, Forgo P, Stine KJ, D'Souza VT (1998) Methods for selective modifications of CDs. Chem Rev 98:1977–1996. https://doi.org/10.1021/cr970012b
8. Sivakumar K, Parinamachivayam G (2022) Review on β-cyclodextrin inclusion complex based chemosensors for heavy metals. J Incl Phenom Macrocycl Chem 102:603–618. https://doi.org/10.1007/s10847-022-01153-2
9. Ross PD, Rekharsky MV (1996) Thermodynamics of hydrogen bond and hydrophobic interactions in cyclodextrin complexes. Biophys J 71:2144–2154. https://doi.org/10.1016/s0006-3495(96)79415-8
10. Ikeda H (2020) Diversely responsive turn-on fluorescent cyclodextrin chemosensors: guest selectivities and mechanism insights. Chemosensors 8:48. https://doi.org/10.3390/chemosensors8030048
11. Wang Z, Xiao S, Chen Y (2006) β-Cyclodextrin incorporated carbon nanotubes-modified electrodes for simultaneous determination of adenine and guanine. J Electroanal Chem 589:237–242. https://doi.org/10.1016/j.jelechem.2006.02.014
12. Healy B, Yu T, C da Silva Alves D, Okeke C, Breslin CB (2021) Cyclodextrins as supramolecular recognition systems: Applications in the fabrication of electrochemical sensors. Materials 14:1668. https://doi.org/10.3390/ma14071668
13. Chen W, Jiang Y, Jiang J, Qian L, Gu W (2019) Determination of 3,3′,4,4′-Polychlorinated biphenyls by an electrochemical sensor based on a methylene blue probe. J Electrochem Soc 166:B908. https://doi.org/10.1149/2.0541912jes
14. Yang Y, Yang X, Yang HF, Liu ZM, Liu YL, Shen GL, Yu RQ (2005) Electrochemical sensor for cinchonine based on a competitive host–guest complexation. Anal Chim Acta 528:135–142. https://doi.org/10.1016/j.aca.2004.10.041
15. Ikeda H, Nakamura M, Ise N, Oguma N, Nakamura A, Ikeda T, Toda F, Ueno A (1996) Fluorescent cyclodextrins for molecule sensing: fluorescent properties, NMR characterization, and inclusion phenomena of N-dansylleucine-modified cyclodextrins. J Am Chem Soc 118:10980–10988. https://doi.org/10.1021/ja960183i
16. Ueno A, Minato S, Suzuki I, Fukushima M, Ohkubo M, Osa T, Hamada F, Murai K (1990) Host-guest sensory system of dansyl-modified. BETA-cyclodextrin for detecting steroidal compounds by dansyl fluorescence. Chem Lett 4:605–608. https://doi.org/10.1246/CL.1990.605
17. Hamasaki K, Ikeda H, Nakamura A, Ueno A, Toda F, Suzuki I, Osa T (1993) Fluorescent sensors of molecular recognition. Modified cyclodextrins capable of exhibiting guest-responsive twisted intramolecular charge transfer fluorescence. J Am Chem Soc 115:5035–5040. https://doi.org/10.1021/ja00065a012
18. Ueno A, Suzuki I, Osa T (1989) Association dimers, excimers, and inclusion complexes of pyrene-appended gamma-cyclodextrins. J Am Chem Soc 111:6391–6397. https://doi.org/10.1021/ja00198a061
19. Wang Y, Ikeda T, Ikeda H, Ueno A, Toda F (1994) Dansyl-β-cyclodextrins as fluorescent sensors responsive to organic compounds. Bull Chem Soc Jpn 67:1598–1607. https://doi.org/10.1246/bcsj.67.1598
20. Udenfriend S, Stein S, Boehlen P, Dairman W, Leimgruber W, Weigele M (1972) Fluorescamine: a reagent for assay of amino acids, peptides, proteins, and primary amines in the picomole range. Science 178:871–872. https://doi.org/10.1126/science.178.4063.871
21. Kalchenko OI, Solovyov AV, Lipkowski J, Kalchenko VI (1999) Study of the complexation of Octakis(diethoxyphosphoryloxy)-tetramethylcalix[4]resorcinarene with benzene derivatives by the RP HPLC method. J Incl Phenom Macrocycl Chem 34:259. https://doi.org/10.1023/A:1008086519227
22. Aoyagi T, Nakamura A, Ikeda H, Ikeda T, Mihara H, Ueno A (1997) Alizarin yellow-modified β-cyclodextrin as a guest-responsive absorption change sensor. Anal Chem 69:659–663. https://doi.org/10.1021/ac960727z
23. James TD, Sandanayake KS, Shinkai S (1995) Recognition of sugars and related compounds by "reading-out"-type interfaces. Supramol Chem 6:141–157. https://doi.org/10.1080/10610279508032530

24. Sasaki K, Nagasaka M, Kuroda Y (2001) New cyclodextrin dimer and trimer: formation of biphenyl excimer and their molecular recognition. Chem Commun 24:2630–2631. https://doi.org/10.1039/B107987E
25. Radi AE, Eissa S (2010) Electrochemistry of cyclodextrin inclusion complexes of pharmaceutical compounds. Open Chem Biomed Methods J 3:74–85. https://doi.org/10.2174/1875038901003010074
26. Baghayeri M, Namadchian M, Karimi-Maleh H, Beitollahi H (2013) Determination of nifedipine using nanostructured electrochemical sensor based on simple synthesis of Ag nanoparticles at the surface of glassy carbon electrode: application to the analysis of some real samples. J Electroanal Chem 697:53–59. https://doi.org/10.1016/j.jelechem.2013.03.011
27. Ferancová A, Labuda J (2001) Cyclodextrins as electrode modifiers. Fresenius J Anal Chem 370:1–10. https://doi.org/10.1007/s002160100752
28. Gao J, Zhang S, Liu M, Tai Y, Song X, Qian Y, Song H (2016) Synergistic combination of cyclodextrin edge-functionalized graphene and multiwall carbon nanotubes as conductive bridges toward enhanced sensing response of supramolecular recognition. Electrochim Acta 187:364–374. https://doi.org/10.1016/j.electacta.2015.11.073
29. Stankovich S, Dikin DA, Dommett GH, Kohlhaas KM, Zimney EJ, Stach EA, Piner RD, Nguyen ST, Ruoff RS (2006) Graphene-based composite materials. Nature 442:282–286. https://doi.org/10.1038/nature04969
30. Khaled E, Kamel MS, Hassan HN, Abd El-Alim SH, Aboul-Enein HY (2015) Novel screen printed potentiometric sensors for the determination of oxicams. RSC Adv 5:12755–12762. https://doi.org/10.1039/c4ra12513d
31. Wang X, Li X, Luo C, Sun M, Li L, Duan H (2014) Ultrasensitive molecularly imprinted electrochemical sensor based on magnetism graphene oxide/β-cyclodextrin/Au nanoparticles composites for chrysoidine analysis. Electrochim Acta 130:519–525. https://doi.org/10.1016/j.electacta.2014.03.039
32. Pingarrón JM, Yanez-Sedeno P, González-Cortés A (2008) Gold nanoparticle-based electrochemical biosensors. Electrochim Acta 53:5848–5866. https://doi.org/10.1016/j.electacta.2008.03.005
33. Guo S, Wang E (2007) Synthesis and electrochemical applications of gold nanoparticles. Anal Chim Acta 598:181–192. https://doi.org/10.1016/j.aca.2007.07.054
34. Hui Y, Ma X, Qu F, Chen F, Yu J, Gao Y (2016) Electropolymerization of carboxymethyl-β-cyclodextrin based on co-electrodeposition gold nanoparticles electrode: electrocatalysis and nonenzymatic glucose sensing. J Solid State Electrochem 20:1377–1389. https://doi.org/10.1007/s10008-016-3119-5
35. Elhag S, Ibupoto ZH, Nur O, Willander M (2014) Incorporating β-Cyclodextrin with ZnO nanorods: a potentiometric strategy for selectivity and detection of dopamine. Sensors 14:1654–1664. https://doi.org/10.3390/s140101654
36. Abd El-Hady D (2007) Selective and sensitive hydroxypropyl-beta-cyclodextrin based sensor for simple monitoring of (+)-catechin in some commercial drinks and biological fluids. Anal Chim Acta 593:178–187. https://doi.org/10.1016/j.aca.2007.05.002
37. Kraft A, Grimsdale AC, Holmes AB (1998) Electroluminescent conjugated polymers—seeing polymers in a new light. Angew Chem Int Ed 37:402–428. https://doi.org/10.1002/(SICI)1521-3773(19980302)37:4%3c402::AID-ANIE402%3e3.0.CO;2-9
38. Ogoshi T, Takashima Y, Yamaguchi H, Harada A (2006) Cyclodextrin-grafted poly (phenylene ethynylene) with chemically-responsive properties. Chem Commun 35:3702–3704. https://doi.org/10.1039/B605804C
39. Hernandez R, Zappi M, Kuo CH (2004) Chloride effect on TNT degradation by zerovalent iron or zinc during water treatment. Environ Sci Technol 38:5157–5163. https://doi.org/10.1021/es049815o
40. Riskin M, Tel-Vered R, Bourenko T, Granot E, Willner I (2008) Imprinting of molecular recognition sites through electropolymerization of functionalized Au nanoparticles: development of an electrochemical TNT sensor based on π-donor−acceptor interactions. J Am Chem Soc 130:9726–9733. https://doi.org/10.1021/ja711278c

41. Fang QL, Geng JL, Liu BH, Gao DM, Li F, Wang ZY, Guan GP, Zhang Z (2009) Inverted opal fluorescent film chemosensor for the detection of explosive nitroaromatic vapors through fluorescence resonance energy transfer. Chem Eur J 15:11507–11514. https://doi.org/10.1002/chem.200901488
42. Sapsford KE, Berti L, Medintz IL (2006) Materials for fluorescence resonance energy transfer analysis: beyond traditional donor–acceptor combinations. Angew Chem Int Ed 45:4562–4589. https://doi.org/10.1002/anie.200503873
43. Feng L, Tong C, He Y, Liu B, Wang C, Sha J, Lü C (2014) A novel FRET-based fluorescent chemosensor of β-cyclodextrin derivative for TNT detection in aqueous solution. J Lumin 146:502–507. https://doi.org/10.1016/j.jlumin.2013.10.039
44. Ikeda H, Ueno A (2009) Fluorescent α-cyclodextrin as a chemosensor for halomethanes. Chem Commun 28:4281–4283. https://doi.org/10.1039/b907629h
45. Crofton KM, Paul KB, DeVito MJ, Hedge JM (2007) Short-term in vivo exposure to the water contaminant triclosan: evidence for disruption of thyroxine. Environ Toxicol Pharmacol 24:194–197. https://doi.org/10.1016/j.etap.2007.04.008
46. Du Preez J, Wenzhan Y (2003) Improving the aqueous solubility of triclosan by solubilization, complexation, and in situ salt formation. J Cosmet Sci 54:537–550
47. Duan MS, Zhao N, Össurardóttir ÍB, Thorsteinsson T, Loftsson T (2005) Cyclodextrin solubilization of the antibacterial agents triclosan and triclocarban: formation of aggregates and higher-order complexes. Int J Pharm 297:213–222. https://doi.org/10.1039/d0em00387e
48. Adhikari BR, Govindhan M, Chen A (2015) Sensitive detection of acetaminophen with graphene-based electrochemical sensor. Electrochim Acta 162:198–204. https://doi.org/10.1016/j.electacta.2014.10.028
49. Blair BD, Crago JP, Hedman CJ, Klaper RD (2013) Pharmaceuticals and personal care products found in the Great Lakes above concentrations of environmental concern. Chemosphere 93:2116–2123. https://doi.org/10.1016/j.chemosphere.2013.07.057
50. Alam AU, Qin Y, Howlader MM, Hu NX, Deen MJ (2018) Electrochemical sensing of acetaminophen using multi-walled carbon nanotube and β-cyclodextrin. Sens Actuators B Chem 254:896–909. https://doi.org/10.1016/j.snb.2017.07.127
51. Morin-Crini N, Crini G (2013) Environmental applications of water-insoluble β-cyclodextrin–epichlorohydrin polymers. Prog Polym Sci 38:344–368. https://doi.org/10.1016/j.progpolymsci.2012.06.005
52. Liao C, Kannan K (2011) Widespread occurrence of bisphenol A in paper and paper products: implications for human exposure. Environ Sci Technol 45:9372–9379. https://doi.org/10.1021/es202507f
53. Ncube P, Krause RW, Mamba BB (2011) Fluorescent sensing of chlorophenols in water using an azo dye modified β-cyclodextrin polymer. Sensors 11:4598–4608. https://doi.org/10.3390/s110504598
54. Liu Y, Shi J, Guo DS (2007) Novel permethylated β-cyclodextrin derivatives appended with chromophores as efficient fluorescent sensors for the molecular recognition of bile salts. J Org Chem 72:8227–8234. https://doi.org/10.1021/jo071131m
55. Cedergren MI, Selbing AJ, Löfman O, Källen BA (2002) Chlorination byproducts and nitrate in drinking water and risk for congenital cardiac defects. Environ Res 89:124–130. https://doi.org/10.1006/enrs.2001.4362
56. Wang J, Ueno A (2000) Naphthol-modified β-cyclodextrins as fluorescent sensors for detecting contaminants in drinking water. Macromol Rapid Commun 21:887–890. https://doi.org/10.1002/1521-3927(20000801)21:13%3c887::AID-MARC887%3e3.0.CO;2-U
57. Fenyvesi É, Jicsinszky L (2009) Cyclodextrin-containing sensors to provide an early warning of contamination. Land Contamin Reclam 17:405. https://doi.org/10.2462/09670513.95
58. Salinas Y, Martínez-Máñez R, Marcos MD, Sancenón F, Costero AM, Parra M, Gil S (2012) Optical chemosensors and reagents to detect explosives. Chem Soc Rev 41:1261–1296. https://doi.org/10.1039/C1CS15173H
59. Morin-Crini N, Winterton P, Fourmentin S, Wilson LD, Fenyvesi E, Crini G (2018) Water-insoluble β-cyclodextrin–epichlorohydrin polymers for removal of pollutants from aqueous

solutions by sorption processes using batch studies: a review of inclusion mechanisms. Prog Polym Sci 78:1–23. https://doi.org/10.1016/j.progpolymsci.2017.07.004

60. Cheng F, An X, Zheng C, Cao S (2015) Green synthesis of fluorescent hydrophobic carbon quantum dots and their use for 2,4,6-trinitrophenol detection. RSC Adv 5:93360–93363. https://doi.org/10.1039/C5RA19029
61. Zhang K, Zhou H, Mei Q, Wang S, Guan G, Liu R, Zhang J, Zhang Z (2011) Instant visual detection of trinitrotoluene particulates on various surfaces by ratiometric fluorescence of dual-emission quantum dots hybrid. J Am Chem Soc 133:8424–8427. https://doi.org/10.1021/ja2015873
62. Valeur B (ed) (2012) Molecular fluorescence: principles and applications. New York, USA
63. López-Nicolás JM, García-Carmona F (2008) Rapid, simple and sensitive determination of the apparent formation constants of trans-resveratrol complexes with natural cyclodextrins in aqueous medium using HPLC. Food Chem 109:868–875. https://doi.org/10.1016/j.foodchem.2008.01.022
64. Ramírez-Galicia G, Garduño-Juárez R, Vargas MG (2007) Effect of water molecules on the fluorescence enhancement of Aflatoxin B1 mediated by Aflatoxin B1: β-cyclodextrin complexes. A theoretical study. Photochem Photobiol Sci 6:110–118. https://doi.org/10.1039/B614107B
65. Delattre F, Cazier F, Cazier F, Tine A (2009) Use a fluorescent molecular sensor for the detection of pesticides and herbicides in water. Curr Anal Chem 5:48–52. https://doi.org/10.2174/157341109787704787
66. Niu X, Mo Z, Yang X, Shuai C, Liu N, Guo R (2019) Graphene-ferrocene functionalized cyclodextrin composite with high electrochemical recognition capability for phenylalanine enantiomers. Bioelectrochem 128:74–82. https://doi.org/10.1016/j.bioelechem.2019.03.006
67. Ji J, Qu L, Wang Z, Li G, Feng W, Yang G (2022) A facile electrochemical chiral sensor for tryptophan enantiomers based on multiwalled carbon nanotube/hydroxypropyl-β-cyclodextrin functionalized carboxymethyl cellulose. Microchem J 175:107133. https://doi.org/10.1016/j.microc.2021.107133
68. Lenik J, Łyszczek R (2022) A potentiometric sensor for ketoprofen based on a β-Cyclodextrin derivative. J Anal Chem 77:246–256. https://doi.org/10.1134/S1061934822020058
69. Amorim CG, Araújo AN, Montenegro MCBSM, Silva VL (2008) Cyclodextrin-based potentiometric sensors for midazolam and diazepam. J Pharm Biomed Anal 48:1064–1069. https://doi.org/10.1016/j.jpba.2008.08.012
70. Khaled E, Hassan HNA, Mohamed GG, Ragab FA, Seleim AEA (2010) β-Cyclodextrin-based potentiometric sensors for flow-injection determination of acetylcholines. Int J Electrochem Sci 5:448–458. https://doi.org/10.1016/S1452-3981(23)15297-7
71. Hasanzadeh M, Pournaghi-Azar MH, Shadjou N, Jouyban A (2014) Determination of lisinopril using β-cyclodextrin/graphene oxide-SO_3H modified glassy carbon electrode. J Appl Electrochem 44:821–830. https://doi.org/10.1007/s10800-014-0689-8
72. Fu L, Lai G, Yu A (2015) Preparation of β-CDs functionalized reduced graphene oxide: application for electrochemical determination of paracetamol. RSC Adv 5:76973–76978. https://doi.org/10.1039/C5RA12520K
73. Shen Q, Wang X (2009) Simultaneous determination of adenine, guanine and thymine based on β-cyclodextrin/MWNTs modified electrode. J Electroanal Chem 632:149–153. https://doi.org/10.1016/j.jelechem.2009.04.009
74. Ramaiah D, Neelakandan PP, Nair AK, Avirah RR (2010) Functional cyclophanes: promising hosts for optical biomolecular recognition. Chem Soc Rev 39:4158–4168. https://doi.org/10.1039/B920032K
75. Hargrove AE, Nieto S, Zhang T, Sessler JL, Anslyn EV (2011) Artificial receptors for the recognition of phosphorylated molecules. Chem Rev 111:6603–6782. https://doi.org/10.1021/cr100242s
76. Fujita K, Fujiwara S, Yamada T, Tsuchido Y, Hashimoto T, Hayashita T (2017) Design and function of supramolecular recognition systems based on guest-targeting probe-modified cyclodextrin receptors for ATP. J Org Chem 82:976–981. https://doi.org/10.1016/j.isci.2020.100927

77. Kanagaraj K, Xiao C, Rao M, Fan C, Borovkov V, Cheng G, Zhou D, Zhong Z, Su D, Yu X, Yao J (2020) A quinoline-appended cyclodextrin derivative as a highly selective receptor and colorimetric probe for nucleotides. iScience 23:(1–18). https://doi.org/10.1016/j.isci.2020.100927
78. Alarcón-Angeles G, Pérez-López B, Palomar-Pardave M, Ramírez-Silva MT, Alegret S, Merkoçi A (2008) Enhanced host–guest electrochemical recognition of dopamine using cyclodextrin in the presence of carbon nanotubes. Carbon 46:898–906. https://doi.org/10.1016/j.carbon.2008.02.025
79. Ramírez-Berriozabal M, Galicia L, Gutiérrez-Granados S, Cortes JS, Herrasti P (2008) Selective electrochemical determination of uric acid in the presence of ascorbic acid using a carbon paste electrode modified with β-Cyclodextrin. Electroanalysis 20:1678–1683. https://doi.org/10.1002/elan.200804232
80. Kumar TN, Ananthi A, Mathiyarasu J, Joseph J, Phani KL, Yegnaraman V (2011) Enzymeless creatinine estimation using poly(3,4-ethylenedioxythiophene)-β-cyclodextrin. J Electroanal Chem 661:303–308. https://doi.org/10.1016/j.jelechem.2011.08.001

Most. israt Jahan is currently a research assistant of Bangladesh Food Safety Authority, Bangladesh at Material Chemistry Research Laboratory, Department of Chemistry at University of Dhaka, Bangladesh. She also served for literature survey at Dhaka University Nanotechnology Center (DUNC) of Dhaka University, Bangladesh; where she contributed her analytical skills and expertise to various projects. She earned her Master's degree in Physical Chemistry. Her research interest is in material science, nanochemistry, supramolecular interactions, and polymer chemistry. One of her research skills is investigation of the effect of several parameters on the enhancement of capacitive behavior of PVDF-hfp. Currently she is conducting a research on developing an electrochemical sensor based on a novel nanocomposite.

Israt Jahan Heme is a graduate student currently pursuing her Master's degree from the Department of Chemistry, University of Dhaka, Bangladesh. She is serving under Student Promotion and Support Unit at Dhaka University Nanotechnology Center (DUNC) of Dhaka University, Bangladesh; where she contributed her analytical skills and expertise to various projects. She is carrying out her research with microemulsion systems at Material Chemistry Research Laboratory, Department of Chemistry at University of Dhaka. Her research interest in material science, supramolecular interactions and specially in microemulsion system. One of her research skills is to investigate the property of ionic liquid based microemulsion system.

Dr. Md. Abu Bin Hasan Susan is a Professor of Chemistry and Director of Dhaka University nanotechnology Center (DUNC) of Dhaka University, Bangladesh. He received his Ph.D. degree from Yokohama National University, Japan in 2000 and was awarded with VBL, JSPS, CREST, and Bridge postdoctoral fellowships. He received the Dean's Award of the Faculty of Science, Dhaka University (2011), the UGC Awards of Bangladesh in 2011 and 2013, the United Group Paper Awards in 2016 and 2017, and the Bangladesh Academy of Sciences Gold Medal in 2012. He received the Silver Medal of the Society of Promotion of Education and Science, India in 2019. His research interest is in the field of Materials Chemistry. He has been a Fellow (2018) of the Bangladesh Academy of Sciences and IUPAC (2016). He has 228 articles including 39 book chapters, 18 proceedings and 3 Books. His citations are 13,470, h-index-36, and i-10 index-106.

Chapter 12
Cyclodextrins: Chemical Modifications and Applications in Sensing

Sheersha Pradhan, Lavanya Lahari Pannuri, and Thangamuthu Mohan Das

Abstract Cyclodextrins (CDs)—a class of water-soluble seminatural cyclic oligosaccharides—comprise α-D-glucopyranose units linked by 1,4-glycosidic linkages. It gained significant attention in the scientific community because of the potential applications in synthetic organic and material chemistry, attributed to its toroidal three-dimensional flexible structures, consisting of a narrow upper rim and a larger bottom rim. Due to several polar carbohydrate moieties, they do not absorb UV–Vis light; nevertheless, they can act as precursors for forming spectroscopically sensitive compounds through several modifications. Synthetic alterations to the cyclodextrin structure can significantly alter the steric layout, hydrophobicity, and charge density, possibly making the formation of host-guest complexes suitable for various applications more effective. This book chapter provides a collection of processes reported in recent literature about constructing cyclodextrins as versatile chemical sensors via molecular recognition phenomena for sensing ions and other small molecular species.

Keywords Cyclodextrins (CDs) · Chemosensors · Inclusion complexes · Nanoprobes · Composites · Metal ion sensing · Small molecule sensing

Abbreviations

CGTase	Cyclodextrin glucanotransferase
RT	Room temperature

S. Pradhan · L. L. Pannuri · T. Mohan Das (✉)
Department of Chemistry, School of Basic and Applied Sciences,
Central University of Tamil Nadu (CUTN), Thiruvarur 610005, Tamil Nadu, India
e-mail: tmohandas@cutn.ac.in

S. Pradhan
e-mail: pradhan.sheersha@gmail.com

L. L. Pannuri
e-mail: lavanyalahari.pannuri@gmail.com

S. Krishnamoorthy (ed.), *Cyclodextrins for Chemosensing*,
https://doi.org/10.1007/978-3-031-89299-8_12

DCC	Dicyclohexyl carbodiimide
DMAP	4-(Dimethylamino)pyridine
DHBC	(*E*)-N'-(4-(diethylamino)-2-hydroxybenzylidene)hydrazinecarbohydrazide
BA-Azo	(*E*)-trihydroxy(4-(4-((4-oxidophenyl)diazenyl)benzamido)phenyl)borate
APBA	*m*-Aminophenylboronic acid
EDC	Ethylene dichloride
NHS	*N*-Hydroxysuccinimide
FL Spectra	Fluorescence spectroscopy
ATP	Adenosine triphosphate
LoD	Limit of detection
LoQ	Limit of quantification
MWCN	Multi-walled carbon nanotubes
FRET	Fluorescence resonance energy transfer
NP	Nanoparticles
CQDs	Carbon quantum dots
DPV	Differential pulse voltammetry
EGDMA	Ethylene glycol dimethacrylate
GCE	Glassy carbon electrode
EP	Electropolymerization
ICD	Induced circular dichroism
ITO	Indium-tin oxide
GO	Graphene oxide

12.1 Introduction

Cyclodextrins (CDs) are a novel class of cyclic oligosaccharides containing, primarily, six to eight 1,4-glycosidically linked glucose units. However, examples of CDs with fewer glucose units have also been reported [1, 2]. α-Cyclodextrin (α-CD) also termed Schardinger's α-dextrin or cyclohexaamylose contains six glucose units, while β-cyclodextrin (β-CD) known as Schardinger's β-dextrin or cycloheptaamylose contains seven glucopyranose units, and γ-cyclodextrin (γ-CD) otherwise termed as Schardinger's γ-dextrin or cyclooctaamylose, comprises eight units of glucopyranose [3]. (Fig. 12.1).

CDs are non-hygroscopic, crystalline, and homogeneous torus-like macrorings [4, 5]. They have a well-defined cup-shaped chemical structure consisting of a hydrophilic exterior and a hydrophobic cavity and are ideal for molecular inclusion complex formation [6, 7]. Guest molecules can thus remain in the hydrophobic cup if their size is comparable to the interior cavity diameter, forming a host-guest complex. In certain circumstances, particular interactions, such as electrostatic interactions and hydrogen bonding, can increase host-guest complex stability. Meanwhile, charge transfer is triggered during the particular contact when guest species enter the cavities of the CD interior. There are

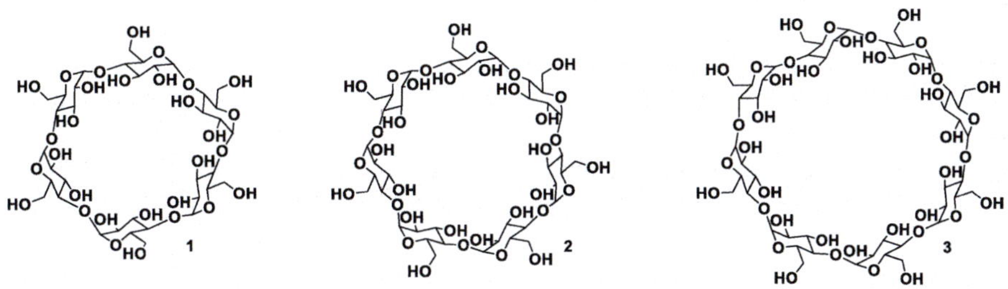

Fig. 12.1 Structural representation of α-cyclodextrin (**1**), β-cyclodextrin (**2**), and γ-cyclodextrin (**3**)

reports of different chemical methods in the literature for developing CD-inclusion complexes, such as paste complexation, slurry complexation, extrusion and dry mixing, damp mixing, and heating and coprecipitation [8–11]. Because of the ease of complexation, they have found applications in agriculture, food processing such as active or smart packaging, textile, medicine, cosmetics, etc. [12–14]. Furthermore, scientific publications regarding research related to cyclodextrins have been abundant, with approximately 2000 publications each year (Fig. 12.2).

In a contemporary manner, these scaffolds have found widespread usage as efficient chemosensors and ecologically acceptable foreign species-detecting tools [15–18]. The design and development of macromolecular hosts for cation and anion binding is an interesting subject in coordination chemistry due to its critical role in the molecular recognition of neutral and ionic species in biological activity studies. Colorimetric and fluorescent chemosensors have gained much interest owing to their

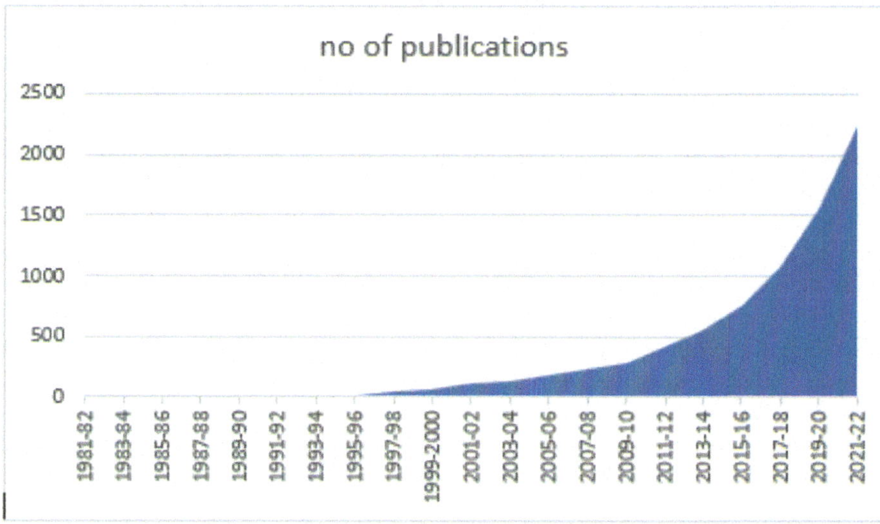

Fig. 12.2 Number of publications per year on cyclodextrin-based chemosensors in 1981–2022 (*Source* https://scholar.google.com/)

low cost, selectivity, sensitivity, and reaction time [19, 20]. This chapter focuses on the discovery, development, and literature concerning synthetically modified cyclodextrins and their application in sensing ions and small molecules in the past decade.

The natural cyclodextrins are formed by the enzymatic breakdown of starch. One of the enzymes is CGTase generated by many bacteria that aids in the breakdown process [21]. The ratio of CD generation is determined by the CGTase sources, which are *Bacillus macerans, Bacillus megatarium, Bacillus circulans, Bacillus ohbensis Thermoanaerobacter sp.*, and *Klebsiella pneumoniae.* This technique involves liquifying starch using α-amylase or heat treatment to produce starch hydrolysate. The new enzyme generated is then treated with CGTase at the optimal temperature and pH for starch breakdown [22, 23]. Flaschel et al. revealed that 1-decanol could be employed as a complexant to improve the synthesis and isolation of α-CD with 98% high purity In 1984 [24]. Rajput and his team in 2016, discovered CGTase from *Microbacterium terrae KNR 9* increased the synthesis of β-dextrin [25]. Larsen and Beeren in 2021 successfully synthesized β-CD in 70% yield from maltose [26].

12.2 Molecular Structure of Natural CDs

The natural dextrins are nonreducing cyclic oligomers made up of 6–8 glucopyranose units connected by α-(1 $\rightarrow$ 4) glycosidic linkage, denoted as α-, β- and γ-CD, respectively, having empirical formula $(C_6H_{10}O_5)_n$, where n = 6–8. Aside from their cyclic configuration, cyclodextrins have a characteristic shape known as a conical frustum [27] (Fig. 12.3). Native CDs have the same height, but their inner and outer diameters grow with the number of glucose units. They possess amphiphilic character due to their conical frustum structure formed by the chair-shaped dextrose units joined via α-(1 $\rightarrow$ 4) linkages [28]. The coupling of dextroses in chair conformation causes CDs to have a truncated cone shape [29, 30]. The general characteristics [31] are given in Table 12.1.

The average cavity diameter grows in proportion to the number of glucose units. As a result, molecules with a comparable size are appropriate for creating host-guest inclusion complexes. Thus, CDs are attractive hosts for sensing, yet these sensors have relatively limited selectivity [32]. Furthermore, CDs are supramolecular hosts that demonstrate higher compatibility with numerous chemical compounds and less toxicity, owing to their natural source [33]. As they comprise carbohydrate units, they do not absorb light in the UV-Vis range (200–800 nm), but they may be transformed into UV-Vis active compounds by adding a chromophore unit. Among the chromophores, fluorophores have excellent sensitivity in analytical applications [34].

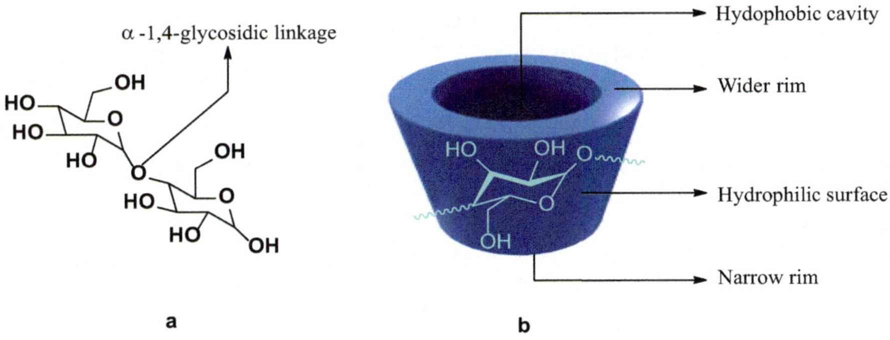

Fig. 12.3 **a** α-(1 → 4) glycosidic bond **b** conical frustum-shape of a cyclodextrin

Table 12.1 Characteristics of α-, β-, and γ-CDs

Properties	α	β	γ
Number of dextrose units	6	7	8
Molecular weight	972	1135	1297
Water solubility (g/100 ml) in RT	14.5	1.85	23.2
$[\alpha]_D$ at 25 °C	150 ± 0.5	162.5 ± 0.5	177.4 ± 0.5
Hollow diameter (in Å)	4.7–5.3	6.0–6.5	7.5–8.3
Torus height (in Å)	7.9 ± 0.1	7.9 ± 0.1	7.9 ± 0.1
Outer diameter (in Å)	14.6 ± 0.4	15.4 ± 0.4	17.5 ± 0.4
Approximate volume of the cavity (in Å^3)	174	262	427
Approximate volume of the cavity for 1 mol of CD (in ml)	104	157	256
Approximate volume of the cavity for 1 g of CD (in ml)	0.1	0.14	0.2
Shape of crystal (from water)	Hexagonal plates	Monoclinic parallelograms	Quadratic prisms
Hydrolysis by A. oryzae amylase	Negligible	Slow	Rapid
pK at 25 °C (by potentiometry)	12.332	12.202	12.081

12.3 Solubility, Cavity Size and Significance to Sensing of Native Cyclodextrins

The presence of hydroxyl groups on the narrow and broad rims determines the water solubility of cyclodextrins. Therefore, the "hydrophilic surface" and "hydrophobic cavity" contribute to their amphiphilicity. Consequently, CDs have a micro-heterogeneous environment in which both hydrophobic and hydrophilic species coexist [35]. Due to their amphiphilic nature, CDs can form inclusion complexes with various species by trapping them in the hollow [36]. Sparingly water-soluble guest species can be best contained in the hydrophobic interior of CDs.

Native CDs have a water solubility order of $\gamma > \alpha > \beta$ at 25 °C [37]. The low hydrophilicity of β-CD is due to a "flip-flop" intramolecular hydrogen interaction. Although this interaction is present in all the natural dextrins, the "flip-flop" type H-bonding in β-CD produces a permanently inflexible intramolecular H-bonding system, making it sparingly soluble in water, and as this is not strong enough in α- and γ-CD, enhances the solubility in water, making them more hydrophilic than β-CD [36].

12.4 Synthetic Modification of Cyclodextrins and Their Significance in Sensing

The relevance of cyclodextrins in chemosensitivity and sensing can be appreciated from the two explanations described in the literature. To begin with, cyclodextrins are preferred to increase the innate sensitivity of organic or inorganic species by encapsulation/tagging [38]. When the chemosensory scaffold is combined with cyclodextrins, the sensitivity and limit of detection (LoD) change dramatically [39]. Second, the target species might be directly captured as a guest molecule within the CD host cavity. β-cyclodextrin (β-CD) is a nontoxic oligosaccharide that may encapsulate organic molecules of specified size and form through noncovalent contact [40, 41].

12.4.1 α-Cyclodextrin-Based Chemosensors

Lai et al. detected and estimated benzene vapors in the atmosphere by using 2,6-per-*O*-(*tert*-butyldimethylsilyl)-α-CD or BDS-α-CD. The BDS-α-CD embedded and plated with gold/silver piezoelectric quartz crystal selectively senses benzene vapors in the air using flow injection and the gas-phase technique (Fig. 12.4). The BDS-α-CD sensor is highly selective for benzene vapors compared to methane, propane, butane, toluene, nitrobenzene, and ammonia with linearity range of 1.01–5120.9 μM [42].

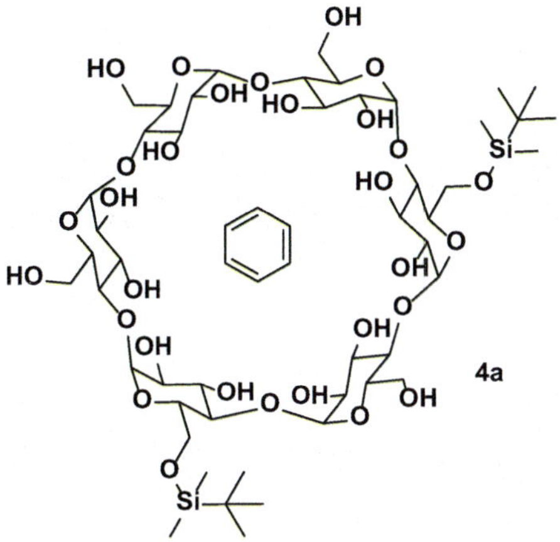

4a

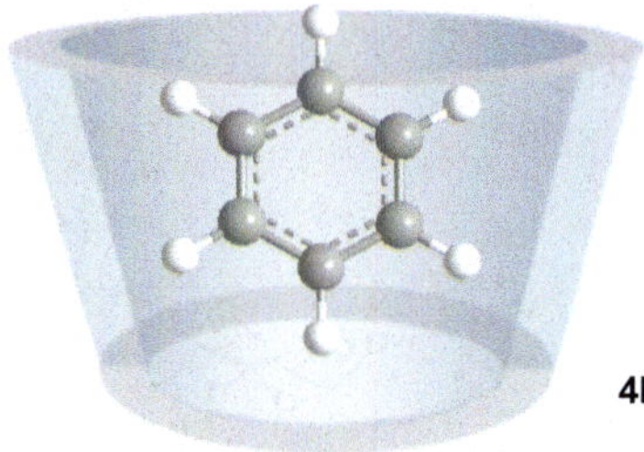

4b

Fig. 12.4 Structural representation of benzene trapped in 2,6-per-*O*-(*tert*-butyldimethylsilyl)-*α*-CD (**4a** and **4b**)

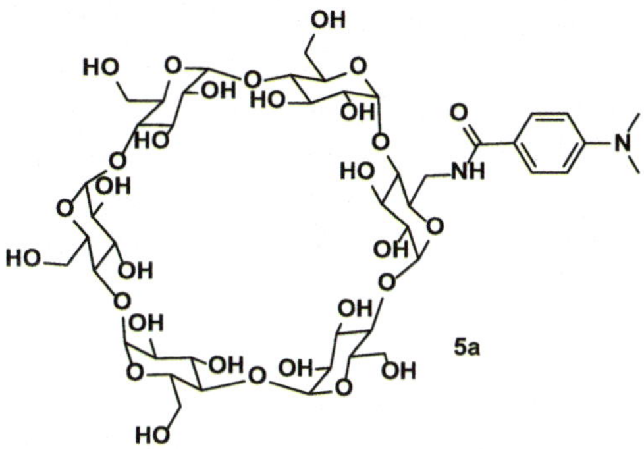

5a

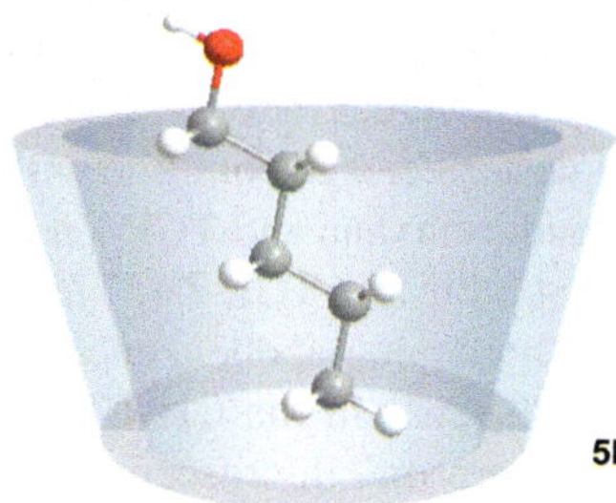

5b

Fig. 12.5 **a** Structural representation of ABDM-*α*-CD, **5a**; **b** inclusion complex of ABDM-*α*-CD, **5a** with *n*-pentanol (**5b**)

Mono-6-deoxy-6-amino-*α*-CD-linked with *p*-dimethylaminobenzoic acid (ABDM-*α*-CD, **5a**) detects aliphatic alcohols like butanol and pentanol derivatives via dual fluorescence (Fig. 12.5). Adding small chain alcohols to the ABDM-*α*-CD cavity (**5b**) reduces typical planar emission intensity while increasing emission because of its distorted intramolecular charge transfer caused by the presence of -$N(CH_3)_2$ group. The detection efficiency order is *n*-butanol < *n*-pentanol < *n*-hexanol [43].

NC0*α*-CD (**6a**) was synthesized by the condensation of 4-chloro-7-nitrobenz-2-oxa-1,3-diazole with mono-6-amino-6-deoxy-*α*-CD. It detects CBr_4 and CCl_4 (halomethanes) through a turn-on fluorescent chemical sensing mechanism at RT and neutral pH in phosphate buffer (0.1 M,) via multisite contacts between the CD scaffold and perhalogenated methane. The absorption intensity of NC0*α*-CD (**6a**)

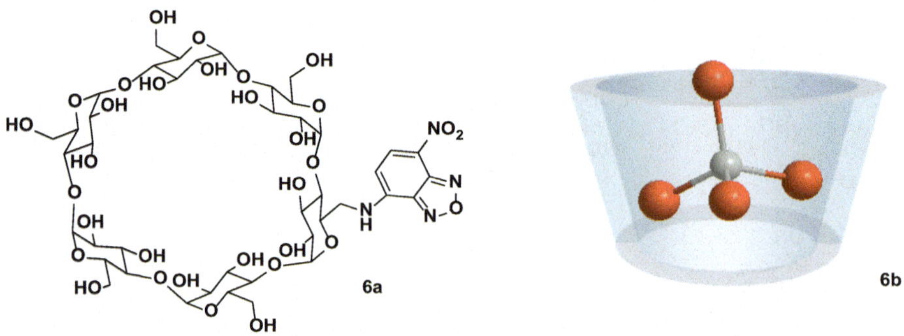

Fig. 12.6 **a** Structural representation of NC0α-CD (**6a**); **b** inclusion complex of NC0α-CD (**6a**) with CBr_4(**6b**)

reduced with the addition of CBr_4 and CCl_4 entering the modified CD's cavity by replacing the attached nitrobenzoxadiazole amine from the inside of the hollow to the outside. The turning-on process happens when CBr_4 is encapsulated, increasing fluorescence intensity [44]. The inclusion complex has been given in Fig. 12.6b.

TCP-Zn (**7**) was employed—where TCP-Zn is (*E*)-4-((4-(1,4,8,11-tetraazacyclotetradecan-1-ylsulfonyl)phenyl)diazenyl)-*N*,*N*-dimethylaniline—as a guest species to combine with α-dextrin (Fig. 12.7) for the selective ATP detection. Pseudorotaxane and TCP-Zn complexes of α-dextrin detected ATP and bound to the guest molecule TCP-Zn (**7**) projected beyond the α-dextrin hollow [45].

Rhodamine-azobenzene-based α-CD functionalized with Fe_3O_4/SiO_2 nanosensor detects Cu^{2+} in the CH_3CN/H_2O medium. Cu^{2+} ions can be used to create selective 'turn-on' fluorescent improvements and an apparent color shift from colorless to yellow-green in fluorescence experiments, as well as light yellow to pink in UV-visible spectrums [46]. Natural α-CD embedded on the microelectrode Ti-W/Au was reported to detect dopamine and L-tyrosine levels. The selectivity and sensing method relies on the variations in surface charge, stereochemistry, and surface affinity of catecholamine derivatives, viz. dopamine and L-tyrosine toward α-dextrin at concentrations ranging from 5 µM to 10 mM [47].

Fig. 12.7 Structural representation of TCP-Zn (**7**) derivative for selective sensing of ATP

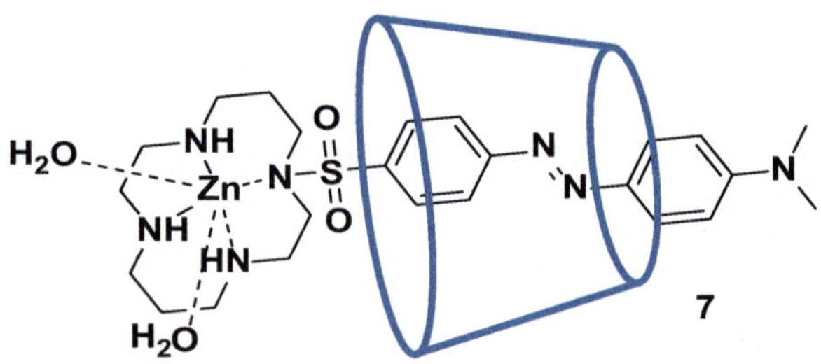

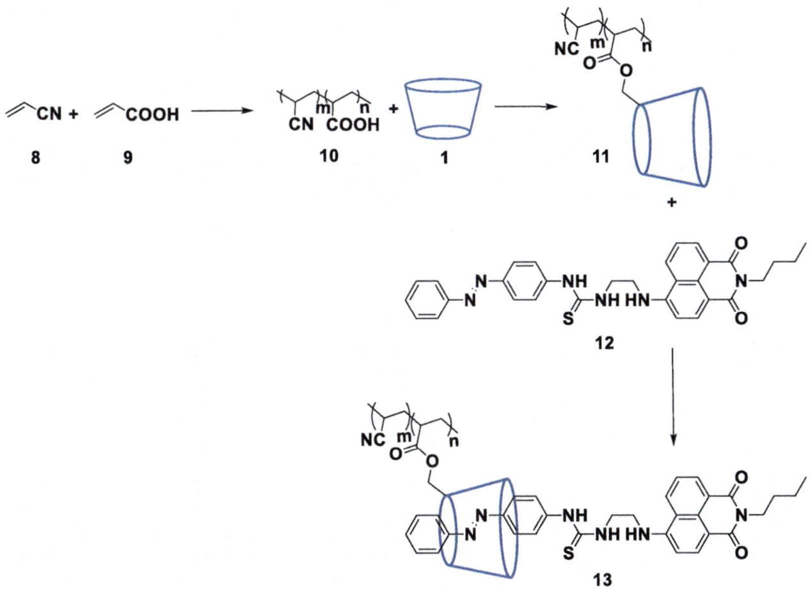

Scheme 12.1 Synthesis of α-CD incorporated *p*-amido-azobenzene (**13**)

An altered *p*-amido-azobenzene (**12**) linked with α-dextrin (**1**) based electrospun nanofiber detects F^- anion in water-acetonitrile mixture. The nanofibrous membrane's enormous surface area allows for excellent adsorption of F^- ions. In the presence of F^- ions, the sensor's fluorescence spectra reached their maximum intensity at 527 nm. Fluorescence intensity gradually diminishes with increasing fluoride ion concentration owing to photoinduced electron transfer influence [48]. The synthesis of α-CD incorporated *p*-amido-azobenzene (**13**) has been given in Scheme 12.1.

12.4.2 β-Cyclodextrin-Based Chemosensors

GO and β-CD functionalized MWCN (**15**) had been utilized in an electrochemical sensor to detect Bisphenol A or 2,2í-bis(4-hydroxyphenyl)propane (BPhA), an endocrine disruptive substance often used in packing of food and beverages. When this electrode device (**15**) got attached to the large surface area of GO, it enhanced the inclusion complex formation of β-CD and resulted in the remarkable detection of BPhA. The chemosensor exhibited a two-step linear response from 0.05–5 μM to 5–30 μM, with an LoD of 6 nM [49]. The synthesis for the preparation of β-CD functionalized multi-walled carbon nanotubes (**15**) has been given in Scheme 12.2.

Pyromellitic anhydride (**16**) was used as a cross-linker to create a β-CD-cross-linked polymer (**17**) that allowed easy fluorescence detection (Scheme 12.3). The

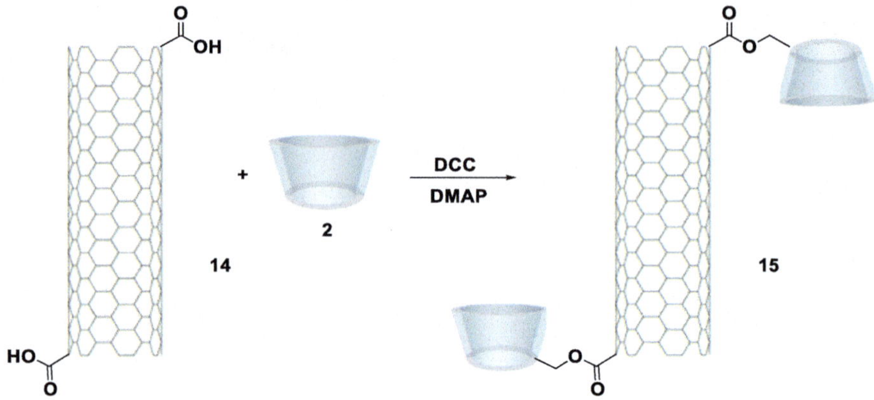

Scheme 12.2 Synthesis of β-CD functionalized MWCN (**15**)

experimental data revealed that the probe had a high affinity for diclofenac and could select it even in the presence of ibuprofen, morphine, amphetamine, and codeine with a 423 nm emission wavelength. The LoD and linear range were 0.92 mM and 1–33 mM, respectively. Pyromellitic anhydride was used to produce β-CD-nanosponge, which was then used to monitor diclofenac levels directly. Diclofenac fluorescence detection was successful even in the presence of competing medicines [50].

A luminous chemosensor was developed using pyrene and two β-cyclodextrin units (Py@2βCD) for aromatic amino acids. Aromatic amino acid D-enantiomers were found to bind more strongly to Py@2βCD than L-enantiomers, resulting in higher complexation energy which was confirmed from the quantum chemical calculations. Using aromatic amino acids as analytes is intriguing since they are chiral; natural aromatic amino acids with a typical L-configuration, whereas D-amino acids are synthetic in nature. As a result, introducing aromatic amino acid enantiomers

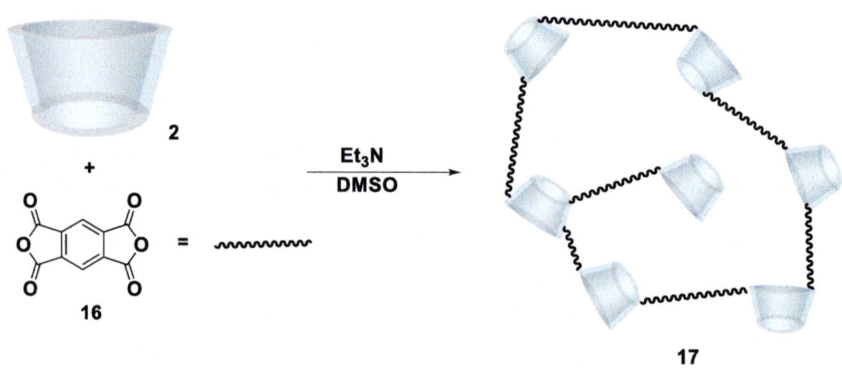

Scheme 12.3 Synthesis of pyromellitic anhydride-based β-cyclodextrin cross-linked polymer (**17**)

into the hollow of the β-CD with D-configuration resulted in the creation of better inclusion complexes [51].

2-hydroxypropyl-β-cyclodextrins (2-HyP-β-CD) was reported as a chemosensor to detect cholesterol, with L-tryptophan serving as a fluorescence probe. The inclusion complex formation stoichiometry of 2-HyP-β-CD with L-tryptophan was determined through a concentration and time-dependent interaction. Upon binding with the amino acid, there was a significant decrease in the fluorescence as 2-HyP-β-CD acted as a fluorescence quencher for L-tryptophan but the gradual addition of cholesterol resulted in a significant increment in fluorescence signals, which suggested the replacement of the amino acid in the 2-HyP-β-CD host. Studies showed that in rats, 2-HyP-β-CD decreases cholesterol levels in plasma rapidly when put intravenously. Its higher water solubility and greater affinity towards cholesterol caused the formation of a soluble complex and increased sensitivity [52].

12.4.3 *β-Cyclodextrin-Based Polymer and Composite Chemosensors*

β-CD linked Ag NPs have been added to ITO electrodes to monitor nitroaromatic isomers like chloro-nitrobenzene and nitroaniline. The lowest detection concentrations were 1×10^{-7}molL^{-1} or below for all isomers. The selectivity was determined by the intensity of the host-guest inclusion complex between β-CD and the corresponding isomers, allowing us to identify the composition one by one from a complex sample with numerous nitroaromatic chemicals. This method appeared unique since resolving the isomers was impossible in any other way [53]. The β-CD derivative modified with silver nanoparticles (βCD-AgNPs, **18**) detected hydrogen peroxide in urine and aqueous media by encapsulation mediated by creatinine (Scheme 12.4), with an LoD of 1.47 nM and a LoQ of 3.76 nM. The sensitivity of H_2O_2 detection was also reported to be affected by the concentration of nanoparticles present, the pH of the media, and temperature. The βCD-AgNPs test showed enhanced sensitivity towards H_2O_2 in urine even compared to the observations obtained from the Horse Radish Assay kit [54].

EP process of β-CD on a GCE creates an electrochemical sensor for detecting fenitrothion (FNT), an organophosphorus pesticide with an LoD of 6 ng/mL. The FNT

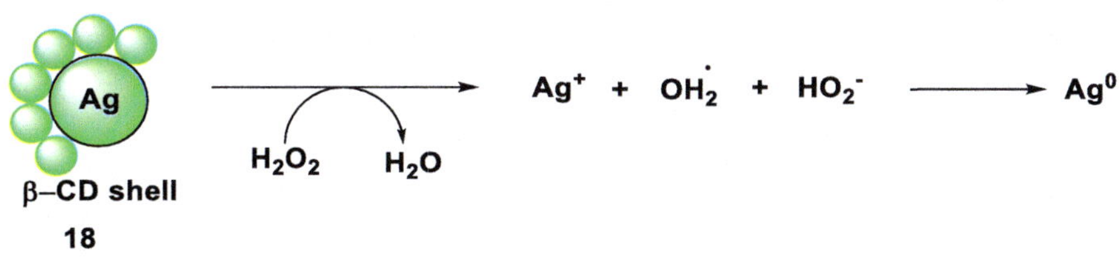

Scheme 12.4 Mechanistic route for the sensing of hydrogen peroxide by βCD-AgNPs (**18**)

detection was influenced by various conditions such as pH, EP cycles, and electrochemical response enrichment time of FNT. The experiment suggested a good linear connection between FNT concentration and the reduction peak current of the nitroso derivative of FNT in the range of 10–150 and 150–4000 ng/mL. Its practicality was studied in tap water and cabbage with a recovery value between 98.43 and 112% [55]. A molecularly engraved poly-β-CD embedded graphite (MEP-β-CD@G) electrode was developed for the electrochemical detection of capsaicin (CP) in chilli pepper samples. The derivative was copolymerized from β-CD, EGDMA, and graphite powder. The lowest LoD for CP was 0.007 μM [56]. A GCE surface was functionalized with a β-CD/CQDs hybrid using electrochemical polymerization to concurrently monitor tryptophan dopamine and uric acid levels. In a system containing all three samples, three well-defined oxidation peaks were obtained with a large separation from each other and also the potential separations of oxidation peaks of dopamine-uric acid and uric acid-tryptophan were up to 150 and 420 mV, respectively, due to the synergistic effect of CQDs with β-CD [57].

Polypyrrole (PPy) is more suitable for preparing of hybrid and smart tools with cyclodextrins because usually it does not degrade under ambient settings and can be developed using simple techniques such as vapor deposition and electrochemical polymerization. A variety of strategies, including surface modification, shape control, and the application of chemicals, are being used to produce and improve the functioning of new sensors [58]. PPy decorated with graphene/CD composite modified screen-printed carbon electrode (SPCE) was utilized to detect Hg^{2+} even in the presence of other metal ions. Chemical oxidation was employed for the preparation of β-CD/PPy composite by taking the PPy monomer in graphene modified β-CD solution in the presence of $FeCl_3$. DPV studies suggested that it can detect Hg^{2+} up to 51.56 μM L^{-1}, with a LoD of 0.47 nM L^{-1} [59]. Chung-Yi Lee et al. worked on the synthesis of hexagonal-shaped plates-like PPy systems using PPy and β-CD. The composite formation depends highly on the amount of the monomeric complex through various intermolecular interactions such as H-bonding and π-π stacking between β-CD and PPy units. The hexagonal-shaped PPy/β-CD plate demonstrated outstanding biosensing behavior, with signal intensity correlated with analyte (dopamine) concentration [60]. A β-CD pyrrole hybrid (Py/β-CD) was synthesized via electrochemical deposition on an electrode surface. The PPy/β-CD polymeric coating over a metallic surface exhibits inhomogeneity, with both highly ordered and extremely disordered polymer chains present. The PPy/β-CD showed better sensitivity towards D-phenylalanine $(138 \pm 15) \times 10^3$ compared to L-phenylalanine $(6 \pm 1) \times 10^3$ in the concentration range of 0.1–0.75 mM [61].

12.4.4 β-Cyclodextrins Metal Complexes for Detection of Metal Ions

The host-gust system created from *β*-CD and DHBC (**21**), a simple chemical consisting of 4-(diethylamino)-2-hydroxybenzaldehyd (**19**) and carbohydrazide (**20**), displayed excellent water solubility during ultrasonic settings. In a clean water environment, the host-guest system exhibited conventional fluorescence behavior towards Al^{3+} (Scheme 12.5). The addition of Al^{3+} caused a considerable increase in blue fluorescence in the host-guest system. The fluorophore of DHBC coordinated with *β*-CD, resulting in a high binding constant. The reaction between *β*-CD-DHBC and Al^{3+} results in a 1:1 stoichiometry molecule [62].

An anthracene derivative of 6-deoxy-6-aminoethylamino-*β*-CD (**26**) produces a modified β-cyclodextrin that enhances fluorescence and serves as a Pb^{2+} (Scheme 12.6) sensor. The spacer between anthracene and *β*-cyclodextrin includes aminoethyl and imino-nitrogens, increasing fluorescence intensity. The compound was yellow, having an absorption band centered at 412 nm. The modified *β*-cyclodextrin: Pb^{2+} ion combination exhibits a log K of 9.2 ± 0.2 at a 1:1 stoichiometry [63].

In a similar work, an anthracene-imino derivative of 3-methyl-2,6-diphenylpiperidin-4-one derivative detects Ca^{2+} ions via turn-on fluorescence. The substance and *β*-cyclodextrin form a yellow host–guest complex with a 1:1 stoichiometric ratio. It detects Ca^{2+} in both free and bound forms, with the cyclodextrin-encapsulated version exhibiting more competitive Ca^{2+} binding in the presence of

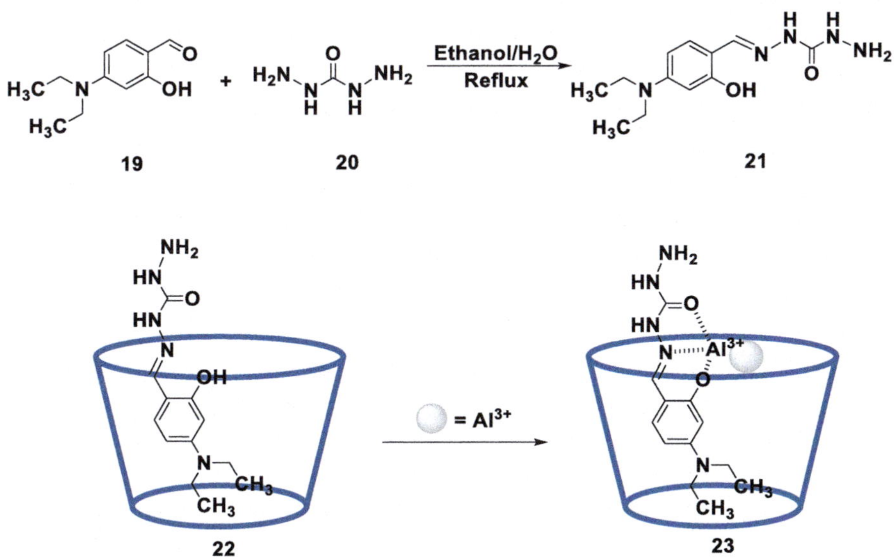

Scheme 12.5 Schematic route for the sensing of Al^{3+} by *β*-CD-DHBC system (**23**)

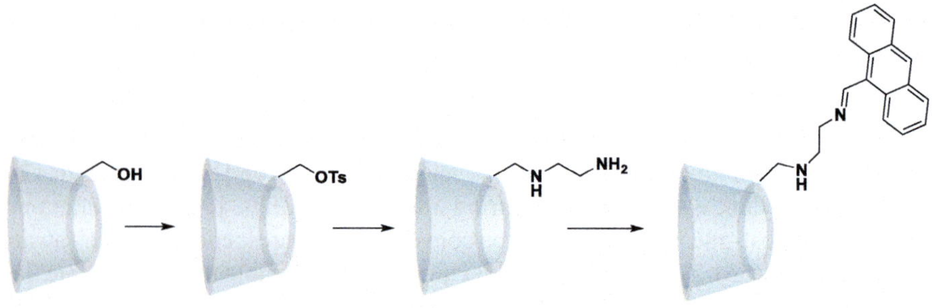

Scheme 12.6 Synthesis of anthracene derivative of 6-deoxy-6-aminoethylamino-*β*-CD (**26**)

different metal ions. In aqueous *β*-CD solution, Ca^{2+} ions attach to the molecule more easily, with a binding constant of 35.71 M^{-1} [64].

Curcumin (L) and *β*-CD formed a supramolecular complex (LC) that selectively bound to Hg^{2+}, resulting in novel fluorescence and absorbance peaks at 502 and 379 nm, in addition to the LC bands at 432 and 535 nm. It also demonstrated an obvious color shift from yellow to colorlessness and fluorescence from strong to weak due to preferential binding with Hg^{2+} ions forming a cationic complex and showed a hypsochromic change in the absorption spectra and quenching in the emission spectra, indicating a significant interaction of the metal ion with LC. Furthermore, studies suggested that deprotonation of the aliphatic -OH group due to complexation in curcumin caused color alterations. The synthesized chemosensor's LoQ and LoD are 16.73 μM and 5.02 μM, respectively [65].

12.4.5 γ-Cyclodextrins-Based Chemosensors

A new podand fluoroionophore/*γ*-CD combination sensor (**28**) exhibited good selectivity for lead (Pb^{2+}) ions in water (Scheme 12.7). The podand fluoroionophore has a pyrene scaffold as the photosignal transducer and pseudo-18-crown-6 system (**27**) as the Pb^{2+} binding site. The wide emission for the pyrene appended podand-*γ*-CD system detected in the higher wavelength area (471 nm) is enhanced, while the pyrene showed emission around 370–410 nm is reduced; hence, *γ*-CD derivative showed better selectivity and sensitivity towards Pb^{2+} [66].

A supramolecular phenylboronic acid azoprobe (B-Azo and BA-Azo)/ *γ*-CD combination (**31**) for detecting D-glucose (**30**) in water had been designed. In addition to the structure of B-Azo, BA-Azo's structure includes a benzamide group as spacer which is a simple phenylboronic acid azoprobe. Only a portion of the spacer differed, yet it had a significant impact on D-glucose selectivity as only BA-Azo (**29**) showed selectivity toward D-glucose in water (Scheme 12.8). Azobenzene chromophores

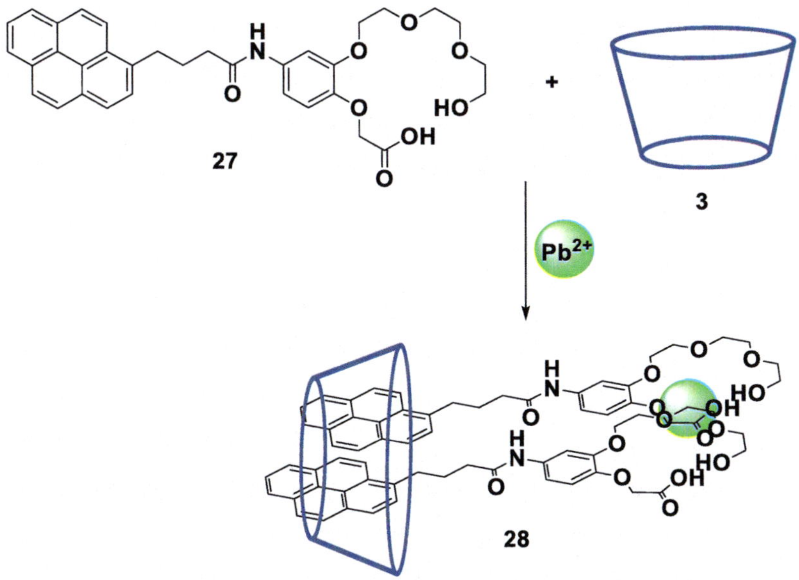

Scheme 12.7 Detection of Pb^{2+} by inclusion complex podand fluoroionophore/γ-CD (**28**)

display significant ICD data by creating an inclusion complex with γ-CD due to the chirality of the CD cavities [67].

Wenlu Wu et al. developed an optical sensor (**35**) for the detection of Fe^{3+} ions (Scheme 12.9) from a Rhodamine B derivative (RB) and Cd/Te quantum dots (Cd/Te-QDs) bridged using γ-CD. The RB derivative could settle into the γ-CD cavity while being affixed onto the Cd/Te-QDs surface. The FRET processed from Cd/Te-QDs to RB in the presence of Fe^{3+} suggested active the response of modified γ-CD towards Fe^{3+} and a notable fluorescence quenching with a linearity between 10 and 60 μM with rising Fe^{3+} concentrations (with a limit of detection of 2.51 μM). The reports suggested the modified quantum dots can be successfully used for sensing of Fe^{3+} in human serum [68].

An inclusion complex (1:1) of γ-CD with a hydrazine-diphenol derivative (4-HBP) was designed and utilized for Al^{3+} ions detection. The 4,4′-((1E, 1′E)-hydrazine-1,2-diylidenebis(methanylylidene)) diphenol (4-HBP) was prepared by condensation of 4-hydroxybenzaldehyde with hydrazine hydrate. Colorimetry, UV, and FL spectral analysis suggested the selective detection of Al^{3+} in competition with different metal ions and in samples like sea water, tap water, etc. [69].

Ridhima Chadha et al. worked on developing and utilizing Ag-γ-dextrin and Au-γ-CD nanoprobes (**40**) by a one-step green synthetic approach (Scheme 12.10) for Raman and colorimetry detection of chlorpyrifos (CPP, **42**). The intensity variation of the Raman bands, 417 (1328) and 411 (773) cm^{-1} as a function of the pesticide concentration ensured to quantify the selectivity and sensitivity of the nanoprobes. The amine and phosphorothioate functional groups present in CPP interacted with

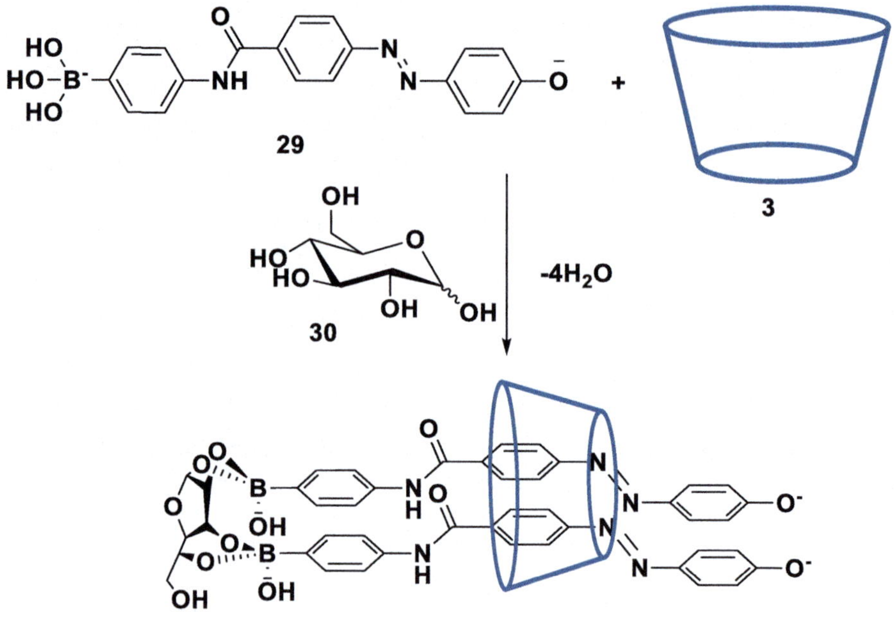

31

Scheme 12.8 Schematic illustration for the detection of D-glucose (**30**) by (B-Azo and BA-Azo)/ γ-CD (**31**)

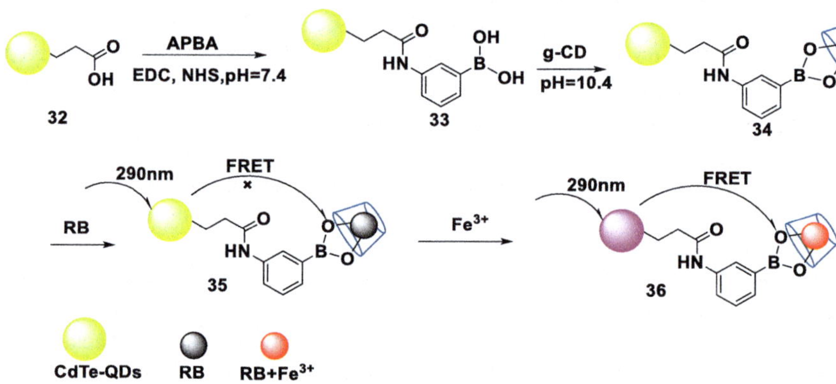

Scheme 12.9 Synthesis and detection of Fe^{3+} by γ-CD bridged Cd/Te-QDs-RB optical chemosensor (**35**)

the metal particles which resulted in the visible color change and increment in the Raman bands even in practical samples like tomatoes and apples. The LoD came out to be 1 and 10 ppb for synthesizing the nanoprobes apple and tomato, respectively, in a water medium [70].

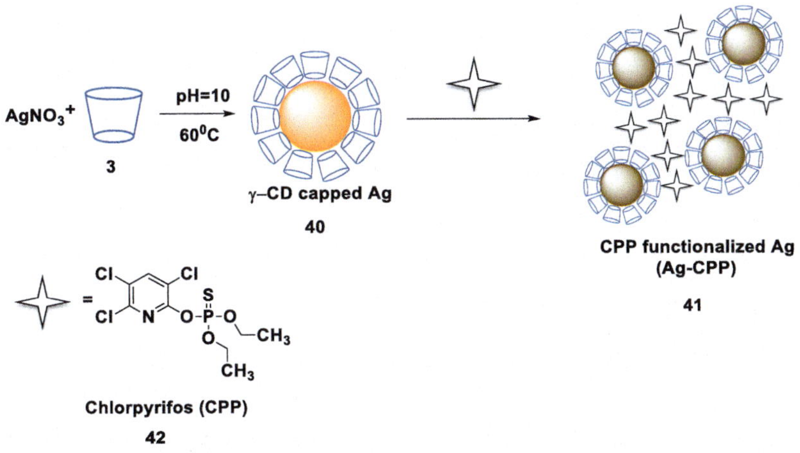

Scheme 12.10 Schematic illustration of detection of CPP by Ag -γ-CD nanoprobes (**40**)

12.5 Conclusion

Chemosensors (CSs) are molecular tools that may convert chemical information into analytical signals—often electrical, magnetic, or optical. The detection limit of CSs is a crucial consideration for determining their future application in practical samples, yet most CSs currently have characteristic fixed detection limits. In recent years, the most contemporary application of CDs has been in sensing and detecting chemical species through various spectroscopic techniques. Even though the native cyclodextrins are not UV-Vis active, chemical modifications have enabled them to be sensitive enough for detection, and the cavity present allows the capture of small molecules and ions. They are also being used to test practical samples because of their binding compatibility with different chemical moieties and nontoxic nature. Currently, many strategies have been reported for the synthetic modification of CDs to tune their characteristics for sensing purposes. However, many areas where cyclodextrins have potential applications are yet to be explored, such as incorporating CDs in the metal-organic framework (MoF) design, which is still growing and has excellent application in anion sensing. Furthermore, precise physicochemical characterization must be given greater importance while synthesizing CD-inclusion complexes, which is vital for chemosensing applications.

This chapter discusses the development, characteristics, and synthetic modification for ionic and neutral species sensing. Even though β-CDs have lower hydrophilicity than α- and γ-CDs, they are the most compatible for chemical modification and synthesizing various composites, nanoprobes, and polymers. The research on γ-CDs has not been as active as that of α- and β-CDs due to the complexity of their modification; however, there have been many reports of various novel works concerning γ-CDs.

References

1. Kitamura S (2000) Cyclic oligosaccharides and polysaccharides, Cyclic polymers. Springer Nature, pp 125–160
2. Matencio A, Caldera F, Cecone C, López-Nicolás JM, Trotta F (2020) Cyclic oligosaccharides as active drugs, an updated review. Pharmaceuticals 13:281
3. Anthony CY, Stapleton LM, Mann JL, Appel EA (2018) Self-assembled biomaterials using host-guest interactions In Self-assembling biomaterials. Woodhead Publishing, pp 205–231
4. Pandey M, Rani P, Adhikari L, Gupta M, Semalty A, Semalty M (2022) Preparation and characterization of cyclodextrin complexes of doxycycline hyclate for improved photostability in aqueous solution. J Incl Phenom Macrocycl Chem 102:271–278
5. Hou J, Zhou W, Bai D, Li S, Han M (2018) Interfacial effect of cyclodextrin inclusion complex on gas adsorption kinetics of dry water emulsion. Colloids Surf A 544:8–14
6. Ma X, Tang H, Malcampo PS, Galoppini E (2017) Organic host-guest complexes for functionalization of nanostructured surfaces. Curr Org Chem 21:2324–2337
7. Ashwin BCMA, Shanmugavelan P, Muthu Mareeswaran P (2020) Electrochemical aspects of cyclodextrin, calixarene and cucurbituril inclusion complexes. J Incl Phenom Macrocycl Chem 98:149–170
8. Astray G, Mejuto JC, Simal-Gandara J (2020) Latest developments in the application of cyclodextrin host-guest complexes in beverage technology processes. Food Hydrocoll 106:105882
9. Del Valle EM (2004) Cyclodextrins and their uses: a review. Process Biochem 39:1033–1046
10. Yang K, Zhang Z, Du J, Li W, Pei Z (2020) Host–guest interaction based supramolecular photodynamic therapy systems: a promising candidate in the battle against cancer. Chem Commun 56:5865–5876
11. Grinberg N, Carr PW (eds) (2020) Advances in chromatography, vol 57. CRC Press
12. Gonzalez Pereira A, Carpena M, García Oliveira P, Mejuto JC, Prieto MA, Simal Gandara J (2021) Main applications of cyclodextrins in the food industry as the compounds of choice to form host–guest complexes. Int J Mol Sci 22:1339
13. Liu Y, Sameen DE, Ahmed S, Wang Y, Lu R, Dai J, Li S, Qin W (2022) Recent advances in cyclodextrin-based films for food packaging. Food Chem 370:131026
14. Fernández MA, Silva OF, Vico RV, de Rossi RH (2019) Complex systems that incorporate cyclodextrins to get materials for some specific applications. Carbohyd Res 480:12–34
15. Hua Z, Qin Q, Bai X, Wang C, Huang X (2015) β-Cyclodextrin inclusion complex as the immobilization matrix for laccase in the fabrication of a biosensor for dopamine determination. Sens Actuators B Chem 220:1169–1177
16. Fang Y, Deng Y, Dehaen W (2020) Tailoring pillararene-based receptors for specific metal ion binding: from recognition to supramolecular assembly. Coord Chem Rev 415:213313
17. Lahiri H, Basu K (2024) Supramolecular sensing platforms: techniques for in vitro biosensing. Chem Eng 8:66
18. Ahmed F, Hussain MM, Khan WU, Xiong H (2024) Exploring recent advancements and future prospects on coordination self-assembly of the regulated lanthanide-doped luminescent supramolecular hydrogels. Coord Chem Rev 499:215486
19. Liu ZH, Devaraj S, Yang CR, Yen YP (2012) A new selective chromogenic and fluorogenic sensor for citrate ion. Sens Actuators B Chem 174:555–562
20. Zhang H, Zhang G, Xu J, Wen Y, Lu B, Zhang J, Ding W (2016) Novel highly selective fluorescent sensor based on electrosynthesized poly (9-fluorenecarboxylic acid) for efficient and practical detection of iron (III) and its agricultural application. Sens Actuators B Chem 230:123–129
21. Allode TDM, Vodounon AC, Dossou AG, Honzounnon NC, Migan ADFM, houenoukpo Hlouedje W (2024) Applications of cyclodextrin-gluconotransferase in the biosynthesis of cyclodextrins: characteristics, sources and production. Am J Microbiol Res 12:63–78
22. Sicard PJ, Saniez MH, Freres R (1987) Cyclodextrins and their industrial uses, pp 77–103

23. Rendleman JA Jr (1993) Enhanced production of γ-cyclodextrin from corn syrup solids by means of cyclododecanone as selective complexant. Carbohyd Res 247:223–237
24. Flaschel E, Landert JP, Spiesser D, Renken A (1984) The production of alpha-cyclodextrin by enzymatic degradation of starch. Ann N Y Acad Sci 434:70–77
25. Rajput KN, Patel KC, Trivedi UB (2016) A novel cyclodextrin glucanotransferase from an alkaliphile Microbacterium terrae KNR 9: purification and properties. 3 Biotech 6:1–11
26. Larsen D, Beeren SR (2021) Building up cyclodextrins from scratch–templated enzymatic synthesis of cyclodextrins directly from maltose. Chem Commun 57:2503–2506
27. Manta K, Papakyriakopoulou P, Chountoulesi M, Diamantis DA, Spaneas D, Vakali V, Naziris N, Chatziathanasiadou MV, Andreadelis I, Moschovou K, Athanasiadou I (2020) Preparation and biophysical characterization of quercetin inclusion complexes with β-cyclodextrin derivatives to be formulated as possible nose-to-brain quercetin delivery systems. Mol Pharm 17:4241–4255
28. Mayer BP, Kennedy DJ, Lau EY, Valdez CA (2023) Evaluation of polyanionic cyclodextrins as high affinity binding scaffolds for fentanyl. Sci Rep 13:2680
29. Pardhi DS, Rabadiya KJ, Panchal RR, Raval VH, Joshi RG, Rajput KN (2023) Cyclodextrin glucanotransferase: fundamentals and biotechnological implications. Appl Microbiol Biotechnol 107:5899–5907
30. Sandilya AA, Natarajan U, Priya MH (2020) Molecular view into the cyclodextrin cavity: structure and hydration. ACS Omega 5:25655–25667
31. Connors KA (1995) Population characteristics of cyclodextrin complex stabilities in aqueous solution. J Pharm Sci 84:843–848
32. Benkovics G, Malanga M, Fenyvesi É (2017) The 'Visualized'macrocycles: chemistry and application of fluorophore tagged cyclodextrins. Int J Pharm 531:689–700
33. Jansook P, Ogawa N, Loftsson T (2018) Cyclodextrins: structure, physicochemical properties and pharmaceutical applications. Int J Pharm 535:272–284
34. Shieh WJ, Hedges AR (1996) Properties and applications of cyclodextrins. J Macromol Sci Part A Pure Appl Chem 33:673–683
35. Szejtli J (1998) Introduction and general overview of cyclodextrin chemistry. Chem Rev 98:1743–1754
36. Betzel CH, Saenger W, Hingerty BE, Brown GM (1984) Circular and flip-flop hydrogen bonding in. beta.-cyclodextrin undecahydrate: a neutron diffraction study. J Am Chem Soc 106:7545–7557
37. Ondo D (2016) Calorimetric study on the interaction of didecyldimethylammonium and decyltrimethylammonium cations with native cyclodextrins in water. J Chem Thermodyn 97:235–243
38. Pinho E, Grootveld M, Soares G, Henriques M (2014) Cyclodextrins as encapsulation agents for plant bioactive compounds. Carbohyd Polym 101:121–135
39. Fukuhara G (2019) Allosteric signal-amplification sensing with polymer-based supramolecular hosts. J Incl Phenom Macrocycl Chem 93127–143
40. Sarkar I, Malini H, Mohan Das T, Mishra AK (2015) Synthesis and evaluation of a glucose attached pyrene, as a fluorescent molecular probe in sugar and non-sugar based micro-heterogeneous media. RSC Adv 5:64604–64613
41. Sarkar I, Hemamalini A, Mohan Das T, Mishra AK (2016) Introduction of an α, β-unsaturated carbonyl conjugated pyrene–lactose hybrid as a fluorescent molecular probe for micro-scale anisotropic media. RSC Adv 6:27933–27943
42. Lai CS, Moody GJ, Thomas JR, Mulligan DC, Stoddart JF, Zarzycki R (1988) Piezoelectric quartz crystal detection of benzene vapour using chemically modified cyclodextrins. J Chem Soc Perkin Trans 2:319–324
43. Hamasaki K, Ueno A, Toda F (1993) A fluorescent α-cyclodextrin as a sensor for detecting aliphatic alcohols by dual fluorescence arising from normal planar and twisted intramolecular charge transfer excited states. J Chem Soc Chem Commun 1:331–333
44. Ikeda H, Ueno A (2009) Fluorescent α-cyclodextrin as a chemosensor for halomethanes. Chem Commun 4281–4283

45. Mahato P, Ghosh A, Mishra SK, Shrivastav A, Mishra S, Das A (2011) Zn (II)– cyclam based chromogenic sensors for recognition of ATP in aqueous solution under physiological conditions and their application as viable staining agents for microorganism. Inorg Chem 50:4162–4170
46. Zhang Y, Li Q, Guo J, Li Y, Yang Q, Du J (2015) A novel magnetic fluorescent chemosensor for Cu 2+ based on self-assembled systems of azobenzene and α-cyclodextrin. RSC Adv 5:66674–66680
47. Yang J-H, Kim HT, Kim H (2014) A cyclodextrin-based approach for selective detection of catecholamine hormone mixtures. Micro Nano Syst Lett 2:1–10
48. Zhang H, Zhou C, Sun J, Li X, Tang J (2021) A nanofibrous membrane fluorescent sensor for fluoride ions prepared by electrospinning and host-guest interaction. Matéria (Rio de Janeiro) 26:e12958
49. Alam AU, Deen MJ (2020) Bisphenol A electrochemical sensor using graphene oxide and β-cyclodextrin-functionalized multi-walled carbon nanotubes. Anal Chem 92:5532–5539
50. Nazerdeylami S, Ghasemi JB, Ziarani GM, Amiri A, Badiei A (2021) Direct monitoring of diclofenac using a supramolecular fluorescent approach based on β-cyclodextrin nanosponge. J Mol Liq 336:116104
51. Avakyan VG, Nazarov VB, Alfimov MV (2018) Fluorescent chemosensor based on the Pyrene@ 2β-cyclodextrin complex: response to aromatic amino acids. High Energy Chem 52:206–211
52. Holkar A, Ghodke S, Bangde P, Dandekar P, Jain R (2022) Fluorescence-based detection of cholesterol using inclusion complex of hydroxypropyl-β-cyclodextrin and l-tryptophan as the fluorescence probe. J Pharm Innov 17:170–179
53. Chen X, Cheng X, Gooding JJ (2012) Detection of trace nitroaromatic isomers using indium tin oxide electrodes modified using β-cyclodextrin and silver nanoparticles. Anal Chem 84:8557–8563
54. Elgamouz A, Nassab C, Bihi A, Mohamad SA, Almusafri AH, Alharthi SS, Abdulla SA, Patole SP (2021) Encapsulation capacity of β-cyclodextrin stabilized silver nanoparticles towards creatinine enhances the colorimetric sensing of hydrogen peroxide in urine. Nanomaterials 11:1897
55. Wang R, Wang S, Qin C, Nie Q, Luo Y, Qin QP, Wang R, Liu B, Luo D (2022) An electrochemical sensor based on electropolymerization of β-cyclodextrin on glassy carbon electrode for the determination of fenitrothion. Sensors 23:435
56. Naskar H, Pradhan S, Ghatak B, Biswas S, Tudu B, Bandyopadhyay R (2021) Electrochemical detection of capsaicin in chili pepper using molecular imprinted poly β-cyclodextrin embedded graphite (MIP-β-CD@ G) electrode. IEEE Sens J 21:17657–17664
57. Chen J, He P, Bai H, He S, Zhang T, Zhang X, Dong F (2017) Poly (β-cyclodextrin)/carbon quantum dots modified glassy carbon electrode: preparation, characterization and simultaneous electrochemical determination of dopamine, uric acid and tryptophan. Sens Actuators B Chem 252:9–16
58. Izaoumen N, Bouchta D, Zejli H, El Kaoutit M, Stalcup AM, Temsamani KR (2005) Electrosynthesis and analytical performances of functionalized poly (pyrrole/β-cyclodextrin) films. Talanta 66:111–117
59. Palanisamy S, Thangavelu K, Chen S-M, Velusamy V, Chang MH, Chen TW, Al-Hemaid FM, Ali MA, Ramaraj SK (2017) Synthesis and characterization of polypyrrole decorated graphene/β-cyclodextrin composite for low level electrochemical detection of mercury (II) in water. Sens Actuators B Chem 243:888–894
60. Lee CY, Hsu DY, Prasannan A, Kalaivani R, Hong PD (2016) Facile synthesis of hexagonal-shaped polypyrrole self-assembled particles for the electrochemical detection of dopamine. Appl Surf Sci 363:451–458
61. Shishkanova TV, Habanová N, Řezanka M, Broncová G, Fitl P, Vrňata M, Matějka P (2020) Molecular recognition of phenylalanine enantiomers onto a solid surface modified with electropolymerized pyrrole-β-cyclodextrin conjugate. Electroanalysis 32:767–774
62. Liu ZC, Zhu WP, Chen YH, Li YX, Ding YJ, Yang WJ, Li K (2015) Water-soluble host–guest system from β-cyclodextrin as a fluorescent sensor for aluminium ions: synthesis and sensing studies. Dalton Trans 44:16528–16533

63. Antony EJ, Raj M, Paulpandi RQ, Paulraj MS, Enoch IV (2015) A highly selective fluorescent sensor for Pb^{2+} based on a modified β-cyclodextrin. J Fluoresc 25:1031–1036
64. Sumithra M, Sivaraj R, Selvan GT, Selvakumar PM, Enoch IV (2017) Ca2+ Ion sensing by a piperidin-4-one derivative and the effect of β-cyclodextrin complexation on the sensing. J Lumin 185:205–211
65. Prabu S, Mohamad S (2020) Curcumin/beta-cyclodextrin inclusion complex as a new "turn-off" fluorescent sensor system for sensitive recognition of mercury ion. J Mol Struct 1204:127528
66. Hayashita T, Qing D, Minagawa M, Lee JC, Ku CH, Teramae N (2003) Highly selective recognition of lead ion in water by a podand fluoroionophore/γ-cyclodextrin complex sensor. Chem Commun 2160–2161
67. Shimpuku C, Ozawa R, Sasaki A, Sato F, Hashimoto T, Yamauchi A, Suzuki I, Hayashita T (2009) Selective glucose recognition by boronic acid azoprobe/γ-cyclodextrin complexes in water. Chem Commun 1709–1711
68. Wu W, Hu Z, Shi C, Xu R, Zhao Y, Ding Y (2023) Construction of CdTe@ γ-CD@ RBD nanoprobe for Fe^{3+}-sensing based on FRET mechanism in human serum. Spectrochim Acta Part A Mol Biomol Spectrosc 296:122645
69. Ganesan V, Mani MK, Narayanan V, Shanmugasundram E, Vellaisamy K, Baskaralingam V, Jeyaraj J, Veerakanellore GB, Rajamohan R, Thambusamy S (2023) Synthesis, characterization of 4,4′-((1E,1′E)-hydrazine-1,2-diylidenebis(methanylylidene)) diphenol and the inclusion complex with γ-cyclodextrin as a fluorescent probe for detection of Al^{3+}. J Photochem Photobiol A 442:114814
70. Chadha R, Das A, Lobo J, Meenu VO, Paul A, Ballal A, Maiti N (2022) γ-Cyclodextrin capped silver and gold nanoparticles as colorimetric and Raman sensor for detecting traces of pesticide "Chlorpyrifos" in fruits and vegetables. Colloids Surf A 641:128558

Sheersha Pradhan has obtained her M.Sc. in Chemistry from Veer Surendra Sai University of Technology, Burla, Odisha, India. She is currently working as a research scholar Under Prof. T. Mohan Das in the Department of Chemistry, School of Basic and Applied Sciences, Central University of Tamil Nadu, Thiruvarur, Tamil Nadu. India. Her current research interest involves design and synthesis of novel *N*- and *C*-glucosides for material and biological applications.

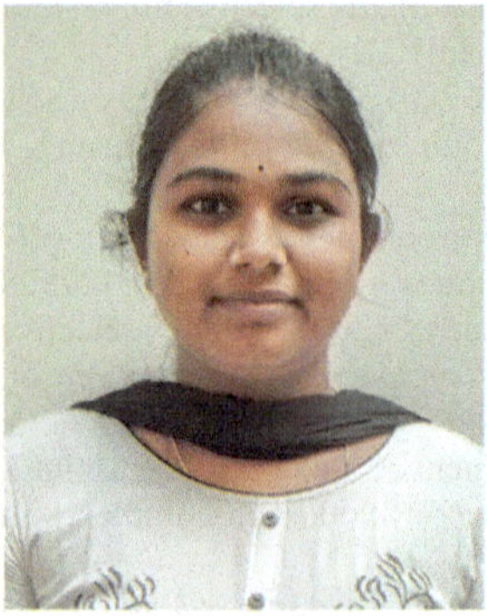

Lavanya Lahari Pannuri has obtained her M.Sc. in Chemistry from the Department of Chemistry, School of Basic and Applied Sciences, Central University of Tamil Nadu, Thiruvarur, Tamil Nadu. India under the supervision of Prof. T. Mohan Das. Her research work involved design and synthesis of novel *N*-glucosides for various metal sensing. Recently she has joined Prof. Santosh Penta at National Institute of Technology, Warangal, Telengana, India for her Ph.D. degree.

T. Mohan Das is currently working as a professor in the Department of Chemistry, School of Basic and Applied Sciences, Central University of Tamil Nadu, Thiruvarur, Tamil Nadu. India. With an experience of over twenty-five years in research and academia, he successfully completed 6 research projects and filed one Indian patent. He has published more than 80 research articles in international peer reviewed journals with h-index 20 and i10-index 42. His research group is working in the area of carbohydrate chemistry and organic synthesis.

Chapter 13
Cyclodextrin Nanoparticles: A New Frontier in Molecular Detection

Pankaj Bhatt, Anasuya Patil, Ranjana, Shashank Soni, and Yudhishther Singh Bagal

Abstract Cyclodextrin (CD)-based nanoparticles (CD-NPs) have gained attention for their potential in sensing applications, particularly in the detection of small molecules. CDs are cyclic oligosaccharides composed of glucose units linked by α-1,4 glycosidic bonds, which allow them to form inclusion complexes with various organic molecules, especially those that are hydrophobic or lipophilic. This property positions CDs as candidates for diverse applications, including drug delivery and biosensing. In electrochemical sensing, CD-based materials have demonstrated utility by providing selectivity and sensitivity in recognizing biomolecules. The mechanism of detection involves the encapsulation of analytes within the cyclodextrin cavities through host–guest interactions. These inclusion complexes enhance the sensitivity and selectivity of sensing processes across various platforms, such as electrochemical, potentiometric, and fluorescent sensors. The optical characteristics of cyclodextrin-based structures ranging from monomers to nanoassemblies have shown promise in identifying a range of physiologically relevant substances. The integration of CDs in sensor systems has been explored for detecting environmental pollutants and biomolecules like proteins and nucleic acids, contributing to the development of biosensors for disease detection and monitoring therapeutic responses. Overall, cyclodextrin-based nanoparticles offer a versatile platform for innovative detection systems across fields such as drug discovery, environmental

P. Bhatt (✉)
Lloyd Institute of Management and Technology, Greater Noida, India
e-mail: pankajbhattb.pharma@gmail.com

A. Patil
KLE College of Pharmacy, Bengaluru, India

Ranjana
School of Pharmacy, Graphic Era Hill University, Dehradun, Uttrakhand, India

S. Soni
Department of Pharmaceutics, Amity Institute of Pharmacy, Amity University Uttar Pradesh, Lucknow, Noida, India

Y. S. Bagal
School of Agriculture, Lovely Professional University, Phagwara, Punjab, India

S. Krishnamoorthy (ed.), *Cyclodextrins for Chemosensing*,
https://doi.org/10.1007/978-3-031-89299-8_13

monitoring, and chiral sensing. This chapter aims to explore the latest advancements in cyclodextrin-based nanoparticles, particularly in their application within pharmaceutical detection technologies.

Keywords Cyclodextrins · Biosensors · Drug delivery · Chemosensors · Nanoparticles · Host–guest interaction

13.1 Introduction

Sensing is an integral part of the scientific and industrial fields as it allows for the detection and measurement of various physical and chemical parameters. It plays a vital role in various applications such as environmental monitoring, health care, food safety, and quality control [1]. It involves the use of devices or systems designed to respond to specific stimuli, converting them into electrical signals or other measurable quantities. Sensing systems can be either passive or active, depending on the method used to acquire and process the sensed information [2]. Cyclodextrins (CDs) have gained significant importance in the realm of chemosensing, largely owing to their distinctive structural features and capacity to form host–guest complexes. Cyclodextrin sensors are devices that utilize the unique properties of cyclodextrins (CDs) to detect and measure specific substances, often referred to as analytes [1]. These naturally occurring oligosaccharides, composed of cyclic glucose units, possess a hydrophilic outer surface and a hydrophobic internal cavity, which allows them to selectively encapsulate a range of small molecules. Their ability to recognize specific molecules is key to their role in chemosensing, facilitating the detection and measurement of analytes through non-covalent interactions [3]. Cyclodextrin is able to form a complex of inclusions with a wide variety of foreign molecules, which subsequently affects their properties and behaviors [62]. Although the three main commercial types of cyclodextrins—α-, β-, and γ-cyclodextrin—are well-known, a variety of modified forms have been created to improve their functional capabilities. These modifications include hydroxypropyl-β-cyclodextrin, methylated-β-cyclodextrin, and sulfobutylated-β-cyclodextrin, as well as more specialized variants like thiolated cyclodextrins and polyrotaxanes. Each of these derivatives possesses distinct properties, broadening their use in diverse industries such as materials science, pharmaceuticals, food safety, and cosmetics. The integration of cyclodextrins into nanoparticle systems amplifies their effectiveness in chemosensing [4]. Cyclodextrin-based nanoparticles exhibit high surface areas and abundant active sites, which increase the effective binding capacity for analytes [5]. This characteristic not only boosts sensitivity but also allows for rapid adsorption kinetics, making these sensors suitable for various applications, including environmental monitoring and medical diagnostics [6]. This chapter focuses on highlighting the latest advancements in cyclodextrin-based nanoparticles, particularly in their application within pharmaceutical detection technologies (Fig. 13.1).

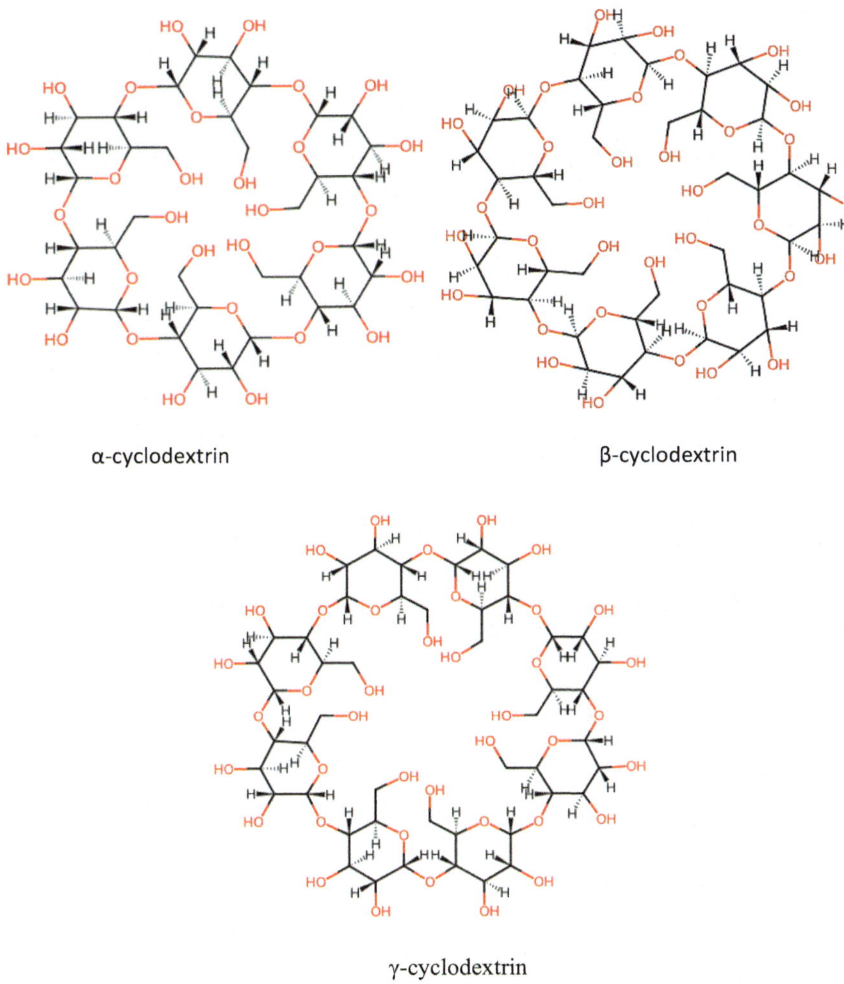

Fig. 13.1 Chemical structures of α-, β-, and γ-cyclodextrin

13.2 Methods of Synthesizing Cyclodextrin-Based Nanoparticles

Synthesis of cyclodextrin-based nanoparticles involves the incorporation of cyclodextrins into nanoparticle matrices through various approaches [7]. These methods aim to control the size, morphology, and surface chemistry of the nanoparticles, thereby influencing their properties and performance in sensing applications [8]. Several techniques have been developed for synthesizing cyclodextrin-based nanoparticles, including.

13.2.1 Encapsulation

Encapsulation is a process by which one substance is enclosed within another substance, forming a distinct entity known as an encapsulated system or encapsulate. In the context of cyclodextrin-based nanoparticles, encapsulation involves the confinement of guest molecules within the hydrophobic cavity of cyclodextrins, which are then incorporated into nanoparticle matrices. This process offers several advantages, including controlled release, enhanced stability, and improved solubility of the encapsulated molecules [5].

13.3 Selection of Cyclodextrins

The first step is to select appropriate cyclodextrins, considering the size, shape, and properties of the foreign molecules to be encapsulated [1]. Cyclodextrins are cyclic oligosaccharides composed of 1,4-glycoside bonds that connect glucose units. These entities have a unique configuration characterized by a hydrophilic exterior surface and a hydrophobic interior, which empowers them to establish complex inclusions with a wide variety of guest molecules [8].

13.4 Formation of Inclusion Complexes

Once the cyclodextrins are chosen, they are dissolved in a suitable solvent, such as water, ethanol, or dimethyl sulfoxide (DMSO), to form a cyclodextrin solution [9]. The guest molecules, which may be hydrophobic or have limited solubility in water, are then added to the cyclodextrin solution. The hydrophobic cavities of cyclodextrins selectively interact with the guest molecules, forming inclusion complexes through non-covalent interactions such as van der Waals forces, hydrogen bonding, and hydrophobic interactions [10]. This process effectively encapsulates the guest molecules within the cyclodextrin cavity, shielding them from the surrounding environment [11].

13.5 Incorporation into Nanoparticle Matrices

The next step involves incorporating the cyclodextrin-guest complexes into nanoparticle matrices to form cyclodextrin-based nanoparticles [6]. This can be achieved through various methods, such as coacervation, emulsification, or nanoprecipitation [9]. In coacervation, the cyclodextrin-guest complexes are dispersed in a polymer solution, followed by the addition of a coacervating agent to induce phase separation and nanoparticle formation. Similarly, in emulsification, the complexes are dispersed in an emulsion system containing surfactants and stabilizers, leading to the formation of nanoparticles upon solvent evaporation or cooling [12].

13.5.1 Surface Modification

Nanoprecipitation is a widely used method for synthesizing nanoparticles with controlled size, shape, and surface properties, particularly suitable for organic nanoparticles, including those containing cyclodextrins, due to its simplicity, versatility, and ability to produce nanoparticles with narrow size distributions [1]. The method involves the rapid mixing of two or more miscible liquids, typically an organic solvent and an aqueous phase, containing the nanoparticle precursors. This rapid mixing induces supersaturation of the solute in the solvent, leading to the formation of nanoparticles through nucleation and subsequent precipitation [13]. The key steps include selecting suitable solvents critical for determining solubility and influencing the resulting nanoparticles' size, morphology, and stability, with organic solvents dissolving hydrophobic components and aqueous phases dissolving hydrophilic ones or acting as a non-solvent for nanoparticle precipitation. Nanoparticle precursors, such as polymers, surfactants, or cyclodextrins, are then dissolved or dispersed in the organic solvent, and their concentration is adjusted to control the resulting nanoparticles' size and concentration [14]. The organic phase containing the nanoparticle precursors is then rapidly injected or dispersed into the aqueous phase under vigorous stirring or agitation, inducing instantaneous nucleation and precipitation of nanoparticles due to the sudden change in solvent conditions [15]. These nanoparticles undergo phase separation and aggregation to form the final particles, their size and morphology are determined by various factors, including precursor concentration, mixing speed, solvent ratio, and temperature. After nanoparticle formation, solvent removal is typically achieved through evaporation or filtration, with additional purification steps like centrifugation or dialysis employed to remove impurities or unreacted precursors, ensuring the nanoparticles' purity [16]. A recent investigation into the surface immobilization of β-cyclodextrin on hybrid silica demonstrated the successful attachment of β-cyclodextrin onto hybrid silica through ethylenediamine as a coupling agent. The synthesized adsorbent exhibited remarkable efficiency in adsorbing p-nitrophenol, achieving equilibrium within only 15 s and a notable adsorption capacity of 69.6 mg/g. This approach underscores the potential of cyclodextrin-based systems for applications in water purification processes [17].

13.5.2 Template-Assisted Synthesis

Template-assisted synthesis is a versatile method used for the fabrication of nanoparticles with controlled size, shape, and structure. This approach involves utilizing templates or scaffolds to guide the formation of nanoparticles, allowing for precise control over their properties [18]. In the context of cyclodextrin-based nanoparticles, template-assisted synthesis offers unique opportunities for the creation of tailored nanostructures with enhanced functionalities for various applications, including sensing [19]. The key steps of template-assisted synthesis include selecting a suitable template or scaffold, depositing or immobilizing nanoparticle precursors onto the template surface, and inducing nanoparticle formation through chemical, physical, or electrochemical processes. The template serves as a spatial guide or mold, dictating the size, shape, and arrangement of nanoparticles during synthesis. After nanoparticle formation, the template can be removed or retained, depending on the desired final product [20]. Template-assisted synthesis enables the fabrication of complex nanostructures, such as nanowires, nanotubes, or hollow nanoparticles, with precise control over dimensions and properties. This method offers advantages such as scalability, reproducibility, and tunability, making it suitable for both research and industrial applications in nanotechnology and materials science [21]. In the field of cyclodextrin-based nanoparticles, template-assisted synthesis holds promise for the development of advanced materials with tailored properties for sensing, drug delivery, catalysis, and other applications [22].

13.5.3 Self-Assembly Method

The self-assembly method is a powerful approach used to fabricate nanoparticles through the spontaneous organization of molecules or nanoscale building blocks into well-defined structures without external intervention [23]. In this process, the inherent interactions between the components drive the assembly process, resulting in the formation of ordered nanostructures. In the context of cyclodextrin-based nanoparticles, self-assembly can be achieved by exploiting the host–guest interactions between cyclodextrin molecules and guest molecules or by utilizing the self-assembly properties of cyclodextrin derivatives as shown in Fig. 13.2 [24]. Cyclodextrins, with their hydrophobic cavities and hydrophilic exteriors, can encapsulate guest molecules within their cavity through non-covalent interactions, such as hydrogen bonding and hydrophobic interactions [1]. By selecting appropriate guest molecules, researchers can control the size, shape, and properties of the resulting nanoparticles. Additionally, cyclodextrin derivatives with specific functional groups can self-assemble into nanostructures through complementary interactions, such as π-π stacking or electrostatic interactions [17]. Self-assembly offers several advantages for nanoparticle synthesis, including simplicity, scalability, and the ability to produce complex structures with precise control over size and morphology [19].

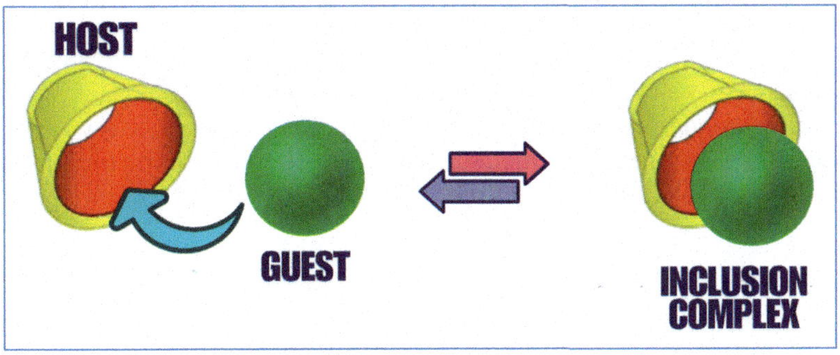

Fig. 13.2 Inclusion complex formation in host–guest chemistry

In the field of cyclodextrin-based nanoparticles, self-assembly holds great promise for the development of functional materials for various applications, including drug delivery, sensing, and catalysis [1].

13.5.4 Crosslinking

Crosslinking techniques involve the crosslinking of cyclodextrins or cyclodextrin derivatives to form nanoparticle-like structures [25]. This can be achieved through chemical crosslinking reactions, such as polymerization or crosslinking of cyclodextrin-containing polymers [26]. Crosslinked cyclodextrin networks can exhibit nanoparticle-like properties, including high surface area, tunable porosity, and controlled release behavior, making them suitable for sensing applications.

13.5.5 Sol–Gel Method

Among the various methods for synthesizing cyclodextrin-based nanoparticles, the sol–gel method stands out as a versatile and widely used approach [27]. In this method, cyclodextrins are incorporated into a sol–gel matrix through a controlled hydrolysis and condensation reaction of precursor molecules, typically alkoxysilanes such as tetraethyl orthosilicate (TEOS). The process begins with the hydrolysis of the precursor molecules in a solvent, forming reactive silanol (Si–OH) groups [28]. Cyclodextrins, either in their pure form or modified with functional groups, are then added to the solution, where they become entrapped within the growing silica network through hydrogen bonding and van der Waals interactions [29]. As the sol–gel reaction progresses, the silica matrix gradually solidifies, entrapping the cyclodextrins and forming nanoparticles with controlled size, morphology, and composition. The resulting cyclodextrin-based nanoparticles exhibit enhanced stability, biocompatibility, and tunable properties, making them suitable for various applications in drug delivery, sensing, and catalysis [30]. The sol–gel method offers advantages such as mild reaction conditions, scalability, and the ability to incorporate a wide range of guest molecules into the nanoparticles, making it a versatile platform for the synthesis of cyclodextrin-based nanoparticles with tailored properties for specific applications [31].

13.6 Sensing Mechanisms and Principles

In addition to health diagnostics, industrial quality control, and environmental monitoring, esoteric technology is essential in many other areas [32]. Cyclodextrin-based nanoparticles are one of the most promising emerging sensing platforms due to their unique properties and functions. Cyclodextrin, which are by nature cyclic oligosaccharides, is an exceptional molecular host for the encapsulation of foreign molecules because of their water-sensitive outer and water-sensitive cavity [33]. The principles underlying sensing with cyclodextrins involve exploiting the selective binding affinity between cyclodextrins and target analytes. When a cyclodextrin-based sensing platform is exposed to a sample containing analytes, the analytes can enter the hydrophobic cavity of cyclodextrins through inclusion complexation [34]. This binding interaction leads to changes in the properties of the cyclodextrin-based nanoparticles, such as their size, stability, or optical properties, which can be measured to detect and quantify the presence of analytes in the sample. The selective binding capability of cyclodextrins allows for highly specific and sensitive detection of target analytes, even in complex sample matrices [35].

13.7 Interaction Mechanisms Between Analytes and Cyclodextrin-Based Nanoparticles

The interaction mechanisms between analytes and cyclodextrin-based nanoparticles depend on various factors, including the chemical nature of the analyte, the structure of the cyclodextrin, and the composition of the nanoparticle. Several interaction mechanisms have been identified, including.

13.7.1 Host–Guest Complexation

The formation of host–guest complexes between cyclodextrins and analytical molecules constitutes the main mechanism of interaction [1]. Inclusion complexes can be formed by analytes that are compatible with the hydrophobic cavity of cyclodextrines by non-covalent interactions, including Van der Waals forces, hydrophobic interactions, and hydrogen bonds [12]. Changes in the characteristics of nanoparticles based on cyclodextrin caused by this selective binding can be detected and quantified, allowing us to deduce the concentration and presence of the analyzer [34].

13.7.2 Surface Adsorption

In addition to inclusion complexation, analyte molecules may also adsorb onto the surface of cyclodextrin-based nanoparticles through physical adsorption or electrostatic interactions [17]. Surface adsorption can further enhance the sensitivity and selectivity of the sensing platform by increasing the effective surface area available for analyte binding and facilitating interactions between analyte molecules and detection probes [19]. Analyte molecules can attach to the surface of cyclodextrin-based nanoparticles through weak van der Waals forces or hydrophobic interactions, enabling a reversible binding process. Moreover, cyclodextrins (CDs) can be chemically modified to incorporate charged groups, promoting electrostatic interactions with oppositely charged analytes, which enhances both binding strength and specificity [11]. Due to their distinctive structure, cyclodextrins can also form inclusion complexes with hydrophobic analytes, thereby stabilizing them and improving their detectability. These interactions expand the available surface area for analyte binding, substantially increasing the sensitivity of chemosensors. Additionally, the adsorption of analytes onto the nanoparticle surface intensifies interactions with detection probes, resulting in amplified signals that elevate sensor performance [35]. Research has shown that these nanoparticles exhibit rapid adsorption kinetics and substantial adsorption capacities for a variety of pollutants, underscoring their effectiveness in environmental monitoring and analytical applications [36].

13.7.3 Induced Aggregation

Analyte binding to cyclodextrin-based nanoparticles can induce aggregation or dispersion of the nanoparticles, depending on the nature of the analyte-cyclodextrin interaction [20]. For example, analytes that disrupt the stability of cyclodextrin-based nanoparticles may induce nanoparticle aggregation, leading to changes in the optical, electrical, or mechanical properties of the sensing platform. These changes can be monitored to detect and quantify analyte concentrations in the sample [37].

13.7.3.1 Signal Transduction Methods

Signal transduction methods are employed to convert the binding events between analytes and cyclodextrin-based nanoparticles into measurable signals that can be detected and quantified. Various signal transduction methods have been developed, depending on the properties of the nanoparticles and the analytical techniques employed [1].

13.7.4 Optical Methods

Optical methods are a group of analytical techniques utilized for detecting and quantifying analytes based on changes in the optical properties of cyclodextrin-based nanoparticles upon analyte binding. These methods leverage the interaction between light and matter to generate measurable signals that correlate with analyte concentration [38]. Optical methods offer several advantages, including high sensitivity, versatility, and the ability to perform measurements in real time. In the context of cyclodextrin-based nanoparticles, optical methods are widely used for sensing applications due to their compatibility with a variety of detection principles and ease of integration into portable and miniaturized devices. Common optical techniques used for sensing with cyclodextrin-based nanoparticles include UV–Visible spectroscopy, fluorescence spectroscopy, and surface plasmon resonance (SPR). In chemosensing, Surface Plasmon Resonance (SPR) offers a significant advantage with its ability to detect molecular interactions without the need for labeling, allowing real-time observation of binding events without employing fluorescent or radioactive markers. The binding of an analyte to the sensor surface induces a change in the local refractive index, which results in a detectable shift in the resonance angle. This measurable shift provides valuable data on binding kinetics, interaction affinity, and specificity, making SPR a highly useful technique for investigating biomolecular interactions, including protein–protein or protein–ligand dynamics [39].

SPR's exceptional sensitivity allows for the detection of analytes in concentrations as low as picomolar to nanomolar levels, making it particularly effective for analyzing trace amounts of substances in complex biological matrices. Moreover, SPR can be coupled with cyclodextrin-based sensors to improve both selectivity and sensitivity. Cyclodextrins can form inclusion complexes with target analytes, enhancing their detection by effectively increasing the concentration of analytes at the sensor interface, thus optimizing the performance of SPR for selective detection [22]. In UV–Visible spectroscopy, changes in the absorption spectrum of the nanoparticles are monitored upon analyte binding, providing information about analyte concentration [40]. Fluorescence spectroscopy involves measuring changes in the fluorescence intensity of the nanoparticles or fluorophore-labeled analytes, which can be used to detect and quantify analytes with high sensitivity. SPR is a label-free technique that detects changes in the refractive index at the nanoparticle surface upon analyte binding, enabling real-time monitoring of binding events [41]. Optical methods offer advantages such as high sensitivity, rapid response times, and the ability to perform multiplexed measurements, making them suitable for a wide range of applications, including environmental monitoring, biomedical diagnostics, and drug discovery [42].

13.7.5 Electrochemical Methods

Electrochemical methods are a class of analytical techniques used to detect and quantify analytes based on changes in electrical properties or signals generated at the surface of electrodes [4]. In the context of cyclodextrin-based nanoparticles, electrochemical methods offer a powerful approach for sensing applications due to their high sensitivity, rapid response times, and compatibility with a wide range of analytes [16]. These methods rely on the electrochemical interactions between analyte molecules and the surface of electrodes modified with cyclodextrin-based nanoparticles. Upon analyte binding, changes in electrical conductivity, electrochemical behavior, or charge transfer kinetics occur at the electrode surface, which can be detected and quantified to infer analyte concentrations in the sample. Common electrochemical techniques used for sensing with cyclodextrin-based nanoparticles include cyclic voltammetry, amperometry, and impedance spectroscopy [27]. In cyclic voltammetry, the current response at the electrode surface is measured as a function of applied potential, providing information about the redox properties of analyte-cyclodextrin interactions. Amperometry involves measuring the current generated by electrochemical reactions at a constant applied potential, with changes in current indicative of analyte binding events [43]. Impedance spectroscopy measures the impedance of the electrode–electrolyte interface, which changes upon analyte binding, providing information about the kinetics and mechanism of analyte-cyclodextrin interactions. Electrochemical methods offer advantages such as high sensitivity, rapid response times, and compatibility with miniaturization and integration into portable devices for on-site analysis [29]. These methods find widespread applications in environmental monitoring, biomedical diagnostics, and food safety, among others. In the context of cyclodextrin-based nanoparticles, electrochemical methods enable the development of sensitive, selective, and cost-effective sensing platforms for diverse applications [1].

13.7.6 Mass-Based Methods

Mass-based methods are analytical techniques that rely on measuring changes in the mass of nanoparticles to detect and quantify analytes in a sample [18]. These methods, which include quartz crystal microbalance (QCM) and surface acoustic wave (SAW) sensors, offer several advantages for sensing applications, including high sensitivity, real-time monitoring, and label-free detection. In mass-based sensing, analyte binding to the surface of cyclodextrin-based nanoparticles induces changes in the mass or mechanical properties of the nanoparticles, which can be detected and quantified using sensitive mass detection techniques. For example, in QCM sensors, analyte binding causes a shift in the resonance frequency of a quartz crystal, proportional to the mass of the bound analyte. The quantification of frequency changes caused by the analyte binding is used to determine the analyte concentration in the sample [44]. Changes in the propagation properties of acoustic waves, including attenuation and speed, induced by the analysis of connections, are also detected and quantified in surface acoustic wave sensors (SAWs) to estimate the concentrations of the analysis [30]. Mass-based techniques offer numerous advantages, such as excellent sensitivity, rapid response times, and compatibility with a wide range of analyzers and sample matrices. These methods are widely implemented in many areas, including food safety, biomedical diagnostics, and environmental monitoring [17]. In terms of nanoparticles composed of cyclodextrin, mass-based detection is a very effective tool for accurate and precise identification of analytes. This in turn accelerates the advancement of the sensor platforms used in a wide range of applications [15].

13.8 Types of Cyclodextrin-Based Nanoparticles for Sensing

Cyclodextrin-based nanoparticles offer a diverse platform for sensing applications due to their unique properties and versatility [1]. Three main types of cyclodextrin-based nanoparticles for sensing include cyclodextrin-modified metal nanoparticles, cyclodextrin-encapsulated quantum dots, and cyclodextrin-coated polymer nanoparticles [33]. Nanoparticles have tailor-made properties suitable for accurate and accurate identification of analytes. As a result, they have immense value in a wide range of disciplines, including drug delivery, biomedical diagnostics, and environmental monitoring [45].

13.8.1 Cyclodextrin-Modified Metal Nanoparticles

Cyclodextrin-modified metal nanoparticles represent a class of nanomaterials extensively utilized in sensing applications due to their unique properties and versatile functionality. In this type of nanoparticles, cyclodextrins, which are cyclic oligosaccharides with a hydrophilic exterior and a hydrophobic cavity, are attached to the surface of metal nanoparticles [27]. This attachment can occur through either covalent bonding or non-covalent interactions, depending on the specific application and desired properties of the nanoparticles. Cyclodextrins act as molecular hosts, capable of selectively binding target analytes within their hydrophobic cavities through inclusion complexation [46]. This selective binding results in changes to the optical properties of the metal nanoparticles, such as their absorbance or scattering characteristics, which can be measured to detect and quantify the presence of analytes in a sample [21]. Gold and silver nanoparticles are commonly used as the metallic core due to their high surface area, excellent biocompatibility, and unique optical properties, including surface plasmon resonance (SPR). Cyclodextrin-modified metal nanoparticles offer several advantages for sensing applications, including high sensitivity, selectivity, and versatility. They can be employed for the detection of various analytes, ranging from small molecules to biomolecules and environmental pollutants [9, 11]. Additionally, these nanoparticles can be integrated into different sensing platforms, including colorimetric, fluorescent, and surface-enhanced Raman spectroscopy (SERS) assays, making them suitable for a wide range of analytical tasks [18]. Cyclodextrin-modified metal nanoparticles hold significant promise for applications in biomedical diagnostics, environmental monitoring, and food safety, among others, due to their ability to provide rapid, sensitive, and reliable detection of analytes in complex sample matrices [47].

13.8.2 Cyclodextrin-Encapsulated Quantum Dots

Cyclodextrin-encapsulated quantum dots are a class of nanomaterials widely utilized in sensing applications for their unique optical properties and versatile functionality [20]. Quantum dots (QDs) are semiconductor nanoparticles with size-tunable optical properties, including high photoluminescence quantum yield and narrow emission spectra. In this type of nanoparticles, cyclodextrins, which are cyclic oligosaccharides with a hydrophilic exterior and a hydrophobic cavity, are used to encapsulate or stabilize quantum dots [2, 7]. The hydrophobic cavities of cyclodextrins provide an ideal environment for encapsulating hydrophobic quantum dots, while the hydrophilic exterior enhances their dispersibility and stability in aqueous solutions [48]. Cyclodextrin-encapsulated quantum dots offer several advantages for sensing applications, including high sensitivity, tunable emission wavelengths, and sensitivity to environmental changes. Analyte molecules can interact with cyclodextrins surrounding the quantum dots through inclusion complexation or surface adsorption, inducing changes in their optical properties [32]. These changes, such as shifts in emission wavelength or intensity, can be monitored to detect and quantify analytes in a sample. Cyclodextrin-encapsulated quantum dots find applications in various fields, including bioimaging, biosensing, environmental monitoring, and drug delivery, due to their ability to provide sensitive, selective, and real-time detection of analytes in complex sample matrices. They offer a versatile platform for developing advanced sensing platforms capable of meeting the diverse analytical needs of different applications [1].

13.8.2.1 Cyclodextrin-Coated Polymer Nanoparticles

Cyclodextrin-coated polymer nanoparticles represent a category of nanomaterials widely employed in sensing applications due to their distinctive characteristics and adaptable functionalities [20]. These nanoparticles consist of cyclodextrins, cyclic oligosaccharides featuring a hydrophilic outer surface and a hydrophobic cavity, coated onto the surface of polymer nanoparticles. Such coating methodologies encompass physical adsorption, covalent bonding, or integration during the synthesis of nanoparticles. Cyclodextrins serve as molecular hosts, capable of selectively binding target analytes within their hydrophobic cavities through inclusion complexation [19]. The polymer nanoparticles provide a stable and biocompatible platform for the immobilization of cyclodextrins and can be functionalized with additional groups for enhanced analyte recognition and binding. Cyclodextrin-coated polymer nanoparticles offer several advantages for sensing applications, including high loading capacity, stability, and biocompatibility. Analyte molecules can interact with cyclodextrins within the polymer nanoparticles, leading to changes in their properties that can be detected using various analytical techniques, including optical, electrochemical, or mass-based methods [49]. These nanoparticles find applications in diverse fields, including drug delivery, chemical sensing, environmental monitoring, and biomedical diagnostics, due to their ability to provide sensitive, selective, and reliable detection of analytes in complex sample matrices. Cyclodextrin-coated polymer nanoparticles offer a versatile platform for developing advanced sensing platforms capable of meeting the diverse analytical needs of different applications [1].

13.9 Applications of Cyclodextrin-Based Nanoparticles in Sensing

Cyclodextrin-based nanoparticles have emerged as versatile tools for sensing applications across diverse fields. Their unique properties, including high surface area, tunable functionality, and selective binding capabilities, enable sensitive and specific detection of analytes [50]. From environmental monitoring to biomedical diagnostics and quality control in food and industrial processes, these nanoparticles offer innovative solutions to address complex analytical challenges [11]. Harnessing the potential of cyclodextrin-based nanoparticles opens avenues for advancements in sensing technology with wide-ranging implications.

13.9.1 Environmental Sensing

Environmental sensing using cyclodextrin-based nanoparticles involves the detection and monitoring of pollutants and contaminants in the environment, including air, water, and soil [19]. Several advantages of CDs and their polymers (e.g., cyclodextrin nanosponges, NSs) are shown in their use for environmental remediation, including the high percentage of pollutants removed from wastewater [33]. When functionalized with different chemical groups or combined with materials like carbon nanomaterials or metal–organic frameworks, CDNs exhibit improved selectivity and sensitivity towards specific pollutants [22]. The mechanism of detection primarily relies on the formation of host–guest complexes, where the pollutant molecules are trapped within the cyclodextrin cavities or adsorbed onto their surfaces. This interaction alters the physical and chemical properties of the CDNPs, which can be monitored through various sensing techniques such as fluorescence, electrochemical methods, and surface plasmon resonance [19]. For instance, when a pollutant binds to the CDNPs, it can induce changes in fluorescence intensity or wavelength, providing a quantifiable signal proportional to the concentration of the contaminant present. During the polymerization reaction, several pores and interstices form in the sponge network, enabling the dispersion of pollution both through host–guest interactions and through pores. Cyclodextrin-based nanoparticles offer several advantages for environmental sensing, including high sensitivity, selectivity, and versatility [51]. These nanoparticles can selectively bind to a wide range of pollutants and contaminants, such as heavy metals, organic pollutants, pesticides, and volatile organic compounds (VOCs), through inclusion complexation or surface adsorption. By incorporating cyclodextrin-based nanoparticles into sensing platforms, such as optical sensors, electrochemical devices, or field-deployable detectors, researchers and environmental agencies can develop sensitive and reliable systems for monitoring environmental pollution [44]. When the interactions between the matrix and the host are reversible, if the CDM arrangement is used as a template for molecular recognition, the molecules can be incorporated and released [28].

13.9.2 Biomedical Sensing

Biomedical sensing using cyclodextrin-based nanoparticles involves the detection and analysis of biomarkers for disease diagnosis and monitoring. Cyclodextrin-based nanoparticles offer unique advantages for biomedical sensing, including high sensitivity, selectivity, and biocompatibility [52]. These nanoparticles can selectively bind to specific biomolecules, known as biomarkers, that are indicative of various diseases or physiological conditions. By incorporating cyclodextrin-based nanoparticles into sensing platforms, such as biosensors or diagnostic assays, researchers can develop sensitive and reliable systems for detecting and quantifying biomarkers in biological samples, such as blood, urine, or saliva [17]. Biomarker detection using cyclodextrin-based nanoparticles enables early diagnosis of diseases, including cancer, infectious diseases, and metabolic disorders, facilitating timely treatment and management strategies [53]. Additionally, these nanoparticles can be functionalized with targeting ligands or detection probes to enhance their specificity and sensitivity for biomarkers, further improving the accuracy of disease diagnosis [8, 15].

13.9.3 Food and Beverage Sensing

Cyclodextrin-based nanoparticles are employed in food and beverage sensing to detect contaminants, additives, and adulterants in food products. Cyclodextrin-based sensors work in detecting food contaminants by leveraging the unique properties of cyclodextrins [1]. These sensors utilize the toroidal structure of cyclodextrins, which have a hydrophobic cavity capable of encapsulating small guest molecules. When a food sample containing contaminants interacts with the sensor, the cyclodextrin molecules selectively capture the target molecules to be detected [54]. This interaction enhances the stability of the contaminants, prevents agglomeration and oxidation, and improves the sensitivity of food detection. By forming inclusion complexes with the contaminants, cyclodextrin-based sensors can effectively detect and quantify food contaminants, contributing to improved food safety and quality control measures in the food industry [2]. These nanoparticles enable rapid and sensitive detection of foodborne pathogens, toxins, and chemical residues, ensuring the safety and quality of food supplies. By integrating cyclodextrin-based nanoparticles into portable and miniaturized sensing devices, food producers and regulatory agencies can perform on-site analysis and quality control to prevent foodborne illnesses and maintain consumer confidence [55].

13.9.4 Industrial Sensing

13.9.4.1 Monitoring of Chemical Processes and Quality Assurance

In industrial sensing, cyclodextrin-based nanoparticles are utilized for monitoring chemical processes, controlling product quality, and ensuring compliance with regulatory standards [19]. These nanoparticles offer high sensitivity and specificity for detecting process intermediates, impurities, and product specifications. Cyclodextrin-based sensors offer distinct advantages compared to other types of sensors in monitoring chemical processes [16]. These sensors leverage the unique properties of cyclodextrins, such as their ability to form inclusion complexes with target molecules, enhancing the selectivity and sensitivity of detection. In contrast to conventional sensors, cyclodextrin-based sensors can provide improved specificity towards particular analytes, leading to more accurate and reliable monitoring of chemical processes [22]. Additionally, the versatility of cyclodextrins allows for the customization of sensors tailored to specific applications, offering a flexible and adaptable sensing solution for diverse industrial needs [56]. Overall, cyclodextrin-based sensors stand out for their enhanced performance in detecting and monitoring chemicals in industrial processes, making them valuable tools for quality assurance and process optimization. By incorporating cyclodextrin-based nanoparticles into online monitoring systems or sensing arrays, industrial manufacturers can optimize production processes, reduce waste, and minimize environmental impact [33].

13.9.5 CDs and CD-Based Polymers in Pesticide Remediation

Cyclodextrins and cyclodextrin-based polymers have garnered significant attention for their potential application in pesticide remediation. The high toxicity of pesticides and agrochemicals poses serious health risks to humans [4]. These materials are widely used around the world to eliminate pests; they are used in parks, schools, residential areas, infrastructure and agricultural areas. Consequently, organic pollutants related to pesticides are widely present in the air, water, and food chains and risk human health by ingestion, inhalation, and contact with the skin. In toxicological studies, the negative effects of pesticides and agricultural chemicals, such as eye irritation and weight loss, are highlighted [5, 40]. This has led to an urgent need for effective ways to minimize these harmful effects on human well-being. These compounds offer unique properties that can aid in the removal or degradation of pesticides from various environmental matrices, including soil and water. CDs have the ability to encapsulate and stabilize a wide range of organic and inorganic compounds, including pesticides [57]. By selectively incorporating the pesticide molecules into the internal cavity of the cyclodextrins, both the solubility and stability of the pesticide are greatly enhanced. Recent studies have developed a solvent platform based on hypercross-linked hydroxypropyl-CD networks to remove polycyclic aromatic hydrocarbons (PAHs) from wastewater. These networks were created using highly concentrated hydroxypropyl-CD solutions with 1,2,3,4 butane tetracarboxylic acid as a crosslinker. The CD-based polymer was synthesized in an environmentally responsible way and showed remarkable sorption capacity [58].

Furthermore, Utzeri et al. have studied the synthesis and characterization of CD nanospheres (NSs) using two rare cross-linkers, dodecane-1,12-diamine (am12) and hexane-1,6-diamine (am6). The physicochemical properties of the synthesized NS were significantly affected by the cross-links selected. The results of the study showed that compared to variables such as imidacloprid concentrations, solution volume, pH, and NSs quantity, am6-NS showed greater sorption capacities of imidacloprid [21, 23].

13.10 Recent Advances and Future Perspectives

13.10.1 Current Trends in Cyclodextrin-Based Nanoparticle Research for Sensing Applications

Cyclodextrin-based nanoparticles have gained significant attention in recent years due to their unique properties and potential for sensing applications. These nanoparticles, which are composed of cyclic oligosaccharides, exhibit remarkable stability and biocompatibility, making them suitable for sensing a broad range of analytes [59].

One of the key trends in the field of cyclodextrin-based nanoparticles is the development of supramolecular architectures. Researchers are exploring various strategies to combine cyclodextrins with other functional molecules or materials to create novel nanoparticles with enhanced sensing capabilities. These supramolecular architectures provide multiple binding sites for analytes, leading to improved sensitivity and selectivity [60].

Research has demonstrated that β-cyclodextrin-based hollow nanoparticles possess notably superior adsorption capacities for pollutants when compared to traditional adsorbents like activated carbon. These nanoparticles have been shown to achieve maximum adsorption capacities as high as 2080.35 mg/g for cationic dyes and heavy metals, which is significantly greater than the capacities of many other materials. Furthermore, their adsorption kinetics are remarkably fast, ranging from 6 to 200 times quicker than other β-CD-based adsorbents, which greatly improves their potential for use in real-time sensing applications [44, 61].

Another significant trend is the use of cyclodextrin-based nanoparticles as fluorescent sensors. These nanoparticles offer excellent optical properties, making them suitable for sensitive detection in biological systems [1]. By incorporating fluorescent dyes or quantum dots into the nanoparticles, researchers have been able to achieve detection limits down to the femtomolar range, opening new possibilities for a wide range of applications [62].

In addition to fluorescent sensing, cyclodextrin-based nanoparticles have also shown promise in electrochemical sensing. The cyclic structure of cyclodextrins allows for the formation of cyclic voltammograms, which can be used to detect and quantify redox-active species. This electrochemical sensing technique offers the advantage of being rapid, portable, and amenable to integration with microfluidic systems, making it suitable for point-of-care applications [63]. Cyclodextrin-based nanoparticles have shown significant promise in electrochemical sensing, particularly through techniques such as cyclic voltammetry (CV) and electrochemical impedance spectroscopy (EIS). In cyclic voltammetry, the cyclic structure of cyclodextrins allows for the formation of cyclic voltammograms, which are instrumental in detecting and quantifying redox-active species. This technique involves sweeping the potential of an electrode and measuring the resulting current, providing insights into the electrochemical behavior of analytes. For instance, when an electroactive guest molecule like irisquinone forms a 1:1 inclusion complex with β-cyclodextrin, variations in peak current and potential can be observed, indicating successful interaction and allowing for the calculation of stability constants [64].

13.11 Challenges and Limitations

While cyclodextrin-based nanoparticles have demonstrated significant potential in sensing applications, several challenges still exist. One key limitation is the limited solubility of analytes in cyclodextrin solutions [1]. To overcome this limitation, researchers have developed encapsulation strategies, such as inclusion complexes or solid-phase extraction, to improve the solubility and stability of analytes within the nanoparticles [65].

Another challenge is the need for specific recognition motifs on the surface of nanoparticles for target analytes. Developing efficient strategies to incorporate specific receptors or antibodies onto the nanoparticle surface is crucial for selectivity and specificity in sensing applications. Additionally, the long-term stability and storage of the nanoparticles are concerns, necessitating the development of suitable packaging materials and storage conditions [66].

13.12 Future Directions and Emerging Technologies

The future direction of cyclodextrin-based nanoparticle research for sensing applications holds great promise [67]. One promising area is the integration of multiple sensing modalities on a single nanoparticle platform. By combining fluorescent sensing, electrochemical sensing, and other sensing techniques, researchers aim to achieve enhanced sensing capabilities and overcome the limitations of individual modalities [68].

Another exciting area of research is the development of multifunctional nanoparticles for sensing applications. These nanoparticles can combine sensing capabilities with other functions such as drug delivery, imaging, or therapy, opening up new possibilities for personalized medicine and combination therapies [69]. Moreover, the integration of nanoparticles with microfluidic systems is also gaining attention. Microfluidic chips can provide controlled environments, sample handling, and data analysis capabilities, enabling the development of portable and point-of-care sensing devices [15]. By integrating cyclodextrin-based nanoparticles with microfluidic chips, researchers aim to develop miniaturized sensing platforms for real-time monitoring of analytes in complex samples [70].

13.13 Conclusion

Recent research on cyclodextrin-based nanoparticles in sensing applications has highlighted their unique properties, including high sensitivity and versatility, which render them valuable tools for detecting and quantifying analytes across diverse applications. Cyclodextrin-based nanoparticles enable sensitive and reliable sensing in complex sample matrices through selective interactions with target molecules. They hold potential for environmental monitoring, biomedical diagnostics, and food safety. Future research efforts should focus on overcoming challenges such as reproducibility, scalability, and standardization to unlock the full potential of these nanoparticles. Moreover, exploring synergies with emerging technologies such as artificial intelligence, nanofabrication, and wearable sensors offers exciting prospects for innovation and impact. By continuing to harness the capabilities of cyclodextrin-based nanoparticles, researchers and industries can propel the development of next-generation sensing platforms with enhanced performance, reliability, and accessibility, ushering in a new era of scientific advancement and societal well-being.

Acknowledgements We would like to express our gratitude to all contributors who provided valuable insights and assistance in crafting this chapter.

References

1. Abdulaziz F, Salah D, Abdulaziz F, Salah D (2021) Gold nanoparticles incorporated with cyclodextrins and its applications. J Biomater Nanobiotechnol 12:79–97. https://doi.org/10.4236/JBNB.2021.124007
2. Ahmad A, Imran M, Ahsan H (2023) Biomarkers as biomedical bioindicators: approaches and techniques for the detection, analysis, and validation of novel biomarkers of diseases. Pharm 2023 15:1630. https://doi.org/10.3390/PHARMACEUTICS15061630
3. Ahmad F, Salem-Bekhit MM, Khan F, Alshehri S, Khan A, Ghoneim MM, Wu HF, Taha EI, Elbagory I (2022) Unique properties of surface-functionalized nanoparticles for bio-application: functionalization mechanisms and importance in application. Nanomaterials 12. https://doi.org/10.3390/NANO12081333
4. Arslan M, Sanyal R, Sanyal A (2020) Cyclodextrin embedded covalently crosslinked networks: synthesis and applications of hydrogels with nano-containers. Polym Chem 11:615–629. https://doi.org/10.1039/C9PY01679A
5. Belenguer-Sapiña C, Pellicer-Castell E, Mauri-Aucejo AR, Simó-Alfonso EF, Amorós P (2021) Cyclodextrins as a key piece in nanostructured materials: quantitation and remediation of pollutants. Nanomaterials 11:1–28. https://doi.org/10.3390/NANO11010007
6. Casulli MA, Taurino I, Carrara S, Hayashita T (2019) Integration methods of cyclodextrins on gold and carbon electrodes for electrochemical sensors. C 5:78. https://doi.org/10.3390/C5040078
7. Celebioglu A, Topuz F, Yildiz ZI, Uyar T (2019) Efficient removal of polycyclic aromatic hydrocarbons and heavy metals from water by electrospun nanofibrous polycyclodextrin membranes. ACS Omega 4:7850–7860. https://doi.org/10.1021/ACSOMEGA.9B00279/ASSET/IMAGES/MEDIUM/AO-2019-00279G_M006.GIF

8. Cengiz B, Gevrek TN, Chambre L, Sanyal A (2023) Self-assembly of cyclodextrin-coated nanoparticles: fabrication of functional nanostructures for sensing and delivery. Mol 28:1076. https://doi.org/10.3390/MOLECULES28031076
9. Chu F, Wei W, Shuaibu NS, Feng H, Wang X, Pan Y (2023) Mass spectrometry-based biosensing and biopsy technology. Chemosensors 11(8):419–419. https://doi.org/10.3390/chemosensors11080419
10. Cova TF, Murtinho D, Aguado R, Pais AACC, Valente AJM (2021) Cyclodextrin polymers and cyclodextrin-containing polysaccharides for water remediation. Polysaccharides 2:16–38. https://doi.org/10.3390/POLYSACCHARIDES2010002
11. Freeman R, Finder T, Bahshi LL, Willner I (2009) β-cyclodextrin-modified cdse/zns quantum dots for sensing and chiroselective analysis. Nano Lett 9:2073–2076. https://doi.org/10.1021/NL900470P/SUPPL_FILE/NL900470P_SI_002.PDF
12. Gadade DD, Pekamwar SS (2020) Cyclodextrin based nanoparticles for drug delivery and theranostics. Adv Pharm Bull 10:166–183. https://doi.org/10.34172/apb.2020.022
13. Gentili A (2020) Cyclodextrin-based sorbents for solid phase extraction. J Chromatogr A 1609:460654. https://doi.org/10.1016/J.CHROMA.2019.460654
14. Ghanizadeh Gerayeli F, Hosseini F, Bagheri Z, Savardashtaki A, Shabaninejad Z, Amani AM, Najafipour S (2020) Colorimetric sensor based on β-cyclodextrin-functionalized silver nanoparticles for zidovudine sensitive determination. Int J Anal Chem 2020. https://doi.org/10.1155/2020/5054864
15. Gharib R, Greige-Gerges H, Fourmentin S, Charcosset C, Auezova L (2015) Liposomes incorporating cyclodextrin–drug inclusion complexes: current state of knowledge. Carbohydr Polym 129:175–186. https://doi.org/10.1016/J.CARBPOL.2015.04.048
16. Guan L, Zeng Z, Liu W, Wang T, Tian S, Hu S, Tian D (2022) Aggregation-induced emission (AIE) nanoparticles based on γ-cyclodextrin and their applications in biomedicine. Carbohydr Polym 298. https://doi.org/10.1016/J.CARBPOL.2022.120130
17. Hadian Z, Maleki M, Abdi K, Atyabi F, Mohammadi A, Khaksar R (2018) Preparation and characterization of nanoparticle β-cyclodextrin: geraniol inclusion complexes. Iran J Pharm Res IJPR 17:39
18. Healy B, Yu T, da Silva Alves DC, Okeke C, Breslin CB (2021) Cyclodextrins as supramolecular recognition systems: applications in the fabrication of electrochemical sensors. Materials (Basel) 14. https://doi.org/10.3390/MA14071668
19. Da HuQ, Tang GP, Chu PK (2014) Cyclodextrin-based host-guest supramolecular nanoparticles for delivery: from design to applications. Acc Chem Res 47:2017–2025. https://doi.org/10.1021/AR500055S/ASSET/IMAGES/MEDIUM/AR-2014-00055S_0007.GIF
20. Ikeda H, Nakamura M, Ise N, Oguma N, Nakamura A, Ikeda T, Toda F, Ueno A (1996) Fluorescent cyclodextrins for molecule sensing: fluorescent properties, NMR characterization, and inclusion phenomena of N-dansylleucine-modified cyclodextrins. J Am Chem Soc 118:10980–10988. https://doi.org/10.1021/JA960183I
21. Kapan B, Kurbanoglu S, Esenturk EN, Soylemez S, Toppare L (2021) Electrochemical catechol biosensor based on β-cyclodextrin capped gold nanoparticles and inhibition effect of ibuprofen. Process Biochem 108:80–89. https://doi.org/10.1016/J.PROCBIO.2021.06.004
22. Karthic A, Roy A, Lakkakula J, Alghamdi S, Shakoori A, Babalghith AO, Bin ET, Sharma R, Lima CMG, Kim B, Park MN, Safi SZ, de Almeida RS, Coutinho HDM (2022) Cyclodextrin nanoparticles for diagnosis and potential cancer therapy: a systematic review. Front Cell Dev Biol 10:984311. https://doi.org/10.3389/FCELL.2022.984311/BIBTEX
23. Lee JU, Lee SS, Lee S, Oh H Bin (2020) Noncovalent complexes of cyclodextrin with small organic molecules: applications and insights into host–guest interactions in the gas phase and condensed phase. Mol 25:4048. https://doi.org/10.3390/MOLECULES25184048
24. Li H, Han C (2008) Sonochemical synthesis of cyclodextrin-coated quantum dots for optical detection of pollutant phenols in water. Chem Mater 20:6053–6059. https://doi.org/10.1021/CM8009176/SUPPL_FILE/CM8009176_SI_002.PDF
25. Li T, Le W (2020) Biomarkers for parkinson's disease: how good are they? Neurosci Bull 36:183–194. https://doi.org/10.1007/S12264-019-00433-1

26. Liu R, Ye X, Cui T (2020) Recent progress of biomarker detection sensors. Research. https://doi.org/10.34133/2020/7949037
27. Liu Y, Liu M, Jia J, Wu D, Gao T, Wang X, Yu J, Li F (2019) β-Cyclodextrin-based hollow nanoparticles with excellent adsorption performance towards organic and inorganic pollutants. Nanoscale 11:18653–18661. https://doi.org/10.1039/C9NR07342F
28. Liu Y, Sameen DE, Ahmed S, Wang Y, Lu R, Dai J, Li S, Qin W (2022) Recent advances in cyclodextrin-based films for food packaging. Food Chem 370:131026. https://doi.org/10.1016/J.FOODCHEM.2021.131026
29. Liu Z, Xu W, Kovaleva EG, Cheng J, Li H (2022) Recent progress in encapsulation and controlled release of pesticides based on cyclodextrin derivative carriers. Adv Agrochem 1:89–99. https://doi.org/10.1016/J.AAC.2022.11.008
30. Marabada D, Li J, Wei S, Huang Q, Wang Z (2023) Cyclodextrin based nanoparticles for smart drug delivery in colorectal cancer. Chem Biol Drug Des 102:1618–1631. https://doi.org/10.1111/CBDD.14344
31. Mariani S, Minunni M (2014) Surface plasmon resonance applications in clinical analysis. Anal Bioanal Chem 406:2303. https://doi.org/10.1007/S00216-014-7647-5
32. Matencio A, Pedrazzo AR, Difalco A, Navarro-Orcajada S, Monfared YK, Conesa I, Rezayat A, López-Nicolás JM, Trotta F (2021) Advances and classification of cyclodextrin-based polymers for food-related issues. Polymers (Basel) 13. https://doi.org/10.3390/POLYM13234226
33. Mendonça MCP, Sun Y, Cronin MF, Lindsay AJ, Cryan JF, O'Driscoll CM (2023) Cyclodextrin-based nanoparticles for delivery of antisense oligonucleotides targeting huntingtin. Pharmaceutics 15:520. https://doi.org/10.3390/PHARMACEUTICS15020520/S1
34. Narita M, Hamada F (2000) The synthesis of a fluorescent chemo-sensor system based on regioselectively dansyl-tosyl-modified β- and γ-cyclodextrins. J Chem Soc Perkin Trans 2:823–832. https://doi.org/10.1039/A907702B
35. Narita M, Koshizaka S, Hamada F (1999) Fluorescent pyrrolinone-modified cyclodextrins as a chemo-sensor for organic guests. J Incl Phenom 35:605–619. https://doi.org/10.1023/A:1008090403417/METRICS
36. Wannasri N, Uppachai P, Senasu T, Nanan S, Katrun P, Vichapong J, Butwong N, Srijaranai S, Mukdasai S (2024) A novel electrochemical sensor based on β-cyclodextrin/bismuth oxybromide/multi-walled carbon nanotubes modified electrode with in situ addition of tetrabutylammonium bromide for the simultaneous detection and degradation of tebuconazole. Microchim Acta 191(11). https://doi.org/10.1007/s00604-024-06797-0
37. Noël S, Léger B, Ponchel A, Sadjadi S, Monflier E (2021) Cyclodextrins as multitask agents for metal nano-heterogeneous catalysis: a review. Environ Chem Lett 19:4327–4348. https://doi.org/10.1007/S10311-021-01298-5
38. Ogoshi T, Harada A (2008) Chemical sensors based on cyclodextrin derivatives. Sensors 8:4961–4982. https://doi.org/10.3390/S8084961
39. Overmyer D, Siegal M, Staton A, Provencio P, Yelton W (1996). Cyclodextrin-based microsensor for volatile organic compounds. https://doi.org/10.2172/920117
40. Ozyilmaz E, Sayin S, Arslan M, Yilmaz M (2014) Improving catalytic hydrolysis reaction efficiency of sol–gel-encapsulated Candida rugosa lipase with magnetic β-cyclodextrin nanoparticles. Colloids Surf B Biointerfaces 113:182–189. https://doi.org/10.1016/J.COLSURFB.2013.08.019
41. Pandey A (2021) Cyclodextrin-based nanoparticles for pharmaceutical applications: a review. Environ Chem Lett 19(6):4297–4310. https://doi.org/10.1007/s10311-021-01275-y
42. Panahi Z, Merrill MA, Halpern JM (2020) Reusable cyclodextrin-based electrochemical platform for detection of trans-resveratrol. ACS Appl Polym Mater 2:5086. https://doi.org/10.1021/ACSAPM.0C00866
43. Pankaj (2020) Anti-cancer cyclodextrin nanocapsules based formulation development for lung chemotherapy. J Pharm Res Int 32:54–63. https://doi.org/10.9734/JPRI/2020/V32I3931024
44. Poulson BG, Alsulami QA, Sharfalddin A, El Agammy EF, Mouffouk F, Emwas AH, Jaremko L, Jaremko M (2021) Cyclodextrins: structural, chemical, and physical properties, and applications. Polysaccharides 3:1–31. https://doi.org/10.3390/POLYSACCHARIDES3010001

45. Rather IA, Hasan A, Ali R, Rather IA, Hasan A, Ali R (2022) Cyclodextrin-based sensors for the recognition of small molecules. Cyclodext Core Concepts New Front. https://doi.org/10.5772/INTECHOPEN.108500
46. Sanità G, Carrese B, Lamberti A (2020) Nanoparticle surface functionalization: how to improve biocompatibility and cellular internalization. Front Mol Biosci 7:587012. https://doi.org/10.3389/FMOLB.2020.587012/BIBTEX
47. Sempionatto JR, Lasalde-Ramírez JA, Mahato K, Wang J, Gao W (2022) Wearable chemical sensors for biomarker discovery in the omics era. Nat Rev Chem 6(12):899–915. https://doi.org/10.1038/s41570-022-00439-w
48. Shen HM, Zhu GY, Bin YuW, Wu HK, Ji HB, Shi HX, Zheng YF, Bin SY (2015) Surface immobilization of β-cyclodextrin on hybrid silica and its fast adsorption performance of p-nitrophenol from the aqueous phase. RSC Adv 5:84410–84422. https://doi.org/10.1039/C5RA15592D
49. Singh R, Singh S (2018) Surface modification of nanomaterials for biomedical applications: strategies and recent advances. Nanobiotechnology 171–217. https://doi.org/10.1201/9781351031585-5
50. Singh RP, Hidalgo T, Cazade PA, Darcy R, Cronin MF, Dorin I, O'Driscoll CM, Thompson D (2019) Self-assembled cationic β-cyclodextrin nanostructures for siRNA delivery. Mol Pharm 16:1358–1366. https://doi.org/10.1021/ACS.MOLPHARMACEUT.8B01307/ASSET/IMAGES/MEDIUM/MP-2018-013079_0008.GIF
51. Sperling RA, Parak WJ (2010) Surface modification, functionalization and bioconjugation of colloidal inorganic nanoparticles. Philos Trans R Soc A Math Phys Eng Sci 368:1333–1383. https://doi.org/10.1098/RSTA.2009.0273
52. Sun HL, Zhang YM, Chen Y, Liu Y (2016) Polyanionic cyclodextrin induced supramolecular nanoparticle. Sci Reports 61(6):1–9. https://doi.org/10.1038/s41598-016-0026-z
53. Syeda SEZ, Nowacka D, Khan MS, Skwierawska AM (2022) Recent advancements in cyclodextrin-based adsorbents for the removal of hazardous pollutants from waters. Polymers (Basel) 14. https://doi.org/10.3390/POLYM14122341
54. Szente L, Szemán J (2013) Cyclodextrins in analytical chemistry: host-guest type molecular recognition. Anal Chem 85:8024–8030. https://doi.org/10.1021/AC400639Y/ASSET/IMAGES/LARGE/AC-2013-00639Y_0005.JPEG
55. Tian BR, Liu YM (2021) Cyclodextrin-active natural compounds in food applications: a review of antibacterial activity. Turkish J Chem 45:1707. https://doi.org/10.3906/KIM-2106-51
56. Tiwari G, Tiwari R, Rai AK (2010) Cyclodextrins in delivery systems: applications. J Pharm Bioallied Sci 2:72. https://doi.org/10.4103/0975-7406.67003
57. Trofymchuk IM, Roik N, Belyakova L (2016) Sol-gel synthesis of ordered β-cyclodextrin-containing silicas. Nanoscale Res Lett 11:1–12. https://doi.org/10.1186/S11671-016-1380-2/TABLES/3
58. Trotta F, Mazzaglia A, Szente L, Esmaeilpour D, Broscheit JA, Shityakov S (2022) Cyclodextrin-Based Polymeric Materials Bound to Corona Protein for Theranostic Applications. Int J Mol Sci 23:13505. https://doi.org/10.3390/IJMS232113505
59. Utzeri G, Matias PMC, Murtinho D, Valente AJM (2022a) Cyclodextrin-based nanosponges: overview and opportunities. Front Chem 10. https://doi.org/10.3389/FCHEM.2022.859406
60. Utzeri G, Murtinho D, Maria TMR, Pais AACC, Sannino F, Valente AJM (2022b) Amine-β-cyclodextrin-based nanosponges. The role of cyclodextrin amphiphilicity in the imidacloprid uptake. Colloids Surf A Physicochem Eng Asp 635:128044. https://doi.org/10.1016/J.COLSURFA.2021.128044
61. Van De Manakker F, Vermonden T, Van Nostrum CF, Hennink WE (2009) Cyclodextrin-based polymeric materials: synthesis, properties, and pharmaceutical/biomedical applications. Biomacromol 10:3157–3175. https://doi.org/10.1021/BM901065F/ASSET/IMAGES/MEDIUM/BM-2009-01065F_0020.GIF
62. Wankar J, Kotla NG, Gera S, Rasala S, Pandit A, Rochev YA, Wankar J, Kotla NG, Rasala S, Pandit A, Rochev YA, Gera S (2020) Recent advances in host-guest self-assembled cyclodextrin carriers: implications for responsive drug delivery and biomedical engineering. Adv Funct Mater 30:1909049. https://doi.org/10.1002/ADFM.201909049

63. Xing C, Zheng X, Deng T, Zeng L, Liu X, Chi X (2023) The role of cyclodextrin in the construction of nanoplatforms: from structure, function and application perspectives. Pharmaceutics 15. https://doi.org/10.3390/PHARMACEUTICS15051536
64. Xu W, Li X, Wang L, Li S, Chu S, Wang J, Li Y, Hou J, Luo Q, Liu J (2021) Design of cyclodextrin-based functional systems for biomedical applications. Front Chem 9:635507. https://doi.org/10.3389/FCHEM.2021.635507/BIBTEX
65. Xu Y, Rashwan AK, Osman AI, Abd El-Monaem EM, Elgarahy AM, Eltaweil AS, Omar M, Li Y, Mehanni AHE, Chen W, Rooney DW (2023) Synthesis and potential applications of cyclodextrin-based metal–organic frameworks: a review. Environ Chem Lett 21:447. https://doi.org/10.1007/S10311-022-01509-7
66. Yang JH, Kim HT, Kim H (2014) A cyclodextrin-based approach for selective detection of catecholamine hormone mixtures. Micro Nano Syst Lett 2:1–10. https://doi.org/10.1186/S40486-014-0001-Z/FIGURES/7
67. Zerkoune L, Angelova A, Lesieur S (2014) Nano-assemblies of modified cyclodextrins and their complexes with guest molecules: incorporation in nanostructured membranes and amphiphile nanoarchitectonics design. Nanomater 4:741–765. https://doi.org/10.3390/NANO4030741
68. Zhang Z, Zhou J, Liu Y, Tang J, Tang W (2015) Cyclodextrin capped CdTe quantum dots as versatile fluorescence sensors for nitrophenol isomers. Nanoscale 7:19540–19546. https://doi.org/10.1039/C5NR06073G
69. Zhao Y, Huang Y, Zhu H, Zhu Q, Xia Y (2016) Three-in-one: sensing, self-assembly, and cascade catalysis of cyclodextrin modified gold nanoparticles. J Am Chem Soc 138:16645–16654. https://doi.org/10.1021/JACS.6B07590/SUPPL_FILE/JA6B07590_SI_001.PDF
70. Zhu G, Yi Y, Chen J (2016) Recent advances for cyclodextrin-based materials in electrochemical sensing. TrAC Trends Anal Chem 80:232–241. https://doi.org/10.1016/J.TRAC.2016.03.022

Pankaj Bhatt is an accomplished academic in the field of Pharmaceutics, holding a Ph.D. from Gurukul Kangri University in Haridwar. With over 12 years of experience in teaching and research, he has made significant contributions to pharmaceutical sciences, evidenced by his impressive portfolio of 63 published papers, which include research articles, reviews, and book chapters. Dr. Bhatt is also a prolific author, having published four books and secured six patents along with one copyright. His research interests primarily focus on nanoparticles and polymer modification, showcasing his commitment to advancing knowledge in these areas. He has achieved notable academic metrics with an h-index of 27 and an i10-index of 38, reflecting his influence and productivity in the scientific community.

Dr. Anasuya Patil M.Pharm, Ph.D., working an as Associate Professor, Department of Pharmaceutics, KLE College of Pharmacy, Bengaluru, Karnataka. She is having more than 21 years of teaching experience and has been guiding PG and Ph.D scholars in the department of Pharmaceutics. She had worked in formulation development at Pharmaceutical industry. She has published several research papers in reputed journals, contributed as author to Advanced drug delivery system book, Regulatory affairs, Industrial Pharmacy, Modern Pharmaceutics, Pharmaceutical Quality Assurance book, etc and in her credit seven Patents.

Dr. Ranjana has done M. Pharma in Pharmaceutical Chemistry from UPTU in 2009 and Ph.D from BHU, Varanasi in 2013.Dr Ranjana has qualified Graduate Aptitude Test in Engineering (GATE) in 2008. she has credited 02 Indian Patents, 01 published Indian Patents,01 UK Patent, 30 international/ National papers and 04 book chapters. She holds experience in academic, research and industry and currently, is a Professor at Graphic Era Hill University, Dehradun. Dr Ranjana, research interest is in biological activities (in vivo& invitro) included Diabetes, Anti-inflammatory, Analgesic, Anti-oxidant etc. and has interest in clinical study and medical writing too.

Dr. Shashank Soni is a highly experienced academician and researcher in the field of Pharmaceutical Sciences. He completed his B. Pharm in 2009 from the Sardar Bhagwan Singh PG Institute of Biomedical Sciences and Research, Balawala, Dehradun, India. He then pursued his M. Pharm in 2011 from the College of Pharmacy, IFTM, Moradabad, India, followed by a Ph.D. in Pharmacy in 2019 from the Faculty of Pharmacy, IFTM University, Moradabad, India. With over 13 years of academic and administrative experience, Dr. Soni currently serves as an Assistant Professor at the Amity Institute of Pharmacy, Amity University Uttar Pradesh Lucknow Campus, Lucknow, India. Throughout his career, he has supervised 18 students in their dissertation work in the field of Pharmaceutics and is currently guiding one Ph.D. student on their dissertation project. Dr. Soni's primary research interests are focused on the development of Gastroretentive Dosage Forms, Ophthalmic Systems, In Situ Gelling Systems for Ophthalmic Applications, Raft Forming Systems, Hydrogels, Sponges, Nanocrystals, and Encapsulation Technologies. His research aims to explore and improve drug delivery systems to enhance therapeutic outcomes in various medical fields.

Dr. Yudhishther Singh Bagal did his Ph.D. in Agricultural Extension and Communication from Sher-e- Kashmir University of Agricultural Sciences and Technology, Jammu, J&K, India. He has published several research articles including book chapters in reputed national as well as international journals and books respectively. Presently he is working as Assistant Professor in School of Agriculture, Lovely Professional University, Phagwara, Punjab, India.

Chapter 14
Applications of β-Cyclodextrin-Linked Carbon Nanodots and Future Perspectives

Ceyda Özen, Ali Zeytünlüoğlu, and Alper Akkaya

Abstract Macrocyclic molecules, like cyclodextrins, are essential in interacting with other molecules and are important actors in supramolecular chemistry. Cyclodextrins, derived from starch enzymatic degradation, exhibit strong stability and excellent complexation abilities. Carbon nanodots (CNDs), especially β-cyclodextrin-based CNDs, offer biocompatibility, non-toxicity, and accessible synthesis advantages. They find applications in diverse fields such as medicine, environmental monitoring, and food safety. Due to their low toxicity and brightness, CNDs have gained prominence in sensing, catalysis, and optoelectronics. β-Cyclodextrin-based biocompatible and easily functionalizable CNDs show promise in biomedical diagnostics, drug delivery, and environmental monitoring. Their potential in medical imaging, formulation of personal care products, and intelligent material development is noteworthy. Looking ahead, β-cyclodextrins and CNDs are poised for significant contributions in drug formulation, nanomedicine, and responsive materials. Continuous research in nanotechnology is likely to unveil new applications and enhance their integration across various industries. The future holds promise for addressing complex challenges in medicine, environmental science, and beyond through the unique properties of β-cyclodextrins and CNDs.

Keywords β-Cyclodextrin · Carbon nanodots · Chemosensing

C. Özen · A. Akkaya
Faculty of Science, Department of Biochemistry, University of Ege, İzmir, Turkey
e-mail: alper.akkaya@ege.edu.tr

A. Zeytünlüoğlu (✉)
Denizli Vocational School of Technical Sciences, Department of Chemistry and Chemical Processing Technologies, University of Pamukkale, Denizli, Turkey
e-mail: azeytun@pau.edu.tr

S. Krishnamoorthy (ed.), *Cyclodextrins for Chemosensing*,
https://doi.org/10.1007/978-3-031-89299-8_14

14.1 Applications of Cyclodextrins with Functionalized Carbon Nanodots (CNDS)

14.1.1 Biosensing

CNDs with β-cyclodextrin can be used in sensing applications as receptors for specific analytes [1]. Due to their excellent optical properties, high fluorescence sensitivity to the environmental medium [2, 3], and the ability to act as influential electron donors, CNDs are proposed to detect various substances [4–7].

A few examples from biosensing applications of functionalized CNDs using cyclodextrin are given below:

Luo et al. worked on developing a new detection method for testosterone. In a study using β-cyclodextrin-functionalized carbon nanoparticles, they demonstrated a reliable and sensitive method for the selective detection of testosterone with high fluorescence quantum yield. They also highlighted that this probe system can monitor testosterone levels in cells due to the low cytotoxicity, high anti-cross-interaction ability, and excellent biocompatibility of β-cyclodextrin CNDs. In addition, the ability of β-cyclodextrin-CNDs to selectively detect testosterone from other steroid hormones was also investigated, indicating that this probe system has the potential to detect testosterone in a wide concentration range selectively [8].

β-glucuronidase (β-Gcu) is a cancer marker that plays a role in the breakdown of glycosaminoglycans. Investigations have shown that an increase in β-glucuronidase (β-Gcu) activity includes many types of cancer. Different methods, such as colorimetric, fluorescence, and electrochemical, monitor β-glucuronidase activity. Among these, although the fluorescence method can provide high sensitivity and real-time detection, the solubility of the β-Gcu substrates is only suitable in organic solvents, reducing the method's effectiveness. Therefore, Yin and colleagues functionalized fluorescent CNDs using cyclodextrin and aminosilane (β-cyclodextrin-SiCNDs) to detect β-glucuronidase (Fig. 14.1). The detection range for β-Gcu was determined as 0.5–60 U L^{-1}. β-cyclodextrin-SiCNDs can detect β-Gcu at the lowest level of 0.14 U L^{-1} [9].

Dopamine is a vital neurotransmitter biomolecule that plays a critical role in various metabolic processes in the body. Sangubotla ve Kim developed fluorescent sensors based on melamine-derived CNDs stabilized with β-cyclodextrin to detect dopamine (Fig. 14.2).

In this study; CNDs can be synthesized from various plant extracts, such as citrus peels, green tea, or other biomass sources, and common solvents for extraction include water or ethanol, which help in dissolving the organic compounds from the plant material. The extraction process typically involves boiling the plant material in the solvent, followed by filtration to obtain the extract, which is then subjected to carbonization or hydrothermal treatment to form CNDs. It is claimed that the developed sensor can detect dopamine at the lowest level of 56 pM [10].

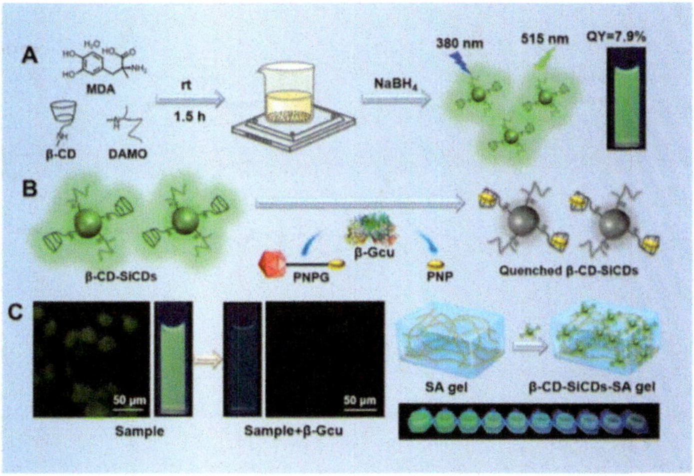

Fig. 14.1 Synthesis strategy of β-cyclodextrin-SiCNDs and detection of β-Gcu [permission from Ref. [9], Copyright, 2024; Elsevier]

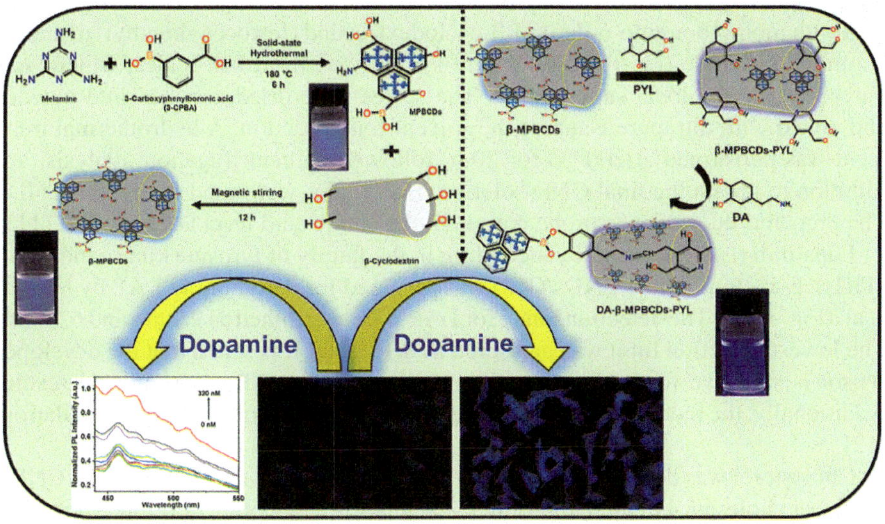

Fig. 14.2 Schematic illustration for the synthesis of MPBCNDs [permission from Ref. [10], Copyright, 2023; Elsevier]

Cholic acid level is an essential diagnostic indicator in liver and minor intestine diseases. The cholic acid levels have limitations such as the equipment being expensive, cholic acid levels being determined with low sensitivity and selectivity, and the long detection times being long. Lu and Fan fabricated a fluorescent "switch" sensing system with high sensitivity and selectivity that is environmentally friendly,

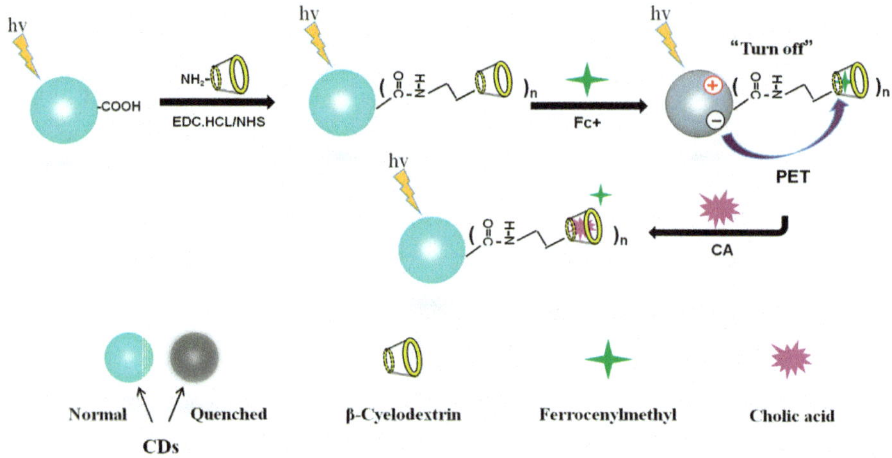

Fig. 14.3 Synthesis of β-cyclodextrin-modified N-doped CND and fluorescent switch detection principle of cholic acid [permission from Ref. [11], Copyright, 2019; Elsevier]

fast, and safe (Fig. 14.3). This system was built using N-doped CNDs functionalized with mono (6-amino-6-deoxy)-β-cyclodextrin and (ferrocenylmethyl) trimethylammonium iodide (Fc +). CNDs were synthesized from fish scales, which were purchased from a local supermarket. The scales were dried, crushed into powder, and mixed with ultrapure water as the solvent for extraction. A hydrothermal treatment was performed at 200 °C for 20 h, followed by centrifugation, dialysis, and filtration to obtain the final CND solution. Fish scales were used to prepare CNDs. The prepared fluorescence probe measured the cholic acid level at 25 nmol L [11].

Lapatinib (LAP) is an anti-cancer drug in the family of tyrosine kinase inhibitors (TKIs). β-cyclodextrin@CNDs/GCE was prepared for determining LAP by Mathad et al. (Fig. 14.4). The detection range for LAP was determined as 9.9 nm and 6.17 um. The lowest detection limit was measured as 0.181 nm. It is stated that the developed sensor is selective for LAP, has good stability, and provides reproducible results. Additionally, the method works in biological fluids and pharmaceutical formulations [12].

Cholesterol was detected in two different studies using β-cyclodextrin and CNDs. To detect cholesterol in human serum, Ganganboina and Doong developed a method using β-cyclodextrin functionalized N-GQD (β-cyclodextrin@N-GQD). Ferrocene (FC) was selected as a redox indicator (Fig. 14.5). The release of FC takes place by taking advantage of the difference in the affinity levels of FC and cholesterol to β-cyclodextrin. The amount of FC released affects the redox signal in the spectrum. This enables the quantitative detection of cholesterol [13].

Second, Sun et al. developed a cholesterol fluorometric detection strategy based on the competition between cholesterol and p-nitrophenol for β-cyclodextrin using carbon quantum dots functionalized with p-nitrophenol and β-cyclodextrin

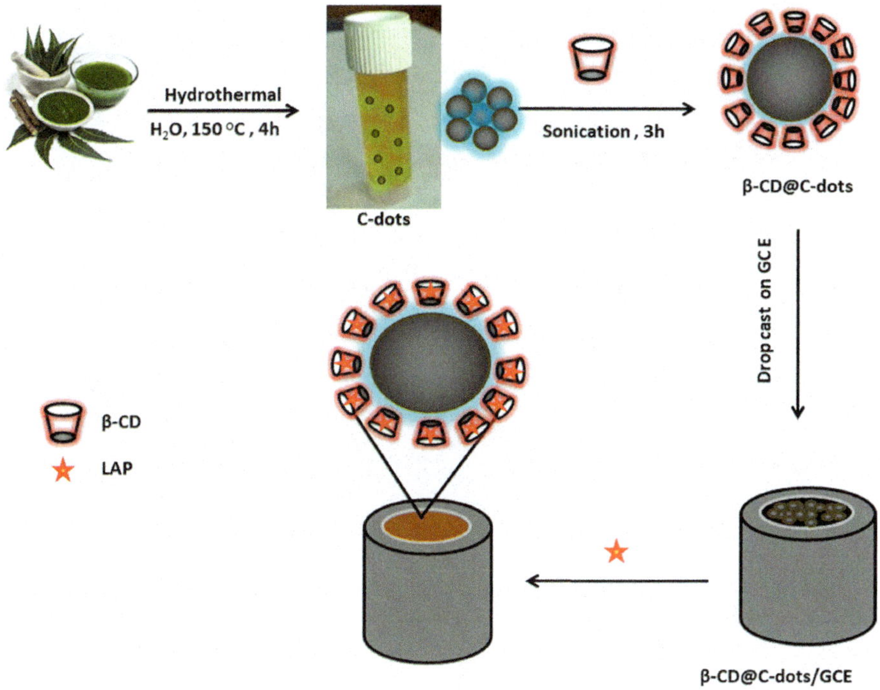

Fig. 14.4 Schematic of β-cyclodextrin@CNDs synthesis [permission from Ref. [12], Copyright, 2022; Elsevier]

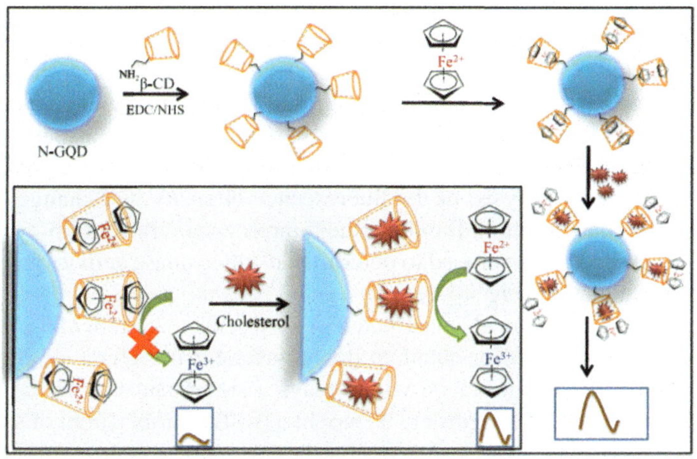

Fig. 14.5 This diagram depicts the blueprint for creating β-cyclodextrin-functionalized nitrogen-doped graphene quantum dots (β-cyclodextrin@N-GQD) and their function as microscopic cholesterol detectors, using ferrocene as a color-changing indicator. [permission from Ref. [13] Copyright, 2018; Springer]

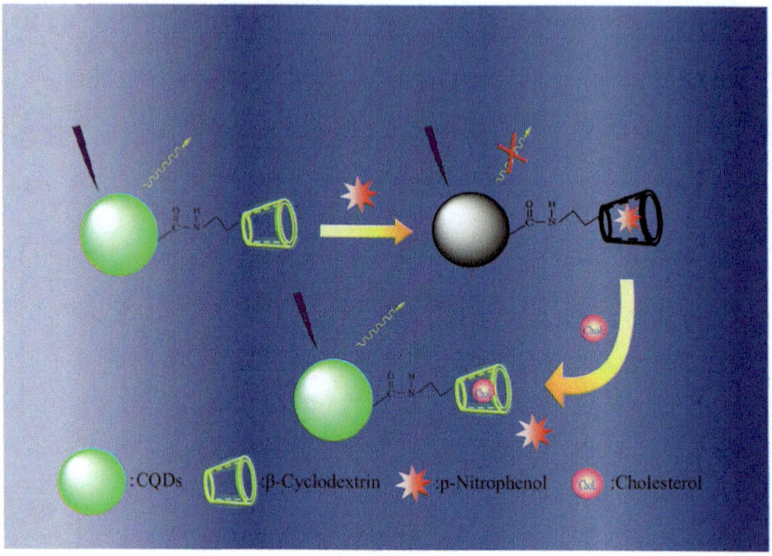

Fig. 14.6 Illustration of detection strategy of cholesterol detection using β-cyclodextrin-CQD [permission from Ref. [14], Copyright, 2017; Elsevier]

(Fig. 14.6). In this study, which takes a similar approach to the strategy of Ganganboina and Doong, the cholesterol level is determined by considering the change in the amount of p-nitrophenol [14].

In another application, 4-nitrophenol will be released after hydrolysis of 4-nitrophenol-α-D-glucopyranoside (NPGlu), which is placed in the hydrophobic cavity of β-cyclodextrin by the α-glucosidase enzyme. In this case, there will be a decrease in fluorescence intensity. The severity of the decrease will be parallel to the increase in the amount of free 4-nitrophenol. Enzyme activity can also be determined according to the decrease in fluorescence intensity. Additionally, the inhibitor will not attend enzyme hydrolysis, or the fluorescence intensity will change depending on the inhibitor concentration. Based on the fluorescence turn-off of β-cyclodextrin-CQDs, this method was developed to determine α-glucosidase activity and screen its inhibitors by Tang et al. (Fig. 14.7) [15].

In another study, Jiang et al. constructed a chiral sensing system of β-cyclodextrin (β-cyclodextrin)-coated sulfur quantum dots (cyclodextrin-SQDs) for chiral recognition of amino acid enantiomers, which plays a significant role in various fields (Fig. 14.8). They could distinguish L-tryptophan (L-Trp) from a pool of amino acids with high selectivity. They defined the L-Trp detection range of the method as 2.3 nM [16].

Quantitative determination of uric acid is significant in clinical evaluations. SiQDs/Cu2-β-cyclodextrin nanoclusters were prepared by Peng and others for the detection of uric acid (UA) and L-cystein as sensitive and selectivity in the presence of interference biomolecules such as ascorbate and bilirubin (Fig. 14.9) [17].

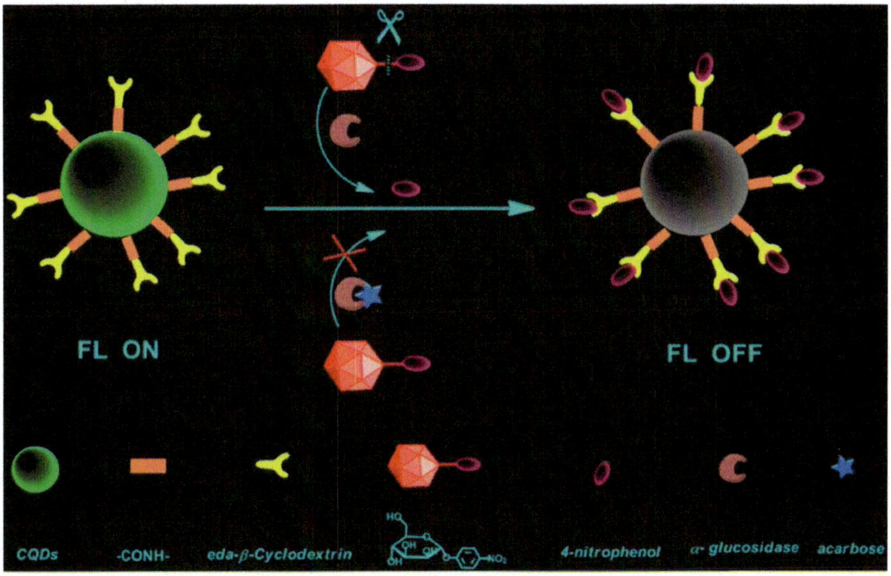

Fig. 14.7 Schematic illustration of a strategy for the detection of α-glucosidase activity and screening of inhibitors using β-cyclodextrin-CQDs [permission from Ref. [15], Copyright, 2017; Elsevier]

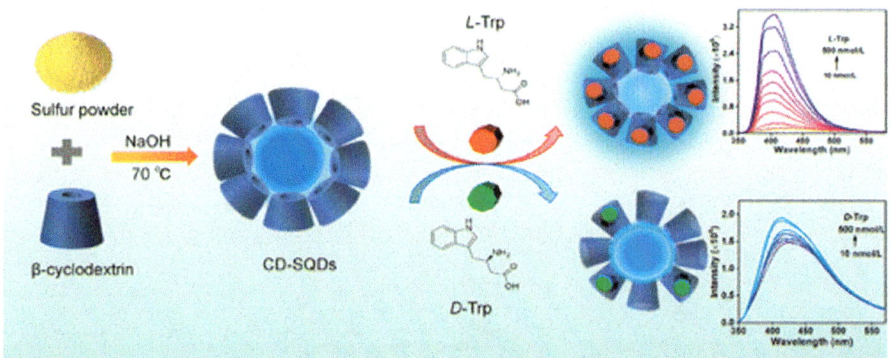

Fig. 14.8 Synthesis of strategy β-cyclodextrin-coated sulfur quantum dots (CD-SQDs) for chiral recognition of amino acid enantiomers [permission from Ref. [16], Copyright, 2023; ACS Publications]

In another study, Dopamine (DA), Uric acid (UA), and tryptophan (Trp) are found together in biological samples; developing a method that can selectively and sensitively determine them is vital for clinical evaluations. β-cyclodextrin/CQDs were prepared for the electrochemical detection of DA, UA, and Trp by Chen and others. The responses were obtained for DA, UA, and Trp, up to 0.14, 0.01, and 0.16 μM concentrations. The detection limit was determined as 0.45, 0.04 and 0.50 [18].

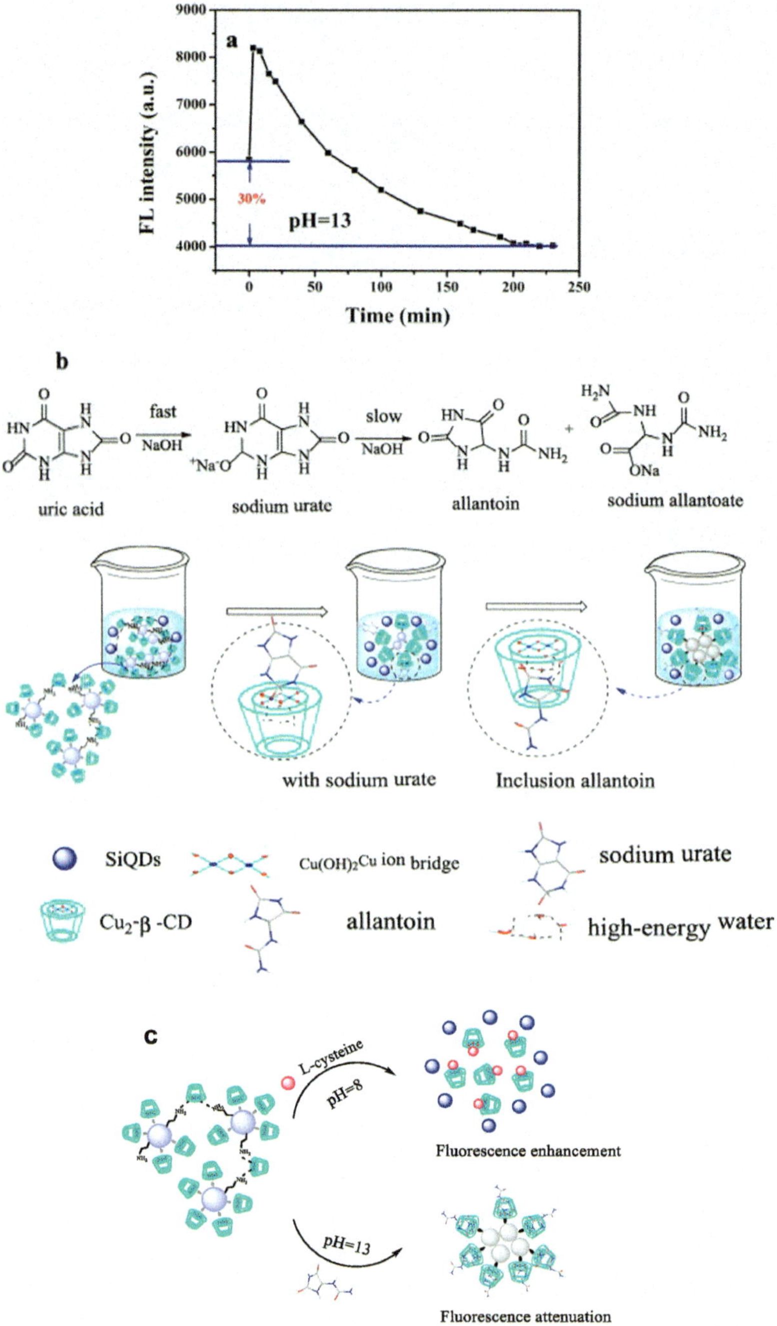

Fig. 14.9 Schematic illustration of the interaction between uric acid (UA) (**a**) L-Cysteine (**b**) and SiQDs/Cu2-β-cyclodextrin (**a**), the graph of change in Fluorescence intensity of SiQDs/Cu2-β-cyclodextrin after adding UA (**c**) [permission from Ref. [17], Copyright, 2022; Elsevier]

14.1.2 Environmental Applications

Cyclodextrin and CNDs frequently detect and remove heavy metals in environmental applications. Cyclodextrin and CNDs can be used as practical materials for the monitoring and removal of heavy metals in water and soil samples. These materials can adsorb and detect heavy metals due to their high surface areas and chemical properties. In addition, these materials have the advantages of being environmentally friendly and economical. Therefore, using cyclodextrin and CNDs to detect and remove heavy metals in environmental applications is an important research topic in environmental analytical chemistry. β-cyclodextrin CNDs are a unique nanomaterial with a high potential for detecting and removing environmental contaminants. Specific advantages of β-cyclodextrin CNDs in metal ion detection include their potential for selective and sensitive detection of metal ions and their ability to serve as effective fluorescent probes due to their unique photoluminescence properties and surface functionalization. These nanomaterials can play an essential role in reducing environmental pollution and creating a more sustainable future. The unique properties of cyclodextrin CDs, such as water solubility, photoluminescence, and fluorescence, make them ideal for various applications. These properties and their biocompatibility, biosensing, and bioimaging capabilities contribute to their potential in metal ion detection and other fields [19].

Cyclodextrin CNDs have been applied to detect metal ions, especially silver (Ag^+). A study demonstrated the one-pot synthesis of CNDs by carbonizing cyclodextrin used to detect Ag^+ ions in aqueous solutions. The CNDs reduced Ag^+ on their surface to elemental silver under sunlight, resulting in a linear change in photoluminescence intensity with Ag^+ concentration, making them effective fluorescent probes for Ag^+ detection [20]. In the study, they were able to reduce Ag^+ ions on their surface to elemental silver (Ag^0) under sunlight (Fig. 14.10), resulting in a change in PL intensity that is linearly related to Ag^+ concentration [20].

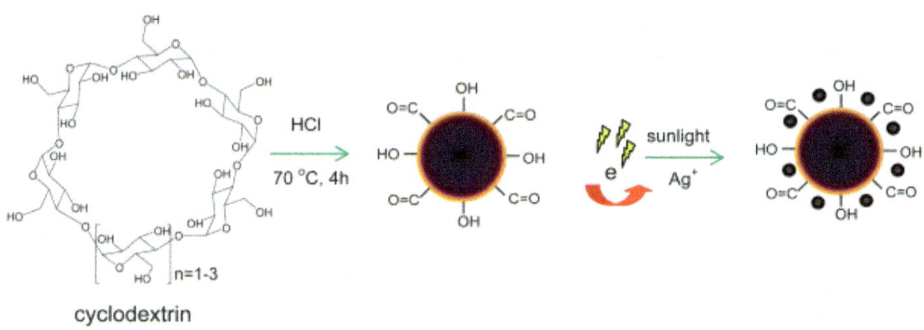

Fig. 14.10 Schematic representation of the synthesis of CNDs from cyclodextrin and the reduction of Ag^+ to Ag^0 by CNDs in the presence of sunlight [permission from Ref. [13], Copyright, 2014; RSC Publishing]

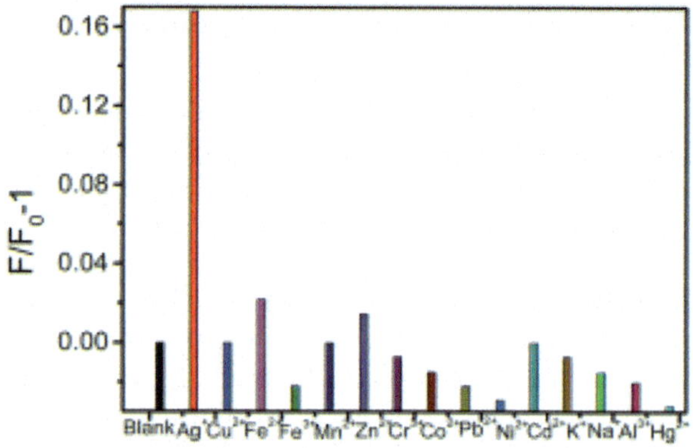

Fig. 14.11 F/F0-1 of aqueous solutions of CNDs (100 mg L^{-1}) with a variety of metal ions (10 mM) [permission from Ref. [20], Copyright, 2014; RSC Publishing]

The document also highlights the CNDs selectivity for Ag^+ detection and their potential application as templates for the *in-situ* synthesis of nano-silver in aqueous solution (Fig. 14.11).

Another study aims to evaluate the composite material's adsorption performance and potential for use in the simultaneous detection and removal of metal ions from water and wastewater. In addition, the study aims to investigate the selectivity of the composite towards different metal ions and optimize its performance for real-world applications.

This research delves into crafting and analyzing supramolecular composites made from poly (β-cyclodextrin-co-citric acid) and carbon quantum dots (QDs). These composites act as both adsorbents and fluorescent probes for divalent metal ions like Ni (II), Cu (II), Cd (II), and Pb (II), which pose environmental threats. Various techniques confirmed the successful synthesis and interaction between cyclodextrins and QDs. Optimal conditions for metal adsorption were identified, leading to impressive capacities (reaching 220 mg/g) and removal efficiencies exceeding 90% for Ni (II) and Cu (II). The composites selectively adsorb metal ions with smaller radii and can be reused with minimal efficiency loss. Additionally, they exhibit quenching luminescence upon metal adsorption, enabling sensing applications.

In conclusion, this novel supramolecular composite holds immense promise for simultaneous water and wastewater monitoring and treatment, making it a valuable tool for environmental protection. This research showcases a material specifically designed to capture all studied metal species. It exhibits unique fluorescence responses for each species, reflecting its selectivity. This dual-functional material can both remove and detect metal ions in water, possessing numerous desirable properties like high light transmittance, exceptional mechanical strength, and thermal/photoluminescent stability. Its outstanding photoluminescence translates to high sensitivity and detection capacity, while its porosity facilitates simultaneous metal adsorption, making it ideal as a filter for rapid decontamination and metal ion removal. The experimental results show that the β-cyclodextrin(CA)/QDs material is the most suitable for the adsorption of Ni(II) and Cu(II) ions. This material showed high removal efficiencies of 96% for Ni(II) and 91% for Cu(II). The removal efficiencies for Cd (II) and Pb (II) ions were 68% and 58%, respectively. These results show that the material is highly selective and successful against Ni (II) and Cu (II) ions [21].

In another study, Lin et al., β-cyclodextrin-modified CNDs-NH_2 utilized the principle of host–guest recognition undergoing a two-step sulfonylation reaction to develop β-cyclodextrin-modified CNDs-NH_2. Due to the unique molecular formation, CNDs-NH_2 is frequently used to determine catechol (CC) and hydroquinone (HQ) analyses. Due to the hydrophilic outer surface and hydrophobic inner cavity, upon entry of analytes into the β-cyclodextrin cavities, a rigid host–guest complex is formed, and intense fluorescence quenching is achieved (Fig. 14.12). Furthermore, the surface modified with CNDs-NH_2 has a high water solubility level and reveals strong fluorescence. Therefore, CC and HQ can be detected in real water samples [22].

The applicability of CND@β-cyclodextrin for hydroquinone and catechol detection in ambient aqueous samples was tested by adding CC or HQ at concentrations of 10, 20, and 50 mM to real water samples. These samples were centrifuged at 12,000 rpm within 30 min and passed through 0.22 μm filters. The data showed that this probe system is suitable for the rapid 30-min determination of phenolic compounds in environmental water samples (recovery rate between 97.4% and 106.7% and RSD less than 4.04%). These data indicate that the synthesized CNDs@β-cyclodextrin nanocomposites are suitable for detecting CC or HQ in real water samples. This method is environmentally friendly, has good biocompatibility,

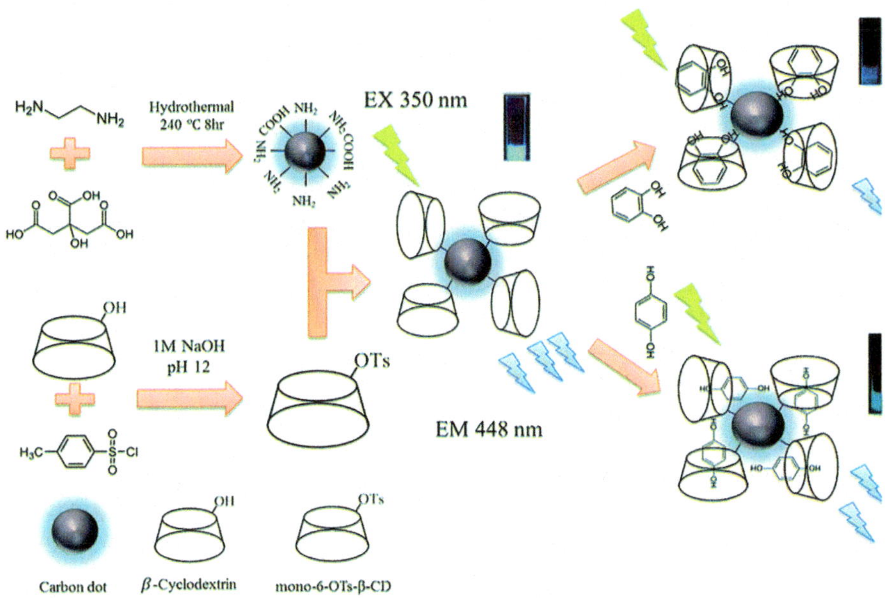

Fig. 14.12 Schematic representation of the fabrication of CNDs @β-cyclodextrin and illustration of the proposed nanosensor [permission from Ref. [22], Copyright, 2018; RSC Publishing]

and may be potentially helpful for cell imaging applications and the recognition of pollutant compounds.

This study aimed to develop an efficient method for detecting hydroquinone and catechol with β-cyclodextrin-modified CNDs. To this end, the synthesis and the characterization of CNDs@β-cyclodextrin nanocomposites and methods for their sensitive detection of hydroquinone and catechol were investigated. The applicability of these nanocomposites for detecting hydroquinone and catechol in ambient aqueous samples was also investigated. This study aims to develop a new approach to detecting phenolic compounds using environmentally friendly and highly biocompatible nanomaterials.

In addition, this method was found to have a lower LOD for CC and HQ detection and a more comprehensive linear range compared to other methods in the literature. In conclusion, CNDs@β-cyclodextrin is a promising nanoprobe for the detection of organic compounds. This method is environmentally friendly, has good biocompatibility, and may also be valuable for cell imaging applications and the recognition of pollutant compounds. This study demonstrated a suitable method for rapidly determining phenolic compounds in environmental water samples.

Cyclodextrin and CNDs, another study (Martínez and Sánchez 2016), which exemplifies the environmental applications of cyclodextrin and CNDs, evaluates the use of electrochemical methods to determine the concentration of glutathione (GSH) in the root extract of the aquatic plant Eichhornia crassipes (water fly) and the potential of β-cyclodextrin modified electrodes in GSH determination. Such research can contribute to understanding aquatic plants' biological and environmental properties and conserving aquatic ecosystems. Therefore, the study can be related to environmental applications. In this study, electrochemical investigation and application of β-cyclodextrin modified carbon electrodes for determining glutathione (GSH) were carried out. G-glutamic acid is a tripeptide derived from the amino acids cysteine and glycine (Fig. 14.13A).

In the first part of this study, the capability of the electrodes to determine GSH was analyzed using different methodologies, and it was determined that the best result was achieved with the carbon paste electrode modified by potential sweeping using circular voltammetry. It was also found that an internal complex was formed between GSH and the deposition of β-cyclodextrin on the surface of the electrode. In the second section of the study, the effect of scan rate and pH on GSH determination was investigated. Electrochemical methods are advantageous, simple, low-cost, and highly sensitive compared to other techniques. GSH exhibits electroactivity due to its ability to lose two protons and two electrons, transforming into glutathione disulfide (GSSG). These two forms coexist in a balanced state within the mixture. In addition, β-cyclodextrin catalyzes the chemical oxidation of GSH [23]. Imagine a ring shaped by seven interconnected sugar molecules (glucopyranose units), as depicted in (Fig. 14.13B(a)). This ring forms a cone-like structure (Fig. 14.13B(b)), with the wider base composed of secondary hydroxyl groups and the narrower opening adorned with primary hydroxyl groups [24]. Interestingly, the inside of this "sugar cone," known as the β-cyclodextrin cavity, prefers to interact with water-fearing molecules (hydrophobic), while the outside, dotted with hydroxyl groups, embraces

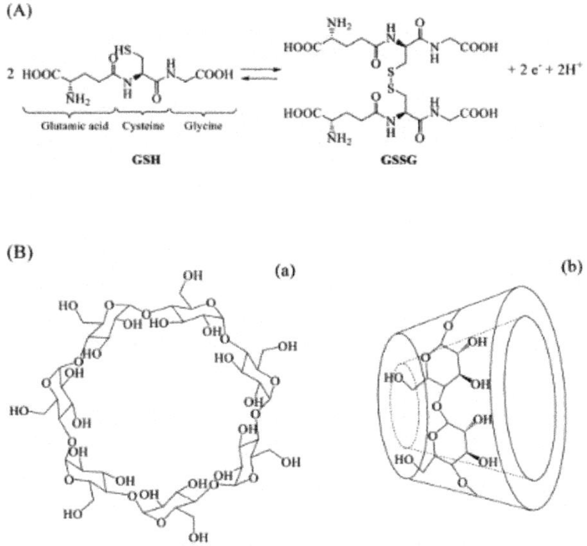

Fig. 14.13 (**A**) Oxidation reaction of GSH. (**B**) Structure of β-cyclodextrin (a) and illustration of its cone-like shape (b). [permission from Ref. [26], Copyright, 2016; Wiley]

water (hydrophilic) [25]. Finally, the electrode successfully determined the GSH concentration in Eichhornia crassipes root extract, and these results were close to the data obtained using the spectroscopic method. This study shows the feasibility of β-cyclodextrin modified electrodes in GSH determination.

Research by Silva et al. delves into the creation and analysis of exceptional composites made from poly (β-cyclodextrin-co-citric acid) and carbon quantum dots (Fig. 14.14). These composites exhibited high adsorption capacities and impressive removal efficiencies for metal ions (Ni (II), Cu (II), Cd (II), and Pb (II)) following successful validation of their synthesis using various characterization techniques. These composites also show selective adsorption of metal ions and can be used for sensing applications by exhibiting luminescence quenching properties with adsorption of metal ions. This research has shown that these composites are promising for monitoring and treating water and wastewater, making them a valuable tool in environmental applications. This study also investigated the adsorption kinetics and isotherm models of specific metal ions and found that these agreed with experimental data. EDX spectroscopy showed that the metal ions were homogeneously distributed on the surface of the adsorbent. These findings indicate that composites of poly (β-cyclodextrin-co-citric acid) and carbon quantum dots are promising for the removal of metal ions from water and wastewater and for sensing applications [21].

This study used citric acid to prepare β-cyclodextrin and carbon quantum dots (QDs) together. This composite material has been used for both adsorption and photoluminescence detection of heavy metal ions (Ni(II), Cu(II), Cd(II), and Pb(II)). The superior properties of this material include high adsorptive capacity, high separation efficiency, selective adsorption ability, and photoluminescence. In addition, this composite material can be used for the rapid removal of metal ions from water and wastewater and detection applications. This study shows that the composites

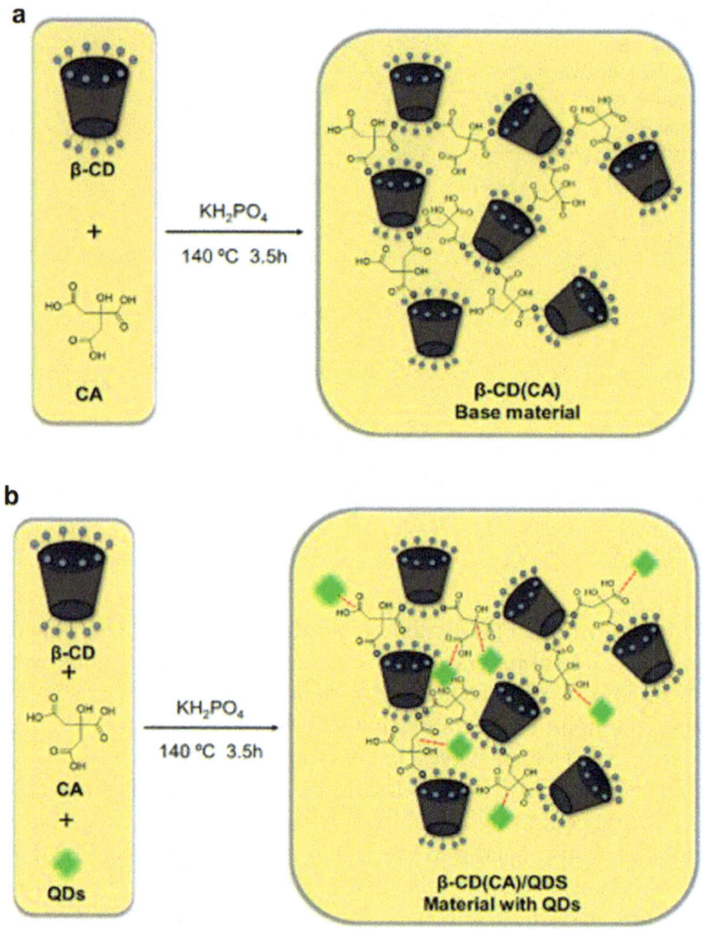

Fig. 14.14 Illustrative schemes for the synthesis of (**A**) β-cyclodextrin(CA) and (**B**) β-cyclodextrin(CA)/QDs [permission from Ref. [21], Copyright, 2024; Elsevier]

formed by combining β-cyclodextrin and QDs are promising tools for environmental applications.

14.1.3 *Food Applications*

Increasing food needs, parallel with the increasing population density today, have revealed the necessity of more food production. However, it has become necessary to develop fast and easy-to-apply measurement methods that ensure control of food production, whether implemented or not in accordance with the law. The basic

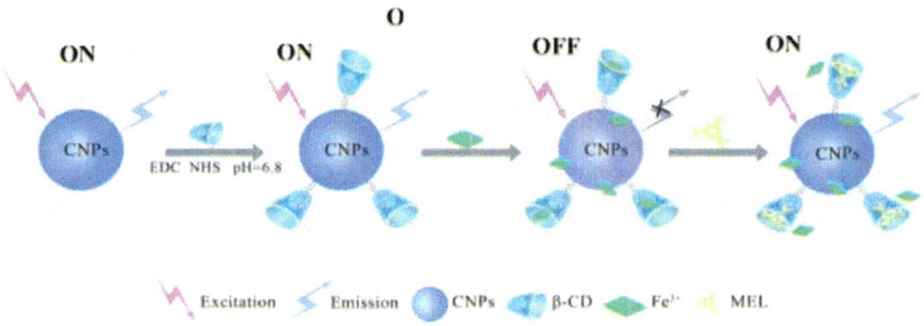

Fig. 14.15 Mechanism illustration for the "ON–OFF-ON" detection of MEL by β-cyclodextrin-CNPs fluorescence probe [permission from Ref. [28], Copyright, 2021; Elsevier]

approach in measurement methods is accurately determining desired or undesirable components in the food product. The use of β-cyclodextrin and CNDs to determine the active ingredients in food or substances that may be harmful to the human body is one of the new approaches that have begun to be applied.

One of the studies by Pinilla-Peñalver et al., β-cyclodextrin-modified graphene quantum dots have been used as a matrix in capillary electrophoresis to detect more than one food analyte. The most important feature of this study is that it can screen a large number of analytes in food samples without the need for organic solvents or surfactants [27].

In another study conducted to detect adulteration in milk, the presence of melanin used for adulteration was determined using β-cyclodextrin modified carbon nanoparticles (Fig. 14.15). Melamine (MEL) is an industrial chemical illegally added to milk to increase protein levels. Detection of MEL in milk was achieved with an intelligent sensor (ON–OFF-ON) designed based on β-cyclodextrin modified carbon nanoparticles (β-cyclodextrin-CNPs) by Lino and staff [28].

There is increasing interest in developing new packaging systems based on biomaterials that control the shelf life, spoilage, and contamination of food products sold in markets. In another study conducted by Ding et al., A composite film surface functionalized with potassium ions was prepared with nitrogen-doped citric acid carbon dot (CND) and hydroxypropyl two β-cyclodextrin/ geraniol (CS/CND@HPCDGer) for simultaneous freshness preservation and quality monitoring of meat products (Fig. 14.16) [29]. This study prepared composite film by casting method, which responds to food spoilage with a pH change. In addition, this film also exhibits antioxidant, antibacterial activity, and UV protection features.

Tert-butylhydroquinone (TBHQ) is a preserved food ingredient that prevents spoilage and rancidity. Use in excessive amounts may cause damage to the immune system, unwanted skin reactions, and visual disturbances. ZnCuMg TMO/ β-cyclodextrin-CB hybrid nanocomposite was prepared by Sebastian et al. for the measurement of TBHQ in food [30].

Vanillin is a flavoring agent used extensively in the food industry. Duran and his colleagues synthesized the β-cyclodextrin–CdSe/ZnS-QDs complex as a vanillin sensor to determine food's vanillin level [31].

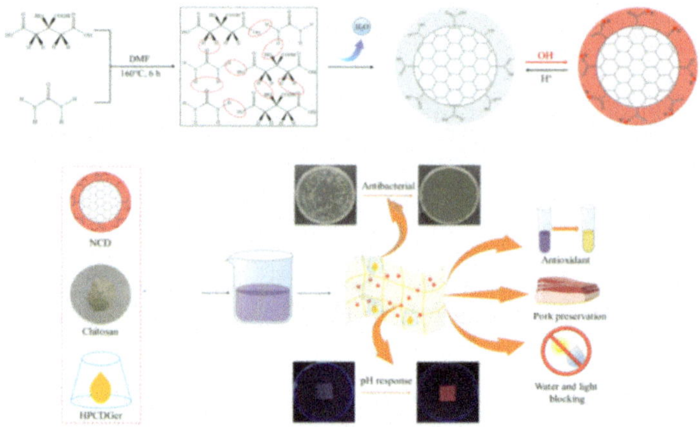

Fig. 14.16 Synthesis of CND, Preparation, and properties of CS/CND@HPCDGer [permission from Ref. [29], Copyright, 2024; SSRN]

14.2 Conclusion

Considering the number of publications in recent years, we find an extreme interest in cyclodextrin and CNDs. These two structures, which we come across in the literature with a new application every day, will be involved in the rapid, sensitive, and selective detection of the presence of viruses, bacteria, and macro and micro components qualitatively or quantitatively in many research applications. Their applications are wider than just these. We may soon see them as structures integrated into the production stage in many sectors that exhibit enzyme-like activities. They can contribute to environmentally sustainable industrial processes by taking part in the disposal of unwanted waste products after production. They can be integrated into systems that require controlled release. Artificial intelligence can be applied more extensively in synthesizing these structures. As a conclusion, the versatility of these structures will offer new perspectives to researchers in the future.

References

1. Xu X, Ray R, Gu Y, Ploehn HJ, Gearheart L, Raker K, Scrivens WA (2004) Electrophoretic analysis and purification of fluorescent single-walled carbon nanotube fragments. J Am Chem Soc 126(40):12736–12737
2. Papaioannou N, Marinovic A, Yoshizawa N, Goode AE, Fay M, Khlobystov A, Sapelkin A (2018) Structure and solvents effects on the optical properties of sugar-derived carbon nanodots. Sci Rep 8(1):6559
3. Wang JC, Violette K, Ogunsolu OO, Hanson K (2017) Metal ion mediated electron transfer at dye–semiconductor interfaces. Phys Chem Chem Phys 19(4):2679–2682

4. Liu C, Tang B, Zhang S, Zhou M, Yang M, Liu Y, Pang DW (2018) Photoinduced electron transfer mediated by coordination between carboxyl on carbon nanodots and Cu^{2+} quenching photoluminescence. J Phys Chem C 122(6):3662–3668
5. Mohammed LJ, Omer KM (2020) Dual functional highly luminescence B, N Co-doped carbon nanodots as nanothermometer and Fe3+/Fe2+ sensor. Sci Rep 10(1):3028
6. Sekar A, Yadav R, Basavaraj N (2021) Fluorescence quenching mechanism and the application of green carbon nanodots in the detection of heavy metal ions: a review. New J Chem 45(5):2326–2360
7. Zhu A, Qu Q, Shao X, Kong B, Tian Y (2012) Carbon-dot-based dual-emission nanohybrid produces a ratiometric fluorescent sensor for in vivo imaging of cellular copper ions. Angew Chem Int Ed 51(29):7185–7189
8. Luo M, Hua Y, Liang Y, Han J, Liu D, Zhao W, Wang P (2017) Synthesis of novel β-cyclodextrin functionalized S, N codoped carbon dots for selective detection of testosterone. Biosens Bioelectron 98:195–201
9. Yin C, Meng Wu M, Sun Q, Su C, Cao S, Niu N, Chen L (2024) Dual-functionalization of fluorescent carbon dots via cyclodextrin and aminosilane for visual detection of β-glucuronidase and bioimaging. Analytica Chimica Acta 1285:341996
10. Sangubotla R, Kim J (2023) Ultra-sensitive turn-on fluorescent sensor based on β-cyclodextrin-stabilized melamine-derived carbon dots for the highly selective dual-detection of dopamine. Ceram Int 49(10):16272–16282
11. Lu X, Fan Z (2019) Determination of cholic acid in body fluids by β-cyclodextrin-modified N-doped carbon dot fluorescent probes. Spectrochim Acta Part A Mol Biomol Spectrosc 216:342–348
12. Mathad AS, Seetharamappa J, Kalanur SS (2022) β-Cyclodextrin anchored neem carbon dots for enhanced electrochemical sensing performance of an anticancer drug, lapatinib via host-guest inclusion. J Mol Liquids 350:118582
13. Ganganboina AB, Doong R (2018) Functionalized N-doped graphene quantum dots for electrochemical determination of cholesterol through host-guest inclusion. Microchimica Acta 185:526
14. Sun Q, Fang S, Fang Y, Qian Z, Feng H (2017) Fluorometric detection of cholesterol based on β-cyclodextrin functionalized carbon quantum dots via competitive host-guest recognition. Talanta 167:513–519
15. Tang C, Qian Z, Qian Y, Huang Y, Zhao M, Ao H, Feng H (2017) A fluorometric and real-time assay for α-glucosidase activity through supramolecular self-assembly and its application for inhibitor screening. Sens Actuators B Chem 245:282–289
16. Jiang W, He R, Lv H, He X, Wang L, Wei Y (2023) Chiral sensing of tryptophan enantiomers based on the enzyme mimics of β-cyclodextrin-modified sulfur quantum dots. ACS Sens 8(11):4264–4271
17. Peng Y, Shao F, Guo K, Zhuo H, Wang Y, Xie X, Tao Y (2022) SiQDs/Cu2-β-CNDnanoclusters: a fluorescence probe for the mutual non-interference detection of uric acid and l-cysteine under alkaline conditions. Inorg Chem Commun 143:109765
18. Chen J, He P, Bai H, He S, Zhang T, Zhang X, Dong F (2017) Poly(β-cyclodextrin)/carbon quantum dots modified glassy carbon electrode: preparation, characterization and simultaneous electrochemical determination of dopamine, uric acid and tryptophan. Sens Actuators B Chem 252:9–16
19. Banger A, Gautam S, Jadoun S, Jangid NK, Srivastava A, Pulidindi IN, Dwivedi J, Srivastava M (2023) Synthetic methods and applications of carbon nanodots. Catalysts 13(5):858
20. Hu M, Yang Y, Gu X, Hu Y, Huang J, Wang C (2014) One-pot synthesis of photoluminescent carbon nanodots by carbonization of cyclodextrin and their application in Ag+ detection. RSC Adv 4(107):62446–62452
21. Silva EC, Gomes CG, Pina J, Pereira RF, Murtinho D, Fajardo AR, Valente AJ (2024) Carbon quantum dots-containing poly (β-cyclodextrin) for simultaneous removal and detection of metal ions from water. Carbohyd Polym 323:121464

22. Lin ZY, Kuo YC, Chang CJ, Lin YS, Chiu TC, Hu CC (2018) Highly sensitive sensing of hydroquinone and catechol based on β-cyclodextrin-modified carbon dots. RSC Adv 8(35):19381–19388
23. Sun W, Liu J, Cui M, Song F, Liu S (1999) Cyclodextrin-catalyzed oxidation of glutathione in solution and in an ion trap. Rapid Commun Mass Spectrom 13(10):950–953
24. Norkus E (2009) Metal ion complexes with native cyclodextrins. An overview. J Incl Phenom Macrocycl Chem 65:237–248
25. Li S, Purdy WC (1992) Cyclodextrins and their applications in analytical chemistry. Chem Rev 92(6):1457–1470
26. Martínez-Sánchez C, Torres-Rodríguez LM, García-de la Cruz RF (2016) Electrochemical studies of inclusion complex formed between glutathione and β-cyclodextrin-modified carbon electrodes and its application for determination of glutathione. Electroanalysis 28(12):3055–3065
27. Pinilla-Peñalver E, Soriano ML, Content AM, Ángel Ríos A (2021) Cyclodextrin-modifed graphene quantum dots as a novel additive for the selective separation of bioactive compounds by capillary electrophoresis. Microchim Acta 188(440):1–12
28. Liao X, Chena C, Shia P, Yue L (2021) Determination of melamine in milk based on β-cyclodextrin modifed carbon nanoparticles via host–guest recognition. Food Chem 338(1–8):127769
29. Ding X, Luo X, Shi W, He X, Yue L, Wang Z (2024) Chitosan film loaded with carbon dots and geraniol/Hydroxypropyl-β-Cyclodextrin inclusion complex: simultaneous freshness preservation and quality monitoring of meat products. https://doi.org/10.2139/ssrn.4682743
30. Sebastian N, Yu WC, Balram D, Fahad S. Al-Mubaddel FS, Noman MT (2022) Nanomolar detection of food additive tert-butylhydroquinone in edible oils based on novel ternary metal oxide embedded β-cyclodextrin functionalized carbon black. Food Chem 377:30, 1–10:131867
31. Duran GM, Contento AM, Rios A (2015) β-Cyclodextrin coated CdSe/ZnS quantum dots for vanillin sensoring in food samples. Talanta 131:286–291

Ceyda Ozen was born in Turkey. She graduated with bachelor's and master's degrees in biochemistry at Ege University. Her master's focused on antimicrobial peptides and cancer studies. She also worked on protein purification, enzymes, and biopolymers. She is currently continuing her academic career in the biochemistry Ph.D. program. Her future research interests include carbon dots and recombinant DNA technology.

URLs: ResearchGate, Googlescholar.

Ali Zeytunluoglu graduated from the biochemistry Ph.D. program at Ege University. He is a lecturer and researcher in the Department of Chemistry and Chemical Processing Technologies of the Vocational School of Technical Sciences at Pamukkale University. His research interests are protein purification and characterization, green synthesis, nanoemulsions, magnetic nanoparticles, and carbon nanodots.

URLs: ResearchGate, Googlescholar.

Dr. Alper AKKAYA is a professor of Biochemistry at Ege University. His main research interests are biochemistry, biotechnology, biomolecule immobilization, and antimicrobial materials. He is working on a TUSEB-funded project investigating antimicrobial and anti-cancer peptides. He plans future works within the framework of carbon nanodots. In his spare time, he enjoys playing chess and listening to music. He lives in İzmir with his wife and two children.

URLs: Googlescholar.

Chapter 15
Cyclodextrin-Based Sensory Molecules for Diverse Sensing Platforms

Sahaya Nadar, Ekta Thakor, and Jegan Nadar

Abstract Cyclodextrins (CDs), due to their discerning identification ability, exceptional of absorptivity, solubility, non-toxic nature, cost-effectiveness, commercial accessibility, and easy functionalization, serve as the primary constituents for constructing elegantly simple yet highly effective supramolecular architectures with fundamental significance. In recent decades, CDs have garnered significant attention in the scientific community, particularly for their role in developing chemical sensors based on molecular recognition phenomena. Recognizing the impressive sensing capabilities of CDs, this discussion aims to present the latest advancements in detecting various biologically relevant small compounds using CDs, employing colorimetric, fluorescence, electrochemical, and potentiometric responses. The sensing events and distinctive optical features in monomeric, dimeric, clustered, and nano-assembled structures derived from cyclodextrin are thoroughly elucidated.

Keywords Cyclodextrins · Chemical sensors · Fluorescence · Potentiometric

15.1 Introduction

Supramolecular chemistry focuses on the interactions and self-assembly of molecules, with cyclodextrins playing a significant role in this area. These host molecules have garnered attention in supramolecular chemistry as they can form inclusion complexes with guest molecules by hydrophobic interaction and hydrogen bonding as non-covalent interaction. Their diverse applications include drug-delivery systems, supramolecular catalysis, sensing organic molecules, and modifying the polarity of molecular chains by forming inclusion complexes. Cyclodextrins'

S. Nadar (✉) · E. Thakor
Department of Pharmaceutical Chemistry, St. John Institute of Pharmacy and Research, Palghar, India
e-mail: sahaya1408@gmail.com

J. Nadar
Department of Pharmacology, St. John Institute of Pharmacy and Research, Palghar, India

S. Krishnamoorthy (ed.), *Cyclodextrins for Chemosensing*,
https://doi.org/10.1007/978-3-031-89299-8_15

capacity to create inclusion complexes has made them highly valuable in this field and essential for developing supramolecular systems. They bind by α-1,4-linkage glucopyranose units and form cyclic oligosaccharides. The most common types are α-cyclodextrin, β-cyclodextrin, and γ-cyclodextrin, consisting of six, seven, and eight glucose units, respectively (Fig. 15.1) [1–3]. The sensing mechanism typically relies on changes in the physical or chemical properties of the cyclodextrin upon guest molecule binding. This could include alterations in fluorescence, absorbance, or conductivity, among other properties. The sensitivity and selectivity of the sensor can be tuned by modifying the cyclodextrin structure or incorporating additional functional groups [2, 4].

In the context of small molecule chemosensing, cyclodextrin-based sensors have been employed to detect various analytes, including drugs, environmental pollutants, and biomolecules. The versatility of cyclodextrins allows for the design of sensors with high specificity and sensitivity for a wide range of applications. Applications of cyclodextrin-based chemosensors include environmental monitoring, medical diagnostics, and pharmaceutical analysis. These sensors can offer advantages such as rapid response, low detection limits, and the ability to discriminate between closely related molecules. Additionally, the biocompatibility of cyclodextrins makes them suitable for biological and medical applications.

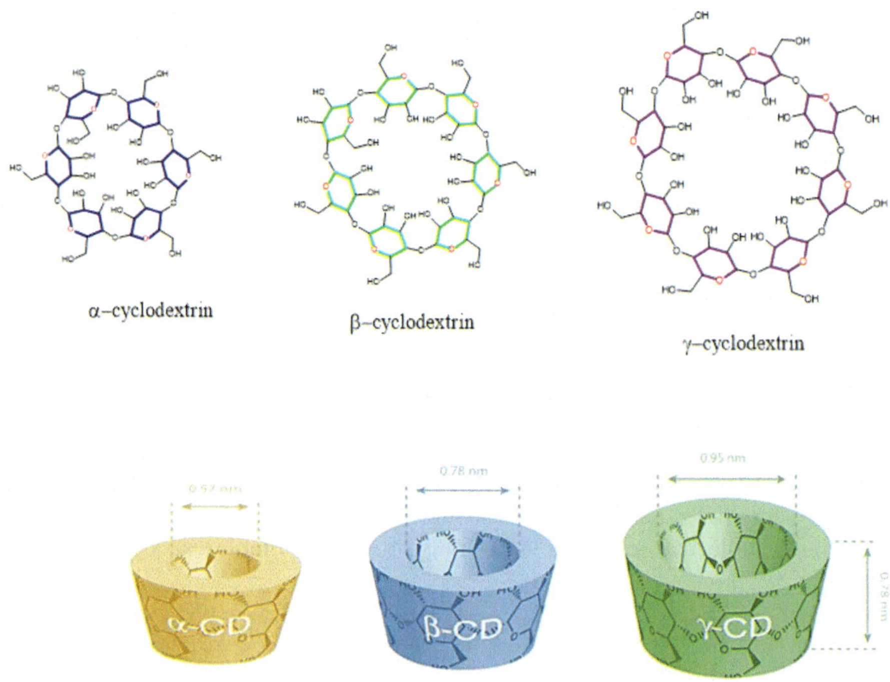

Fig. 15.1 2D and 3D structure of αCD (left), βCD (middle), and γCD (right)

15.2 Chromophore-Appended Cyclodextrins as Classical Chemical Sensors

The color and fluorescence properties of chromophores are generally influenced by their microenvironment. Consequently, fluorescence compounds and dye on entering into cyclodextrins aid in the detection of the inclusion of guest molecules into the cyclodextrins cavity. The exploration of chromophore-modified cyclodextrins has emerged as an extensive area of research, yielding a plethora of reported sensors. Notably, Ueno and colleagues have been pioneers in the development of chromophore-appended cyclodextrins chemical sensors [5].

This includes widely synthesized "turn-off" fluorescent chemosensors, where the fluorescence intensity decreases upon complexation with guest molecules. In aqueous media, fluorophore-cyclodextrins complex forms a self-inclusion conjugate through an adjustable linker. Adding a competitive guest to the solution leads to the exclusion of the fluorophore from the inside to the outside of the cyclodextrins cavity. The fluorescent cyclodextrins demonstrate enhancement of emission in the self-inclusion complex owing to the cyclodextrins cavity's hydrophobic environment, and the fluorophore's exclusion to the prominent aqueous media weakens its fluorescence intensity. Various "turn-off" fluorescent chemical sensors have been synthesized, incorporating cyclodextrins derivatives with fluorophore molecules like pyrene, naphthyl, anthracene, dansyl, and dimethylaminobenzoyl components (Fig. 15.2) [6].

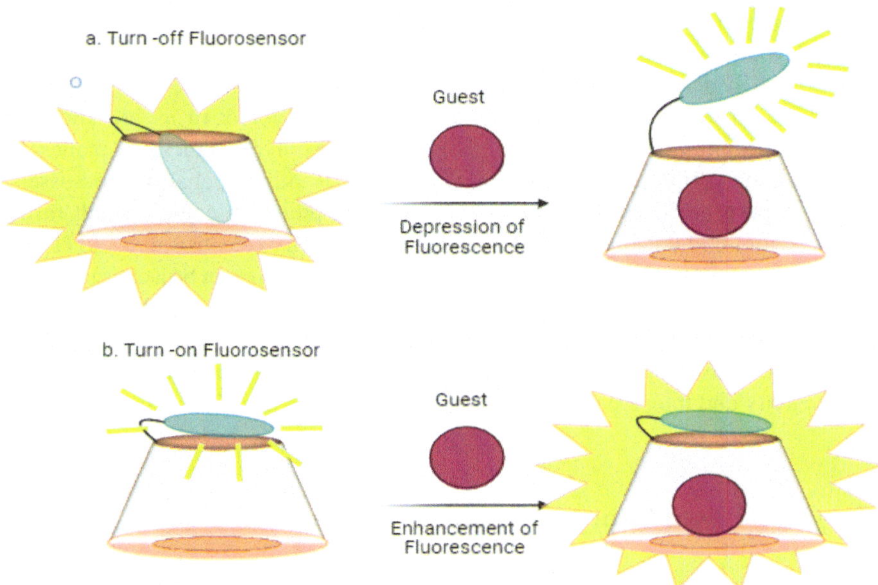

Fig. 15.2 **a** Turn-off fluorosensors. **b** Turn-on fluorosensors

Additionally, a novel category of "turn-on" fluorescent chemical sensors has been reported, where the fluorescence intensity increases upon the formation of host–guest complexes. Using a rigid spacer between cyclodextrins and fluorophore prevents the formation of self-inclusion complexes in fluorophore-cyclodextrins conjugates. Consequently, the fluorophore moiety is surrounded by a hydrophilic environment, leading to the suppression of fluorescence [7]. By introducing a hydrophobic guest molecule into the cavity of the fluorophore-cyclodextrins conjugate, the fluorophore is situated in a more hydrophobic environment, increasing in fluorescence intensity. Various "turn-on" type fluorescent chemical sensors, including 4-amino-7-nitrobenz-2-oxa-1,3-diazole-cyclodextrins [8, 9], naphthol-cyclodextrins [10, 11], and hydroxyquinoline-cyclodextrins conjugates [12, 13], have been successfully reported. Recognizing that pH indicators like p-nitrophenol, phenolphthalein, and methyl red exhibit color changes based on pH conditions, they designed cyclodextrin derivatives appended with these dye molecules [14–16]. In aqueous environments, the dye moiety was included within the hydrophobic cyclodextrins cavity, forming a self-inclusion complex and isolating it from the surrounding aqueous media.

15.3 Cyclodextrin-Based Supramolecular Systems as Chemical Sensors

For effective supra-molecular structure cyclodextrins that are biocompatible, cheap, non-toxic, acceptable solubility in water are considered essential building blocks in the development of chemical sensors. Naturally occurring cyclodextrins are torus-shaped structures composed of cyclic oligosaccharides present as possessing hydrophilic surface and hydrophobic central cavity. Cyclodextrins are employed for building supramolecular structures of rotaxanes, polyrotaxanes, catenanes and supramolecular polymeric materials [17, 18]. CDs oligosaccharide possesses a unique property to form inclusion complexes with other molecules which further can act as nontoxic sensing media. Few interactions such as electrostatic and H-bond are responsible for the host–guest stable complex. Also, there can be a charge transfer upon the occupancy of the guest into the pocket of cyclodextrins [19].

To form an inclusion complex with large guest molecules cyclodextrins-dimer and cyclodextrins-trimer are covalently coupled as they provide a huge hydrophobic central cavity and thus are ideal structures for designing chemical sensors.

15.4 CD-Based Supramolecule with Dansyl Group

Ueno et al. have proposed a β-cyclodextrins-dimer fluorescent sensor used for the recognition of steroids. Primary faces of two β-cyclodextrins molecules are linked through the dansyl group (Fig. 15.3). The mechanism included in this detection

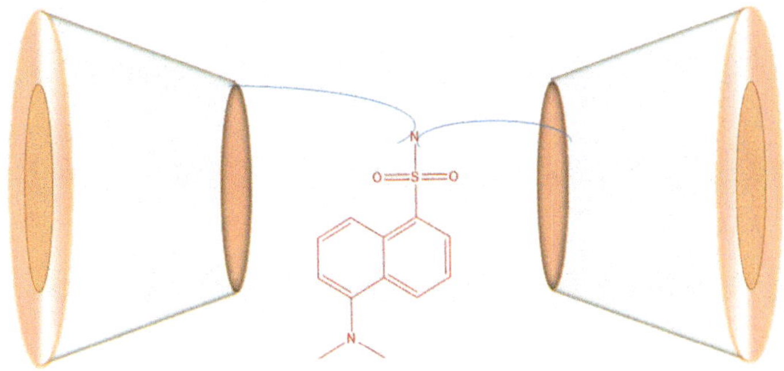

Fig. 15.3 Primary faces of two β-cyclodextrins molecules linked through the dansyl group

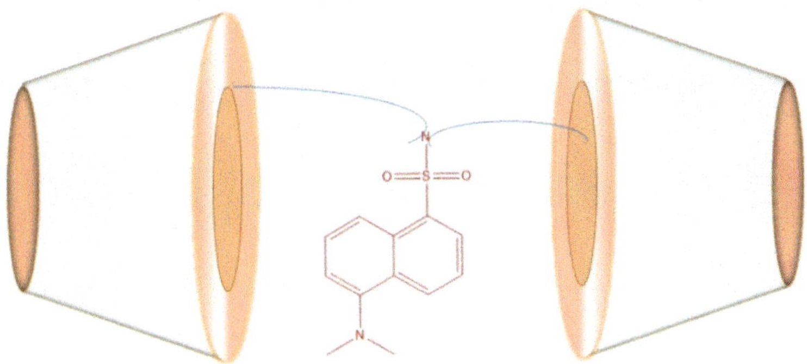

Fig. 15.4 Secondary faces of two β-cyclodextrins molecules linked through the dansyl group

of steroid molecules is by the formation of inclusion complex of steroids as guest molecules in hydrophobic cavities which leads to the exclusion of dansyl moiety space to aqueous hydrophilic from hydrophobic media resulting in fluorescence quenching effect of dansyl molecules [3, 20, 21].

Reinhoudt and the group conducted similar experiments but here they linked the secondary face of two β-cyclodextrins-dimmers through the dansyl group [22] (Fig. 15.4).

Kikuchi et al. worked on β-cyclodextrins linear trimer where dansyl moieties act as linkers and examined various bile acids by host–guest chemistry [23] (Fig. 15.5).

Sasaki et al. constructed a fluorescent cyclic trimer with permethylated β-cyclodextrins linked by biphenyl moieties. The experimental studies revealed fluorescence modulations for anthracene derivatives possessing two alkyls [23] (Fig. 15.6).

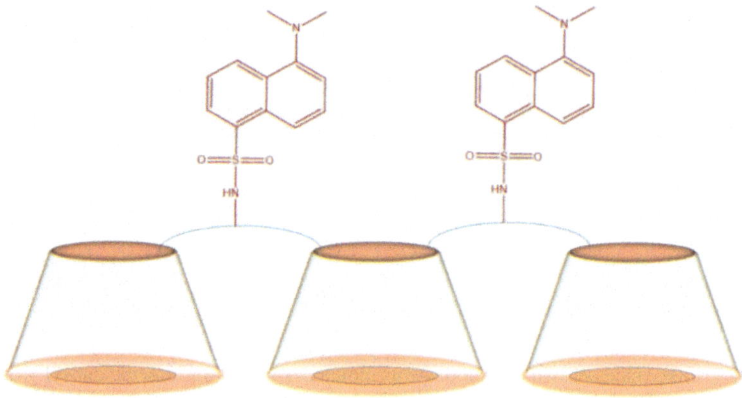

Fig. 15.5 Dansyl moieties linked β-cyclodextrins trimer

Fig. 15.6 Biphenyl linked β-cyclodextrins trimer

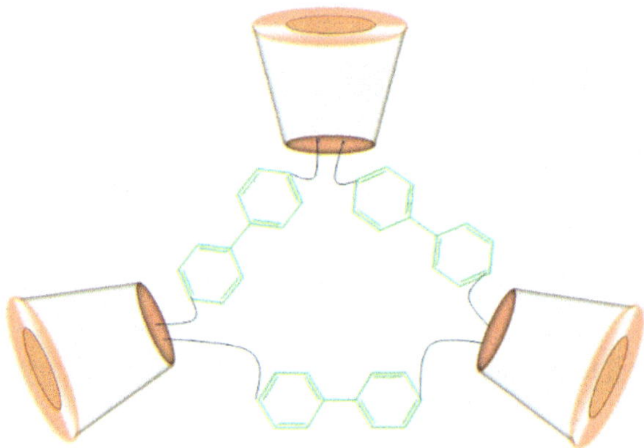

15.5 Cyclodextrins-Based Supramolecule with Pyrene

A molecular hybrid conjugate of 2:1 host–guest sandwiched complex of β-cyclodextrins and pyrene crown ether was developed by Hayashita et al. There is the disappearance of emission of pyrene monomer in the presence of K^+ ion leads to excimer emission [24–26] (Fig. 15.7).

Enhancement in fluorescence emission by attaching boronic acid fluorophore between cyclodextrins and pyrene has been noticed by Tong et al. in the fabrication of conjugate sugar sensing systems [27–29] (Fig. 15.8).

Suzuki et al. synthesized pyrene by using a triamine spacer to double threaded γ-cyclodextrins dimer which exhibited strong excimer fluorescence in aqueous solution [30, 31] (Fig. 15.9).

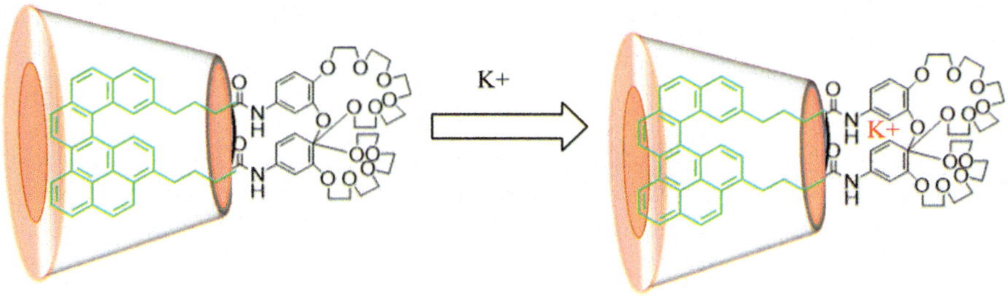

Fig. 15.7 β-Cyclodextrins and pyrene crown ether

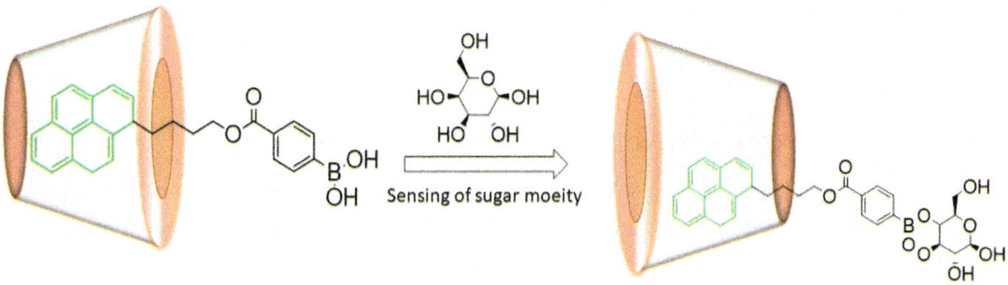

Fig. 15.8 β-Cyclodextrins in the presence of conjugate sugar sensing systems

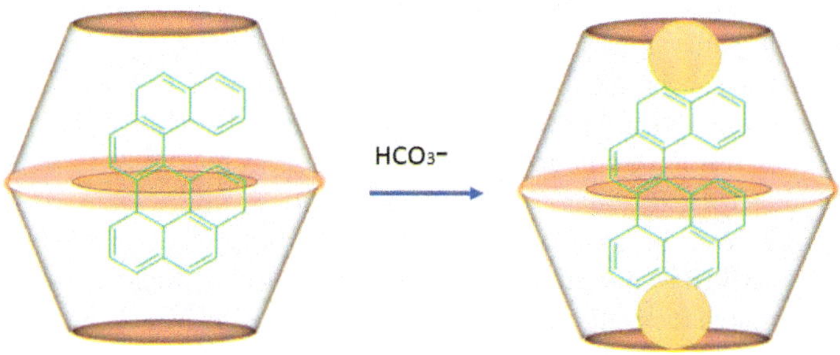

Fig. 15.9 Double threaded γ-cyclodextrins dimer

Anderson and coworkers fabricated rotaxane with stilbene (Fig. 15.10). The stilbene in the hydrophobic cavity of γ-cyclodextrins provide a hydrophobic environment for the guest molecule the stilbene acts as the fluorophore in the complex formation between rotaxane and the guest molecule [32, 33].

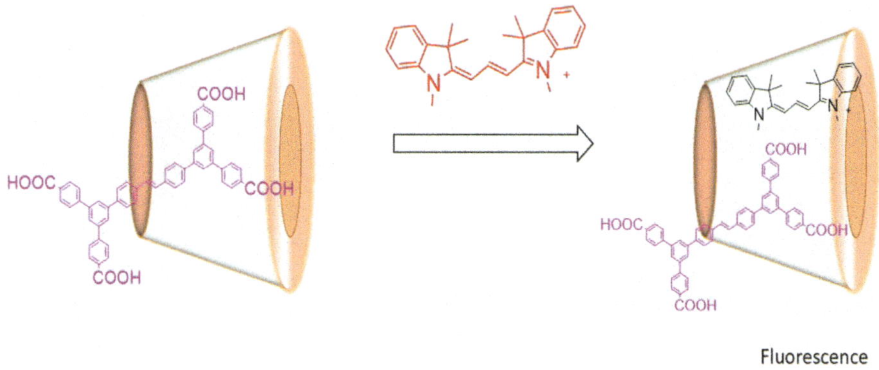

Fig. 15.10 γ-Cyclodextrins with rotaxane and stilbene

15.6 Cyclodextrin-Based Electrochemical Sensors

The field of cyclodextrin-enhanced chemical sensors, or biosensors, has seen significant expansion in research over the last ten years, as seen by the wide range of papers that have emerged in this field. When it comes to accurately and quantitatively identifying medications and biomolecules in complex samples, such as bodily fluids, and tablets, electrochemical sensors have shown to be quite successful. Because of their straightforward design, the sensors provide several benefits, such as repeatable and reversible readings that guarantee precise and accurate analytical data. Their extensive development can be attributed to their wide range of linearity, low detection limits, superior selectivity, and relative simplicity and affordability.

Because of their unique characteristics, cyclodextrins have been thoroughly investigated as models for molecular recognition, inclusion complex formation, and selective analyte pre-concentration at the electrode surface. They are becoming more widely acknowledged as modifiers for a range of electrochemical sensors, helping to improve membrane electrode selectivity in enantiomer discrimination and affecting the activity of organic electrode processes. Several investigations have shown that the peak currents of organic molecules are strongly influenced by the amount of β-cyclodextrins in the electrode [34–37].

The detection of biomolecules and environmentally dangerous contaminants has drawn a lot of interest to electrochemical sensing in recent years due to its high selectivity and sensitivity, fast analysis, and sophisticated and portable gear. In the last ten years, conducting polymers, carbon nanotubes (CNTs), graphene, and other cyclodextrins-based carbon nanomaterials have shown to be extremely successful in the field of electrochemical sensing. Because cyclodextrins-based conducting polymers can provide high stability, good catalytic activity, and electrical characteristics when attached to electrodes by the electrooxidation process of monomers, they are currently the focus of research attention for electrochemical sensing.

In this context [38], Luong et al. doped a composite film of polypyrrole and poly(*N*-acetyltyramine) with boron and sulfobutylether functionalized β-cyclodextrins to enable the selective electrochemical determination of the neurotransmitter dopamine as well as other interfering analytes such as uric acid and ascorbic acid [39–41].

Regarding cyclodextrins-based carbon nanomaterials, Huang's et al. used the electrochemical impedance approach to detect 3,30,4,40-tetrachlorobiphenyl by modifying a glassy carbon electrode (GCE) with pyrene-functionalized β-cyclodextrins and single-walled carbon nanotubes and [42–44] (Fig. 15.11).

Recognizing chiral enantiomers in the fields of medicine and pharmacy, using CD-based electrochemical sensors is of prime significance. Recently, Yang and colleagues altered the GCE surface using copper ions, multi-walled CNTs (MWCNTs), and hydroxypropyl β-cyclodextrins grafted cellulose to create a sensitive electrochemical sensor for identifying the chiral enantiomers of tryptophan. The electrochemical sensor that was created demonstrated a greater inclination towards L-Trp in contrast to D-Trp [45, 46]. It was also able to track the amount of D-Trp present in a racemic combination. These investigations open the door to the creation of practical chiral enantiomer for the discerning of numerous chiral molecules.

On the contrary, by pasting MWCNTs and graphene oxide on a screen-printed carbon electrode, a sensitive electrochemical sensor based on β-cyclodextrins was reported to be able to identify the endocrine-disrupting substance bisphenol A

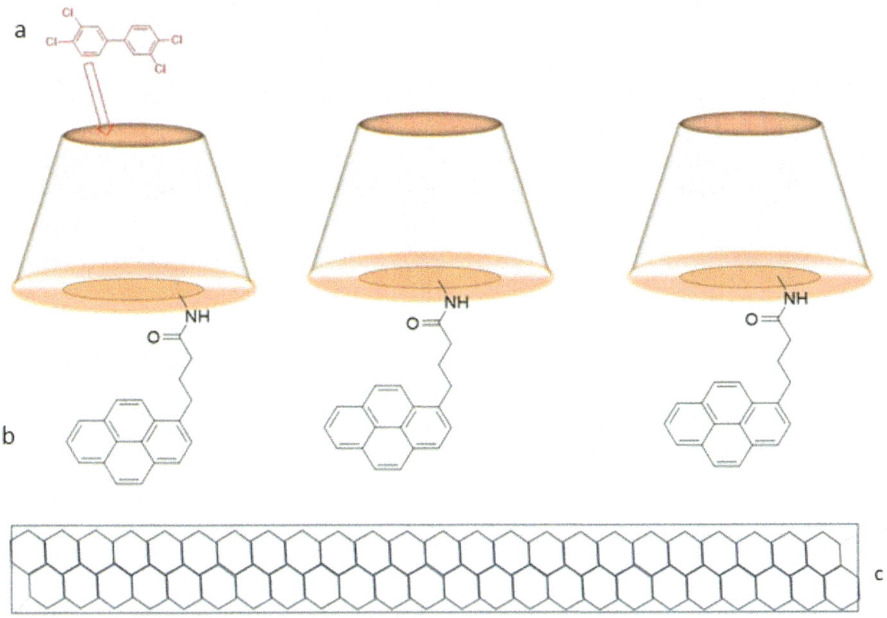

Fig. 15.11 **a** 3,30, 4,40-Tetrachlorobiphenyl, **b** guest molecule and the π–π-stacking adsorption of pyrene functionalized β-CD, **c** onto the side wall of SWCNT

in hydrosoluble solution. This versatile system employs a diffusion-controlled method to detect bisphenol A in drinking water with great sensitivity, with a detection minimum quantity of 6 nM. Bisphenol A monitoring has great potential for determining the quality of water [47, 48].

Many cyclodextrins-based potentiometric sensors that use electrochemical processes have been effectively created over the past few years, and they have found important uses in the pharmaceutical sciences, industries, environmental monitoring, medical, and agricultural fields. Lenik and associates created potentiometric sensors based on functionalized β-cyclodextrins in this context to identify medications such as ketoprofen and naproxen. It was concluded that the enclosure of naproxen guest molecule was partial or full within the cavity of functionalized β-cyclodextrins that act as host. Furthermore, Amorim et al. created potentiometric sensors for psychiatric medication compounds like midazolam and diazepam using functionalized β-cyclodextrins. Khaled's research team created carbon paste electrodes using polyvinyl chloride based on β-cyclodextrins to identify several derivatives of acetylcholine, such as butyrylcholine, acetylthiocholine, and acetylmethylcholine.

15.7 Cyclodextrin-Based Polymers as Chemical Sensors

Guest molecules play an important role in the preparation of polymeric networks by triggering a change in their structural properties and their ability to bind to cyclodextrins. Takashima et al. combined pAAm polymer with Azobenzene (AZOB) moieties as guest molecules in water with cyclodextrins-derivatized poly(allylamine). They studied the physical characteristics of the polymer in the presence of different isomers of AZOB [49]. The viscoelastic gel was obtained with pAAm-trans-AZOB with the cyclodextrins-polymer while when the same was bombarded with UV light (360 nm) led to the formation of pAAm-cis-AZOB with the cyclodextrins-polymer which provided polymer of low viscosity (Fig. 15.12).

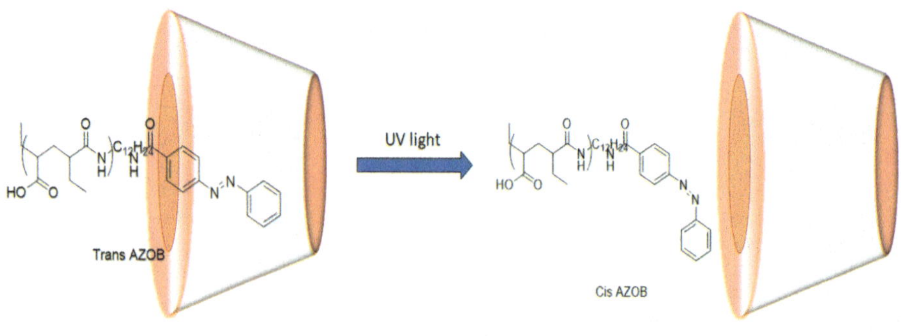

Fig. 15.12 pAAm polymer with azobenzene (AZOB) moieties as guest molecules in water with cyclodextrins-derivatized poly(allylamine)

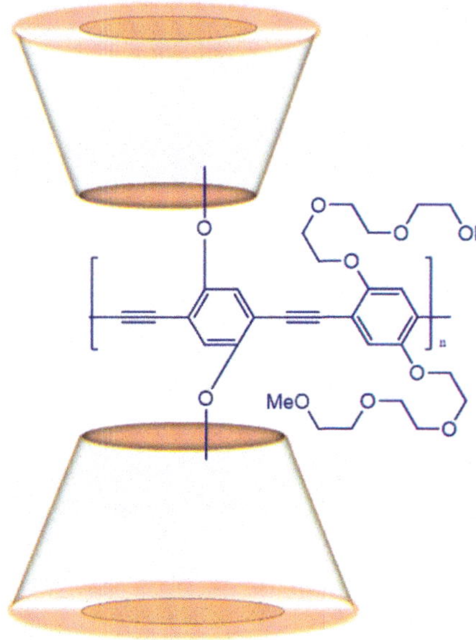

Fig. 15.13 Poly(phenylene ethynylene) (Polym-1) linked with two β-cyclodextrins moiety

Tomoki et al. synthesized poly(phenylene ethynylene) (Polym-1) linked in between two β-cyclodextrins moiety. Blue fluorescence of Polym-1 was observed in *N*,*N*-dimethylformamide while in aqueous media it showed green fluorescence. Polym-1 is not completely soluble in monomeric form in aqueous solution. Polym-1 can show a change in fluorescence or quench fluorescence by the addition of guest compounds aliphatic and acceptor guests, respectively [50, 51] (Fig. 15.13).

Yashima et al. synthesized helical polyacetylene-carrying cyclodextrins moieties (Polym-2). Depending on the inclusion complex formed by Polym-2 and guest molecules, visible color change was observed. It was observed that the chirality of the guest molecule also affected complexation. This was explained by using an example of (*S*)-1-phenylethylamine which shows a color change from yellow to red but (*R*)-1-phenylethylamine did not show any color change (Fig. 15.14).

Ueno and coworker synthesized a cyclodextrins-peptides conjugate, (Polym-3) possessing a pyrene donor and a coumarin acceptor in the side chains which forms a covalent bond. They observed that strong fluorescence is indicated by intramolecular fluorescence resonance energy transfer (FRET) at the absorption wavelength of pyrene to excitation caused by the coumarin group. In the presence of a guest molecule, the fluorescence from the coumarin moiety decreased resulting from the exclusion of the coumarin group from β-cyclodextrins cavity [52–54] (Fig. 15.15).

Various research groups have created metallocyclodextrin-based cyclodextrins, which are used in the chemical sensors sector. Accordingly, successful reports of ligands made up of metal binding sites have been made. Examples of such ligands include crown ether, diethylenetriaminepentaacetate (DTPA), and ethylenediaminetetraacetate (EDTA). As lanthanide metal ions (Eu^{2+} and Tb^{2+}) possess

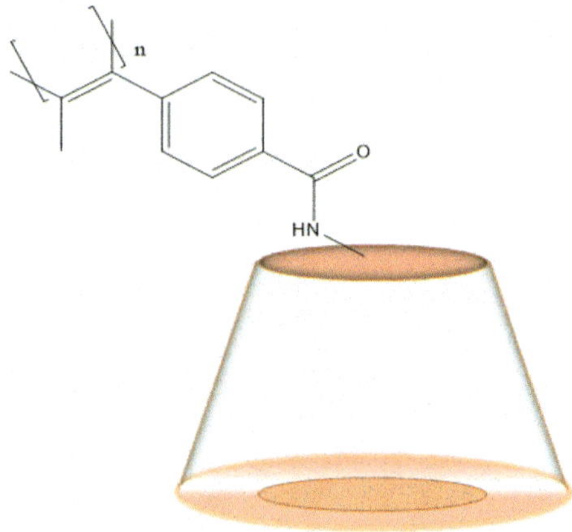

Fig. 15.14 Helical polyacetylene-embedded cyclodextrins moieties (Polym-2)

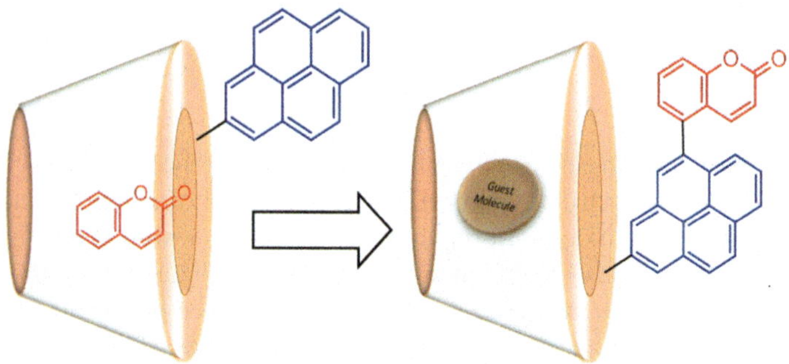

Fig. 15.15 Helical polyacetylene-embedded cyclodextrins moieties (Polym-2)

intense fluorescence and extended lives, they are often utilized in the employment of creating metallocyclodextrin-based chemical sensors among all metal ions Among the different sensing mechanisms, it has been determined that the one responsible for sensing metallocyclodextrins is absorption energy transfer emission (AETE). The first step in this sensing system is the absorption of photon energy, which excites the light-harvesting guest molecule. Next, the energy is transferred to a photoactive metal ion (Eu^{2+}/Tb^{2+}), and these metal ions emit light. Faint fluorescence in aqueous solution is observed due to the complexes of Eu^{2+} and Tb^{2+} ions with the attached CDs such as crown ether-cyclodextrins (Fig. 15.16a) and DTPA-CD (Fig. 15.16b) as they do not possess aromatic hydrocarbons serving as light-harvesting groups.

Toluene, biphenyl, and benzene, for example, are examples of aromatic guest/ light-harvesting molecules that can be added to the inner hydrophobic cavity of metallocyclodextrins. This allows the molecules to be included and exhibit fluorescence

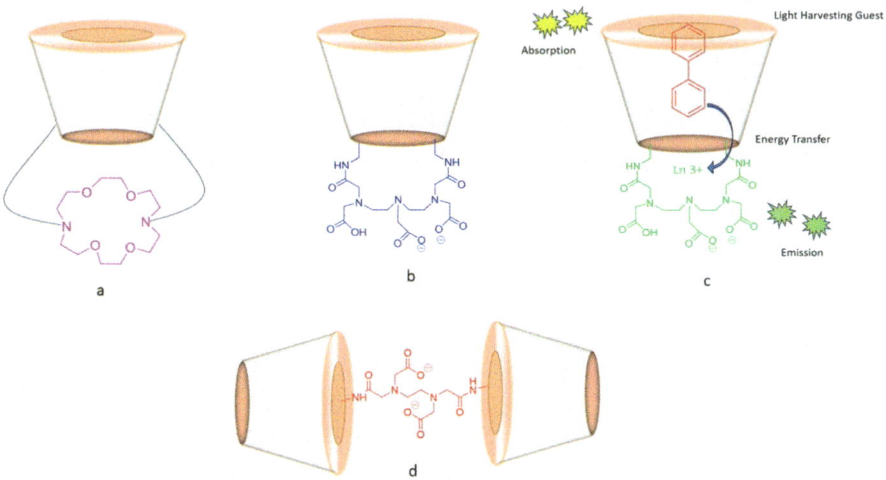

Fig. 15.16 **a** Crown ether appended CD, **b** DTPA appended CD, **c** AETE sensing phenomenon's mechanical, **d** EDTA-linked β-CD dimer

enhancement via AETE, providing a novel method for developing fluorescent chemical sensors that can be turned on (Fig. 15.16). However, two β-cyclodextrins units are connected by EDTA in Reinhoudt and colleagues' creation of β-cyclodextrins dimer (d). An experiment performed on the complex between (Fig. 15.16d) and lanthanide metal ions leads to adamantyl ends forming inclusion complex in the hydrophobic cavities of β-cyclodextrins dimer (Fig. 15.16d).

As a result, AETE has been seen from the lanthanide metal ion complexed with β-cyclodextrins dimer to the biphenyl group of adamantane dimer. The entire lanthanide metallocyclodextrin-based assembly works similarly to a fluorescent chemical sensor that can be turned on and off. Furthermore, via the AETE sensing phenomenon, polypyridine and hepta bipyridine-appended cyclodextrins, ramify as chemical sensors for specific guest molecules like metal ions of lanthanide [55].

The use of CDs with transition metal cation ligands attached as fluorescent chemical sensors has been investigated by Liu et al. β-CD dimer has been synthesized, where dimer of β-cyclodextrin units are connected by the biquinoline molecule. This group allows the dimer of β-cyclodextrin to form a complex with transition metal ion of the Cu(II) (Fig. 15.17a). An inclusion complex was formed upon the subsequent addition of the steroid guest molecule, exhibiting an increase in fluorescence and functioning as an effective fluorescent chemical sensor.

Additionally, a similar group has created a β-CD-based selective fluorescence sensor functionalized with quinoline for Zn(II) ion as well as various other interfering metal ions, including Mg(II) and Ca(II) (Fig. 15.17b). The sensor may be essential for Zn(II) imaging compounds in living organisms in the actual world. Yang and colleagues have reported on the development of a sensitive and selective β-CD-based fluorescence sensor (Fig. 15.17c) that can identify Zn(II) ions. Tetraphenylporphyrin

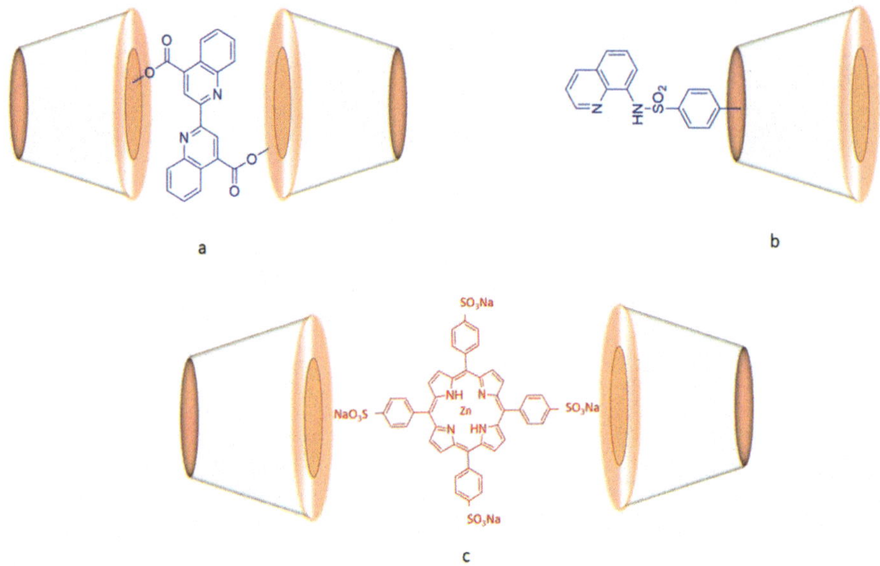

Fig. 15.17 **a** Biquinoline group linked β-CD dimer, **b** quinoline functionalized β-CD sensor, c tetraphenylporphyrin and Zn(II) ions

units and alkylated β-CD units make up the developed fluorescence sensor in a 2:1 stoichiometric ratio. Here, specific complex formation between Zn(II) and meso-tetraphenylporphyrin and other interfering metal ions in an aqueous media has shown fluorescence amplification.

15.8 Summary

The realm of chemical sensors utilizing cyclodextrins derivatives holds significant importance, despite its history spanning less than three decades. Chemical sensors utilizing cyclodextrins derivatives have emerged as a vital area of research in the last few decades, presenting promising opportunities for various applications. Traditionally, chemical sensors with chromophore-modified cyclodextrins were crafted. However, advancements in various disciplines including supramolecular chemistry, polymer chemistry, biochemistry, inorganic chemistry, and nanotechnology have prompted a shift toward developing sophisticated sensing systems. The unique host–guest chemistry of cyclodextrins makes them promising candidates for the development of chemosensors capable of selectively detecting small molecules. As research in this field progresses, further advancements in the design and application of cyclodextrin-based sensors are likely to contribute to improved analytical techniques and diagnostic tools. The current research emphasis lies in integrating ideas from these diverse fields to construct innovative chemical sensors. By combining cyclodextrins chemistry with insights from other disciplines, novel chemical sensors boasting high sensitivity, extensive versatility, and advanced functionality have been realized. Continued progress in these related research areas is expected to facilitate the creation of new cyclodextrins chemical sensors with enhanced attributes. The authors anticipate that this chapter will contribute novel perspectives to the field of supramolecular sensors in general, and specifically enhance the understanding of sensors based on cyclodextrins.

15.9 Further Reading

For an overview of the chemosensing of small molecules using cyclodextrins refer Fluorescent Cyclodextrins for Molecule Sensing: Fluorescent Properties, NMR Characterization, and Inclusion Phenomena of *N*-Dansylleucine-Modified Cyclodextrins.

References

1. Ikeda H, Nakamura M, Ise N et al (1996) Fluorescent cyclodextrins for molecule sensing: fluorescent properties, NMR characterization, and inclusion phenomena of N-dansylleucine-modified cyclodextrins. J Am Chem Soc 118:10980–10988. https://doi.org/10.1021/ja960183i
2. Sarabia-Vallejo Á, Caja MdM, Olives AI et al (2023) Cyclodextrin inclusion complexes for improved drug bioavailability and activity: synthetic and analytical aspects. Pharmaceutics 15:2345. https://doi.org/10.3390/pharmaceutics15092345
3. Xu W, Li X, Wang L et al (2021) Design of cyclodextrin-based functional systems for biomedical applications. Front Chem 9:635507. https://doi.org/10.3389/fchem.2021.635507
4. Zhou J, Jia J, He J et al (2022) Cyclodextrin inclusion complexes and their application in food safety analysis: recent developments and future prospects. Foods 11:3871. https://doi.org/10.3390/foods11233871
5. Ma X-K, Cheng Q, Zhou X, Liu Y (2023) Macrocycle γ-cyclodextrin confined polymeric chromophore ultralong phosphorescence energy transfer. JACS Au 3:2036–2043. https://doi.org/10.1021/jacsau.3c00255
6. Liu Y, Shi J, Guo D-S (2007) Novel permethylated β-cyclodextrin derivatives appended with chromophores as efficient fluorescent sensors for the molecular recognition of bile salts. J Org Chem 72:8227–8234. https://doi.org/10.1021/jo071131m
7. Garrido E, Pla L, Lozano-Torres B et al (2018) Chromogenic and fluorogenic probes for the detection of illicit drugs. ChemistryOpen 7:401–428. https://doi.org/10.1002/open.201800034
8. Mistri T, Alam R, Dolai M et al (2013) A 7-nitrobenz-2-oxa-1,3-diazole based highly sensitive and selective turn-on chemosensor for copper(II) ion with intracellular application without cytotoxicity. Org Biomol Chem 11:1563–1569. https://doi.org/10.1039/C3OB27405E
9. Ikeda H, Murayama T, Ueno A (2005) Skeleton-selective fluorescent chemosensor based on cyclodextrin bearing a 4-amino-7-nitrobenz-2-oxa-1,3-diazole moiety. Org Biomol Chem 3:4262–4267. https://doi.org/10.1039/b508477f
10. Wang J, Ueno A (2000) Naphthol-modified β-cyclodextrins as fluorescent sensors for detecting contaminants in drinking water. Macromol Rapid Commun 21:887–890. https://doi.org/10.1002/1521-3927(20000801)21:13%3c887::AID-MARC887%3e3.0.CO;2-U

11. Hamasaki K, Ikeda H, Nakamura A et al (1993) Fluorescent sensors of molecular recognition. Modified cyclodextrins capable of exhibiting guest-responsive twisted intramolecular charge transfer fluorescence. J Am Chem Soc 115:5035–5040. https://doi.org/10.1021/ja00065a012
12. Oliveri V, Sgarlata C, Vecchio G (2016) Cyclodextrins 3-functionalized with 8-hydroxyquinolines: copper-binding ability and inhibition of synuclein aggregation. Chem –Asian J 11:2436–2442. https://doi.org/10.1002/asia.201600824
13. Chu HM, Zhang RX, Huang Q et al (2017) Chemical conjugation with cyclodextrins as a versatile tool for drug delivery. J Incl Phenom Macrocycl Chem 89:29–38. https://doi.org/10.1007/s10847-017-0743-3
14. Samuelsen L, Holm R, Schönbeck C (2020) Cyclodextrin binding constants as a function of pH for compounds with multiple pKa values. Int J Pharm 585:119493. https://doi.org/10.1016/j.ijpharm.2020.119493
15. Szente L, Szemán J (2013) Cyclodextrins in analytical chemistry: host–guest type molecular recognition. Anal Chem 85:8024–8030. https://doi.org/10.1021/ac400639y
16. Thurein SM, Lertsuphotvanit N, Phaechamud T (2018) Physicochemical properties of β-cyclodextrin solutions and precipitates prepared from injectable vehicles. Asian J Pharm Sci 13:438–449. https://doi.org/10.1016/j.ajps.2018.02.002
17. Ponchel A, Monflier E (2023) Application of cyclodextrins as second-sphere coordination ligands for gold recovery. Nat Commun 14:1283. https://doi.org/10.1038/s41467-023-36700-z
18. Wüpper S, Lüersen K, Rimbach G (2021) Cyclodextrins, natural compounds, and plant bioactives—a nutritional perspective. Biomolecules 11:401. https://doi.org/10.3390/biom11030401
19. Jansook P, Ogawa N, Loftsson T (2018) Cyclodextrins: structure, physicochemical properties and pharmaceutical applications. Int J Pharm 535:272–284. https://doi.org/10.1016/j.ijpharm.2017.11.018
20. Hamasaki K, Usui S, Ikeda H et al (1997) Dansyl-modified cyclodextrins as fluorescent chemosensors for molecular recognition. Supramol Chem 8:125–135. https://doi.org/10.1080/10610279708233976
21. Xu C, Wu Y-L, Li Z, Loh XJ (2019) Cyclodextrin-based sustained gene release systems: a supramolecular solution towards clinical applications. Mater Chem Front 3:181–192. https://doi.org/10.1039/C8QM00570B
22. van Dienst E, Snellink BHM, von Piekartz I et al (1995) Selective functionalization and flexible coupling of cyclodextrins at the secondary hydroxyl face. J Org Chem 60:6537–6545. https://doi.org/10.1021/jo00125a045
23. Kikuchi T, Narita M, Hamada F (2001) Synthesis of bis dansyl-modified β-cyclodextrin liner trimer having multi-recognition sites and high hydrophobic environment. Tetrahedron 57:9317–9324. https://doi.org/10.1016/S0040-4020(01)00935-8
24. Bai H, Wang J, Phan CU et al (2021) Cyclodextrin-based host-guest complexes loaded with regorafenib for colorectal cancer treatment. Nat Commun 12:759. https://doi.org/10.1038/s41467-021-21071-0
25. Sayed M, Pal H (2021) An overview from simple host–guest systems to progressively complex supramolecular assemblies. Phys Chem Chem Phys 23:26085–26107. https://doi.org/10.1039/D1CP03556H
26. Roy MN, Ekka D, Saha S, Roy MC (2014) Host–guest inclusion complexes of α and β-cyclodextrins with α-amino acids. RSC Adv 4:42383–42390. https://doi.org/10.1039/C4RA07877B
27. Tong A-J, Yamauchi A, Hayashita T et al (2001) Boronic acid fluorophore/β-cyclodextrin complex sensors for selective sugar recognition in water. Anal Chem 73:1530–1536. https://doi.org/10.1021/ac001363k
28. Fang G, Wang H, Bian Z et al (2018) Recent development of boronic acid-based fluorescent sensors. RSC Adv 8:29400–29427. https://doi.org/10.1039/C8RA04503H
29. Wang R, Zhan D, Fang G et al (2022) A novel boronic acid-based fluorescence turn on sensor for specific detection of adenosine in urine. Dyes Pigm 197:109859. https://doi.org/10.1016/j.dyepig.2021.109859

30. Suzuki I, Ui M, Yamauchi A (2006) Supramolecular probe for bicarbonate exhibiting anomalous pyrene fluorescence in aqueous media. J Am Chem Soc 128:4498–4499. https://doi.org/10.1021/ja055772f
31. Wang Q, Zhang Q, Zhang Q-W et al (2020) Color-tunable single-fluorophore supramolecular system with assembly-encoded emission. Nat Commun 11:158. https://doi.org/10.1038/s41467-019-13994-6
32. Klotz EJF, Claridge TDW, Anderson HL (2006) Homo- and hetero-[3]rotaxanes with two π-systems clasped in a single macrocycle. J Am Chem Soc 128:15374–15375. https://doi.org/10.1021/ja0665139
33. Man JYH, Au-Yeung HY (2019) Synthesis of a [6]rotaxane with singly threaded γ-cyclodextrins as a single stereoisomer. Beilstein J Org Chem 15:1829–1837. https://doi.org/10.3762/bjoc.15.177
34. Dawson RE, Lincoln SF, Easton CJ (2008) The foundation of a light driven molecular muscle based on stilbene and α-cyclodextrin. Chem Commun 3980–3982. https://doi.org/10.1039/B809014A
35. Zhu G, Yi Y, Chen J (2016) Recent advances for cyclodextrin-based materials in electrochemical sensing. TrAC, Trends Anal Chem 80:232–241. https://doi.org/10.1016/j.trac.2016.03.022
36. Lenik J (2017) Cyclodextrins based electrochemical sensors for biomedical and pharmaceutical analysis. Curr Med Chem 24:2359–2391. https://doi.org/10.2174/0929867323666161213101407
37. Liu R, Li B, Li F et al (2022) A novel electrochemical sensor based on β-cyclodextrin functionalized carbon nanosheets@carbon nanotubes for sensitive detection of bactericide carbendazim in apple juice. Food Chem 384:132573. https://doi.org/10.1016/j.foodchem.2022.132573
38. Bouchta D, Izaoumen N, Zejli H et al (2005) A novel electrochemical synthesis of poly-3-methylthiophene-γ-cyclodextrin film: application for the analysis of chlorpromazine and some neurotransmitters. Biosens Bioelectron 20:2228–2235. https://doi.org/10.1016/j.bios.2004.12.004
39. Sravanthi M, Manjunatha KG (2021) Synthesis and characterization of conducting polypyrrole with various dopants. Mater Today: Proc 46:5964–5968. https://doi.org/10.1016/j.matpr.2020.11.762
40. Borges MHR, Nagay BE, Costa RC et al (2023) Recent advances of polypyrrole conducting polymer film for biomedical application: toward a viable platform for cell-microbial interactions. Adv Coll Interface Sci 314:102860. https://doi.org/10.1016/j.cis.2023.102860
41. Trung VQ, Hung HM, Van Khoe L et al (2022) Synthesis and characterization of polypyrrole film doped with both molybdate and salicylate and its application in the corrosion protection for low carbon steel. ACS Omega 7:19842–19852. https://doi.org/10.1021/acsomega.2c01561
42. Healy B, Yu T, C. da Silva Alves D et al (2021) Cyclodextrins as supramolecular recognition systems: applications in the fabrication of electrochemical sensors. Materials (Basel) 14:1668. https://doi.org/10.3390/ma14071668
43. Wang S-Y, Li L, Xiao Y, Wang Y (2019) Recent advances in cyclodextrins-based chiral-recognizing platforms. TrAC, Trends Anal Chem 121:115691. https://doi.org/10.1016/j.trac.2019.115691
44. Maduraiveeran G, Jin W (2021) Carbon nanomaterials: Synthesis, properties and applications in electrochemical sensors and energy conversion systems. Mater Sci Eng, B 272:115341. https://doi.org/10.1016/j.mseb.2021.115341
45. Qian J, Yi Y, Zhang D, Zhu G (2019) Electrochemical recognition of tryptophan enantiomers using a multi-walled carbon nanotube@polydopamine composite loaded with copper(II). Microchim Acta 186:358. https://doi.org/10.1007/s00604-019-3469-7
46. Mounesh RKRV (2021) Decorated CoPc with appliance of MWCNTs on GCE: sensitive and reliable electrochemical investigation of heavy metals. Microchem J 160:105610. https://doi.org/10.1016/j.microc.2020.105610
47. Monisha RS, Mani RL, Sivaprakash B et al (2023) Remediation and toxicity of endocrine disruptors: a review. Environ Chem Lett 21:1117–1139. https://doi.org/10.1007/s10311-022-01455-4

48. Gonsioroski A, Mourikes VE, Flaws JA (2020) Endocrine disruptors in water and their effects on the reproductive system. Int J Mol Sci 21:1929. https://doi.org/10.3390/ijms21061929
49. Przybyla MA, Yilmaz G, Remzi Becer C (2020) Natural cyclodextrins and their derivatives for polymer synthesis. Polym Chem 11:7582–7602. https://doi.org/10.1039/D0PY01464H
50. Zhu P, Li W, Zhang Y et al (2023) β-Cyclodextrin derived full-spectrum fluorescent carbon dots: the formation process investigation and biological applications. Chin Chem Lett 34:108239. https://doi.org/10.1016/j.cclet.2023.108239
51. Danquah MK, Wang S, Wang Q et al (2019) A porous β-cyclodextrin-based terpolymer fluorescence sensor for in situ trinitrophenol detection. RSC Adv 9:8073–8080. https://doi.org/10.1039/C8RA06192K
52. Ling Q, Cheng T, Tan S et al (2020) Fluorescence-resonance energy transfer (FRET) within the fluorescent metallacycles. Chin Chem Lett 31:2884–2890. https://doi.org/10.1016/j.cclet.2020.08.020
53. Hossain MA, Mihara H, Ueno A (2003) Novel peptides bearing pyrene and coumarin units with or without β-cyclodextrin in their side chains exhibit intramolecular fluorescence resonance energy transfer. J Am Chem Soc 125:11178–11179. https://doi.org/10.1021/ja036427y
54. Masuko M, Ohuchi S, Sode K et al (2000) Fluorescence resonance energy transfer from pyrene to perylene labels for nucleic acid hybridization assays under homogeneous solution conditions. Nucleic Acids Res 28:e34–e40. https://doi.org/10.1093/nar/28.8.e34
55. Rajbanshi B, Saha S, Das K et al (2018) Study to probe subsistence of host–guest inclusion complexes of α and β-cyclodextrins with biologically potent drugs for safety regulatory dischargement. Sci Rep 8:13031. https://doi.org/10.1038/s41598-018-31373-x

Sahaya Nadar is an Assistant Professor in the Department of Pharmaceutical Chemistry at St. John Institute of Pharmacy and Research [SJIPR] (Autonomous), Palghar, Maharashtra, India. She has over 12 years of experience in teaching and research. She has published several research and review papers. She has over 200 citations. She is also a Reviewer for esteemed journals under Taylor & Francis and Bentham Science. Her current research focuses on the computer-aided design of novel anticancer small molecules and the formulation of topical nanostructured lipid carriers (NLCs).

Ekta Thakor is currently working as an Assistant Professor, Department of Pharmaceutical Chemistry at St. John Institute of Pharmacy and Research, Palghar affiliated to University of Mumbai. She has 6 years of academic experience. She has pursued her M. Pharm from Dr. Bhanuben Nanavati College of Pharmacy, University of Mumbai in the year 2018. She is also pursuing her Ph.D. in Pharmaceutical Chemistry from the Institute of Pharmacy, Nirma University. Her recent research work focuses on drug designing as potent anti-tubercular drugs.

Jegan Nadar is an Assistant Professor in the Department of Pharmacology at St. John Institute of Pharmacy and Research [SJIPR] (Autonomous), Palghar, Maharashtra, India. He has over 10 years of experience in teaching and research. He has published several research and review papers. His current research focuses on Neuropharmacology and Neuropsychopharmacology.

Chapter 16
Cyclodextrins as Fluorescent Chemosensors for the Heavy Metal Ions and Volatile Organic Carbons

Priya

Abstract Heavy metals such as lead, chromium, mercury, and bismuth are being utilized in various processes such as industrial preparation, mining, batteries, being a part of vehicular exhaust, agricultural waste is polluting the environment. Hence, their detection and removal is a big concern in the research area. On the other hand, toxic volatile organic carbons (VOCs) also pose serious threats to human life as they are present in air. In this book chapter, we have discussed Cyclodextrins and the role of their host–guest chemistry in sensing toxic heavy metal ions and VOCs. Cyclodextrins are macrocyclic molecules having a unique structure comprising of glucopyranose molecules attached through alpha-1, 4 glycosidic linkages in a cyclic or ring form resulting in a helical shape. Cyclodextrin-based rigid ring or cavity can act as a host for the analyte and it exhibits host–guest chemistry through a coordination mechanism. They are highly efficient, biocompatible, water soluble, cost effective, and easily available. Metal ions can form chelates or complexes with the suitable functional groups present inside the moiety. As a result, the metal ion is encapsulated inside and there is an increase (Turn on chemosensor) or decrease (Turn off chemosensor) in the luminescence properties of the host cyclodextrin moieties. Similarly, VOCs can also interact with the hydrophobic cavity of cyclodextrins. It forms the basis of sensing and is studied through spectrofluorometric systems. The main feature of fluorescence sensing is selectivity w.r.t. different analytes which plays a pivotal role. Sensitivity towards the specific analyte, mechanism of interaction, and medium of coordination are the other key features of the sensing mechanism. Sensing can be observed with or without a blue/red shift. Nowadays, researchers are also working on modified cyclodextrins which is also a part of the given chapter. Hence, it's a vast area of research and currently in demand.

Priya (✉)
School of Chemical Engineering and Physical Sciences, Lovely Professional University, Phagwara, India
e-mail: priya.28594@lpu.co.in; priyabhalla90@gmail.com

S. Krishnamoorthy (ed.), *Cyclodextrins for Chemosensing*,
https://doi.org/10.1007/978-3-031-89299-8_16

Graphical Abstract

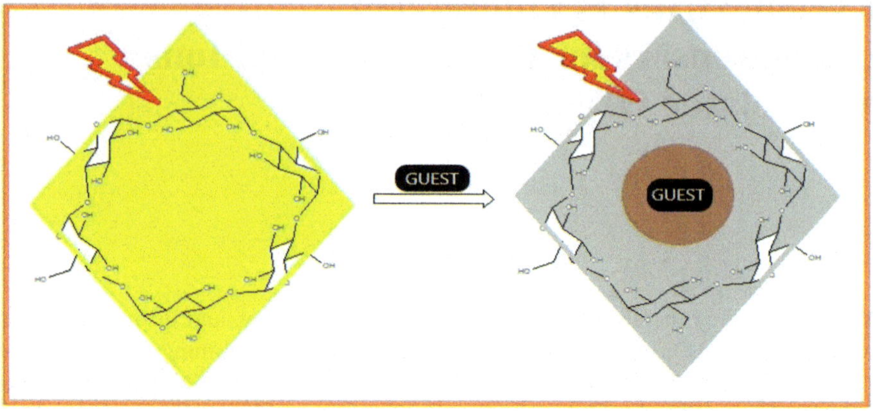

Keywords Cyclodextrins · Host–guest chemistry · Chemosensors · Heavy metal ions · VOCs · Selectivity · Sensing

16.1 Introduction

16.2 Toxic Heavy Metal Ions and Volatile Organic Carbons (VOCs)

As per the World Health Organization (WHO), most of the transition and heavy metal ions have certain permissible limits (Table 16.1) and if their concentration increases above the safety limit, they become toxic. The concentration limit may vary depending upon the type of metal ions, sample type, environment, or whether we are talking about local or international standards. The limits may be revised over time as the conditions and toxicity of the metal ions change. One must refer to a suitable standard or regulatory body [1]. For a few of the metal ions, the permissible limits are given below.

Volatile organic carbons such as methanol, ethanol, toluene, xylene, and acetone are one of the major air pollutants we have these days. These are very commonly used in cosmetics, varnishes, room fresheners, cleaning materials, and printers and are easily evaporated at room temperature leading to a toxic environment. For example, acetone poses several health issues to humans such as breathing problems, irritation in the nose or inflammation. Hence, their detection is of prime importance to us [4].

Table 16.1 Permissible limits of some metal ions in sample water and their health effects

Heavy metal ion	Permissible limits in water (mg/L)	Health effects	Refs.
Pb	0.1	Central nervous system and kidneys	[2]
Cr	0.05	DNA defects and asthma	[2]
Hg	0.002	Brain and respiratory issues	[2]
Bi	0.0007	Skin-related issues and depression	[2]
Co	0.05	Vision loss, nausea, heart issues, and thyroid	[2]
Zn	5	Gastrointestinal disorders, kidney and liver-related issues	[2]
Fe	0.03	Lung tumor	[2]
Ag	0.05	Brain damage, kidney damage. Eyes, lungs, and liver-related issues	[3]
Ni	0.2	Lung and nose cancer, heart problems	[3]
As	0.05	Cancer, brain, and heart-related issues	[2]
Sb	0.02	Irritation to eyes and skin	[3]
Mn	0.1	Memory loss, attention problem, and motor skills are not properly developed	[3]
Cd	0.01	Neural damage	[2]

16.3 Parameters of Fluorescent Chemosensors

Any fluorescent chemosensor is evaluated on the basis of its performance in terms of Limit of Detection (LOD) and Limit of Quantification (LOQ). They give an idea about the lowest concentration of the analyte which can sensed and quantified by the given chemosensor. Every chemosensor has its own values of LOD and LOQ. The lower the values, the more precise and accurate will be the detection and quantification. LOD is the lowest concentration of analyte which can be detected and differentiated from the background or blank signal [1]. It is calculated using the formula given in Eq. (16.1).

$$\mathrm{LOD} = \frac{3.3 \times \sigma}{S} \tag{16.1}$$

σ and S are the standard deviation and slope of the calibration curve, respectively.

LOQ is the next step to LOD and is a parameter to evaluate precision factors at very low concentrations. It is calculated as in Eq. (16.2):

Table 16.2 LOD and LOQ values of some fluorescent metal ion detectors

Fluorescent chemosensors	Heavy metal ion	LOD	LOQ	Refs.
1-Hydroxy-2-naphthoic acid: β-cyclodextrin	Chromium	30 μM	90 μM	[5]
Curcumin: β-cyclodextrin	Mercury	5.02 μM	16.73 μM	[6]
Benzoguanamine: β-cyclodextrin	Cerium	671	2034	[7]
Cyclodextrin functionalized photoluminescent metal nanoclusters	Iron	1.33×10^{-7} M	4.55×10^{-7} M	[8]

$$\text{LOQ} = \frac{10 \times \sigma}{S} \tag{16.2}$$

LOD and LOQ values of some cyclodextrin-based fluorescent heavy metal ion detectors have been given below (Table 16.2).

16.4 Properties of Cyclodextrins as Fluorescent Chemosensors

Sensitivity

It is a very important feature of chemosensors. The better the sensitivity towards the analyte, the better the chemosensor. Due to the rigid structure imposed by the cyclodextrins, we observe enhanced emission in terms of the intensity (hyperchromic or hypochromic shift) with or without wavelength shift as a result of ordered structural behavior. In addition, there is a change in the environment of the analyte as it is shifted from a polar aqueous environment to a slightly less polar cavity which makes its response better [9].

Selectivity

It means the shape and size of the analyte is important. Every moiety acts as an enzyme that can only take up a unique entity in its moiety. Cyclodextrin-based inclusion complexes further provide rigidity that increases the chance of selectivity towards a unique analyte.

Medium

The type of medium (polar or organic) and pH of the medium play an important role in deciding the sensing behavior. Most of the chemosensors prefer an aqueous neutral medium with pH 7 around to show their best emission results. Few of them show their best results in acidic or basic conditions and organic solvents [10].

16.5 Sensing Behaviors

Spectrophotometric (for colored changes) and spectrofluorometric (for luminescence-based changes) techniques have been used so far for the detection of heavy metal ions and VOCs using cyclodextrins.

Change in Fluorescence Intensity Without Shift

Increase in Fluorescence Intensity (Hyperchromic Shift)

Various inclusion complexes of cyclodextrin have been observed in which on analyte complexation an increase in the fluorescence intensity is observed. It is because of the rigidity offered by the cyclodextrins which does not allow nonradiative decay as a consequence of binding or chelation towards the given analyte (Fig. 16.1).

Decrease in Fluorescence Intensity (Hypochromic Shift)

As a result of the inclusion complex formed, cyclodextrins offers rigidity to the molecule inside undergoing complexation with the analyte that enables charge transfer between the donor site and analyte. As a result, emission quenching is favored (Fig. 16.1).

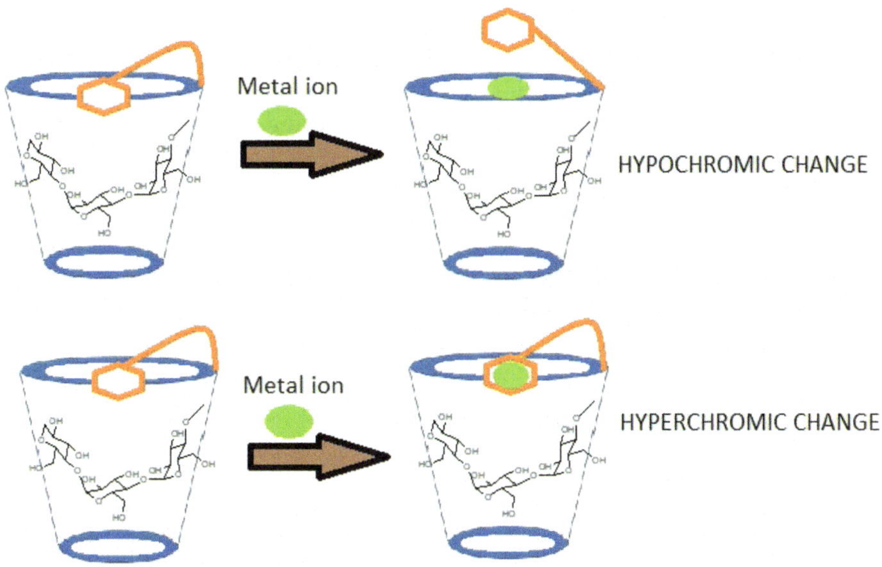

Fig. 16.1 Hypochromic and hyperchromic shifts observed as fluorophore unit is excluded and included inside cyclodextrin

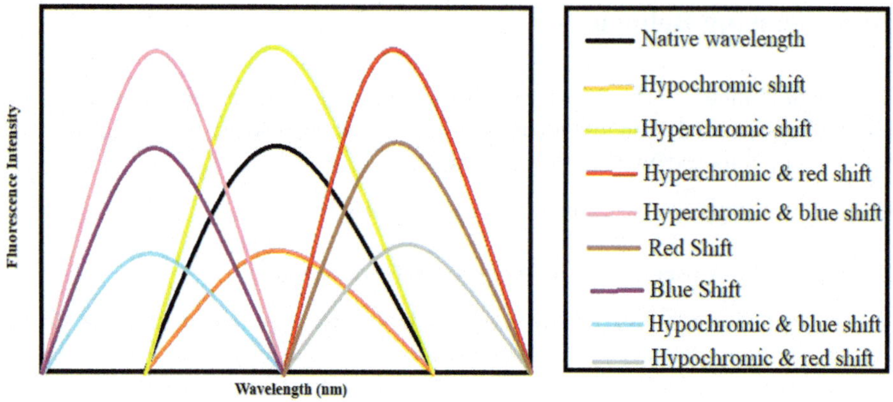

Fig. 16.2 Possible sensing behaviors of fluorescent chemosensors w.r.t. analytes

Change in Intensity with Wavelength Shift

In some of the inclusion complexes, it has been observed that with change intensity there is a shift in wavelength too. It can be a red shift (increase in wavelength) or blue shift (decrease in wavelength).

No Change in Intensity with Wavelength Shift

In some of the inclusion complexes, no change in intensity was observed. Instead, a shift in wavelength (red or blue shift) has been observed.

Various mechanisms involved in fluorescent sensing of analytes are given in (Fig. 16.2).

16.6 Sensing Mechanisms

Various sensing mechanisms are followed as a result of the interaction of the analyte with the chemosensor which are given below.

Photo-Induced Electron Transfer (PET)

In the presence of photons, electron transfer occurs from the donor site to the fluorophore present in the molecule. Once this donor site is shielded by an analyte-forming complex, no electron transfer occurs. Hence, fluorescence intensity is enhanced. In the presence of a polar solvent, electron transfer is easier and leads to fluorescence quenching of the fluorophore unit [11] (Fig. 16.3a).

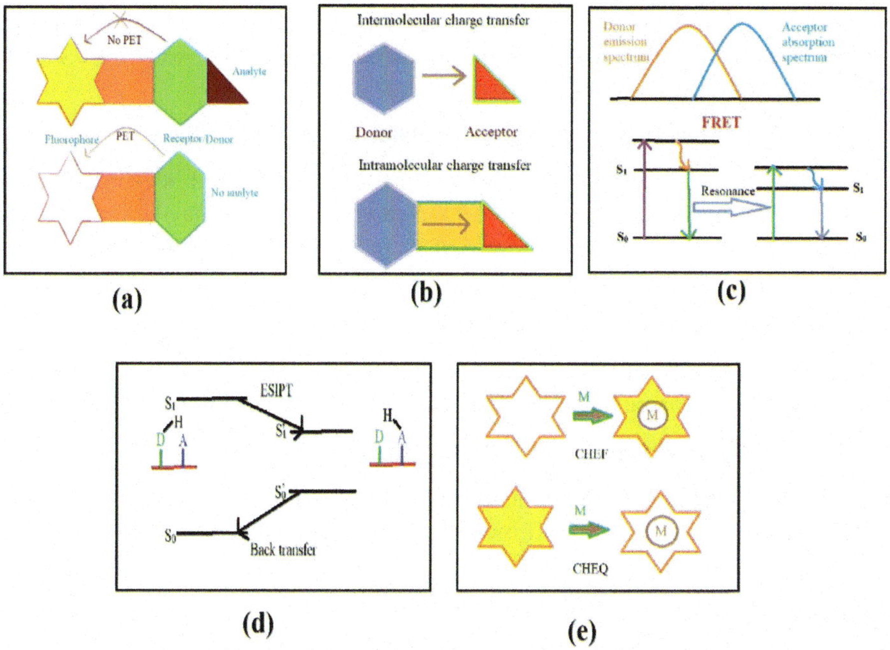

Fig. 16.3 Sensing mechanisms. **a** Photo-induced electron transfer, **b** intermolecular and intramolecular charge transfer, **c** fluorescence resonance energy transfer, **d** excited state intermolecular proton transfer, **e** chelation enhanced effects

Photo-Induced Charge Transfer (PCT)

Charge transfer occurs between donor and acceptor as per this mechanism in the presence of light. In conjugated systems, the charge transfer is partial. But in donor–acceptor systems, there is a full charge transfer. The parameters or factors affecting the charge transfer are the nature of donor binding with analyte, complexation type, nature of analyte, and fluorophore [12].

Intramolecular and Intermolecular Charge Transfer (ICT)

The charge transfer may occur from the donor to the acceptor site in different molecules called intermolecular charge transfer. It may occur from the donor site to fluorophore within the same molecule and it creates a dipole moment as a result of charge rearrangement. The charge transfer is accompanied by red shift for wavelength. As the analyte binds with the donor site, the charge transfer is prohibited. It is represented by a blue shift [13] (Fig. 16.3b).

Fluorescence Resonance Energy Transfer (FRET)

The donor in its excited state can transfer its energy to the acceptor in its ground state. The basic conditions to be followed are the closeness of the donor and acceptor sites, overlapping of the emission spectrum of donor and absorption spectrum of acceptor and parallel orientation of both sites. This process is nonradiative in nature. The acceptor should absorb at the emission wavelength of donor which will lead to resonance phenomenon [14] (Fig. 16.3c).

Excited State Intermolecular Proton Transfer (ESIPT)

In this mechanism, a proton is transferred between donor and acceptor through intramolecular hydrogen bonding. In certain cases, the groups may be far apart and hydrogen transfer occurs through solvent or medium. Net exchange is of proton (Fig. 16.3d).

Chelation Enhanced Fluorescence Effect (CHEF)

As the analyte binds with the donor site and forms a chelate or complex, it will enhance the fluorescence intensity of fluorophore-containing donor site (Fig. 16.3e).

Chelation Enhanced Quenching Effect (CHEQ)

As a result of chelation between analyte and donor site of fluorophore, its fluorescence effect may get quenched. This effect is called Chelation Enhanced Quenching (Fig. 16.3e).

Ligand to Metal Charge Transfer

If ligand-based molecular orbitals are completely filled and metal contains empty or partially filled molecular orbitals, it leads to charge transfer from ligand to metal. They form very intense bands.

Excimer Formation

Excimer is a complex formed between excited and ground states of two fluorophores of the same category. It leads to luminescence. However, if the donor site is chelating with analyte, the complex formation is hindered. As a result, fluorescence is quenched.

16.7 Cyclodextrins as Fluorescent Chemosensors

Cyclodextrins are quite efficient chemosensors or hosts because of the rigidity they provide to guest molecules through various dynamic interactions such as H-bonding, hydrophobic interactions or van der Waals interactions. The guest molecule can be an organic molecule containing functional groups such as aldehyde, ketone, amine, hydroxyl, or carboxylic acid. The inclusion complexes so formed have wide applications in the fields of supramolecular chemistry, materials chemistry, biochemistry, electronics, polymer chemistry, nanoscience, catalysis, drug delivery, designing cosmetics, environmental field, food technology, biotechnology, textile industries, and pharmaceutical sector. The compound encapsulated inside the cavity gets a rigid fit and that increases its selectivity and sensitivity. They are different from the other chemosensors in terms of their better selectivity/recognition of the analyte, extraordinary biocompatibility, non-toxic nature, non-carcinogenic, aqueous solubility, cost-effectiveness, and availability and scope for derivatization. Cyclodextrins being a combination of hydrophobic and hydrophilic moieties prefer less polar molecules compared to water. The physical properties of guest molecules such as stability, solubility, bioavailability, volatility, sublimation, and rate of dissolution change as they form inclusion complexes with cyclodextrins. Zhang et al. have prepared a fluorescent probe based on beta-cyclodextrin (4-[4-(acetylamino)styryl]-1-cyclodextrin pyridinium *p*-toluenesulfonate) for the detection of toxic organic pollutant 4-nitro phenol. It was used as a "Turn off type" sensor and LOD was 0.01 μM [15]. Silva et al. have synthesized a composite from poly(α-cyclodextrin) and carbon dots in which carbon dots were embedded within polymer poly[(α-cyclodextrin)-v-(citric acid)]. It has been used in the removal and detection of volatile organic carbons such as toluene and xylene [16].

16.8 Fluorescence-Based Heavy Metal Ion Sensing Using Cyclodextrins

Water is a universal solvent and can take up almost all useful compounds and pollutants existing naturally or toxic effluents coming from various sources such as industries, mining, agricultural areas, and domestic level. It is an essential part of our body and is required for the survival of humans, flora, and fauna. It is a medium of existence for the aquatic animals and plants. It covers 70.9% of the area occupied by the earth out of which only 2.5% comes in the freshwater category. The percentage of quality fresh water for drinking is further very low. A lot of people are facing health issues in rural and urban areas because of the poor quality of water supplied to them. The pollutants present may include inorganic compounds, organic compounds, or microorganisms [17]. The sources of pollutants are various industries such as leather, electrochemicals, electronics, automobiles, construction, military, laboratories, surgical equipment, dairy, food, pulp manufacturing, textiles, agricultural fields, mining, and nuclear plants. Some of the effluent chemicals are degraded over time, the rest stay in their original form for years or are degraded very slowly. Heavy metal ions such as lead, mercury, cadmium, arsenic, cobalt, nickel, chromium, zinc, manganese, palladium, antimony fall in that category, and their degradation rate is very slow compared to their accumulation rate in our bodies. Hence, people are facing a lot of health issues including short-term and long-term impacts [18]. Vomiting, tumors, depression, diarrhea, pneumonia, gastrointestinal issues, stomatitis, paralysis, lung issues, bone defects and may even lead to death. Hence, it becomes important to detect these toxic analytes if they are present in a given water sample so that they can be removed further and water can be purified and become safe for drinking [19].

Selective determination of metal ions has become a concern of the researchers. Every metal ion has its required concentration range above which it becomes toxic. If that level is crossed, we need to take action for their removal. The range may vary as per the origin and various regulatory bodies dealing with the concern are WHO, EPA, and USEPA. Various methods are being used for the quantification and removal of the metal ions such as Atomic Absorption Spectroscopy, inductively coupled plasma emission spectroscopy, coagulation, Ion exchangers, precipitation, atomic emission spectroscopy, electrochemical methods, reverse osmosis, separation and voltammetry. Fluorescence-based detection is better than these methods as it is cost-effective, efficient, quick methods, easier to practice, more selective, and sensitive. It does not require any other reference materials. Fluorophore concentration by observing the changes in its fluorescence intensity, wavelength, or lifetime of the fluorophore as a result of its interaction with the guest molecule. The main points deciding the changes are the structure of the receptor site at fluorophore, type of interaction with analyte, binding affinity of analyte, selectivity in the presence of other analytes, lowest concentration of the ions which can be detected, medium of interaction, nature of the solvent, pH and ionic strength. There can be an increase in the intensity of fluorescence or a decrease in the intensity of fluorescence as a result of the interaction. A shift in wavelength (either blue or red shift) can also be observed [20]. Narayanan et al. performed spectral analysis of inclusion complexes prepared from syringaldehyde azine and β-cyclodextrin and used it for further detection of copper, lead, and aluminum ions. The mechanism of detection was Chelation Enhanced Fluorescence mechanism and Ligand to Metal Charge Transfer [21]. Almahri et al. have prepared β-Cyclodextrin implanted lipophobic polymer dots which were found to be antimicrobial and anticancer laborers. They have been also utilized in the estimation of silver ions [22]. In the research article of Ganesan et al., 4, 4′-((1E,1′E)-hydrazine-1,2-diylidenebis(methanylylidene))diphenol and γ-Cyclodextrin based chemosensor has been utilized for the detection of aluminum ions. HOMO–LUMO band gap value and DFT analysis have confirmed the formation of an inclusion complex [23].

Cyclodextrins have multiple –OH groups which can be substituted with other groups to enhance their selectivity and sensitivity towards the analyte. Otherwise, the primary hydroxyl groups are more reactive as compared to the secondary hydroxyl groups. It can be mono, di, tri, or multiple times substituted by simple organic reactions [24]. Cyclodextrins modified with ligand sites such as Ethylene diaminetetraacetate, diethylenetriaminepentaacetate, crown ethers, etc. can also act as fluorescent sensors for metal ions. It involves the absorption energy transfer-based emission in which energy is transferred from a light-emitting guest molecule such as an aromatic moiety (biphenyl, toluene) to the metal ion such as lanthanide ions with longer lifetime that results in luminescence [25] (Fig. 16.4).

Supramolecular systems designed from cyclodextrins also help in metal ion sensing in an effective way. They can either form dimers/trimers or can be conjugated with other species such as crown ethers forming hybrid structures [26].

Cyclodextrin-based chemosensors reported in the literature have been given (Table 16.3).

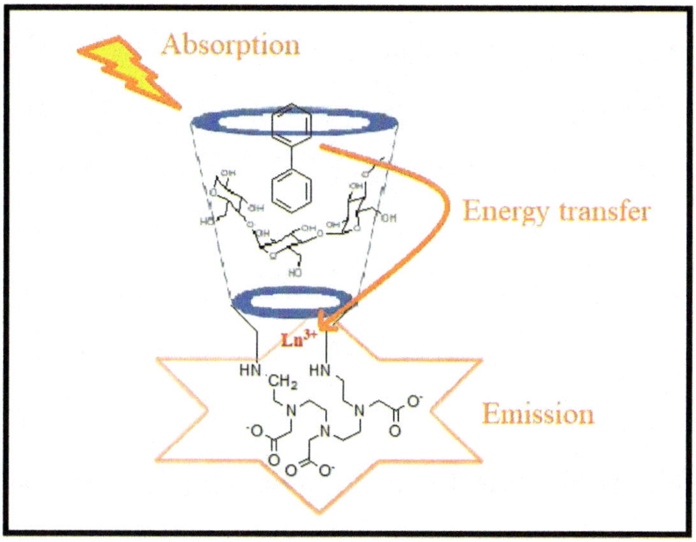

Fig. 16.4 Modified cyclodextrin with light emitting guest molecule showing energy transfer mechanism

Table 16.3 Cyclodextrin-based chemosensors reported in the literature

Cyclodextrin-based chemosensor	Heavy metal ion sensed	Change observed	Mechanism	Refs.
β-Cyclodextrin-driven inclusion complexation with phenylalanine-capped AuNPs	Cr(III) and Fe(III)	Redshift and color change	Coordination complex	[27]
β-Cyclodextrin (β-CD) and thiosemicarbazide	Cu^{2+}	Turn off sensor with color change	Fluorescence quenching	[28]
β-Cyclodextrin (β-CD) based probes containing naphthalimide group	Hg^{2+}/Hg^{+}	Bright yellow fluorescence emitted	Micelle formed as a result of complexation	[29]
Amidoxime-functionalized non-porous β-cyclodextrin polymer	U(VI)	Turn off type	Photo-induced electron transfer	[30]
CdTe@γ-cyclodextrin@rhodamine derivative	Fe^{3+}	Quenching	Fluorescence resonance energy transfer	[31]
Carboxymethyl chitosan crosslinked β-cyclodextrin containing hydrogen bonded NC QDs	Mn^{2+}	Quenching	Chelation and aggregation	[32]
β-Cyclodextrin-enhanced Eu^{3+} luminescence aggregates	Fe^{3+}	Quenching	Metal ion-ligand chelation	[32]

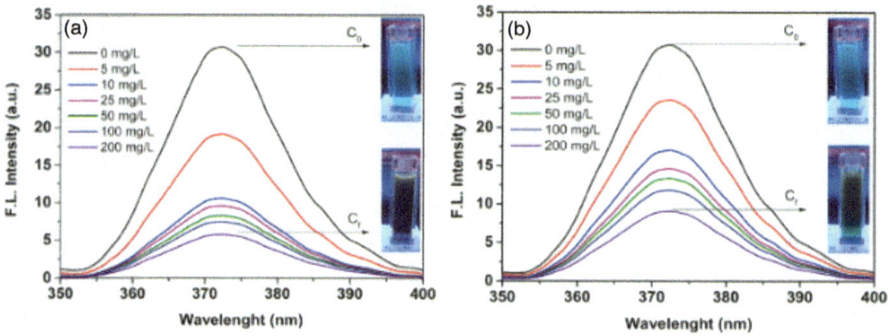

Fig. 16.5 Effect of variable concentrations of **a** toluene and **b** xylene on fluorescence spectra of poly(α-cyclodextrin)/carbon quantum dots [16]

16.9 Cyclodextrins for Fluorescent Sensing of Toxic Gases

In addition, the cavity of Cyclodextrins can also act as a molecular recognition site for analytes such as Volatile Organic Carbons (VOCs). They have limited selectivity but can be modified as per the requirement. They exhibit physical interactions with the analyte which is reversible in nature. Becuwe et al. have prepared Indolizone-β-cyclodextrin based fluorescent derivatives for the detection of 1-Adamantol, benzene, toluene, phenol, and *p*-cresol [33]. Liu et al. have synthesized perylene bisimide-bridged-bis-(permethyl-β-cyclodextrin) which acts as a fluorescent detector in the solid state and helps in the detection of organic amines as well as nitro-based compounds [34]. Hu et al. have prepared β-cyclodextrin functionalized graphene-based fluorescent sensor for the detection of tetrahydrofuran [35]. Silva et al. have synthesized a fluorescent composite probe based on poly(α-cyclodextrin)/carbon quantum dots which was further used for the removal and detection of toluene and xylene [16] (Fig. 16.5).

16.10 Role of Computational Analysis

Computational analysis is a technique by which we can determine the sensing behavior of a chemosensor towards the analyte at a given external. The various analyses involved are Density Functional Theory (DFT), Molecular Dynamics (MD), and Monte Carlo (MC). Without the use of computational studies, it becomes difficult to pick and choose a metal ion that will generate a response from the newly prepared chemosensor. While performing computational analysis, we need to define the application for which the analysis is to be carried out in terms of the type of analyte and conditions of its interaction with the chemosensor. It also helps in optimizing various parameters. The reported research carried out on the computational studies of Cyclodextrins has been given (Table 16.4).

Table 16.4 Literature review of computational analysis

Computational analysis performed	Sample	Metal ion detected	Refs.
Density functional theory (DFT) and molecular dynamics (MD)	3,3′-Dihydroxybenzidine: α-cyclodextrin	Hg^{2+} and Fe^{3+}	[36]
DFT	Antipyrine derivative in aqueous and β-cyclodextrin media	Vanadyl ion	[37]
DFT	Imidazolylpiperazine derivative and β-cyclodextrin	Al^{3+}	[38]
DFT	*N*-Phenyl-*p*-phenylenediamine into β-CD	Fe^{2+}	[39]
DFT	Beta-cyclodextrin inclusion complex with benzoguanamine	Ce^{4+}	[7]

16.11 Conclusion

In this book chapter, details about cyclodextrins as Fluorescent chemosensors and their application in sensing of heavy metal ions and volatile organic carbons has been given. The importance of metal ion and VOCs sensing has also been discussed. Permissible limits of heavy metal ions and parameters deciding the quality of fluorescent chemosensors have also been reported. Factors deciding the properties of chemosensors have also been part of the discussion. Sensing behavior and corresponding mechanisms such as PET, PCT, ICT, FRET, CHEF, CHEQ, excimer formation and ligand-to-metal charge transfer have been explained. Modified forms of cyclodextrins can also be utilized in fluorescence sensing of different analytes. Computational analysis such as DFT, MD, and MC which is currently in demand, and its role in studying the fluorescence mechanism has also been elaborated.

References

1. Khan SA, Alam MZ, Mohasin M, Ahmad S, Salma U, Parveen H, Mukhtar S, Al-Anazi M, Alotaibi FA, Abdelaziz MA (2023) Ultrasound-assisted synthesis of chalcone: a highly sensitive and selective fluorescent chemosensor for the detection of Fe^{3+} in aqueous media. J Fluoresc 34:723–728. https://doi.org/10.1007/s10895-023-03317-w
2. Jyothi NR (2021) Heavy metal sources and their effects on human health. Intech Open eBooks. https://doi.org/10.5772/intechopen.95370
3. Kumar P, Gacem A, Ahmad MT, Yadav VK, Singh S, Yadav KK, Alam MM, Dawane V, Piplode S, Maurya P, Ahn Y, Jeon B, Cabral-Pinto MM (2022) Environmental and human health implications of metal(loid)s: source identification, contamination, toxicity, and sustainable clean-up technologies. Front Environ Sci 10. https://doi.org/10.3389/fenvs.2022.949581
4. Nagare A, Dhadage A, Ghare P, Naval R, Athare A, Gogoi P, Datir A, Navgire M (2024) Room temperature acetone sensing activity of β-cyclodextrin coated MoO_3-ZrO_2 nanocomposite. Surf Interfaces 105068. https://doi.org/10.1016/j.surfin.2024.105068

5. Sivakumar K, Bhakyajothi V, Parameswari M, Prema D, Stalin T (2013) Spectral studies on the supramolecular assembly of 1H2NA: β-CD complex and its analytical application as chemosensor for the selective sensing of Cr^{3+}. Polycycl Aromat Compd 33:221–235. https://doi.org/10.1080/10406638.2013.781044
6. Prabu S, Mohamad S (2020) Curcumin/beta-cyclodextrin inclusion complex as a new "turn-off" fluorescent sensor system for sensitive recognition of mercury ion. J Mol Struct 1204:127528
7. Sivakumar K, Parinamachivayam G, Murali Krishnan M, Chakravarty S, Bharathi A (2018) Preparation, characterization and molecular modeling studies of the beta-cyclodextrin inclusion complex with benzoguanamine and its analytical application as chemosensor for the selective sensing of Ce^{4+}. Spectrochim Acta Part A Mol Biomol Spectrosc 200:212–225
8. Mohandoss S, Palanisamy S, You SG, Lee YR (2022) Synthesis of cyclodextrin functionalized photoluminescent metal nanoclusters for chemoselective Fe^{3+} ion detection in aqueous medium and its applications of paper sensors and cell imaging. J Mol Liq 356:118999
9. Kalogiouri NP, Papatheocharidou C, Samanidou VF (2024) Recent advances towards the use of deep eutectic solvents and cyclodextrins in the extraction of food contaminants: from traditional sample pretreatment techniques to green microextraction and beyond. Trends Anal Chem 173:117649
10. Kavitha R, Stalin T (2014) A highly selective chemosensor for colorimetric detection of Hg^{2+} and fluorescence detection of pH changes in aqueous solution. J Lumin 149:12–18. https://doi.org/10.1016/j.jlumin.2013.11.044
11. Banerjee A, Sahana A, Das S, Lohar S, Guha S, Sarkar B, Mukhopadhyay SK, Mukherjee AK, Das D (2012) A naphthalene exciplex based Al^{3+} selective on-type fluorescent probe for living cells at the physiological pH range: experimental and computational studies. Analyst 137(9):2166–2175
12. Valeur B, Leray I (2000) Design principles of fluorescent molecular sensors for cation recognition. Coord Chem Rev 205(1):3–40
13. Xu Z, Xiao Y, Qian X, Cui J, Cui D (2005) Ratiometric and selective fluorescent sensor for Cu II based on internal charge transfer (ICT). Org Lett 7(5):889–892
14. Serin JM, Brousmiche DW, Frechet JM (2002) A FRET-based ultraviolet to near-infrared frequency converter. J Am Chem Soc 124(40):11848–11849
15. Zhang Z, Fang F, Zhang P, Zhang X, Ma H, Wei Y (2024) Synergistic effect of inner filtering effect and host-guest interaction based on β-cyclodextrin fluorophore for promoted 4-nitrophenol detection. Colloids Surf, A 695:134087
16. Silva EC, Pina J, Pereira RF, Murtinho D, Valente AJ, Fajardo AR (2024) Synthesis and characterization of fluorescent poly(α-cyclodextrin)/carbon quantum dots composite for efficient removal and detection of toluene and xylene from aqueous media. Environ Pollut 347:123778. https://doi.org/10.1016/j.envpol.2024.123778
17. Sader KM, Padilla AG, Realpe A, Morantes MA, Soares JOB (2019) Removal of heavy metal water pollutants (Co^{2+} and $Ni^{2+)}$ using polyacrylamide/sodium montmorillonite (PAM/Na-MMT) nanocomposites. ACS Omega 4(6):10834–10844
18. Zhang X, Wei H, Guan Q, Yang X, Yu Q, Zhang M, Xia Y (2023) Maternal exposure to trace elements, toxic metals, and longitudinal changes in infancy anthropometry and growth trajectories: a prospective cohort study. Environ Sci Technol 57(32):11779–11791
19. Cheng S (2003) Effects of heavy metals on plants and resistance mechanism, heavy metal pollution in China. Environ Sci Pollut Res 10:256–264. https://doi.org/10.1065/espr2002.11.141.2
20. Potes MDL, Nakadi FV, Frois CFG, Vale MGR, Da Silva MM (2019) Investigation of the conditions for selenium determination by photochemical vapor generation coupled to graphite furnace atomic absorption spectrometry. Microchem J 147:324–332
21. Narayanan V, Ganesan V, Shanmugasundram E, Durganandini S, Vellaisamy K, Amirthalingam H, Rajamohan R, Thambusamy S (2023) Spectral studies of synthetic symmetric syringaldehyde di-Schiff azine with β-cyclodextrin inclusion complexes as a fluorescent probe for the detection of metal ions. J Photochem Photobiol A, Chem 445:115069. https://doi.org/10.1016/j.jphotochem.2023.115069

22. Almahri A, Al-Bonayan AM, Attar RMS, Karkashan A, Abbas B, Al-Qahtani SD, El-Metwaly NM (2023) Multifunctional lipophobic polymer dots from cyclodextrin: antimicrobial/anticancer laborers and silver ions chemo-sensor. ACS Omega 8(19):16956–16965. https://doi.org/10.1021/acsomega.3c00873
23. Ganesan V, Mani MK, Narayanan V, Shanmugasundram E, Vellaisamy K, Baskaralingam V, Jeyaraj J, Veerakanellore GB, Rajamohan R, Thambusamy S (2023) Synthesis, characterization of 4,4′-((1E,1′E)-hydrazine-1,2-diylidenebis(methanylylidene)) diphenol and the inclusion complex with γ-cyclodextrin as a fluorescent probe for detection of Al^{3+}. J Photochem Photobiol, A 442:114814. https://doi.org/10.1016/j.jphotochem.2023.114814
24. Khan AR, Forgo P, Stine KJ, D'Souza VT (1998) Chem Rev 98:1977–1996
25. Kameda T, Takaizumi M, Kumagai S, Saito Y, Yoshioka T (2021) Preparation of Zn-Al layered double hydroxide intercalated with carboxymethyl-β-cyclodextrin by anion exchange method and its Ni^{2+} adsorption property. Soft Mater 19:139–147. https://doi.org/10.1080/1539445X.2020.1779088
26. Rather IA, Ali R (2021) Indicator displacement assays: from concept to recent developments. Org Biomol Chem 19:5926–5981. https://doi.org/10.1039/D1OB00518A
27. Trivedi MU, Sharma P, Kaur A, Sharma RK (2024) Unraveling the β-cyclodextrin-driven inclusion complexation with phenylalanine-capped AuNPs for colorimetric detection of Cr(III) and Fe(III). ACS Appl Nano Mater 7(4):4377–4387
28. Wang J, Yang XQ, Li N, Wang LL, Xu XY, Zhang C (2023) A cyclodextrin-based turn-off fluorescent probe for naked-eye detection of copper ions in aqueous solution. Spectrochim Acta Part A Mol Biomol Spectrosc 287:122069
29. Liu J, Qin Y, Li X, Wang R, Guo H, Tang J, Liu J, Liu L (2023) One-step synthesis of graphene quantum dots by laser-induced polymer. Mater Sci Semicond Process 153:107146. https://doi.org/10.1016/j.mssp.2022.107146
30. Sun Y, Leng R, Ma X, Zhang J, Han B, Zhao G, Ai Y, Hu B, Ji Z, Wang X (2023) Economical amidoxime-functionalized non-porous β-cyclodextrin polymer for selective detection and extraction of uranium. Chem Eng J 459:141687
31. Wu W, Hu Z, Shi C, Xu R, Zhao Y, Ding Y (2023) Construction of CdTe@γ-CD@RBD nanoprobe for Fe^{3+}-sensing based on FRET mechanism in human serum. Spectrochim Acta Part A Mol Biomol Spectrosc 296:122645
32. Wang R, Gu W, Liu Z, Liu Y, Ma G, Wei J (2021) Simple and green synthesis of carbonized polymer dots from nylon 66 waste fibers and its potential application. ACS Omega 6(48):32888–32895. https://doi.org/10.1021/acsomega.1c04808
33. Becuwe M, Landy D, Delattre F, Cazier-Dennin F, Fourmentin S (2008) Fluorescent indolizine-b-cyclodextrin derivatives for the detection of volatile organic compounds. Sensors 8(6):3689–3705. https://doi.org/10.3390/s8063689
34. Liu Y, Wang K, Guo D, Jiang B (2009) Supramolecular assembly of perylene bisimide with β-cyclodextrin grafts as a solid-state fluorescence sensor for vapor detection. Adv Func Mater 19(14):2230–2235. https://doi.org/10.1002/adfm.200900221
35. Hu H, Xin J, Hu H, Wang X, Lu X (2014) Organic liquids-responsive β-cyclodextrin-functionalized graphene-based fluorescence probe: label-free selective detection of tetrahydrofuran. Molecules 19(6):7459–7479. https://doi.org/10.3390/molecules19067459
36. Maniyazagan M, Rameshwaran C, Mariadase R, Jeyakanthan J, Premkumar K, Stalin T (2017) Fluorescence sensor for Hg^{2+} and Fe^{3+} ions using 3,3′-dihydroxybenzidine: α-cyclodextrin supramolecular complex: characterization, in-silico and cell imaging study. Sens Actuators, B Chem 242:1227–1238
37. Selvan GT, Poomalai S, Ramasamy S, Selvakumar PM, Muthu Vijayan Enoch IV, Lanas SG, Melchior A (2018) Differential metal ion sensing by an antipyrine derivative in aqueous and β-cyclodextrin media: selectivity tuning by β-cyclodextrin. Anal Chem 90(22):13607–13615
38. Kaliyamoorthi K, Maniraj S, Govindaraj TS, Ramasamy S, Selvakumar Paulraj M, Enoch IV, Melchior A (2020) Unusual fluorescence quenching-based Al^{3+} sensing by an imidazolylpiperazine derivative, β-cyclodextrin encapsulation-assisted augmented sensing. J Fluoresc 30:445–453

39. Prema D, Thamaraiselvi S, Yamuna R (2020) Encapsulation of N-phenyl p-phenylenediamine into β-CD: spectral, molecular modelling studies and sensor application for detecting Fe^{2+} ion. J Mol Liq 319:113990

Dr. Priya is working as Assistant Professor in Chemistry for the Department of Chemical Engineering and Physical Sciences, Lovely Professional University, Phagwara, Punjab, India. She has received her Ph.D. degree from Dr. B. R. Ambedkar National Institute of Technology, Jalandhar (Punjab), India. She has also qualified CSIR-UGC NET exam with 37th rank in all over India. Her research areas include synthesis of polymer based nanomaterials with their applications in waste water treatment (adsorption and photo catalysis), in chemosensing and in biomedical field. She has 7 years of experience in teaching as well as research. Her h-index is 14, and i10 index is also 14. She has published 14 research publications in various International Journals of high repute. In addition, she is also an author in 3 books from various National and International publishers. She is also a member of renowned Asian Polymer Association (APA).

Chapter 17
Advances in Cyclodextrin-Based Sensors for Recognition of Small Molecules

Aditi Boruah, Priyanka Nath, and Pranjal K. Baruah

Abstract Cyclodextrins (CDs) play a significant role in chemical sensors, particularly in molecular recognition-based analytical techniques like separation science, electroanalysis, single-molecule sensing, spectroscopy, and clinical diagnostics hence, received growing interest among scientists during the past few decades. This chapter focuses on the recent development in the use of CDs for the sensing of several biologically relevant small compounds using fluorescence, electrochemical, and potentiometric response methods. It primarily emphasizes chromophore-linked CDs, metallocyclodextrins, CD-based supramolecular systems, polymer-based CDs, and their role in electrochemical sensing. Additionally, advances in nanotechnology have facilitated the integration of CDs into nanomaterial-based sensor systems, enhancing their functionality and expanding their applications. The chapter also discusses the advantages, disadvantages of CD, and the unique detection mechanisms of CD-functionalized materials in electrochemical sensors, along with recent developments in CD-based electrochemical sensors.

17.1 Introduction

Cyclodextrin (CD) is a cyclic polysaccharide molecule characterized by a sizable ring created through the end-to-end linkage of α-1,4 glycosidic bonds of D-(+)-glucose units. Typically, CD exhibits steric repulsion, resulting in a glucose unit count exceeding 6. Specifically, α-, β-, and γ-CDs consist of 6, 7, and 8 glucose units, respectively [1]. CD's three-dimensional configuration resembles a truncated cone, featuring hydroxyl groups distributed across the molecule's surface, facilitating water solubility [2]. Conversely, the interior cavity of CD exhibits relative hydrophobicity, enabling the inclusion of lipophilic molecules through host–guest interactions, provided they possess an appropriate size [3]. The distinct features of CDs have led to extensive research into the development of novel sensors for detecting various analytes, including small molecules [4], nucleic acids, and proteins [1]. Due to

A. Boruah · P. Nath · P. K. Baruah (✉)
Department of Applied Sciences, GUIST, Gauhati University, Guwahati, India
e-mail: baruah.pranjal@gmail.com; pranjal@gauhati.ac.in

S. Krishnamoorthy (ed.), *Cyclodextrins for Chemosensing*,
https://doi.org/10.1007/978-3-031-89299-8_17

their selective recognition capability, exceptional biocompatibility, water solubility, non-toxic nature, cost-effectiveness, commercial availability, and easy functionalization, CDs serve as pivotal building blocks for the creation of elegantly simple yet highly effective supramolecular architectures of fundamental significance [5]. The cytotoxic effects of cyclodextrins are minimal, hence CDs have been extensively researched in various fields, including enzyme simulation, drug solubilization and modification, cosmetics, agriculture, catalysis, food, spices, textiles, extraction and separation of bioactive compounds, environmental protection, organic synthesis, etc. [6]. β-CD is applied in the development of nanogels, mild oral insulin administration systems, taste-masking agents, nanofibers, nanoparticles (NPs), siRNA delivery carriers, gene delivery systems, and nanosponges for drug delivery [7]. Its biocompatibility, biodegradability, and low immunogenicity make it an excellent candidate for use as a drug carrier [8–10]. β-CD-based nanoarchitectures, formed through structural combinations, have emerged as a remarkable platform in nanostructured materials and nanotechnology due to their unique physicochemical properties and capacity to interact with biological elements [7]. Ongoing research focuses on developing efficient CD-based nano-carriers, especially for anticancer drugs. In recent decades, CDs have garnered significant attention in the scientific community, particularly for their role in developing chemical sensors through molecular recognition phenomena [5].

17.2 Molecular Recognition

The process of reversible encapsulation of guest molecules inside the hydrophobic cavity of CD molecules is known as molecular recognition in CDs [11]. As cyclodextrin research progresses, it is essential to clearly define key concepts like inclusion complexation, molecular recognition, and complexation energy. Understanding these concepts will improve host–guest chemistry research and drive advancements in supramolecular chemistry [12]. The primary driving forces for CD inclusion complex formation are van der Waals, steric effect, and hydrophobic interactions. While electrostatic and hydrogen bonding interactions are not dominant, they can significantly affect the complex's conformation. Steric effects play a crucial role here—if the guest molecule is too large or has bulky substituents, it may not fit properly into the cavity. Conversely, if it is too small, the interaction may not be optimal. The classical view suggests that hydrophobic association is driven by an increase in entropy, as the organized water molecules surrounding the solute become more disordered when they move into the bulk solution [13]. It is noteworthy that the driving forces typically coexist or work together synergistically. The strength of each force often depends on the specific inclusion system involved. Their relative strengths depend on factors like the size and modification of the CD cavity, the shape, volume, polarity, and number of substituents on the guest, as well as the reaction medium, temperature, and ionic strength [14].

17.3 Study of the Complexation of Structure

CDs have a hydrophobic cavity that enables the inclusion of molecules—referred to as "invited guests"—whose hydrophobic properties and size match the cavity's dimensions. Meanwhile, the hydroxyl groups on CDs enhance the solubility of these complexes in water. Encapsulation can occur one or more times, depending on the size of the guest, with complex formation mainly governed by geometric compatibility. Meier et al. found that the association constant for the C10-β-cyclodextrin complex is ten times higher than that for the C10-γ-cyclodextrin complex, highlighting that a stable inclusion complex requires the guest molecule to fit the CD cavity size [15]. For instance, naphthalene is too large for α-CD but can form a stable complex with β-CD, while anthracene can fit into γ-CD. Propionic acid does not fit well in the big cavities, but it is compatible with α-CD. To maximize interaction between the hydrophobic portion of the guest and the CD cavity, the contained molecules are often orientated in the host in this manner. To the greatest extent feasible, the hydrophilic portion of the guest molecule stays at the complex's exterior face. This guarantees optimal interaction between the solvent and the host's hydroxyl groups [16]. Host–guest complexes form because host molecules have specialized structures, like pores or hydrophobic regions, that interact with guest molecules. The host's internal structure can selectively accommodate guests of various sizes and shapes, ensuring high selectivity and specificity [17].

17.4 Cyclodextrin-Based Sensors

CD-based sensors detect inclusion complexes by monitoring changes in optical properties, such as fluorescence or absorption. For example, a fluorophore linked to the guest molecule may exhibit altered fluorescence upon complex formation with CD [18]. They are highly selective and sensitive, with mass spectrometry detecting changes in mass-to-charge ratios and NMR spectroscopy observing signal shifts. Their ability to recognize and bind molecules makes them popular in chemical sensing [19]. In chemical and biosensor research, accurately measuring small molecules in aqueous solutions with surface-bound receptors is essential. Effective sensors require sensitive transduction mechanisms and proper surface functionalization. Supramolecular structures like CDs and calixarenes are important for forming inclusion complexes and detecting small molecules [20]. CD-based sensors are particularly effective for molecular recognition of small molecules, offering a versatile and sensitive platform for various detection applications [5].

17.5 Chromophore-Linked Cyclodextrins for Chemical Sensing

Optical chemosensors are practical and intriguing, utilizing visible changes in color, absorption, or emission to indicate successful recognition of a target analyte by a host molecule. These chemosensors utilize chromophores attached to basic sensing structures to detect these optical variations, facilitating the determination of successful recognition events. Researchers globally have functionalized CDs with various dyes and fluorescent moieties to optimize the output, as changes in the microenvironment of chromophores can alter color and fluorescence patterns. The use of chromophore-appended CDs has been extensively explored for detecting a variety of hydrophobic guest molecules within hydrophobic cavities in recent years [5, 21].

17.6 Metallocyclodextrin for Chemical Sensing

Metallocyclodextrin-based CDs have been created and applied by multiple research teams in the realm of chemical sensors [22, 23]. Metallocyclodextrins generally use absorption energy transfer emission (AETE) as their sensing mechanism. Photon absorption is the first step in this system, which excites a guest molecule that harvests light. The energy is then transferred to a photoactive metal ion (Eu^{2+}/Tb^{2+}), which causes these metal ions to emit. Complexes of Eu^{2+} and Tb^{2+} ions with appended CDs like crown ether-CD and DTPA-CD show minimal fluorescence in aqueous solutions due to the lack of aromatic hydrocarbons, which usually act as light-harvesting groups [24]. However, incorporating aromatic guests into the hydrophobic cavity of metallocyclodextrins enhances fluorescence through AETE, providing a unique approach for developing turn-on fluorescent chemical sensors.

17.7 Supramolecular System Based on Cyclodextrin for Chemical Sensing

Supramolecular structures from CD units are of great interest due to their unique properties and applications, such as in chemical sensors. Effective detection relies on cooperative recognition and dynamic modifications of these structures [25]. The cyclic permethylated β-CD trimer effectively captured an anthracene derivative with two hydrophobic alkyl chains, as indicated by fluorescence changes [26]. Some of the recent works on supramolecular systems based on CD for chemical sensing are mentioned here. Çubuk et al. developed a CD-based polymeric fluorescence sensor employing UV-curing for the detection of Cd(II) ions. The sensor underwent testing for multiple parameters, such as pH, duration, and coexisting ions, utilizing excitation and emission wavelengths of 374 nm and 422 nm, respectively. The measurement

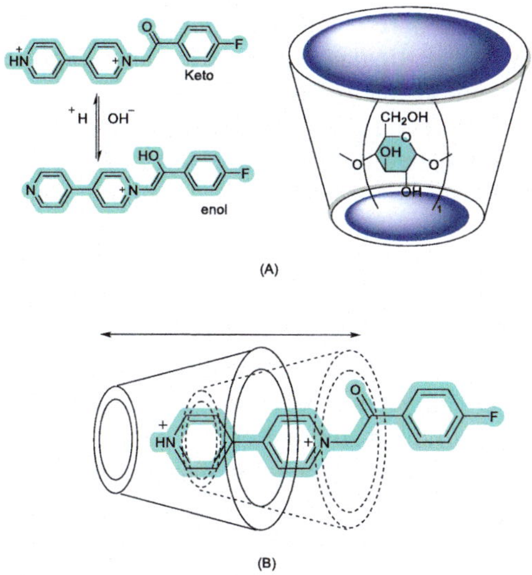

Fig. 17.1 **a** The molecular structures of the host β-CD (right) and the guest molecule 1-(4-fluoroacetophenone)-4,4′-bipyridinium dichloride (left) utilized in the study. **b** A schematic diagram of the inclusion complex FOV2+@β-CD

of Cd(II) ions occurred within the range of 4.45×10^{-9} to 4.45×10^{-8} mol L^{-1}, exhibiting a detection limit of 6.23×10^{-10} mol L^{-1}. The sensor exhibited selectivity, remaining unaffected by prevalent metal ions such as Pb(II) and Zn(II), even at concentrations 1000 times greater. It was effectively utilized to identify Cd(II) ions in biscuit and tap water samples [27]. Multistimuli-responsive materials are in high demand for applications in smart windows, sensors, and bionic manufacturing. Wang and colleagues developed a guest molecule, 1-[2-(4-fluorophenyl)-2-oxoethyl]-4,4-bipyridinium dichloride ($FOV\text{-}Cl_2$) (Fig. 17.1a), which undergoes pH-dependent ketone-enol tautomerism in aqueous solutions. The solid-state inclusion complex formed with β-CD and FOV2+ (ketone form) displayed distinct chromic behaviors when exposed to light, heat, and ammonia vapors, as analyzed using UV-vis and ESR spectroscopy. The chromic behavior of the β-CD inclusion complex with FHV^+ (enol form) was also studied for comparison. Furthermore, the fluorescence emission of the FOV2+@β-CD complex can be reversibly controlled using a photochromic method. These characteristics suggest potential applications in anti-counterfeiting, ammonia sensing, and inkless printing, highlighting the intriguing practical uses of these multistimuli-responsive inclusion complexes [28]. Figure 17.1b shows the mechanism of the β-CD inclusion complex.

17.8 Polymer Based on Cyclodextrin for Chemical Sensing

Researchers worldwide are actively exploring ways to enhance the electrical and optical properties of π-conjugated polymers, which are extensively employed in electroluminescence, LEDs, electrical conductivity, and chemical sensors. These polymers are receiving significant attention for their roles in various applications, with a particular emphasis on adjusting their optical and electrical traits through stimuli such as oxidation–reduction, pH changes, and metal cations [29]. Harada and coworkers have developed a water-soluble conjugated fluorescent polymer, based on poly(phenylene ethynylene) with CD functionalization. This polymer exhibits blue fluorescence in DMF and green fluorescence in aqueous solutions [30]. However, In an aqueous solution, this polymer doesn't fully dissolve in its monomeric state; rather, it forms partially quenched aggregates of the poly(phenylene ethynylene) backbone. The introduction of guest molecules leads to alterations in the fluorescence. Specifically, the presence of aliphatic and acceptor molecules causes the polymer to undergo fluorescence quenching and shifts in color [31]. This phenomenon can be attributed to the interaction between 1-adamantanecarboxylic acid and the β-CD units, which leads to the disruption of various intermolecular π-stacking interactions within the polymeric backbone. Coumarin and pyrene are important in biological systems and sensing. Ueno et al. developed a β-CD-peptide hybrid polymeric conjugate by combining coumarin as an acceptor and pyrene as a donor [32, 33].

17.9 Cyclodextrin for Electrochemical Sensing

Electrochemical sensing detects chemical substances via electrical signals from their interaction with electrodes. Its sensitivity, selectivity, rapid response, and potential for miniaturization have made it valuable in fields like environmental monitoring, biomedical diagnostics, and food safety [33, 34]. Over the past decade, CD-based carbon nanomaterials, which include carbon nanotubes (CNTs), graphene, and conducting polymers, have demonstrated significant utility in the realm of electrochemical sensing [35]. Studies show that β-CD can adsorb onto carbon-based substrates like graphene, forming water-dispersible nanocomposites. Methylene blue easily enters the β-CD cavity, creating a stable host–guest complex. β-CD polymer (β-CDP), cross-linked with epichlorohydrin, offers improved water solubility and higher binding affinity, making it a promising candidate for electrochemical sensors in biomarker detection [1]. Many CD-based potentiometric sensors utilizing electrochemical techniques have been developed in recent years, boasting significant applications across various fields including pharmaceutical sciences, industries, environmental monitoring, medicine, and agriculture [36]. Taking into account these advantages, Lenik et al. have pioneered the development of functionalized β-CD-based potentiometric sensors for determining pharmaceutical drugs like naproxen and ketoprofen, wherein the naproxen molecule is observed to be partially or completely encapsulated within the cavity of the functionalized β-CD host [37]. Gong et al. developed an electrochemical sensing method to detect DNA, RNA, and protein targets using the host–guest interaction between MB-labeled probes and a β-CD-based nanocomposite. The formation of target-probe complexes reduces probe binding to the electrode, leading to a detectable signal drop. After optimization, the platform demonstrated excellent sensitivity for cancer-related analytes—p53 DNA, microRNA-21, and thrombin—with low detection limits. The assay separates detection into two stages: target recognition and signal transduction, overcoming challenges seen in traditional sensors and advancing host–guest chemistry in biomarker detection [1].

Chirality is an essential property of biological systems [38], described as the inability of an entity to fully superimpose onto its mirror image [39]. While chiral substances have equal chemical structures and physical properties among their enantiomers, they demonstrate unique biological interactions, pharmacological effects, and toxicities in a chiral environment [40]. For example, L-tryptophan, an important amino acid, is a critical protein-building ingredient and serves as a biomarker of chronic disorders [41]. Zang and colleagues developed a rapid electrochemical sensor for detecting tryptophan (Trp) enantiomers using a CD-modified microporous organic network (CD-MON) as the chiral recognition agent. The CD-MON, stabilized on the electrode surface by BSA, enhances chiral recognition through steric hindrance, host–guest interactions, and hydrogen bonding. The sensor effectively distinguishes between Trp enantiomers, with a detection limit of 2.6 μM for L-Trp. UV-Vis spectroscopy confirmed CD-MON's ability to adsorb Trp enantiomers, aligning with the electrochemical results [41].

17.10 β-Cyclodextrin-Based Nanocomposite

β-CDs play a key role in nanocomposites and metallic nanoparticles by stabilizing NPs and enhancing their catalytic properties. They improve NP attributes like drug loading, solubility, stability, and targeted delivery, offering potential for controlled release, increased efficacy, and safety, making them ideal for simultaneous diagnosis and treatment. This work focuses on leveraging CDs to address issues in NP composition and boost their performance [7]. Table 17.1 lists various β-CD-based nanocomposites along with their applications.

17.11 Nanocarbon Hybrids Based on Cyclodextrin for Chemical Sensing

Researchers from all over the world have been very interested in carbon nanomaterials in recent years because of their special electrical, structural, and mechanical capabilities. These materials include carbon nanotubes, fullerene, nanodiamonds, and other polyaromatic hydrocarbons. This interest has led to the challenge of constructing CD-based sensors through hybridization with carbon nanomaterials [44]. Grüner and coworkers used field-effect transistors (FETs) hybridized with pyrene-functionalized β-CDs painted on SWCNTs to detect common organic guest molecules and the FET properties of these hybrid nano-conjugates were shown to be extremely sensitive

Table 17.1 β-CD based nanocomposites identify various templates through molecular recognition, resulting in line arrangement, load distribution, and applications

Sl. No.	β-CD-based nanocomposites	Template molecule detected	Line range	LOD	Application in real sample
1	ILs(2)-βCD	Fluoroquinolones (FQs)	0.5–100 g/L	0.43–1.83 g/L	For separation and identification of trace FQs in complicated matrices [7]
2	SMIP/MA@PCNS	Thiophene	22.6 mg S/g	16.2 mg S/g	Effectively used to separate thiophene from oil directly [7]
3	β-CD base nanocomposite	Bisphenol A	0.02–1.0 mM	8 nM	Proven for recognizing bisphenol A in contaminated water and potentially applicable to certain other natural contaminants as well [42]
4	β-CD Fe_3O_4 NPs	Erythromycin	0.51 × 0005 μmol/mL	97.34% recovery	Recovery as well as segregation of ERY from samples of industrial waste [43]

and dependent on the association constants between the competitive organic guest molecules and β-CDs [45].

17.12 Nanoparticle Hybrids Based on Cyclodextrin for Chemical Sensing

Kaifer and colleagues developed CD-functionalized gold nanoparticles and discovered that adding a ferrocene dimer caused nanoparticle aggregation, resulting in a red shift and solid formation. This color change signals a sensing event with specific guest molecules [46]. They later even detected the well-known carbon nanomaterial, fullerene, by leveraging the aggregation phenomenon with the help of β-CD and gold nanoparticles [47]. Elgamouz and his team developed silver nanoparticles with β-CD to create a nanoprobe capable of encapsulating creatinine, which can be detected through a colorimetric response. This β-CD-silver nanoparticle-based nano-colorimetric probe has proven to be an efficient method for detecting reactive oxygen species in human urine samples, particularly H_2O_2 [48].

17.13 Advantages and Limitations of Cyclodextrin-Based Sensors

CD-based sensors offer significant advantages, including high selectivity, enhanced sensitivity, functionality, biocompatibility, and wide versatility across different applications. These features make them a valuable tool in fields such as environmental monitoring, food safety, and medical diagnostics [25, 49, 50]. However, they also have several limitations they do face challenges such as non-specific binding, environmental sensitivity, and limited solubility. These limitations can affect their performance in terms of selectivity, sensitivity, and practical applicability. CDs mainly bind hydrophobic molecules, limiting their ability to detect polar or water-soluble analytes. This reduces their versatility, particularly for larger molecules, and can result in slow sensor responses, making them unsuitable for real-time monitoring. These limitations hinder the widespread use of CD-based sensors in applications requiring high selectivity or real-time detection [51, 52].

17.14 Conclusion

Chemical sensors utilizing CD derivatives have gained significance in less than three decades. This chapter summarizes the development of CD-based materials, from monomers to nano-assemblies, that detect chemicals at μM/nM levels. The future of CD-based sensors is promising, with potential for innovation in environmental protection, healthcare, food safety, and smart technology. Key areas of focus include enhancing selectivity, minimizing interference, and making sensors scalable and affordable for industrial and consumer use. Future research may integrate CDs with nanomaterials like graphene, MOFs, or quantum dots to create multifunctional sensors. Advancements will lead to more sophisticated, sensitive, and selective CD-based sensors, enabling real-time, on-site, and non-invasive monitoring across various industries.

This chapter covers molecular recognition, inclusion complex formation, and encapsulation in CD chemistry, with applications in supramolecular chemistry, biochemistry, polymer chemistry, pharmaceuticals, and nanotechnology. It highlights the role of steric effects in small molecule encapsulation and explores optical and electrochemical sensing mechanisms, emphasizing the unique features of CD-based sensors.

References

1. Gong H, Bao C, Guo X, Tian F, Qiu L, Yang W (2024) A universal electrochemical sensor for detection of nucleic acids and protein based on host-guest recognition of β-cyclodextrin polymer. Microchem J 204:110987

2. Liu Z, Liu Y (2022) Multicharged cyclodextrin supramolecular assemblies. Chem Soc Rev 51:4786–4827
3. Zhang Y, Liu Y, Liu Y (2019) Cyclodextrin-based multistimuli-responsive supramolecular assemblies and their biological functions. Adv Mater 32:1806158
4. Suzuki Y, Mizuta Y, Mikagi A, Misawa-Suzuki T, Tsuchido Y, Sugaya T, Hashimoto T, Ema K, Hayashita T (2023) Recognition of D-glucose in water with excellent sensitivity, selectivity, and chiral selectivity using gamma-cyclodextrin and fluorescent boronic acid inclusion complexes having a pseudo-diboronic acid moiety. ACS Sens 8:218–227
5. Rather IA, Hasan A, Ali R (2023) Cyclodextrin-based sensors for the recognition of small molecules. In: Cyclodextrins—core concepts and new frontiers
6. Xu Y, Rashwan AK, Osman AI, El-Monaem EMA, Elgarahy AM, Eltaweil AS, Omar M, Li Y, Mehanni AE, Chen W, Rooney DW (2022) Synthesis and potential applications of cyclodextrin-based metal–organic frameworks: a review. Environ Chem Lett 21:447–477
7. Tarannum N, Kumar D, Kumar N (2022) β-Cyclodextrin-based nanocomposite derivatives: state of the art in synthesis, characterization and application in molecular recognition. ChemistrySelect 7:e202200140
8. Zafar N, Fessi H, Elaissari A (2014) Cyclodextrin containing biodegradable particles: from preparation to drug delivery applications. Int J Pharm Res 461:351–366
9. Qiu N, Li X, Liu J (2017) Application of cyclodextrins in cancer treatment. J Incl Phenom Macrocycl Chem 89:229–246
10. Gidwani B, Vyas AA (2015) Comprehensive review on cyclodextrin-based carriers for delivery of chemotherapeutic cytotoxic anticancer drugs. Biomed Res Int 2015:1–15
11. Rather IA, Wagay SA, Hasnain MS, Ali R (2019) New dimensions in calix[4] pyrrole: the land of opportunity in supramolecular chemistry. RSC Adv 9:38309–38344
12. Song LX, Bai L, Xu XM, He J, Pan SZ (2009) Inclusion complexation, encapsulation interaction and inclusion number in cyclodextrin chemistry. Coord Chem Rev 253:1276–1284
13. Sandilya AA, Natarajan U, Priya MH (2020) Molecular view into the cyclodextrin cavity: structure and hydration. ACS Omega 5:25655–25667
14. Omari MMA, Zughul MB, Davies JED, Badwan AA (2006) Effect of buffer species on the complexation of basic drug terfenadine with β-cyclodextrin. J Incl Phenom Macrocycl Chem 58:227–235
15. Meier MM, Bordignon Luiz MT, Farmer PJ, Szpoganicz BJ (2001) The influence of β-and γ-cyclodextrin cavity size on the association constant with decanoate and octanoate anions. J Incl Phenom 40:291–295
16. Davis ME, Brewster ME (2004) Cyclodextrin-based pharmaceutics: past, present and future. Nat Rev Drug Discovery 3:1023–1035
17. Schluep T, Cheng J, Khin KT, Davis ME (2006) Pharmacokinetics and biodistribution of the camptothecin–polymer conjugate IT-101 in rats and tumor-bearing mice. Cancer Chemother Pharmacol 57:654–662
18. Sinn S, Biedermann F (2018) Chemical sensors based on cucurbit[n]uril macrocycles. Isr J Chem 58:357–412
19. Wang Y, Ng SC (2012) HPLC enantioseparation on cyclodextrin-based chiral stationary phases. In: Chiral separations. Methods in molecular biology, pp 69–79
20. Janshoff A, Steinem C, Michalke A, Henke C, Galla H-J (2000) Monofunctionalized β-cyclodextrins as sensor elements for the detection of small molecules. Sens Actuators, B Chem 70:243–253
21. Krämer J, Kang R, Grimm L, De Cola L, Picchetti P, Biedermann F (2022) Molecular probes, chemosensors, and nanosensors for optical detection of biorelevant molecules and ions in aqueous media and biofluids. Chem Rev 122:3459–3636
22. Haider JM, Pikramenou Z (2005) Photoactive metallocyclodextrins: sophisticated supramolecular arrays for the construction of light activated miniature devices. Chem Soc Rev 34:120
23. Kameda T, Takaizumi M, Kumagai S, Saito Y, Yoshioka T (2020) Preparation of Zn–Al layered double hydroxide intercalated with carboxymethyl-β-cyclodextrin by anion exchange method and its Ni^{2+} adsorption property. Soft Mater 19:139–147

24. Mortellaro MA, Nocera DG (1996) A supramolecular chemosensor for aromatic hydrocarbons. J Am Chem Soc 118:7414–7415
25. Ogoshi T, Harada A (2008) Chemical sensors based on cyclodextrin derivatives. Sensors 8:4961–4982
26. Sasaki K, Nagasaka M, Kuroda Y (2001) New cyclodextrin dimer and trimer: formation of biphenyl excimer and their molecular recognition. Chem Commun 2001:2630–2631
27. Çubuk S, Yılmaz Ö, Yetimoğlu EK, Kahraman MV (2017) Determination of Cd(II) ions by using cyclodextrin based polymeric fluorescence sensor. J Turk Chem Soc, Sect A: Chem 4:537–548
28. Wang X, Xu C, Lin R, Sun W, Ye M, Xu L, Liu J (2024) Acid–base regulated inclusion complexes of β-cyclodextrin with 1-[2-(4-fluorophenyl)-2-oxoethyl]-4,4′-bipyridinium dichloride displaying multistimuli-responsive chromic behaviors and photomodulable fluorescence. J Mater Chem C 12:2764–2771
29. Mousazadeh H, Pilehvar-Soltanahmadi Y, Dadashpour M (2021) Cyclodextrin based natural nanostructured carbohydrate polymers as effective non-viral siRNA delivery systems for cancer gene therapy. J Control Release 330:1046–1070
30. Ogoshi T, Takashima Y, Yamaguchi H, Harada A (2006) Cyclodextrin-grafted poly(phenylene ethynylene) with chemically-responsive properties. Chem Commun 2006:3702–3704
31. Hossain MA, Mihara H, Ueno A (2003) Novel peptides bearing pyrene and coumarin units with or without β-cyclodextrin in their side chains exhibit intramolecular fluorescence resonance energy transfer. J Am Chem Soc 125:11178–11179
32. Hossain MA, Mihara H, Ueno A (2003) Fluorescence resonance energy transfer in a novel cyclodextrin-peptide conjugate for detecting steroid molecules. Bioorg Med Chem Lett 13:4305–4308
33. Simoska O, Gaffney EM, Minteer SD, Franzetti A, Cristiani P, Grattieri M, Santoro C (2021) Recent trends and advances in microbial electrochemical sensing technologies: an overview. Curr Opin Electrochem 30:100762
34. Lu S, Peng Y, Ying Y, Long Y (2020) Electrochemical sensing at a confined space. Anal Chem 2:5621–5644
35. Zhu G, Yi Y, Chen J (2016) Recent advances for cyclodextrin-based materials in electrochemical sensing. Trends Anal Chem 80:232–241
36. Mostafa SM (2020) Novel potentiometric sensors based on β-cyclodextrin and dibenzo 18-crown-6 ionophores/mesoporous silica nanoparticles for clidinium determination. Int J Electrochem Sci 15:3347–3364
37. Lenik J, Łyszczek R (2022) A potentiometric sensor for ketoprofen based on a β-cyclodextrin derivative. J Anal Chem 77:246–256
38. Niu X, Zhao R, Liu Y, Yuan M, Zhao H, Li H, Yang X, Xu H, Wang K (2023) Adaptive host–guest chiral recognition in nanoarchitectonics with biomimetic MOF mimicking DNA. J Mater Chem A 11:23376–23386
39. Niu X, Zhao R, Yan S, Pang Z, Li H, Yang X, Wang K (2023) Chiral materials: progress, applications, and prospects. Small 19:2303059
40. Karimian M, Dashtian K, Zare-Dorabei R, Norouzi S (2023) Paper-based microfluidic system and chiroptical functionalized gold nano-oval for colorimetric detection of L-tryptophan. Anal Chim Acta 1285:342022
41. Zhang X, Wang F, Chen Z (2024) Electrochemical chiral sensor for recognition of amino acid enantiomers with cyclodextrin-based microporous organic networks. Anal Chim Acta 1316:342879
42. Ali H, Mukhopadhyay S, Jana NR (2019) Selective electrochemical detection of bisphenol A using a molecularly imprinted polymer nanocomposite. New J Chem 43:1536–1543
43. Zhang Y, Li J, Wang F, Wu G, Qv X, Hong H, Liu C (2016) Recovery and separation of erythromycin from industrial wastewater by imprinted magnetic nanoparticles that exploit β-cyclodextrin as the functional monomer. J Sep Sci 39:450–459
44. Ali R, Alvi S (2020) The story of π-conjugated isotruxene and its congeners: from syntheses to applications. Tetrahedron 76:131345

45. Zhao Y, Hu L, Stoddart JF, Grüner G (2008) Pyrenecyclodextrin-decorated single-walled carbon nanotube field-effect transistors as chemical sensors. Adv Mater 20:1910–1915
46. Liu J, Ong W, Román E, Lynn MJ, Kaifer ÁE (2000) Cyclodextrin-modified gold nanospheres. Langmuir 16:3000–3002
47. Liu J, Álvarez JC, Ong W, Kaifer ÁE (2000) Network aggregates formed by C60 and gold nanoparticles capped with γ-cyclodextrin hosts. Nano Lett 1:57–60
48. Elgamouz A, Nassab C, Bihi A, Mohamad SAI, Almusafri AHSA, Alharthi SS, Abdulla SAE, Patole SP (2021) Encapsulation capacity of β-cyclodextrin stabilized silver nanoparticles towards creatinine enhances the colorimetric sensing of hydrogen peroxide in urine. Nanomaterials 11:1897
49. Ahmad Rather I, Hasan A, Ali R (2023) Cyclodextrin-based sensors for the recognition of small molecules. In: Ali R (ed) Cyclodextrins—core concepts and new frontiers. IntechOpen, p 108500
50. Matencio A, Rubin Pedrazzo A, Difalco A, Navarro-Orcajada S, Khazeai Monfared Y, Conesa I, Rezayat A, López-Nicolás JM, Trotta F (2021) Advances and classification of cyclodextrin-based polymers for food-related issues. Polymers 13:4226
51. Wang SY, Li L, Xiao Y, Wang Y (2019) Recent advances in cyclodextrins-based chiral-recognizing platforms. TrAC, Trends Anal Chem 121:115691
52. Karadurmus L, Gumustas M, Bakirhan NK, Ozkan SA (2023) Chiral sensing as a future challenge in electroanalytical chemistry: cyclodextrin-based chiral sensors. Crit Rev Anal Chem 53:498–519

Aditi Boruah was born in the Sivasagar district of Assam. She received her M.Sc. with a specialization in Organic Chemistry in 2019 from the University of Science and Technology, Meghalaya (USTM). Currently, she is pursuing a Ph.D. as a DST-WISE Junior Research Fellow under the supervision of Prof. Pranjal Kumar Baruah in the Department of Applied Sciences (Chemical Science Division) at Gauhati University, Guwahati, Assam. Her research interests focus on C-H functionalization and the synthesis of heterocyclic compounds using multi-component reactions.

Priyanka Nath was born in Hojai district of Assam, India. She completed her M. Sc. in Chemistry from, the University of Science and Technology Meghalaya (USTM). She qualified the State Level Eligibility Test Commission (SLET), Assam (NE Region) Year:2023. She is currently pursuing her PhD under the guidance of Prof. Pranjal K. Baruah in the Department of Applied Sciences (Chemical Science Division), Gauhati University, Guwahati, Assam. Her research work broadly focuses on developing novel methodologies via the C-H functionalization of heterocyclic compounds.

Pranjal K. Baruah began his academic journey at Dibrugarh University, earning his B.Sc. and M.Sc. with 1st class and 2nd rank. He completed his PhD at NCL, Pune, in 2007. As a postdoc, he was involved in the research of antiepileptic compounds with Prof. Harold Kohn at UNC Chapel Hill and later earned a Marie Curie Fellowship at University of Oxford, UK working with Prof. Martin D. Smith on foldamers and asymmetric catalysis. Joining Gauhati University in 2011, he became a Professor in 2022. His research spans C-H functionalization, organic synthetic methodology, catalysis, nanomaterials, and food preservatives. He has >75 publications in international journals and few book chapters. Currently he serves an editorial role at Discover Chemistry journal. His h-index is 24, and i10 index is 48. He has received funding from various Indian funding agencies, like DST and DBT.

Chapter 18
Sensing Reactive Oxygen and Nitrogen Species with Cyclodextrins and Their Nanostructured Systems

Álvaro Sarabia-Vallejo, Marta García-Cerrato, Pilar López-Alvarado, M. Antonia Martín, and J. Carlos Menéndez

Abstract Reactive oxygen (ROS) and nitrogen species (RNS) exist in a delicate equilibrium within living organisms. Their physiological role as messengers and signaling intermediates makes them of unparalleled relevance for the correct regulation of their levels inside cells requiring precise generation and quenching systems to achieve this fine-tuning task. If these mechanisms are evaded and excessive levels of ROS and RNS are produced and/or accumulated, their function turns into a highly toxic role and devastating consequences arise in the diseased state such as neurological, cardiovascular, or neoplastic disorders. Therefore, it is of great interest to develop detection and monitoring systems for ROS and RNS. Cyclodextrins are organic scaffolds with the ability to encapsulate small molecules inside, improving some of their properties such as hydrosolubility and stability. In addition, they are highly biocompatible exhibiting a benign safety profile. Thus, cyclodextrins have been widely used for different biological applications due to their privileged potential in biomedicine. In this context, cyclodextrins are interesting components for the design of detection systems for ROS and RNS. From designs barely containing a certain cyclodextrin derivative to highly complex structures with numerous components like polymers or nanoparticles, different approaches have proved successful in detecting and monitoring the concentration of oxidative species. In this chapter, the most relevant strategies involving cyclodextrin-based sensing systems for ROS and RNS are revised, and representative examples have been selected to illustrate this dynamic and innovative research field.

Á. Sarabia-Vallejo (✉) · M. García-Cerrato · P. López-Alvarado · J. Carlos Menéndez (✉)
Unidad de Química Orgánica y Farmacéutica, Departamento de Química en Ciencias Farmacéuticas, Facultad de Farmacia, Universidad Complutense, Madrid, Spain
e-mail: alsarabi@ucm.es

M. Antonia Martín
Unidad de Química Analítica, Departamento de Química en Ciencias Farmacéuticas, Facultad de Farmacia, Universidad Complutense, Madrid, Spain

S. Krishnamoorthy (ed.), *Cyclodextrins for Chemosensing*,
https://doi.org/10.1007/978-3-031-89299-8_18

18.1 An Overview of the Relevance of Reactive Oxygen and Nitrogen Species

18.2 Introduction to ROS and RNS

Reactive oxygen (ROS) and nitrogen (RNS) species are unstable, highly reactive chemical entities. They exhibit reduced lifetimes and limited operating radii due to their rapid quenching by reaction with surrounding molecules. These species are commonly found in biological systems, which have evolved to respond and interpret their signals as part of intrinsic and environmental information. Thus, ROS and RNS are part of various signaling pathways and perform important physiological functions, contributing to the maintenance of homeostasis. However, an undue generation of ROS and RNS results in an imbalance between oxidative species and antioxidant mechanisms and hence oxidative stress, which damages biomolecules and translates into cytotoxicity. Moreover, the sustained excessive production of ROS and RNS leads to chronic inflammatory processes, which are dysfunctional immune responses, driving cellular and tissular damage in a wide range of diseases.

Reactive oxygen species (ROS) are highly unstable molecules prone to react with electron-rich species and include singlet oxygen ($^1O_2$), superoxide radical anion ($O_2^{\cdot -}$), hydroxyl radical ($HO^{\cdot}$), and hydrogen peroxide (H_2O_2). Reactive nitrogen species include nitric oxide ($^{\cdot}NO$) and peroxynitrite ($ONOO^-$) (Fig. 18.1). The most relevant ones, due to their abundance and oxidative potential in biological systems, are superoxide radical anion, hydroxyl radical, and hydrogen peroxide.

Superoxide radical anion ($O_2^{\cdot -}$) has key physiological functions and participates in cellular signaling and processes such as senescence, apoptosis, and aging. Phagocytic cells resort to the generation of superoxide radical anions to eliminate pathogens [1]. Also, excessive levels of superoxide radical anion can lead to states of oxidative stress [2]. Superoxide can be detoxified through enzymatic processing by superoxide dismutase to yield O_2 and H_2O_2, which is a less reactive ROS and thus less toxic (Fig. 18.2a). Superoxide can also react with H_2O_2 to yield hydroxyl radical ($HO^{\cdot}$),

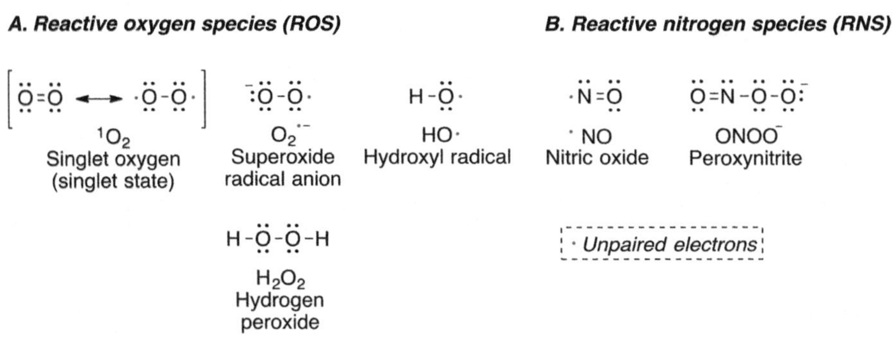

Fig. 18.1 Chemical structures of reactive oxygen (ROS) and nitrogen species (RNS)

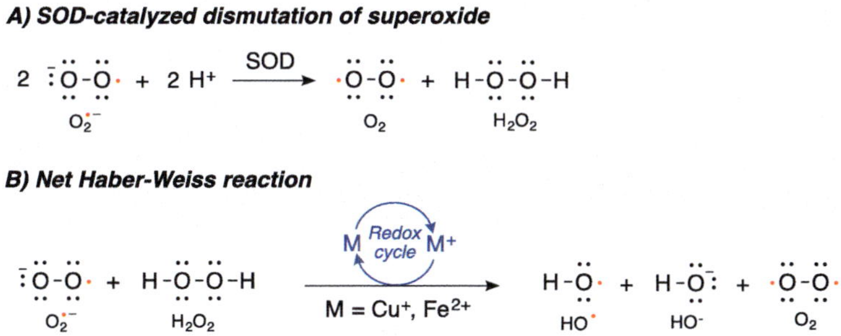

Fig. 18.2 Detoxification of superoxide radical anion by a superoxide dismutase (SOD)-mediated-enzymatic route (**a**) or by non-enzymatic processes (**b**)

a process that is mediated by certain metal cations (iron, copper) via the so-called Haber–Weiss reaction (Fig. 18.2b). This radical, one of the most potent oxidants known, reacts more readily than superoxide with biomolecules such as proteins, lipids, sugars, and nucleic acids.

H_2O_2 is a key signaling molecule involved in redox metabolism and plays a role in a variety of processes like cell survival and proliferation and the recruitment of immune cells, working by causing oxidative modifications in oxidizable amino acids such as cysteine residues in targets like transcriptional factors, MAP kinases, and protein tyrosine phosphatases. Since the translocation of hydrogen peroxide is part of the finely tuned redox metabolism and signaling within cells, there are specific aquaporins, named peroxiporins, allowing for the distribution of H_2O_2 along cellular compartments [3]. H_2O_2 is enzymatically produced in the respiratory chain, by oxidases like NADPH oxidases and by the dismutation of superoxide radical anion by the superoxide dismutase enzyme (SOD), as shown in Fig. 18.2. The homeostasis of hydrogen peroxide is essential for correct cellular functionality and several mechanisms monitor and maintain the levels of H_2O_2 within physiological ranges inside the cell, typically between 1 and 100 nM [4]. These mechanisms include the activity of catalase, an enzyme that decomposes hydrogen peroxide into molecular oxygen and water, and peroxidases like glutathione peroxidase that reduce hydrogen peroxide to water (Fig. 18.3).

Hydroxyl radical ($HO^{\cdot}$) is a highly reactive ROS capable of causing extensive oxidative damage to a wide variety of biomolecules. Its very high reactivity makes

Fig. 18.3 Hydrogen peroxide decomposition by **a** catalase (CAT) and **b** glutathione peroxidase (GPx)

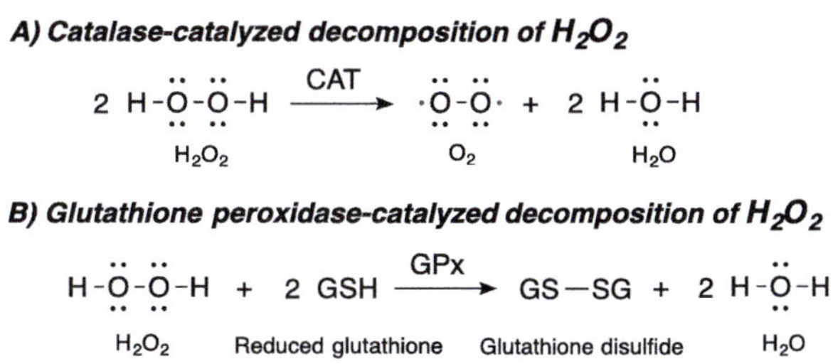

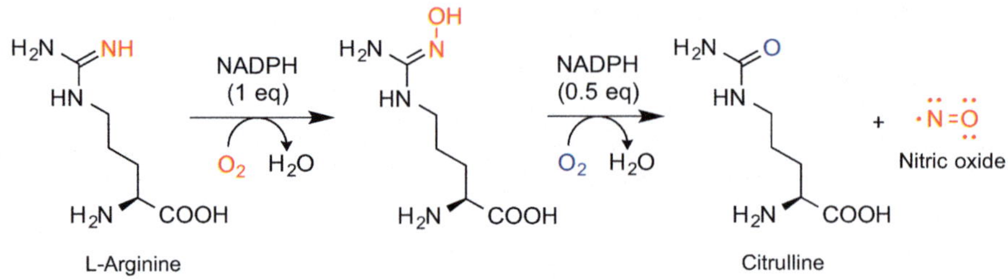

Fig. 18.4 Nitric oxide production catalyzed by nitric oxide synthases (NOS)

HO˙ an unlikely player in physiological processes due to the risk of uncontrolled oxidation reactions. Instead, hydroxyl radical is a key player in oxidative stress because it damages the DNA, causes lipid peroxidation compromises the integrity of biological membranes, and readily reacts with sugars and amino acids [5, 6]. Because in biological environments the hydroxyl radical is mainly produced through the Haber–Weiss reaction, the control of the concentration of catalytic metals, especially iron, is crucial for the regulation of its production.

Nitric oxide (˙NO) is a reactive nitrogen species with key functions in blood pressure regulation by promoting vasodilation. It also works as a neurotransmitter, an antimicrobial agent produced by macrophages and an apoptosis regulator, among other functions [7, 8]. However, an excess of nitric oxide leads to oxidative stress and induces cell death. Thus, a rigorous balance in the production of ˙NO is required to ensure its correct physiological function while not compromising cell viability. Nitric oxide is synthesized from L-arginine by nitric oxide synthases (NOS) across tissues (neurons, macrophages, endothelial cells, erythrocytes, among others) (Fig. 18.4).

18.3 ROS and RNS Overproduction and Imbalance. Oxidative Stress and Inflammation

An adequate equilibrium between the generation and the elimination of ROS and RNS results in a physiological situation known as oxidative eustress. On the other hand, an exacerbated production of oxidative species can lead to the insufficient scavenging activity of enzyme-based antioxidant mechanisms and the depletion of low molecular weight antioxidants so that the remaining ROS and RNS can oxidize nearby biomolecules, altering their structures through oxidative modifications and impairing their functionality. This dysregulation results in a situation of oxidative distress (Fig. 18.5).

The overproduction of ROS and RNS triggers the innate immune system and elicits inflammatory processes. Inflammation is an adaptive response intended to combat early-stage cell and tissue damage of varied etiologies; however, its sustained instauration represents a maladaptive situation. Long-term inflammation also unleashes the

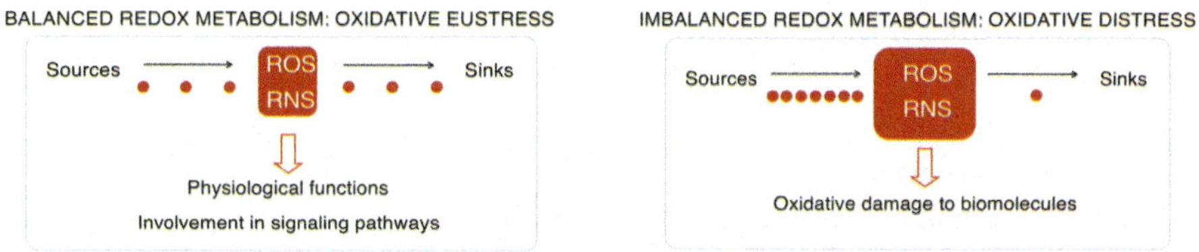

Fig. 18.5 The balance of the redox metabolism results in a physiological oxidative eustress, while the imbalanced accumulation of ROS and RNS leads to oxidative stress and disease states

overproduction of oxidative species, making the whole process enter a self-feeding cycle leading to chronic diseases associated with an imbalance in ROS and RNS [9]. Cancer, diabetes, neurodegenerative and cardiovascular conditions, and autoimmune diseases exhibit a key inflammatory component accompanied by a dysregulation of the redox metabolism. Although the main cause of these diseases is generally not an oxidative stress state, the instauration of a highly oxidant environment in cells and tissues worsens the underlying pathology and contributes to the progression and extension of the disease.

Cancer cells present a higher concentration of ROS as a consequence of an impaired redox metabolism. A higher concentration of ROS translates into several phenomena promoting the tumorigenic features of cancer cells: it causes damage to DNA and activates pro-tumorigenic signaling pathways together with an enhancement of cell survival and the ability to escape from apoptotic mechanisms [10], allowing cancer cells to survive even with compromised genetic integrity. To counteract the pro-oxidant environment, which threatens their viability, cancer cells upregulate their antioxidant defense mechanisms, ensuring their survival.

Neurodegenerative diseases are age-related conditions of multifactorial aetiology. Among the multiple processes involved in disease appearance and progression, oxidative stress, and sustained neuroinflammation are known to contribute to neurodegeneration. In Alzheimer's disease, oxidative stress is triggered by neurotoxic events, most of them linked to the toxicity of misfolded proteins such as the amyloid beta and tau proteins. The resulting chronic neuroinflammation is driven by a dysregulated microglial response and, far from having a healing effect, it causes the disease to progress and spread [11]. In Parkinson's disease, for instance, oxidative stress is responsible for the degeneration of dopaminergic neurons [12].

Sustained hyperglycaemia in diabetes increases the levels of ROS while also promoting inflammatory responses. Oxidative stress has an impact on the complications derived from diabetes by causing damage to nerves, blood vessels and kidneys [13, 14]. In addition, metabolic syndrome patients suffer from a noticeable impairment in the antioxidant defense system [15].

18.4 Cyclodextrins for Sensing ROS and RNS

An early example of the use of cyclodextrins for improving the detection of hydrogen peroxide is based on the chemiluminescence from its reaction of 10,10′-dimethyl-9,9′-biacridinium dinitrate (lucigenin), which was found to be improved in organized media, generated by addition of surfactants or β-cyclodextrin (Fig. 18.6) [16]. Several cyclodextrin derivatives have also been exploited to promote the chemiluminescence of luminol [17, 18].

A simple electrochemical sensor for ROS was designed based on the redox properties of coenzyme Q_{10} (CoQ_{10}), a natural quinone. This compound was immobilized at the surface of glassy carbon electrodes, combined with α- or β-CD, whose role is to ensure the formation of a well-dispersed CoQ_{10} film, and a Nafion® membrane to ensure sensor stability. This sensor was applied to the detection of oxidant molecules, hydrogen peroxide, and the superoxide radical using fixed potential chronoamperometry and square wave voltammetry (SWV), based on the redox equilibrium between CoQ_{10} and its reduced hydroquinone form (Fig. 18.7) [19].

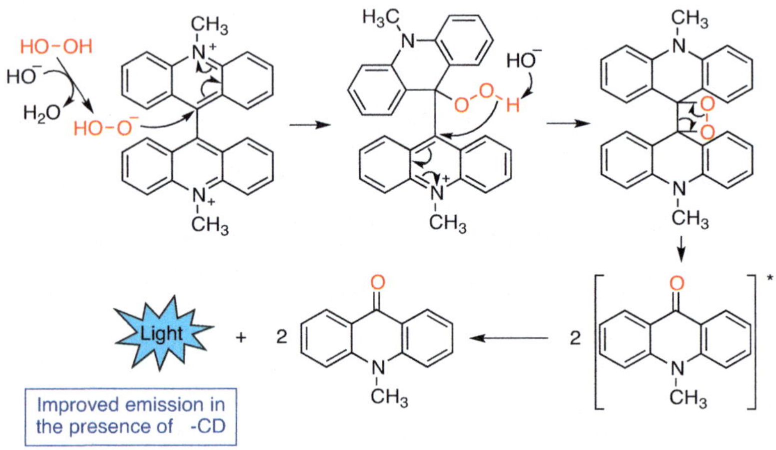

Fig. 18.6 β-cyclodextrin improves the lucigenin chemiluminescence reaction

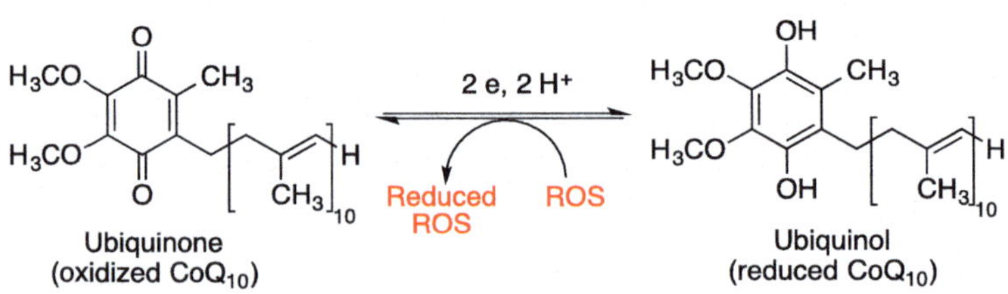

Fig. 18.7 The ubiquinone-ubiquinol redox equilibrium

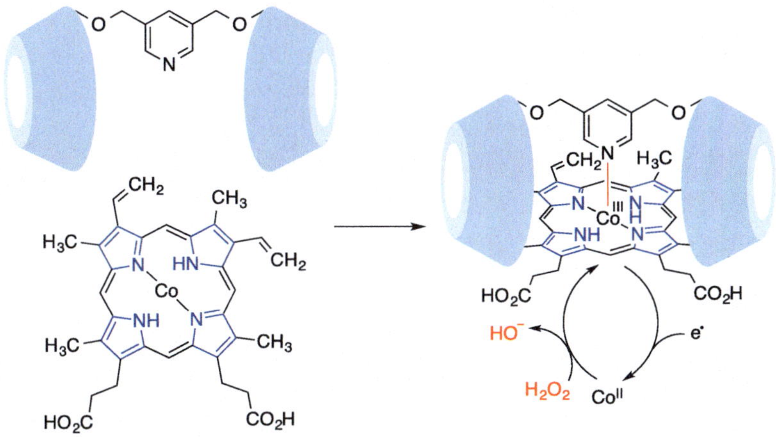

Fig. 18.8 Formation of an inclusion complex with cobalt protoporphyrin (Co^{III}-PPIX@Py2CD) and cobalt-catalyzed reduction of hydrogen peroxide

A per-*O*-methylated-cyclodextrin dimer, functionalized with 3,5-bis(bromomethyl)pyridine, formed an inclusion complex with cobalt protoporphyrin (Co^{III}-PPIX@Py2CD) where the pyridine nitrogen is linked to Co^{III}. This complex was applied as a catalyst for the electrochemical reduction of hydrogen peroxide (H_2O_2), achieving a wide linear range for H_2O_2 detection and a detection limit in the 10^{-7} M range (Fig. 18.8) [20].

A cyclodextrin-based activatable duplex reporter was reported that allowed real-time in vivo imaging of contrast-induced acute kidney injury. This molecule, which has a high renal clearance thanks to its cyclodextrin-based structure, contains a chemiluminescence signaling moiety that is activated by oxidative stress (e.g. superoxide anion). Additionally, it has a near-infrared fluorescence (NIRF) that is responsive to lysosomal damage (releasing *N*-acetyl-β-D-glucosaminidase, NAG). It therefore detects sequentially the upregulation of superoxide and NAG in the kidneys of living mice, outperforming prior assays (Fig. 18.9) [21].

In order to measure ROS production inside living cells, radical trapping using cyclic nitrones has some advantages over other methods based on fluorescence or chemiluminescence, particularly its specificity for superoxide anion and hydroxyl radical. These traps, such as 5,5-dimethyl-1-pyrroline *N*-oxide (DMPO), are able to form stable radical adducts used in electron spin resonance spectroscopy (ERS) to detect and study free radicals. To overcome problems such as the low capture rate of these traps and their spontaneous, cell-enhanced decomposition, new cyclic nitrones containing a permethylated β-cyclodextrin moiety (CD-DIPPMPO) have been synthesized and shown to be particularly sensitive and specific of extracellular superoxide (Fig. 18.10). This stability is attributed to the extracellular localization of the nitroxyl adduct, which prevents the reduction of the adducts by cellular antioxidants such as ascorbic acid and glutathione [22].

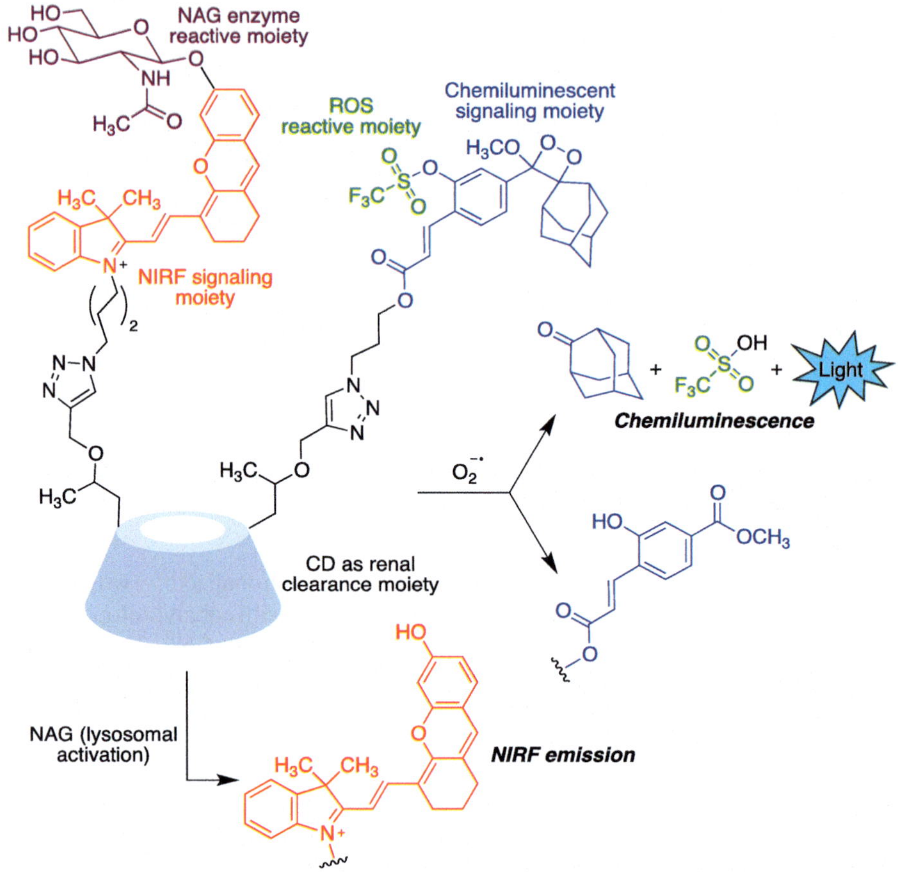

Fig. 18.9 Dual chemiluminescent and fluorescent sensor for the sequential detection of superoxide radical anion and lysosomal damage in contrast-induced acute kidney injury

Nitric oxide is the primary reactive nitrogen species. One traditional method for its determination involves its reaction with 2,3-diaminonaphthalene (DAN) to give a 2,3-naphthotriazole, which needs to be deprotonated with a strong alkali to its anion, which has a much higher quantum yield. The formation of a DAN-β-cyclodextrin complex was found to still allow the formation of the 2,3-naphthotriazole derivative, which remains in the cyclodextrin. This increases its fluorescence quantum yield even at neutral pH, allowing a greater generality of the method (Fig. 18.11) [23].

In another approach, an electrochemical assay for sensing $NO^{\cdot}$ in biological systems has been described, based on its electrochemical reduction to NO^{-} by ferrocene, facilitated by a gold nanocomposite-modified glassy carbon electrode. In this method, ferrocene was introduced as a hydroxypropyl-β-cyclodextrin complex in order to solubilize it. This protocol achieved good sensitivity (0.03 nA/nM), detection limit (25.75 nM), and response time (<5 s) [24].

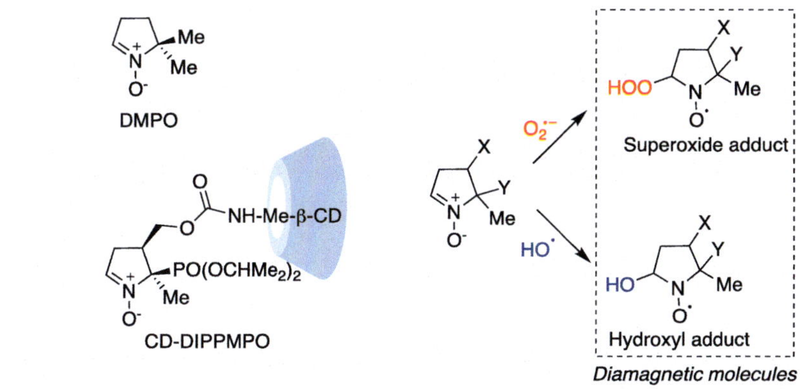

Fig. 18.10 Structures of DMPO and CD-DIPPMPO and their spin-trapping reactions with superoxide and hydroxyl adducts

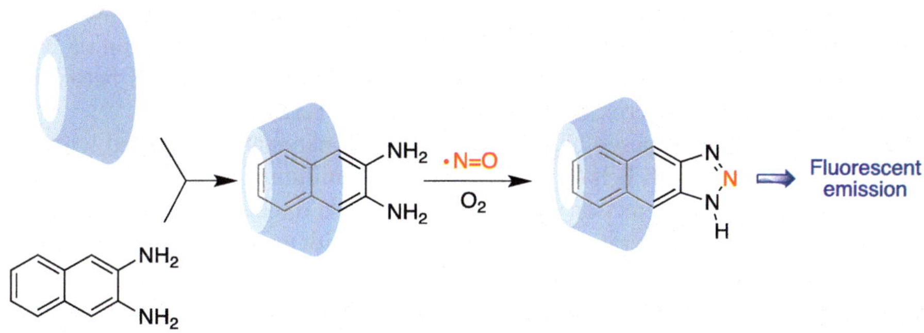

Fig. 18.11 Inclusion complex of 2,3-diaminonaphthalene with β-cyclodextrin for the sensitive detection of nitric oxide

18.5 Nanostructured Systems Based on Cyclodextrins for Sensing ROS and RNS

Nanostructured systems for the detection of ROS

Some CD-based H_2O_2 sensors are based on nanoparticles of certain metals or polymers with cyclodextrins. Sometimes ferrocene is included inside the cyclodextrins in order to improve the stability of the nanoparticle, to which a fluorescent (fluorescein) or chemiluminescent (luminol) compound is added for H_2O_2 detection. In recent clinical practice, smart electrochemical biosensors in which a bioactive molecule (cytochrome-c, catalase, peroxidase) is responsible for ROS detection have emerged as promising analytical tools. Some examples of nanostructured systems for ROS detection are discussed below.

One simple example of such nanostructures are silver nanoparticles stabilized with a β-cyclodextrin shell, which can act as sensitive and low-cost colorimetric

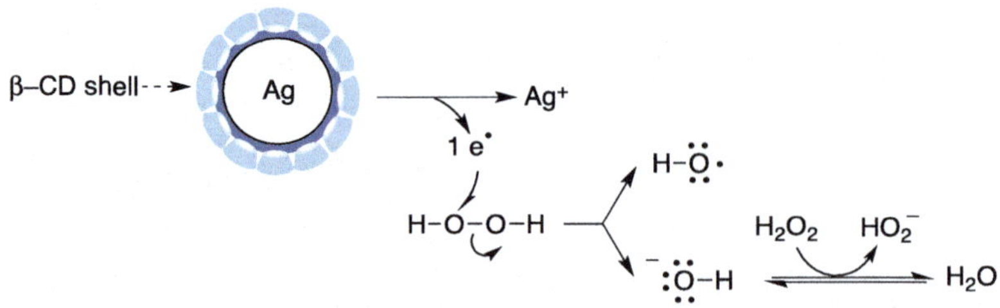

Fig. 18.12 Mechanism of β-cyclodextrin-Ag nanoparticle reaction with hydrogen peroxide

sensors for urine samples [25]. The detection of H_2O_2 is based on the fact that, when the nanoparticles react with this species, an oxidative dissolution of the nanoparticles occurs, releasing silver ions (Ag^+) into the medium. This process causes a change in the color of the solution from olive green to colorless, which correlates directly with the concentration of H_2O_2. The change in absorbance, measured at 384 nm by UV-Vis spectroscopy, serves as a quantifiable signal to determine H_2O_2 levels with high sensitivity and reproducibility. The limits of detection and quantification of H_2O_2 by this method are much lower than those of commercial horseradish peroxidase (HRP)-based hydrogen peroxide detection kits (Fig. 18.12).

Inclusion complexes based on magnetic $CoFe_2O_4$ nanoparticles coated with a β-cyclodextrin shell, which behave as catalysts in chemiluminescence systems, have been proposed for the detection of hydrogen peroxide under basic conditions. The cyclodextrin moieties generate inclusion complexes with the luminol anion, bringing together the catalytic metal and the chemiluminescence promoter. The nanoparticles decompose hydrogen peroxide into hydroxyl radicals ($HO^{\cdot}$) and also generate superoxide anions ($O_2^{\cdot-}$), which react with luminol to generate the excited 3-aminophalate anion (3-APA*), responsible for light emission (Fig. 18.13). This method offers high sensitivity and reproducibility and is suitable for the detection of hydrogen peroxide in water samples with a detection limit of 2.0×10^{-8} mol/L [26].

Following the implementation of organic superbases such as 1,5,7-triazabicyclo[4.4.0]dec-5-ene (TBD) as a way to increase chemiluminescent emission of luminol [27], the same authors described a self-reporting supramolecular system based on luminol-generated chemiluminescence in biological environments. It involves the formation of host–guest complexes between luminol and a copolymer decorated with TBD, which is included in randomly methylated β-cyclodextrin units (Me-β-CD). These complexes encapsulate luminol and the superbase into the β-cyclodextrin, leading to the subsequent controlled release of luminol under oxidative stress. The host–guest interaction breaks down in the presence of H_2O_2, leading to the oxidation of luminol to the blue-emitting excited state of 3-aminophthalic acid (3-APA) (Fig. 18.14). This design does not require additional bases or catalysts for strong light emission, facilitating real-time detection of ROS in biological environments [28].

Another example of nanoparticles designed for hydrogen peroxide (H_2O_2) detection are supramolecular fluorescent nanoparticles that self-assemble via non-covalent

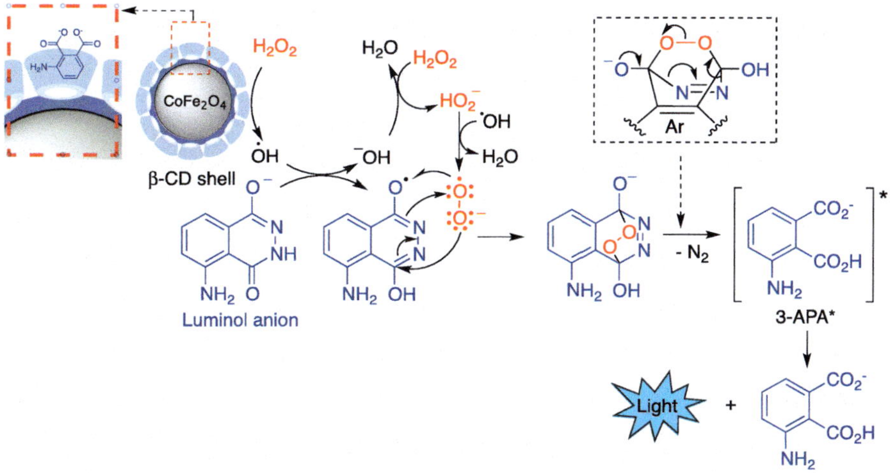

Fig. 18.13 Proposed chemiluminescence process of luminol-H_2O_2 in the presence of inclusion complexes between β-CD and magnetic $CoFe_2O_4$ nanoparticles

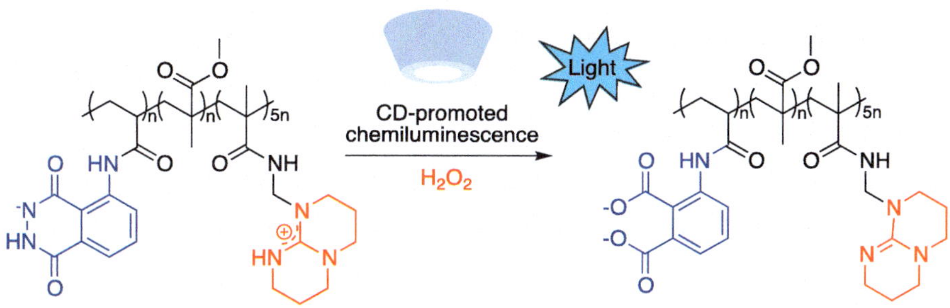

Fig. 18.14 General reaction pathway of the host–guest complexation reaction and oxidation process, triggered by the addition of H_2O_2

interactions between two key components, fluorescein isothiocyanate (FITC) bound to β-cyclodextrin (β-CD) and ferrocene (Fc)-modified rhodamine B (RB). The interaction between the host (FITC-β-CD) and the guest (Fc-RB) allows the creation of a fluorescence resonance energy transfer (FRET) effect that gives the particle red coloration. This fluorescence changes from red (RB) to green (FITC) in the presence of H_2O_2, as the FRET effect is disrupted by decomposition of the structure owing to loss of affinity of ferrocene for the cyclodextrin cavity upon oxidation of Fe^{2+} to Fe^{3+} [29]. This method provides a sensitive method (10 μM), for the detection of H_2O_2 in cancer cells, which usually have higher concentrations of this compound compared to normal cells (Fig. 18.15). In vitro studies demonstrated that the nanoparticles could be efficiently internalized by cancer cells (HeLa) [30].

Atherosclerosis is the result of endothelial injury, progressive local inflammation, and excessive lipid accumulation. Atherosclerotic plaques rich in foam cells

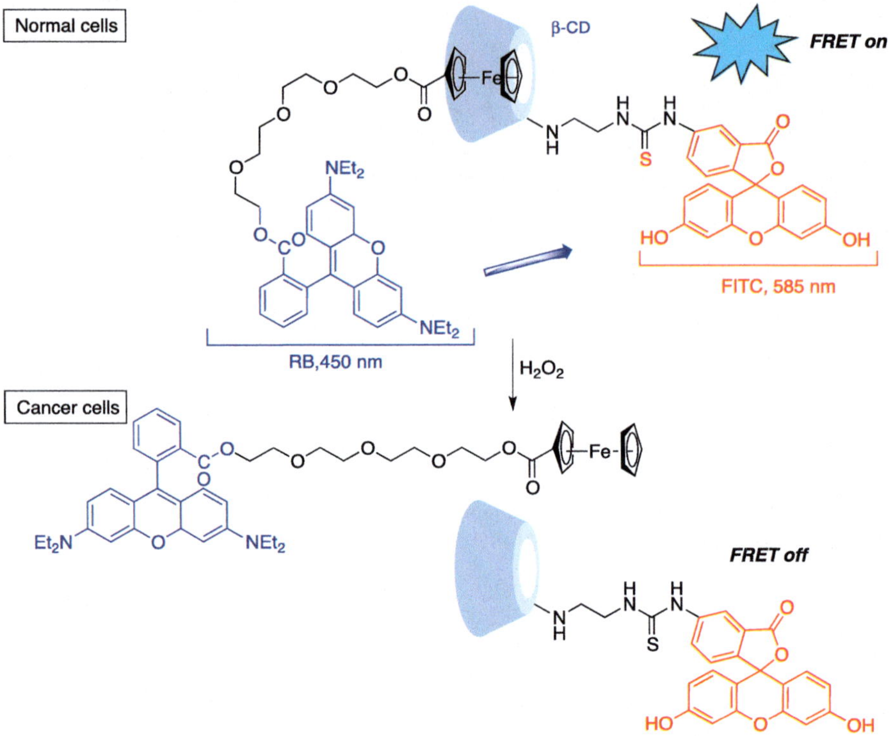

Fig. 18.15 A cyclodextrin-based nanoparticle system that detects H_2O_2 based on FRET quenching upon cyclodextrin-ferrocene complex disruption subsequent to Fe^{2+} oxidation

are prone to rupture and form thrombi, which can cause heart attacks and strokes. Early assessment of the vulnerability of these plaques is crucial to reduce cardiovascular disease-related mortality. To this end, nanoparticles have been designed that respond to reactive oxygen species (ROS) and have dual targeting capabilities: they use hyaluronic acid (HA) to bind to the CD44 receptor, present on endothelial cells under inflammation, and a specific peptide (PTP) to bind to phosphatidylserine, a marker on the surface of apoptotic foam cells. These nanoparticles also include a ferrocene-derived triphenylamine derivative (FC-TPA) fluorescent probe that is selectively released in the presence of ROS and produces lipid-specific imaging enhancement. This probe, which has the ability to switch between fluorescent states when protonated or unprotonated, is prepared by encapsulation in PTP-modified cyclodextrins (CD), and the entire complex is coated with hyaluronic acid (HA) to form the nanoparticles (NP). The design allows the nanoparticles to accumulate in atherosclerotic plaques and be released in response to elevated ROS levels, providing targeted, lesion-site-specific release (Fig. 18.16). Both in vitro, with foam cells, and

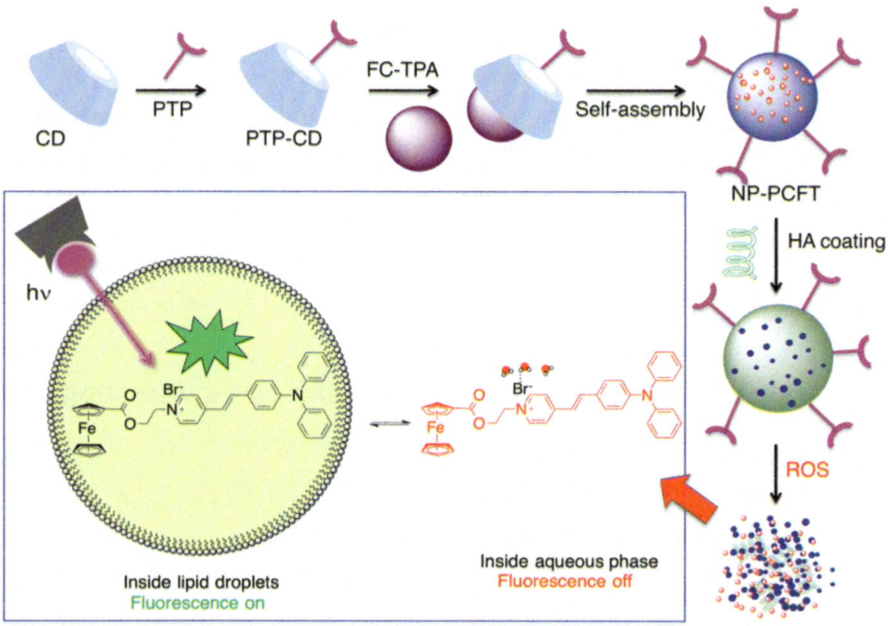

Fig. 18.16 Cyclodextrin-containing nanoparticles for the detection of ROS in atherosclerotic lesions

in vivo, in mice with atherosclerosis, the nanoparticles achieved selective and effective accumulation in these cells in atherosclerotic plaques, allowing detection of plaque lipid burden, highlighting their potential as a diagnostic tool [31].

Other sensors using advanced imaging techniques such as electron paramagnetic resonance (EPR), magnetic resonance imaging (MRI), and optical imaging to determine the total redox state (TRC) and oxidative stress in cells and tissues are based on quantum dots (QD) coated with cyclodextrins functionalized with multiple nitroxide groups and triphenylphosphonium (TPP).

Two sensors have been designed, the TRC sensor (QD@CD-TEMPO) and the OxiStress sensor (QD@CD-TEMPOH). The former is in a paramagnetic form that generates an EPR signal and MRI contrast, while the latter is diamagnetic and fluorescent. Both sensors can switch between these two forms depending on the redox environment, allowing accurate tracking of changes in the balance between oxidants and reductants. The fluorescence of the QDs is quenched by nitroxide radicals in the TRC form but is restored when the nitroxides are reduced to hydroxylamines in the OxiStress form (Fig. 18.17). The sensors were tested in mice with hypercholesterolaemia-induced renal dysfunction, allowing them to track changes in the redox state of renal tissues by MRI. The results showed that healthy tissues have a high reducing capacity, which is lower in tissues affected by oxidative stress. Their ability to combine fluorescence and magnetic resonance imaging offers a powerful

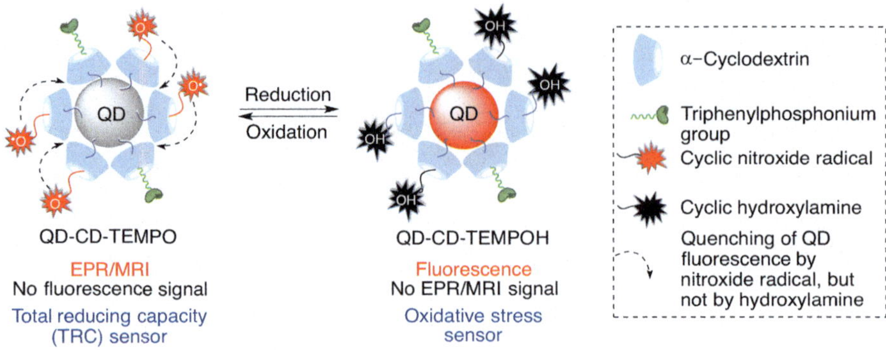

Fig. 18.17 Schematic representation of quantum sensors: TRC sensor (QD@CD-TEMPO) and OxiStress sensor (QD@CD-TEMPOH)

platform for the diagnosis and monitoring of oxidative stress-related diseases such as cancer, atherosclerosis, and neurodegeneration [32].

A biosensor for electrochemical detection of H_2O_2 was prepared on a glassy carbon electrode with catalase, an H_2O_2 decomposing enzyme, immobilized by chitosan (CTS-CAT) and ferrocene, which was included in β-cyclodextrin to increase its stability (CD-FE complex) (Fig. 18.18). The CTS-CAT electrode produced a strong reduction peak current in response to H_2O_2 and the CD-FE complex amplified the current, allowing the detection of H_2O_2 concentrations ranging from 1.0×10^{-7} to 6.0×10^{-3} M [33].

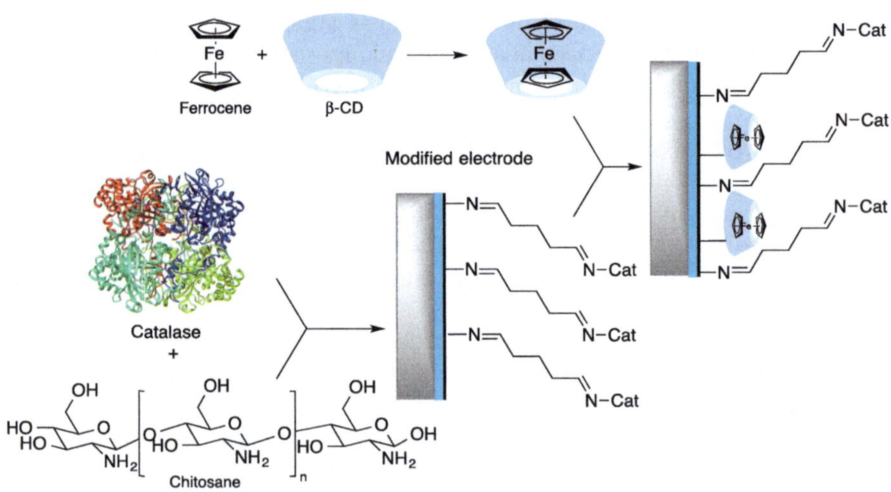

Fig. 18.18 Schematic diagram of the CTS-CAT/CD-FE composite electrode. The catalase molecule was generated from PDB 1dgf

Another electrochemical biosensor model uses biocompatible hybrid hydrogels integrated with gold nanocubes (AuNCs) for the detection of H_2O_2. It is based on a hydrogel combining chitin (CS) and β-cyclodextrin (β-CD), and was obtained by graft copolymerization. Its porous structure allows for adequate immobilization of biomolecules and nanoparticles. By integrating AuNCs into this matrix, due to their cubic shape and plasmonic properties, a significant improvement in the electrical conductivity of the hydrogel was achieved, which increased the sensor's ability to transfer electrons and also provided a larger surface area for the immobilization of biomolecules. As a biological receptor for H_2O_2 detection, cytochrome c was used, a redox-active protein that plays an essential role in electron transfer reactions within the mitochondrion. Cytochrome c was immobilized on the hydrogel nanocomposite matrix by cross-linking with thiol bonds of 3-mercaptopropionic acid (MPA). The reduction of hydrogen peroxide to water, driven by the redox activity of cytochrome c, generates a detectable electrochemical signal with a high sensitivity and a low detection limit (15×10^{-9} M) for H_2O_2. The hydrogels improve biocompatibility and preserve enzymatic activity in vitro when immobilizing bioactive molecules against other compounds present in biological environments, such as ascorbic acid and lactic acid, which makes them potentially interesting for the development of implantable diagnostic devices in the future [34].

Nanostructured systems for the detection of RNS

Several cyclodextrin-based strategies are available to detect RNS, specifically to detect nitric oxide (NO) in living cells, both in humans and plants. One of them is based on the phenomenon of plasmon resonance energy transfer (PRET) between plasmonic nanoparticles, such as gold nanoparticles (GNPs), and organic dyes, such as rhodamine B-derived molecules (RdBMs). A PRET nanosensor contains a GNP bound to several β-CDs by hydrophobic interactions. A rhodamine derivative is then inserted into the CD cavity via hydrophobic interactions. The non-fluorescent rhodamine reacts with NO to form rhodamine B (RdB), which is fluorescent. Under such conditions, the PRET effect between the GNP and RdB occurs (Fig. 18.19). This implies that the absorption peak of RdB overlaps with the scattering peak of GNP@CD and this causes the scattering intensity of GNP@CD to decrease. This strategy is sensitive to NO concentration, and allows quantitative determination of NO in solution and single-particle imaging of exogenous and endogenous NO in living cells [35].

Another technique uses a de novo host–guest self-assembled nanosensor based on β-CD polymers, which the authors have named CTA, to detect NO in cells and in inflammatory mice models. Overproduction of NO is known to be related to the development of inflammatory diseases. The CTA nanosensor consists of an organic molecule (RN-adH) containing a naphthalimide luminescent group, a rhodamine B fluorophore (internal reference for self-calibration) and an adamantyl group whereby RN-adH is inserted into the β-CD polymer. By self-assembly, the CTA nanosensor is generated. RN-adH has an *o*-phenylenediamine in its structure, which reacts with NO to furnish a triazole derivative. This structural modification causes changes in its absorption spectra, and hence in the color of the CTA nanosensor, and a change

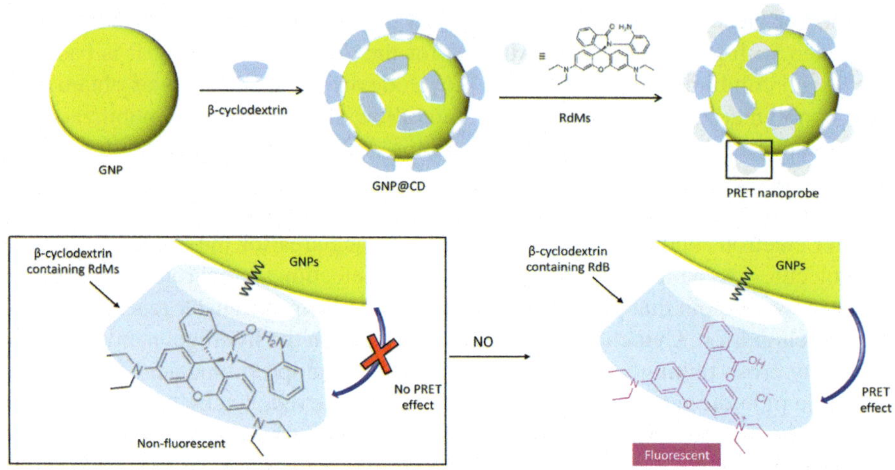

Fig. 18.19 Synthesis of the PRET nanoprobe. Reaction of RdMs (non-fluorescent) with NO to form RdB (fluorescent)

in its fluorescence emission via the FRET effect (Figs. 18.20 and 18.21). Since the fluorescence intensity is directly proportional to the NO concentration, it enables its quantitative determination. The measurements need to be performed under acidic conditions since the fluorescence intensity of the CTA nanosensor is weak in the presence of base [36].

The nanoprobe BDNA@HPβCD is used to detect NO in animals and plants, as well as to localize liver damage in mouse models. It consists of a molecular probe (BDNA) and 2-hydroxypropyl-β-cyclodextrin (HPβCD). BDNA has a benzothiadiazole core with a *o*-phenylenediamine fragment and, at each end of the core, two phenylnaphthynamine groups are attached. The naphthyl groups of BDNA are accommodated into the hydrophobic cavity of HPβCD, thus forming the BDNA-HPβCD complex. This complex self-assembles into nanoaggregates in aqueous media to become the activatable nanoprobe BDNA@HPβCD, which is almost non-emissive in the absence of NO. However, in the presence of NO, the *o*-phenylenediamine group of BDNA reacts with NO to form a triazole, leading to the activated probe (BDNANO). This results in strong fluorescence emission at 938 nm ($\lambda_{exc} = 808$ nm) and an intense absorption signal at 680 nm, which allows for near-infrared II (NIR-II) fluorescence imaging and optoacoustic imaging of NO, respectively. Both fluorescence emission and absorption increase with increasing NO concentration (Fig. 18.22) [37].

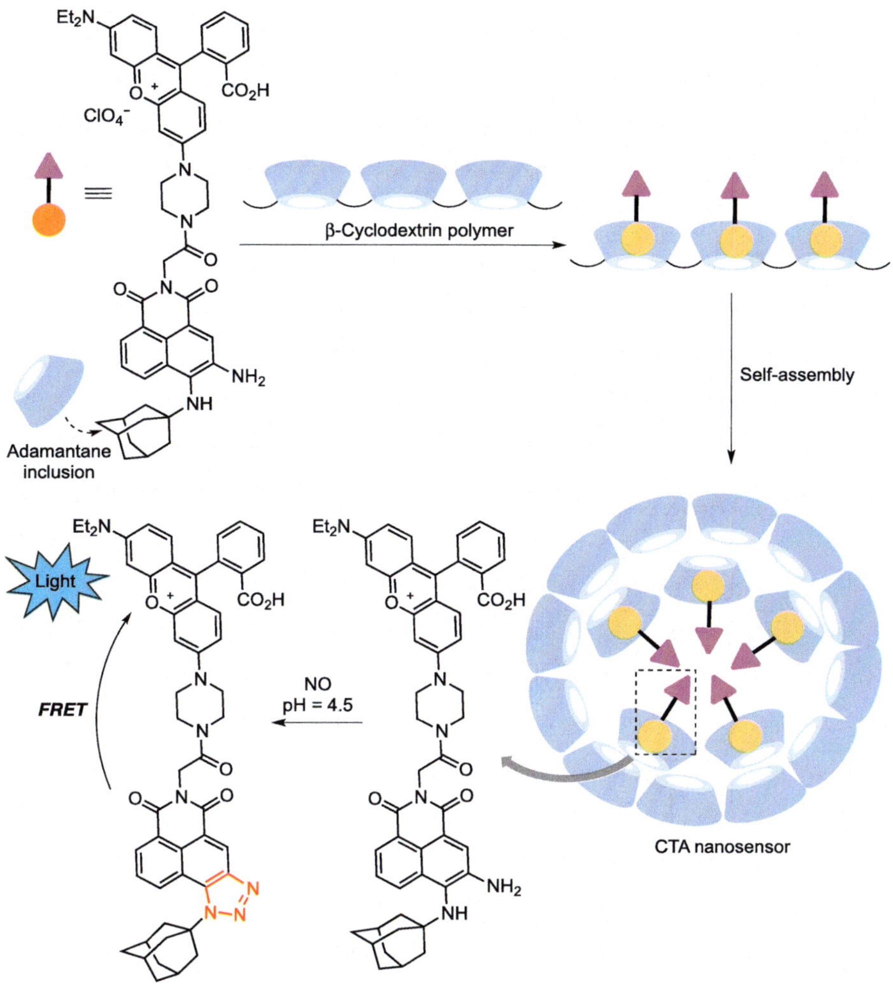

Fig. 18.20 Construction of the CTA nanosensor and change in the chemical structure of RN-adH upon reaction with NO

18.6 Miscellaneous Cyclodextrin-Based, ROS-Responsive Diagnostic Nanosystems

Biosensors

Some biosensors have been designed for the electrochemiluminescent (ECL) recognition of biomolecules such as proteins based on electrodes coated with cyclodextrin nanoparticles that amplify their signal by the presence of ROS. One example is an electrochemiluminescent (ECL) immunosensor for the detection in human serum

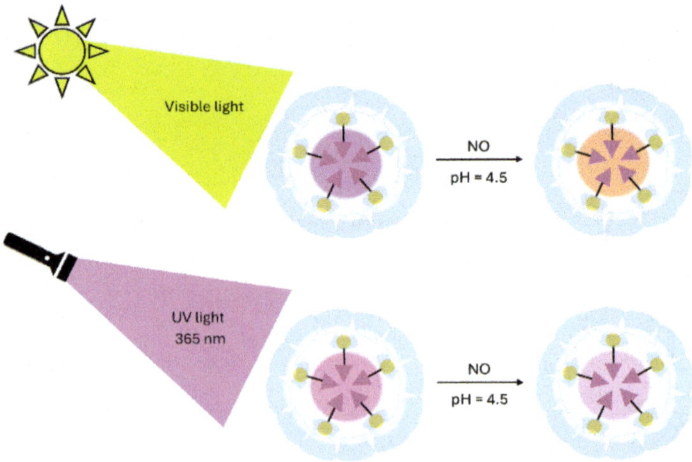

Fig. 18.21 Color change of the CTA nanosensor before and after reacting with NO

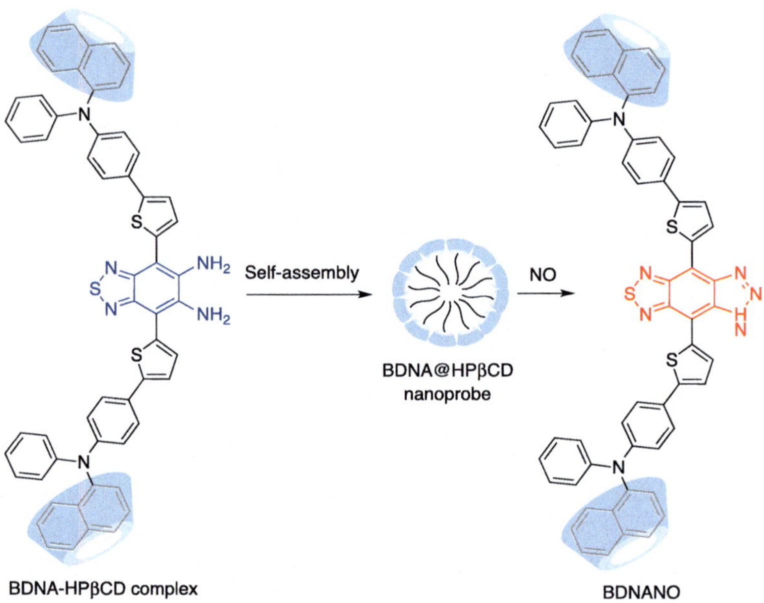

Fig. 18.22 Assembly of the activatable nanoprobe BDNA@HPβCD and its reaction with NO to form the activated probe BDNANO

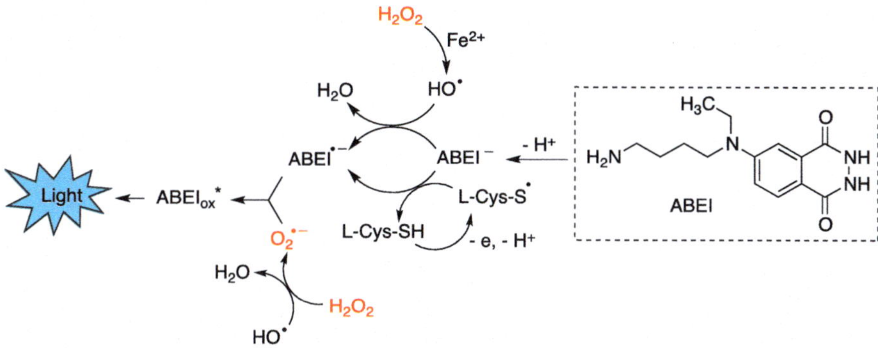

Fig. 18.23 Chemoluminescence generated from ABEI (see Fig. 18.13 for additional chemical details)

samples of β_2-microglobulin (β_2-M), a clinically relevant biomarker for the detection of chronic kidney disease. In this sensor, β-cyclodextrin and ferrocene form a functional host–guest complex and the luminophore used is *N*-(aminobutyl)-*N*-ethylisoluminol (ABEI), a luminol derivative that is integrated into the immunosensor structure to increase electron transfer efficiency and enhance the ECL signal (Fig. 18.23). In addition, hollow gold nanoparticles (HAuNPs) are used to load cysteine (L-Cys), which acts as a coreactant and signal amplifier. The ferrocene not only catalyzes the decomposition of hydrogen peroxide, generating hydroxyl and superoxide radicals, but also promotes a higher signal intensity in the detection of β_2-M. In the immunosensor, β_2-M is captured on the surface of the modified electrode using specific antibodies. Subsequently, the ABEI-β-CD-Fc-L-Cys complex binds to β_2-M, which reduces the distance between the luminophore and the coreactant, increasing the efficiency of the ECL signal. This process allows accurate detection of β_2 with a significantly lower detection limit compared to other analytical methods and with excellent selectivity and stability [38].

An additional biosensor, designed for the detection of methicillin-resistant *Staphylococcus aureus* (MRSA), is based on a sandwich methodology where the electrode is coated with β-cyclodextrin-conjugated luminol nanoparticles, which act as anodic luminophores. The biosensor includes a cathodic luminophore based on an organic material with a metal center, MOF, zinc-modified porphyrin (PCN 224/Zn), which, in conjunction with a recombinant phage cell binding domain, captures MRSA. This dual design allows the system to detect and distinguish MRSA signals across two different potentials: an anodic and a cathodic potential. The mechanism of the biosensor works through competition between reactive oxygen species (ROS) produced in ECL reactions. Both the anodic and cathodic luminophore consume ROS in different electrochemical processes, which generate opposite signals. The PCN 224/Zn-based luminophore accelerates the reduction of $S_2O_8^{2-}$ to produce sulfate radicals, while the H_2O_2-activated luminol emits in the anodic process. The system also uses resonance energy transfer (FRET) between the two luminophores, which improves the sensitivity of the biosensor [39].

Theranostics

Theranostic agents, i.e., multifunctional systems capable of exerting a therapeutic effect and at the same time providing diagnostic information, are highly valuable in the search for pharmaceuticals active against oxidative stress. This double potential results in a synergistic approach where the therapeutic effect of the system can be tracked yielding key information in preclinical stages.

An example of this approach can be found in the design of mesoporous silica nanoparticles (MSNP) functionalized on their surface with a non-fluorescent fluorescein derivative capable of reacting with hydrogen peroxide and undergoing a chemical reaction that yields a highly fluorescent compound (Fig. 18.24) [40]. Moreover, these nanoparticles are loaded with captopril, a heart failure drug, and further modified by the addition of α-cyclodextrin that forms a host–guest complex with the non-fluorescent fluorescein derivative located at the surface of the nanoparticles. The addition of α-cyclodextrin prevents captopril from being released. When in the presence of high concentrations of hydrogen peroxide, this ROS reacts with the non-fluorescent fluorescein derivative yielding a strongly fluorescent molecule and, at the same time, breaks the host–guest complex releasing the α-cyclodextrin scaffolds and, thus, letting captopril out of the nanoparticles. This system proved its potential in biological systems in a transgenic zebrafish model of heart failure with promising therapeutic outcomes. Thus, these modified nanoparticles represent a dual functional system capable of providing diagnostic information on the concentration of hydrogen peroxide, a key player in the redox metabolism and a potentially harmful oxidant, and releasing a heart failure drug, captopril, right at the site of higher oxidative stress.

Nanogels are convenient delivery systems for diagnostic and therapeutic purposes consisting of polymeric chains capable of carrying drugs within them. One example

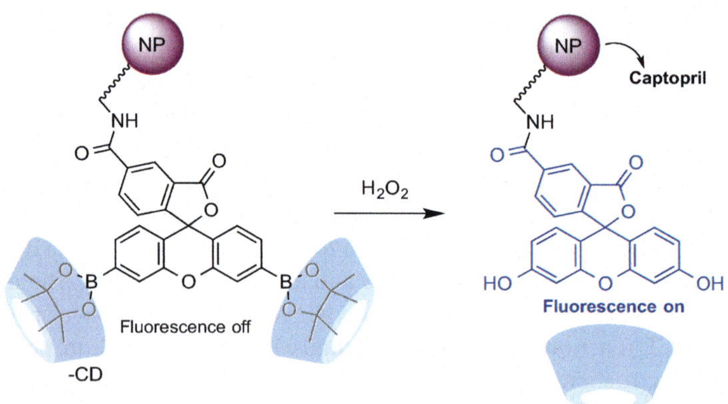

Fig. 18.24 Multifunctional cyclodextrin-based system for the simultaneous detection of hydrogen peroxide and targeted release of captopril in heart failure. NP: nanoparticle

representing this approach consists in the design of cyclodextrin-based nanogels loaded with luminol and doxorubicin [41]. In these nanogels, luminol undergoes a quenching of its native fluorescence when in the presence of hydrogen peroxide due to its decomposition. This way, the system is able to detect high levels of hydrogen peroxide, typical of inflammation-related processes, by the detection of fluorescence quenching. In addition, the nanogel exhibits an enhanced release of doxorubicin, an anticancer drug, in the presence of high levels of hydrogen peroxide. Thus, this dual-function diagnostic and therapeutic system is a potential theranostic tool against cancer cells, which show higher concentrations of ROS, including hydrogen peroxide, than healthy cells.

The use of cyclodextrins in systems to detect and scavenge ROS and RNS is highly interesting for diagnostic and therapeutic purposes in diseases where oxidative stress is present. Conversely, complementary approaches where cyclodextrin-based systems produce ROS are interesting for different applications like the detection of analytes or the use of localized oxidative stress as a therapeutic tool. In this context, a nanostructured system of mannose-modified β-cyclodextrin (CDM) supported on graphene oxide (GO) and graphitic carbon nitride (g-C_3N_4, CN) was designed as an antibacterial agent (Fig. 18.25) [42]. CDM works as a bait to capture *Escherichia coli* bacteria, GO is responsible for the generation of ROS upon light irradiation (photodynamic therapy) and the CN fraction releases heat upon exposure to light (photothermal therapy). This way, a dual and synergistic mechanism based on ROS and heat is selectively targeted to kill bacteria.

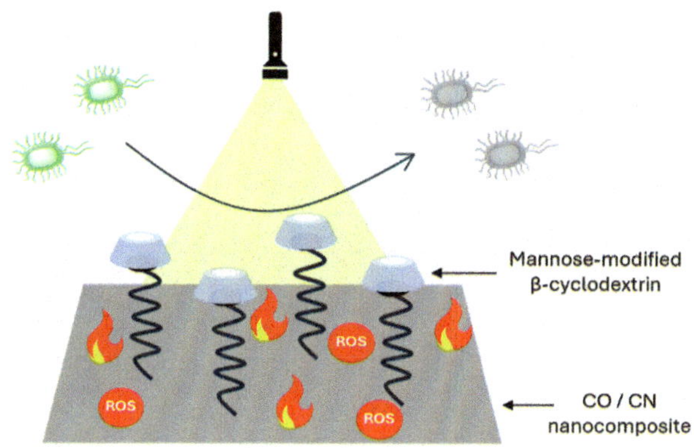

Fig. 18.25 Mannose-modified β-cyclodextrin-based nanosystem for the selective capture and destruction of bacteria through ROS and heat production upon light exposure

18.7 Future Perspectives in Cyclodextrin-Based Nanostructured Materials for ROS Sensing

Nanofibers based on cyclodextrins are particularly attractive nanostructures, since they are nontoxic and incorporate only cyclodextrins and antioxidant or sensing molecules, being free of polymeric materials [43, 44]. Moreover, for sensing purposes, they do not require performing additional reactions, since the sensor is incorporated into the cyclodextrin forming the nanofiber. Nanofibers can be produced by the association of cyclodextrin units into filaments by the electrospinning technique, which involves the application of an electrical field to the tip of a needle inside which flows a dispersed/jetting viscous solution producing a "Taylor cone" and developing the filaments that are solidified and collected on a surface by electrical forces (Fig. 18.26) [45]. This method generates nanofibers with are highly reproducible in terms of morphological, porosity, and solubility characteristics, providing a decisive advantage over [46]. As an example, an important increase (tenfold) in the antioxidant capacity of resveratrol was observed for the resveratrol/HPβ-CD and resveratrol/HPγ-CD nanofibrous films compared to the resveratrol/pullulan nanofibrous entities, a difference that can be attributed to increased solubility and chemical stability. The application of CD-based nanofibers to ROS sensing is still in its infancy and will undoubtedly constitute one of the main topics of progress in this area.

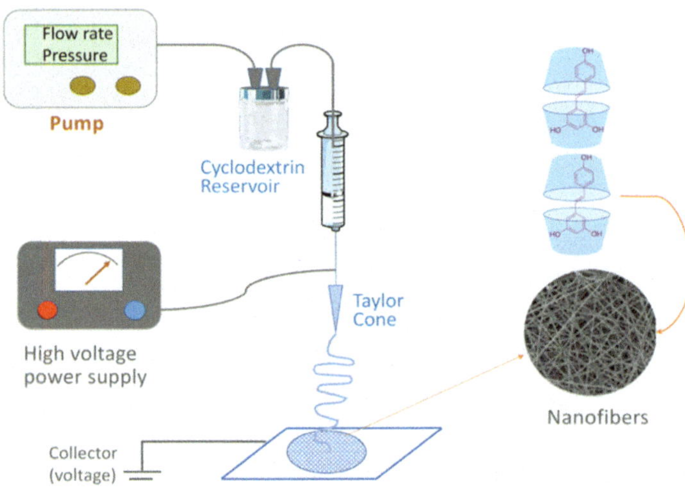

Fig. 18.26 Scheme of the electrospinning technique for the synthesis of resveratrol-cyclodextrin nanofibers

18.8 Cyclodextrin-Based Systems for the Determination of ROS Scavenging Activity

The non-polar cyclodextrin cavities constitute a very favorable environment capable of accommodating, in a confined space, a large variety of molecules, altering their chemical behavior and reactivity. Thus, CDs serve as vehicles for drugs and bioactive substances and protect them against possible ROS and RNS species. Native and chemically modified CDs, as well as a variety of nanostructured materials containing CDs, facilitate reactions that are not possible in other environments due to the proximity of reagents inside CDs cavities. Several reviews have been devoted to CD-mediated catalysis [47], and in particular to CD-promoted oxidation and reduction reactions [48, 49].

The antioxidant activity of compounds such as polyphenols, vitamins C and E, and flavones is significantly increased when they are found forming inclusion complexes with CDs or with CD-based nanoparticles and nanofibers. This effect can be explained by the increase in the solubility of the bioactive compounds in the presence of CDs, as well as by the possible influence of the CDs on the outcome of the reactions with the different probes used (e.g. DPPH). However, the benefits and drawbacks that the use of cyclodextrins may pose in the determination of antioxidant capacity need to be considered and critically analyzed. As it is well known, the inclusion in cyclodextrin cavities increases fluorescence, and therefore, in those reactions where fluorimetry is used as a method for quantification of antioxidant activity such as the case of ORAC and ABTS, it should be taken into consideration that interferences may occur if CDs are used to encapsulate bioactive compounds [50]. Table 18.1 summarizes the modifications in antioxidant capacity for a set of bioactive compounds encapsulated in CDs or their derivatives.

18.9 Conclusions

Cyclodextrins are versatile organic molecules capable of allocating small molecules within their hydrophobic cavity, resulting in an enhancement of their stability and water solubility. In addition, cyclodextrins can be included as part of supramolecular nanosystems such as polymers, nanoparticles, and nanofibers, leading to novel and enhanced capabilities. These features have been exploited for the design of tools for the detection and monitoring of ROS and RNS with different purposes. Different strategies are employed in the design and implementation of tools for the sensing of these oxidative species, including a broad variety of mechanisms and proving the versatility and complexity of existing approaches. This attractive and dynamic field will continue to evolve to the design of novel cyclodextrin-based sensing systems for oxidative species with improved capabilities.

Table 18.1 Modifications in the antioxidant activity (AA) of ROS scavengers in the presence of cyclodextrins (CDs) and nanosystems based on CDs

Sample	Evaluation method	Cyclodextrin/ nanosystem	Comments	Refs.
Rosa damascena essential oil: β-citronellol, geraniol, nonadecane, linalool, and heneicosane	$ABTS^{\bullet+}$. The AA is evaluated by comparison to ascorbic acid and rutine Spectrophotometric determination	β-CD IC*	Increase in 2–3 times the AA of the IC* with regard to free components of the essential oil. AA is increased in long time periods (40 days)	[51]
Naringenin, a natural flavonoid present in citrus fruits	DPPH and $ABTS^{\bullet+}$. The AA is evaluated by comparison to trolox in both cases Spectrophotometric determination	β-CD IC* and carbon quantum dots based on the β-CD IC*	AA for both IC* (33.5% enhancement) and carbon quantum dots IC* (56.9% enhancement) with regard to free naringenin. Potential delivery vehicle for bioactive compounds	[52]
Daidzein, an aglycone-type isoflavone	DPPH method Spectrophotometric determination	Cyclodextrin-based metal–organic frameworks	AA of the IC increased 1.3 times with regard to free daidzein	[53]
Trans-cinnamaldehyde	DPPH, $ABTS^{\bullet+}$, and FRAP methods. Spectrophotometric determination. The AA is evaluated by comparison to vitamin C in all cases	β-CD IC*	AA of the IC is significantly increased with regard to free *trans*-cinnamaldehyde. The AA enhancement is higher for DPPH and $ABTS^{\bullet+}$ assays than for FRAP assay	[54]
Ellagic acid	DPPH method. The AA is evaluated by comparison to vitamin C Spectrophotometric determination	Nanosponges based on β-CD using diphenylcarbonate as crosslinker	No significant differences are observed in the AA of free ellagic acid and the ellagic acid β-CD nanosponges	[55]
Vitamin C	DPPH method Spectrophotometric determination	HPβ-CD nanofibers	AA is increased from 52.61% for free vitamin C to 95.66% for the complexed form of vitamin C in HPβ-CD nanofibers	[56]

(continued)

Table 18.1 (continued)

Sample	Evaluation method	Cyclodextrin/ nanosystem	Comments	Refs.
Resveratrol	DPPH method Spectrophotometric determination	Sulfobutylether β-CD IC and nanoparticles based in sulfobutylether β-CD IC with poly(lactic-co-glycolic acid)	Non-statistically significant differences were observed in the AA of free resveratrol and the resveratrol nanoparticles based on sulfobutylether β-CD	[57]
Resveratrol	DPPH method Spectrophotometric determination	HPβ-CD and HPγ-CD nanofibers	AA of resveratrol is significantly improved for resveratrol/CD nanofibers compared to free resveratrol	[58]
β-Carotene	DPPH method Spectrophotometric determination	HPβ-CD and HPγ-CD nanofibers	AA of the CDs-based nanofibers is slightly decreased with regard to free β-Carotene. Nanofibers enhance the photostability of β-carotene for long periods	[59]

AA: antioxidant activity; $ABTS^{\bullet+}$: 2,2′-azino-bis(3-ethylbenzothiazoline-6-sulfonic acid) CD: cyclodextrin; DPPH: 2,2-diphenyl-1-picrylhydrazyl; FRAP: ferric reducing antioxidant power; HPβ-CD: hydroxypropyl-beta-cyclodextrin; HPγ-CD: hydroxypropyl-gamma-cyclodextrin; IC: inclusion complex; Ref.: literature references

References

1. Kettle AJ, Ashby LV, Winterbourn CC, Dickerhof N (2023) Superoxide: the enigmatic chemical chameleon in neutrophil biology. Immunol Rev 314:181–196
2. Andrés CMC, Pérez de la Lastra JM, Andrés Juan C, Plou FJ, Pérez-Lebeña E (2023) Superoxide anion chemistry—its role at the core of the innate immunity. Int J Mol Sci 24:1841
3. Medraño-Fernández I, Bienert GP, Sitia R (2023) Peroxiporins. CRC Press, Boca Raton
4. Di Marzo N, Chisci E, Giovannoni R (2018) The role of hydrogen peroxide in redox-dependent signaling: homeostatic and pathological responses in mammalian cells. Cells 2018(7):156
5. Halliwell B, Adhikary A, Dingfelder M, Dizdaroglu M (2021) Hydroxyl radical is a significant player in oxidative DNA damage in vivo. Chem Soc Rev 50:8355–8360
6. Prior RL (2015) Oxygen radical absorbance capacity (ORAC): new horizons in relating dietary antioxidants/bioactives and health benefits. J Funct Foods 18:797–810
7. Andrabi SM, Sharma NS, Karan A, Shahriar SMS, Cordon B, Ma B, Xie J (2023) Nitric oxide: physiological functions, delivery, and biomedical applications. Adv Sci 10:202303259
8. Rosselli M (1998) Role of nitric oxide in the biology, physiology and pathophysiology of reproduction. Hum Reprod Update 4:3–24
9. Kim Y-W, West XZ, Byzova TV (2013) Inflammation and oxidative stress in angiogenesis and vascular disease. J Mol Med 91:323–328
10. Arfin S, Jha N K, Jha S K, Kesari K K, Ruokolainen J, Roychoudhury S, Rathi B, Kumar D (2021) Oxidative stress in cancer cell metabolism. 10:642
11. Cai Y, Liu J, Wang B, Sun M, Yang H (2022) Microglia in the neuroinflammatory pathogenesis of Alzheimer's disease and related therapeutic targets. Front Immunol 13:856376

12. Dias V, Junn E, Mouradian MM (2013) The role of oxidative stress in Parkinson's disease. J Parkinson's Dis 3:461–491
13. Jha JC, Ho F, Dan C, Jandeleit-Dahm K (2018) A causal link between oxidative stress and inflammation in cardiovascular and renal complications of diabetes. Clin Sci 132:1811–1836
14. Caturano A, D'Angelo M, Mormone A, Russo V, Mollica MP, Salvatore T, Galiero R, Rinaldi L, Vetrano E, Marfella R, Monda M, Giordano A, Sasso FC (2023) Oxidative stress in type 2 diabetes: impacts from pathogenesis to lifestyle modifications. Curr Issues Mol Biol 45:6651–6666
15. Vona R, Gambardella L, Cittadini C, Straface E, Pietraforte D (2019) Biomarkers of oxidative stress in metabolic syndrome and associated diseases. Oxid Med Cell Longev 2019:1–19. https://doi.org/10.1155/2019/8267234
16. Malehorn CL, Riehl TE, Hinze WL (1986) Improved determination of hydrogen peroxide or lucigenin by measurement of lucigenin chemiluminescence in organised assemblies. Analyst 111:941–947
17. Yuan D-Q, Lu J, Atsumi M, Yan J-M, Kai M, Fujita K (2007) Cerium complexes of cyclodextrin dimers as efficient catalysts for luminol chemiluminescence reactions. Org Biomol Chem 5:2932–2939
18. Maeztu R, Tardajos G, Gonzélez-Gaitano G (2010) Natural cyclodextrins as efficient boosters of the chemiluminescence of luminol and isoluminol: exploration of potential applications. J Phys Chem B 114:2798–2806
19. Barsan MM, Diculescu VC (2019) New electrochemical sensor based on CoQ10 and cyclodextrin complexes for the detection of oxidative stress initiators. Electrochim Acta 302:441–448
20. Fan MF, Wang H-M, Nan L-J, Wang AJ, Luo X, Yuan P-X, Feng JJ (2020) The mimetic assembly of cobalt prot-porphyrin with cyclodextrin dimer and its application for H_2O_2 detection. Anal Chim Acta 1097:78–84
21. Huang J, Lyu Y, Li J, Cheng P, Jiang Y, Pu KA (2019) Renal-clearable duplex optical reporter for real-time imaging of contrast-induced acute kidney injury. Angew Chem Int Ed 58:17796–17804
22. Abbas K, Hardy M, Poulhès F, Karoui H, Tordo P, Ouari O, Peyrot F (2014) Detection of superoxide production in stimulated and unstimulated living cells using new cyclic nitrone spin traps. Free Rad Biol Med 71:281–290
23. Esquembre R, Pastor I, Mallavia R, Mateo CR (2005) Fluorometric detection of nitric oxide using 2,3-diaminonaphthalene incorporated in β-cyclodextrin. J Photochem Photobiol A 173:384–389
24. Varatharajan S, Sathish Kumar K, Berchmans S, Amutha R, Kiruthiga PV, Pandima Devi K (2010) Synergistic effect of hydroxypropyl-β-cyclodextrin encapsulated soluble ferrocene and the gold nanocomposite modified glassy carbon electrode for the estimation of NO in biological systems. Analyst 135:2348–2354
25. Elgamouz A, Nassab C, Bihi A, Mohamad SAI, Almusafri AHSA, Alharthi SS, Abdulla SAE, Patole SP (2021) Encapsulation capacity of β-cyclodextrin stabilized silver nanoparticles towards creatinine enhances the colorimetric sensing of hydrogen peroxide in urine. Nanomaterials 11:1897
26. He S, Shi W, Zhang X, Li J, Huang Y (2010) β–Cyclodextrins-based inclusion complexes of $CoFe_2O_4$ magnetic nanoparticles as catalyst for the luminol chemiluminescence system and their applications in hydrogen peroxide detection. Talanta 82:377–383
27. Geiselhart CM, Schmitt CW, Jöckle P, Mutlu H, Barner-Kowollik C (2019) A guanidine-based superbase as efficient chemiluminescence booster. Sci Rep 9:14519
28. Geiselhart CM, Mutlu H, Tzvetkova P, Barner-Kowollik C (2020) Chemiluminescent self-reporting supramolecular transformations on macromolecular scaffolds. Polym Chem 11:4213–4220
29. Naresh Kumar T, Sivabalan S, Chandrasekaran N, Phani KLN (2014) Ferrocene-functionalized polydopamine as a novel redox matrix for H_2O_2 oxidation. J Mater Chem B 2:6081–6088

30. Wei X, Dong R, Wang D, Zhao T, Gao Y, Duffy P, Zhu X, Wang W (2015) Supramolecular fluorescent nanoparticles constructed via multiple non-covalent interactions for the detection of hydrogen peroxide in cancer cells. Chem Eur J 21 11427–11434
31. He Z, Chen Q, Duan X, Zhong Y, Zhu L, Mou N, Yang X, Cao Y, Han Z, He H, Wu S, Wang G, Qin X, Qu K, Zhang K, Liu J, Wu W (2024) Reactive oxygen species-responsive nano-platform with dual-targeting and fluorescent lipid-specific imaging capabilities for the management of atherosclerotic plaques. Acta Biomateralia 181:375–390
32. Lazarova D, Semkova S, Zlateva G, Tatsuya H, Aoki I, Bakalova R (2021) Quantum sensors to track total redox-status and oxidative stress in cells and tissues using electron-paramagnetic resonance, magnetic resonance imaging, and optical imaging. Anal Chem 93:2828–2837
33. Dong W, Wang K, Chen Y, Li W, Ye Y, Jin S (2017) Construction and characterization of a chitosan-immobilized-enzyme and β-cyclodextrin-included-ferrocene-based electrochemical biosensor for H_2O_2 detection. Materials 10:868
34. Manickam P, Vashist A, Madhu S, Sadasivam M, Sakthivel A, Kaushik A, Nair M (2020) Gold nanocubes embedded biocompatible hybrid hydrogels for electrochemical detection of H_2O_2. Bioelectrochemistry 131:107373
35. Wang SM, Wang H, Gao H, Zhou J, Zhao W, Chen HY, Xu JJ (2023) Single-particle plasmonic sensing of nitric oxide in living cells. Anal Chem 95:7062–7069
36. Yang Q, Zhou Y, Tan L, Xie C, Luo K, Li X, Zeng M, Zhou L (2023) Rationally constructed *de novo* fluorescent nanosensor for nitric oxide detection and imaging in living cells and inflammatory mice models. Anal Chem 95:2452–2459
37. Chen J, Chen L, Fang Y, Zeng F, Wu S (2022) Refashioning benzothiadiazole dye as an activatable nanoprobe for biomarker detection with NIR-II fluorescence/optoacoustic imaging. Cell Rep Phys Sci 3:100570
38. Xie X, Wang H, Zhang L, Liu Y, Chai Y, Yuan Y, Yuan R (2018) A novel electrochemiluminescence immunosensor based on functional β-cyclodextrin-ferrocene host-guest complex with multiple signal amplification. Sens Actuators B Chem 258:1146–1151
39. Ma Y, Huang J, Xue J, Liu L, Ouyang H, Guo T, Fu Z (2024) Dual-mechanism-driven ratiometric electrochemiluminescent biosensor for methicillin-resistant *Staphylococcus aureus*. Anal Chem 96:2702–2710
40. Tan SY, Teh C, Ang CY, Li M, Li P, Korzh V, Zhao Y (2017) Responsive mesoporous silica nanoparticles for sensing of hydrogen peroxide and simultaneous treatment toward heart failure. Nanoscale 9:2253–2261
41. Soleimani K, Beyranvand S, Souri Z, Ahmadian Z, Yari A, Faghani A, Shams A, Adeli M (2023) Ferrocene/β-cyclodextrin based supramolecular nanogels as theranostic systems. Biomed Pharmacother 166:115402
42. Wang T, Bai Z, Wei W, Hou F, Guo W, Wei A (2022) β-Cyclodextrin-derivative-functionalized graphene oxide/graphitic carbon nitride composites with a synergistic effect for rapid and efficient sterilization. ACS Appl Mater Interfaces 14:474–483
43. Celebioglu A, Uyar T (2010) Cyclodextrin nanofibers by electrospinning. Chem Commun 46:6903–6905
44. Dodero A, Schlatter G, Hébraud A, Vicini S, Castellano M (2021) Polymer-free cyclodextrin and natural polymer-cyclodextrin electrospun nanofibers: a comprehensive review on current applications and future perspectives. Carbohyd Polym 26:118042
45. Topuz F, Uyar T (2019) Electrospinning of cyclodextrin functional nanofibers for drug delivery applications. Pharmaceutics 11:6
46. Topuz F, Uyar T (2020) Electrospinning of cyclodextrin nanofibers: the effect of process parameters. J Nanomater 7529306
47. Molnár A (2021) Synthetic application of cyclodextrins in combination with metal ions, complexes, and metal particles. ChemCatChem 13:1424–1474
48. Payamifar S, Poursattar Marjani A (2023) Recent advances in β-cyclodextrin-based catalysts for reducing toxic nitroaromatic: an overview. Appl Organomet Chem 37:7287
49. Noël S, Léger B, Ponchel A, Sadjadi S, Monflier E (2021) Cyclodextrins as multitask agents for metal nano-heterogeneous catalysis: a review. Environ Chem Lett 19:4327–4348

50. Navarro-Orcajada S, Conesa I, Matencio A, Rodríguez-Bonilla P, García-Carmona F, López-Nicolás JM (2022) The use of cyclodextrins as solubility enhancers in the ORAC method may cause interference in the measurement of antioxidant activity. Talanta 243:123336
51. Hadiana Z, Kamalabadi M, Phimolsiripol Y, Balasubramanian B, Lorenzo-Rodriguez JM, Khaneghah AM (2023) Preparation, characterization, and antioxidant activity of β-cyclodextrin nanoparticles loaded *Rosa damascena* essential oil for application in beverage. Food Chem 403:134410
52. Liang Y, Hou D, Ni Z, Cao M, Cai L (2022) Preparation, characterization of naringenin, β-cyclodextrin and carbon quantum dot antioxidant nanocomposites. Food Chem 375:131646
53. Inoue Y, Yoshida M, Ezawa T, Tanikawa T, Arce F Jr, Lee See G, Tomita J, Suzuki M, Oguchi T (2022) Inclusion complexes of daidzein with cyclodextrin-based metal–organic framework-1 enhance its solubility and antioxidant capacity. AAPS PharmSciTech 23:1–14
54. Jo Y-J, Cho H-S, Chun J-Y (2021) Antioxidant activity of β-cyclodextrin inclusion complexes containing trans-cinnamaldehyde by DPPH, ABTS and FRAP. Food Sci Biotechnol 30:807–814
55. Sharma K, Kadian V, Kumar A, Mahant S, Rao R (2022) Evaluation of solubility, photostability and antioxidant activity of ellagic acid cyclodextrin nanosponges fabricated by melt method and microwave-assisted synthesis. J Food Sci Technol 59:898–908
56. Khan N, Singh AK, Saneja A (2023) Preparation, characterization, and antioxidant activity of l-ascorbic acid/HP-β-cyclodextrin inclusion complex-incorporated electrospun nanofibers. Foods 12:1363
57. Wang X, Parvathaneni V, Shukla SK, Kulkarni NS, Muth A, Kunda NK, Gupta V (2020) Inhalable resveratrol-cyclodextrin complex loaded biodegradable nanoparticles for enhanced efficacy against non-small cell lung cancer. Int J Biol Macromol 164:638–650
58. Celebioglu A, Tekant D, Kilic ME, Durgun E, Uyar T (2022) Orally fast-disintegrating resveratrol/cyclodextrin nanofibrous films as a potential antioxidant dietary supplement. ACS Food Sci Technol 2:568–580
59. Yildiz ZI, Topuz F, Kilic ME, Durgun E, Uyar T (2023) Encapsulation of antioxidant beta-carotene by cyclodextrin complex electrospun nanofibers: solubilization and stabilization of beta-carotene by cyclodextrins. Food Chem 423:136284

Álvaro Sarabia-Vallejo obtained a Degree in Pharmacy from the Faculty of Pharmacy from Complutense University of Madrid where he is now a predoctoral researcher in the Ph.D. program of Medicinal Chemistry. His research interests involve synthetic organic chemistry, UV-Vis spectroscopy and fluorescence spectroscopy and their application to medicinal chemistry. His current research work addresses the design and synthesis of novel small-molecule theranostics against Alzheimer's disease, and the analytical spectroscopic characterization and pharmacological evaluation of the compounds using different techniques and resources including cyclodextrin inclusion complexes, fluorescence lifetime imaging and photodynamic therapy.

Marta García-Cerrato is a predoctoral researcher in the Ph.D. program of Medical Chemistry in the Department of Chemistry in Pharmaceutical Sciences at the Faculty of Pharmacy of the Complutense University of Madrid, Spain. She has studied a Bachelor's degree in Chemistry at the Autonomous University of Madrid and a Master's degree in Drug Discovery at the Complutense University of Madrid. She is in a research group focused on the development of new drugs to treat neurodegenerative and neglected diseases. Her doctoral thesis is focused on the search, synthesis and pharmacological evaluation of new compounds with potential antischistosomal activity.

M. Pilar López-Alvarado is a Professor in the Department of Chemistry in Pharmaceutical Sciences at the Faculty of Pharmacy of the Complutense University of Madrid, Spain. She has more than 28 years of teaching and research experience. She is co-author of 65 research papers. Her h-index is 23 and her i10 index is 43. She has participated in research projects funded by several national agencies, such as MEC-MCIIN, UCM-BSCH, CAM and international AEI-EU. Her current research interest is the development of bioactive molecules for the treatment of neurodegenerative diseases.

María Antonia Martín is a Lecturer in Analytical Chemistry, in the Department of Chemistry in Pharmaceutical Sciences at Faculty of Pharmacy of the Complutense University of Madrid. Since 1989 she is working on molecular recognition by cyclodextrins and their analytical applications. Her main research areas are fluorescence chemo-sensors for studying alterations of proteins and DNA and environmentally friendly spectroscopic and chromatographic analysis, thanks to the financial support of Spanish Organisms (MEC, MICINN; MINECO, UCM) She is author of 93 research and review papers and 6 book chapters. Her *h*-index is 23 and her i10 index is 50.

José Carlos Menéndez is a Professor in Organic and Medicinal Chemistry in the Department of Chemistry in Pharmaceutical Sciences at the Faculty of Pharmacy of the Complutense University of Madrid. His main research areas are in the areas of medicinal chemistry (neurodegenerative diseases, neglected tropical diseases, cancer, diagnostics and theranostics) and synthetic methodology, with an emphasis in green synthetic methods and mechanochemical synthesis. He has authored or co-authored about 300 research and review articles, 40 book chapters and four books. His *h* and i10 indexes are 51 and 228, respectively.

The manufacturer's authorised representative in the EU is Springer Nature Customer Service Centre GmbH, Europaplatz 3, 69115 Heidelberg, Germany. If you have any concerns regarding our products, please contact ProductSafety@springernature.com

Printed and bound by CPI Group (UK) Ltd, Croydon, CR0 4YY

30/09/2026

02225563-0013